Energy and Power

Energy and Power

Germany in the Age of Oil, Atoms, and Climate Change

STEPHEN G. GROSS

OXFORD
UNIVERSITY PRESS

Oxford University Press is a department of the University of Oxford. It furthers the University's objective of excellence in research, scholarship, and education by publishing worldwide. Oxford is a registered trade mark of Oxford University Press in the UK and certain other countries.

Published in the United States of America by Oxford University Press
198 Madison Avenue, New York, NY 10016, United States of America.

© Oxford University Press 2023

All rights reserved. No part of this publication may be reproduced, stored in a retrieval system, or transmitted, in any form or by any means, without the prior permission in writing of Oxford University Press, or as expressly permitted by law, by license, or under terms agreed with the appropriate reproduction rights organization. Inquiries concerning reproduction outside the scope of the above should be sent to the Rights Department, Oxford University Press, at the address above.

You must not circulate this work in any other form
and you must impose this same condition on any acquirer.

Library of Congress Cataloging-in-Publication Data
Names: Gross, Stephen G., 1980– author.
Title: Energy and power : Germany in the age of oil, atoms, and climate change / Stephen G. Gross.
Description: New York, NY : Oxford University Press, [2023] | Includes index.
Identifiers: LCCN 2023006046 (print) | LCCN 2023006047 (ebook) | ISBN 9780197667712 (hardback) | ISBN 9780197667736 (epub)
Subjects: LCSH: Energy policy—Germany—History—20th century. | Renewable energy sources—Germany—History—20th century.
Classification: LCC HD9502.G32 G767 2023 (print) | LCC HD9502.G32 (ebook) | DDC 333.790943—dc23/eng/20230310
LC record available at https://lccn.loc.gov/2023006046
LC ebook record available at https://lccn.loc.gov/2023006047

DOI: 10.1093/oso/9780197667712.001.0001

Printed by Integrated Books International, United States of America

To Rachel, Duncan, and Lucy

CONTENTS

Introduction: The Paradoxes of German Energy 1

PART I   THE OLD ENERGY PARADIGM

1. Energy Price Wars and the Battle for the Social Market Economy: The 1950s 19
2. The Coupling Paradigm: Conceptualizing West Germany's First Postwar Energy Transition 46
3. Chains of Oil, 1956–1973 71
4. The Entrepreneurial State: The Nuclear Transition of the 1950s and 1960s 99
5. Shaking the Coupling Paradigm: The 1973 Oil Shock and Its Aftermath 125

PART II   THE NEW ENERGY PARADIGM

6. Green Energy and the Remaking of West German Politics in the 1970s 155
7. Reinventing Energy Economics after the Oil Shock: The Rise of Ecological Modernization 183
8. Energetic Hopes in the Face of Chernobyl and Climate Change: The 1980s 211

9. The Energy Entanglement of Germany and Russia: Natural Gas, 1970–2000   238

10. Unleashing Green Energy in an Era of Neoliberalism: The *Energiewende*   267

Coda: German Energy in the Twenty-First Century   295

Acknowledgments   313
Archives and Abbreviations   315
Notes   317
Index   389

Introduction

The Paradoxes of German Energy

The resource base is far more fundamental to economic development than questions of political and social order. The old dispute of capitalism versus socialism pales into insignificance before the life-or-death choice of renewables versus non-renewable resources.[1]

An Energy Miracle?

On June 1, 2004, before 3,000 delegates from over 150 countries, Germany's flamboyant environmental minister declared that "the age of renewables has now begun." Jürgen Trittin, a stalwart of the Green Party known for his acerbic wit, was presiding over an international conference on renewable energy in his nation's former capital of Bonn. Reflecting on Germany's recent achievements, Trittin saw a model for the world. Energy efficiency and solar and wind power, he proclaimed, were the key to fighting global warming and bringing prosperity to the world. No more were these technologies a "niche market," he concluded. "They are our future."[2]

Trittin had good reason for optimism. Since 1998 he had helped craft a groundbreaking political experiment: the first coalition of the Left in a large country that combined a Social Democratic Party with a new political force, the Greens. Between 1998 and 2002 these former rivals passed a remarkable array of legislation that aimed to rebuild their nation's infrastructure on an ecological foundation. "The end of the oil age," so they hoped, "is in sight."[3] After 1998 this Red-Green government began placing solar panels on 100,000 roofs. It levied an ecological tax on energy use. It reformed the power grid by guaranteeing profits to anyone who produced electricity from renewables. Its development bank aggressively financed energy efficiency enhancements in Germany's buildings and homes. For Hermann Scheer, Trittin's Social Democratic partner, known as the "solar king" by his allies, this was "an economic revolution of the most far-reaching kind."[4] Its architects aspired to smash the power of fossil fuel conglomerates and erect a more democratic economy. They aimed to turn Germany into a "global leader" of green technology, where the "export markets are huge." And they promised to liberate the nation from the "energy conflicts" that loomed as the world burned through its conventional petroleum.[5]

When Trittin addressed the renewable conference on the banks of the Rhine, these promises seemed on the cusp of realization. Germany boasted a third of global wind power capacity. Solar was growing even faster. By 2004 the Federal Republic accounted for 80 percent of all solar photovoltaics in Europe. And by 2006 it was installing nearly half of all new solar panels in the world, despite a cloudy climate that gets as much annual sunlight as Alaska. New production techniques developed by Germany's *Mittelstand* companies made them global leaders in solar and wind technology, unleashing a revolutionary fall in the cost of renewables and bringing carbon-free energy into the realm of the economically possible. The country, moreover—the third largest manufacturer in the world after the United States and China—had taken great strides in reducing its energy footprint. While America, France, and Britain were all reaching a new historical peak in the total volume of energy they consumed, Germany's energy use had fallen by 15 percent from its 1979 height even as its economy expanded (see Figures I.1 and I.2).[6]

Against the backdrop of this success, Scheer and Trittin heralded the Red-Green agenda as humanity's best chance to stop global warming. In retrospect many agreed, lauding Germany as "the most successful large advanced economy

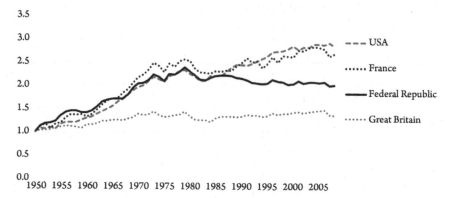

Figure I.1. Total primary energy consumption in the Federal Republic, 1950–2008. Indexed from 1950 = 1. *Source:* European energy data from Energy History: Joint Center for History and Economics. https://sites.fas.harvard.edu/~histecon/energyhistory/; American energy data from U.S. Energy Information Administration. www.eia.gov; GDP and price index from the OECD. https://data.oecd.org/.

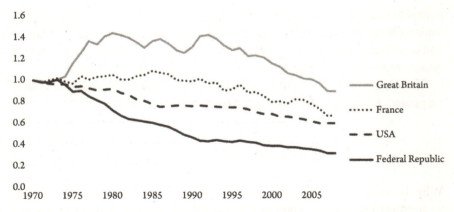

Figure I.2. Energy intensity of the Federal Republic's economy (primary energy consumption divided by real GDP), 1970–2008. Indexed to 1970 = 1. *Source:* European energy data from Energy History: Joint Center for History and Economics. https://sites.fas.harvard.edu/~histecon/energyhistory/; American energy data from U.S. Energy Information Administration. www.eia.gov; GDP and price index from the OECD. https://data.oecd.org/.

to date in developing its clean energy economy," as a "world leader in energy efficiency," as the nation that lit the "touchpaper" that would forever change the state of renewable power.[7]

Today the Federal Republic is continuing this energy transition—its *Energiewende*—in the hopes of combating global warming. But this transition was never only, or even primarily, about climate change. The promise of high-tech exports and skilled jobs, of a more democratic society, of a geopolitically secure nation—all of these dreams have been as much an engine of energy policy as has the quest for a stable climate. Many of these hopes date to the 1970s, before global warming was fully understood, or earlier still, to the 1950s. Only by unearthing this deeper history, and looking beyond climate change, can we understand how energy came to be the central problem of German economics and politics in the twenty-first century.

The path toward a greener energetic future, moreover, has been littered with contradictions, advances, and backtracking. At nearly the same moment that Tritten was touting his nation's accomplishments, German companies were embarking on a new round of investment into power plants that ran on dirty, lignite coal. These would give Germany the notorious reputation as home to eight of the twelve largest sources of carbon in Europe. The country was stumbling into ever greater reliance on natural gas from a dangerous nation that would eventually threaten peace in Europe: Vladimir Putin's Russia. The Federal Republic, moreover, increasingly found itself at odds with the rest of the European Union, which was struggling to build a common energy agenda. And despite decoupling energy use from growth, by the 2010s Germany seemed poised to miss its carbon targets. Critics spoke of an "Ecocide," the title of a dystopian film in which citizens of the Global South sue the Federal Republic for endangering the right to life. In some ways, the very success of Germany's green vision entrenched the need for conventional fuels, at least for a time.[8]

The longer history behind these energy successes and failures, behind Germany's energy dreams and paradoxes, is the subject of this book.

Putting Energy in European History

Why do we use the energy we do? How do we shift from one energy system to another? How do our choices about energy shape society and politics, and the other way around? In our era of global warming, these questions have taken on an urgency once reserved for issues of war and peace. Today carbon dioxide in the atmosphere is creeping toward a level the Earth has not experienced for several million years. As a consequence, the world is warming at an astonishing pace. In 2015 the Paris Climate Accords aspired to hold temperature increases to 1.5 degrees Celsius. Three years later the International Panel on Climate Change (IPCC) all but accepted that humanity would smash through this limit by 2030. Even if we do not, we watch wildfires burn through California and Southern Europe and hurricanes rip through the Atlantic bringing once-in-a-century storms every few years. The "climate catastrophe," as West Germans named this future in the 1980s, is upon us.[9]

Perhaps the most wicked aspect of this problem is that carbon emitted anywhere—Germany, the United States, China—flows into the same atmosphere, such that those suffering the greatest impact from warming are not those who emitted this carbon in the first place. Global warming embodies the pernicious logic of a new sort of modernity, where the risks of industrialization transcend space as well as time, making it inordinately difficult to hold accountable those who are responsible.[10]

The prime source of human atmospheric carbon is the world's fossil energy system. Today a mind-boggling network of coal mines and oil fields, offshore rigs, fracking wells, pipelines, compressor stations, tankers, and power plants prepares nearly 160,000 terawatt hours (TWh) of energy each year to make a lifestyle of iPhones and automobiles possible. In 1950 the world consumed a small fraction of the energy it does today. Transforming this energy into usable form has been accomplished through one of the most stupendous outlays of capital in human history. Each year investors spend nearly two trillion dollars upgrading the "second crust" of the earth that is our energy system.[11]

In one sense, the cost seems worth it. This energy leverages human ingenuity, providing a panoply of goods and experiences that were unimaginable to our grandparents. Energy is the ability to do work, and one can translate its power from one form to another. A "cup of gasoline," for instance, has the same energy as fifty people "pulling a Fiat for 2 hours." With the energy from our fossil fuel system, it is as though each person today has nearly 90 people working for them around the clock, providing light, motion, and heat. But these averages obscure vast differences across space and culture: an average American consumes 80,000 kilowatt hours a year, a German 44,000, a citizen of the Democratic Republic of Congo just 489.[12]

This fossil energy, however, is wrecking the natural foundations of our society. It has been for decades. Our hydrocarbon supply chain is characterized by immense waste. Only a miniscule fraction of the energy embodied in a lump of coal or a barrel of oil is actually used to cool milk in the fridge or move people in cars: over 97 percent is lost through the multilayered process of extracting, transforming, and distributing the energy that comes from fossil fuels. With this immense waste, burning fossil fuels has accounted for over two-thirds of all the carbon humanity has put into the air. And the pace is accelerating: since the first IPCC report in 1990, humanity has burned more carbon than in all previous recorded history.[13]

Understanding why so much of the world has adopted an economy rooted in oil, coal, and natural gas is more urgent than ever. To identify the causes behind our fossil system and the actors who "lit this fire" of global warming, we must look to the specific choices made by particular groups, classes, businesses, governments, political parties, and experts.[14] For one of the paradoxes of global warming is the incredible sway of unintended consequences. Many, if not most, of the decisions that placed humanity on a trajectory of global warming were made before climate was even politicized, and for reasons that had little to do with carbon and everything to do with jobs, power, geopolitics, or profits.

In fact, energy is everywhere in history, shaping our understanding of why industrialization began in a cold, damp corner of Europe or how the United States came to exercise a unique sort of global power in the twentieth century. Yet too often energy remains a side note in the narratives that have shaped our understanding of the contemporary world. The short twentieth century made famous by Eric Hobsbawm, which began with World War I and the Russian Revolution and ended with the collapse of Communism in 1989, is defined through politics and ideas; its central experience the "religious wars" fought between capitalism and socialism. The most influential narratives of twentieth-century Europe revolve around the "ideological struggle for Europe's future," and how liberal democracy emerged changed from its contest with fascism and communism.[15] Above all, it is the shadow cast by the violence of World War II and the Holocaust that orients the masterworks of twentieth-century Europe.[16]

These narratives understandably focus on the ideas and social structures that spawned war and genocide, alongside the paradox that these catastrophes occurred as Europeans were experiencing the most radical material changes since the dawn of agriculture. Energy belongs to both sides of the coin, the catastrophic and the uplifting. Yet it hardly appears in stories that privilege politics, ideas, or diplomacy. Some histories note how "ridiculously cheap" energy underpinned the golden years of growth after 1945. Energy briefly rears its head again to explain the rise of ecological movements in the 1960s and 1970s. It makes a final appearance with the oil shocks of 1973 and 1979. That Europe changed its approach to fossil fuels after this pivotal decade "speaks for itself." But does it?[17]

Energy carries more heft in histories of twentieth-century Germany, but only recently. In 2003 two preeminent scholars could interpret this nation's "shattered" past without mentioning energy at all.[18] Even more than for Europe, postwar history for Germany is "a history of attempts to come to terms with the 'German catastrophe,'" that is, the Third Reich, and to understand how National Socialism and the Holocaust were followed by the rise of a liberal democracy in the Western half of the country.[19] The prime questions of postwar Germany all stem from this effort to fit the two halves of the twentieth century together: How did the Federal Republic forge democracy after the Third Reich? How did it achieve an Economic Miracle after world war? How did it come to terms with its Nazi past? How did Germans navigate the division of their once powerful nation?[20]

Now new challenges associated with globalization, neoliberalism, and ecology are leading historians to refocus their gaze away from mid-century and look for the origins of the twenty-first century in the trauma of the 1970s, which witnessed everything from the decline of heavy industry to domestic terrorism, oil shocks, and the collapse of monetary systems.[21]

As the 1970s have gained prominence, so too has energy. One goal of this book is thus to integrate energy into the arc of postwar history.[22] European historians can no longer treat energy as mere background. We must shine a spotlight onto what has become the most pressing question of our time: how the consumption of massive amounts of energy has become naturalized, and with it the accompanying volumes of carbon emissions that are changing our environment. Historians must identify those actors who caused the emergence of high-energy society, understand the rationale behind their actions, and explain why some individuals, organizations, and states revolted against this system. But historians must also explore how the consumption of energy is entangled with landmark developments that define the conventional narrative of modern Europe, like the rise and fall of governments, the success or failure of ideologies like neoliberalism, the spread of new cultural mores, or the ebb and flow of geopolitical frictions connected to the Cold War and decolonization.

Indeed, energy's place in history is far from self-explanatory, and it can open an entirely new perspective on the prime topics of the twentieth century. The end of World War II marked a watershed by dividing Germany and reconfiguring its political system. But in the sphere of energy and consumer life, the greater transformation occurred in the decade after 1958, as the nation became a high-energy society that ran on petroleum and electricity. Cheap energy may have fueled the Economic Miracle, but *why* was energy inexpensive in the first place? What social, political, and geopolitical *work* went into making it so? West Germany's social market economy looks much different once one appreciates how regimented the flow of energy was. After 1949 Germany may have been a divided power in the heart of Europe. But equally important is understanding its place in a global hydrocarbon network that was in a profound state of flux. The Federal Republic is also known for supporting the integration of Europe. But energy often put it at loggerheads with its allies and with the European Community, revealing a deeply nationalist orientation at key moments in history. Energy even lends a new perspective on how West Germans grappled with their terrible past. For as a new generation in the 1960s and 1970s began questioning their parents' participation in the Nazis' crimes, many also turned to questions of ecology, energy, and nuclear power to criticize the continuity of the Federal Republic with the Third Reich, calling for a decentralized, more democratic form of governance.

Germany's Lessons for Energy Transitions

Energy belongs to the history of twentieth-century Europe and Germany now more than ever since historians' "vanishing point" is shifting, and with it the questions we ask.[23] Like the 1970s, the turn of the millennium has become a new focal point of research. For by then, at the very latest, the global community knew with utter clarity that it was headed toward a carbon catastrophe. "If no effective countermeasures are taken," so Germany's parliamentary inquiry into climate change announced in the 1990s, "then we must expect dramatic consequences for all regions of the world."[24] Put differently, it was clear the world needed a transition away from fossil fuels. But nearly everywhere, political and corporate leaders were failing miserably to make headway.

Why was this transition stalled? And how have energy transitions happened throughout history? Herein lays the second goal of the book: to see what Germany's postwar experience can teach us about the nature of energy transitions: why they occur, how they are experienced, and how they are connected to other types of change. Understanding past transitions, one hopes, can offer landmarks for appreciating the sorts of actors, processes, or arguments that influence energy outcomes, guide our expectations for what is feasible, suggest the range of challenges and opposing forces that lay ahead, and not least, reveal how energy is entangled with so many other spheres of life. A new corpus of energy histories has tackled these questions, illustrating how historians have much to contribute to the fight against global warming.[25]

Yet the conventional understanding remains dominated by other disciplines that explain transitions through changes in technology and prices. That petroleum became the prime global fuel after 1945 has everything to do with efficiency and little to do with politics or social formations, we are told. As Bruce Usher puts it, "basic economic principles, primarily cost, are the main drivers of energy transitions. Cost is key."[26] Marxian scholars have pushed against this price-is-king paradigm by portraying transitions as class conflicts in which the cost of energy hardly matters. Instead, new energy arrangements triumph when they enable those with capital to better control labor. Britain's industrial revolution to coal-fired steam engines occurred, in this rendering, "in spite of the persistently superior cheapness of water," or put differently, because steam overcame "the barriers to procurement not of energy, but of *labour*."[27]

Both views have merit, but both err in elevating a single dimension of transitions: price versus social conflict. Cost and efficiency certainly matter. Energy shifts, however, are far more complex than this. Prices and markets are themselves embedded in cultural practices, social institutions, political arrangements, and geopolitical constellations. The price of a commodity, especially something as laden with power as energy, is determined as much by struggles between interest groups, political actors, and experts as it is by supply and demand.[28]

To integrate these two views we must cast a wider net, one that captures changes in technology and prices while paying attention to how these shape, and are in turn shaped by, social and political conflicts.

Postwar Germany offers fertile ground for this. With its succession of energy upheavals, the history of the Federal Republic of Germany can illuminate why transitions succeed or fail, and how such shifts are tied to changes in other spheres. The country's experience illustrates how three types of forces make energy transitions happen. First, the market for energy—or rather, the diverse markets for different energies—profoundly influence whether consumers and producers adopt a new fuel. Can a new energy or technology provide the same or even better services at a lower price than existing ones? But many forces shape prices, including states and competing social actors.[29] Throughout history, and particularly in postwar Germany, the price of energy has been a political football manipulated by a range of actors. To understand why transitions unfold, one must look beyond markets to states and the social actors who shape markets.

States, the second force, have powerful reasons for desiring one energy over another, which range from geopolitics to jobs and ecology. States set the legal framework that determines how energies are distributed. States often favor particular fuels through instruments like taxation, import restrictions, land-use rights, or monopoly privileges. States have even acted as entrepreneurs by investing in particular technologies and creating new energy markets where none existed before. State policies, in other words, can transform a new energy from a minor commodity into a full-fledged system. But states themselves are complex entities riven by divisions. In a democracy like the Federal Republic, political parties channel these conflicts, and thus exploring parties and their factions is necessary for understanding how transitions unfold.

Third, social actors outside the state also influence the rise and fall of energy systems, including corporations, unions, experts, and grassroots movements. Firms, above all the multinational oil conglomerates that spanned the globe, shaped policy by fielding operations outside the jurisdiction of any one state. But other social forces influenced the policy landscape as well, through lobbying and protest, by changing the terms of debate, and by mobilizing new stakeholders. In postwar Germany every energy transition was profoundly accelerated, slowed, or stalled by social actors.

Energy transitions, put simply, are not the result of an invisible hand, of technological enhancements to the "performance" of an energy, or of an inherent "quest" for greater efficiency.[30] Rather, they are actively made by the interaction of markets, states, and social groups. But to understand why energy transitions succeed or fail, and how they transform other spheres of life, we must add a second axis, that of mobilization. Germany's experience illustrates how transitions require a mixture of four factors to succeed: policy linkages, political coalitions, compelling ideas about the future, and crises.

When transitions unfolded in Germany they did so, first, because advocates of the rising energy clearly linked their agenda to other priorities. Oil, nuclear power, energy efficiency, natural gas, and renewables all advanced because their proponents skillfully connected these to other goals that were dear to Germany's political elite and public.

Second, energy transitions succeeded when their advocates forged coalitions with other groups. All of West Germany's transitions became conduits through which political parties and social actors could build alliances to pursue agendas that were much broader than questions of what energy to consume. Third, ideas about the future have propelled or held back new energy systems. Those actors who could define the problem facing society or tell compelling stories about what the future would hold *if* their desired energy were adopted proved the most successful in building coalitions and advancing their cause. These future ideas came in many different guises: stories about a technological gap with rival nations; predictions about the demise of democracy; forecasts about the limits of fossil fuels. But whatever their form, these future narratives shaped Germany's transitions because they were often the glue holding coalitions together and strengthening linkages.

Crises, lastly, drove new energies forward, and German energy history is impossible to understand without them. Like ideas, crises came in many forms, from moments of stupendous oversupply or shortages to human disasters. But in every instance, they helped break down older ideas, coalitions, and linkages and create the space for new ones to form.

Energy transitions, put differently, are profoundly historical processes that cannot be reduced to a single cause. They are defined by contingency, the possibility of alternative pathways, and human choice. They happen *in time*, in which outside events and decisions intervene, often in quite unexpected ways, to reorient the way societies produce and consume energy. They are embedded in a welter of micro and macro developments. And they are deeply contested affairs, shaped by the collision of interest groups and political actors advocating one form of energy over another.

German Divergence

This book shows how the Federal Republic walked a different energy path than other large, industrial countries like the United States, Britain, and France. Yet that very difference can help us understand patterns common to energy transitions. The story that follows is a mixture of peculiarities and general developments, with Germany's energetic road both an inspiration and a warning.

As in other countries, policy in the Federal Republic was informed by two distinct energy paradigms. The first, older paradigm held that energy must be cheap and abundant for growth, and that these goals were best achieved through markets. But gradually a new energy paradigm arose to question this, suggesting that societies could achieve welfare without consuming ever more energy; that energy prices should be forcefully guided by the state; and that the very concept of growth must be reimagined.

The history of German energy is the history of the struggle between these two paradigms. The idiosyncratic thread in Germany's story are those decisions that led the new paradigm, during crucial moments, to overcome the older one, in contrast to the United States, Britain, and France. Strikingly, the new paradigm strengthened just when neoliberalism as a philosophy of governance advanced around the world and decried the very policies that might kickstart a green transition. Even as climate change was creating an urgent need for decisive state action, faith in the efficiency of free prices, skepticism of government action, and a desire to insulate markets and private capital from democratic politics, voters, and parties spread through powerful global institutions, from the US Treasury to the European Commission and the World Trade Organization. That a newly reunified Germany bucked the global trend toward market fundamentalism to pursue a state-guided drive into renewables stands out as a puzzle.[31]

Germany's idiosyncratic energy trajectory grew partly from the distinctive circumstances it faced after World War II. Whereas the United States pioneered high-energy society in the early twentieth century, powering growth with its rich endowment of all fossil fuels, Germany had preciously little hydrocarbons in its own territory. Nor, after their failed bid for continental hegemony under the Nazis, did Germans own any of the multinational companies that held commanding stakes in the great oil fields of the Middle East, in contrast to Americans, British, and French. The Federal Republic came to depend on foreign-owned hydrocarbons, and this would make the possibility of a new, greener energy system seem attractive and viable earlier in Germany than elsewhere.

But more importantly, Germany's unique path grew from specific decisions made after 1945. Both the United States and Britain responded to the economic malaise of the 1970s with market fundamentalism and the construction of a "neoliberal order." Under Margaret Thatcher and Ronald Reagan, both great powers shifted energy ever more into the hands of the market, their leaders believing that deregulation, low taxes, privatization, and market pricing would revive growth and enrich key stakeholders, while also managing challenges of energy scarcity and environmental damage. Even the European Union, through its Commission in Brussels, began gravitating toward neoliberal ideas as it tried to reinvigorate growth through energy market reform. France, meanwhile, followed a dirigiste policy that prioritized nuclear power for reasons of national prestige, which eventually made it an alternative model for combating global warming.[32]

Germany, by contrast, chose a different energetic route from that of the United States, Britain, or France, a route that historians often explain by pointing to the country's powerful environmental movement. What Paul Nolte has called Germany's Green *Sonderweg*—or special path—is embodied in the rise of an influential Green Party far earlier than in these other nations.[33] The product of an anti-nuclear grassroots movement in the 1970s, the Green Party broke into national Parliament in 1983 and altered West Germany's political culture. This narrative explains Germany's energy pathway as the product of outsiders—citizens' initiatives, left-wing student groups, renewable enthusiasts—who hoped not only to reform the energy system, but to transform West German political culture by building a small-scale, more democratic, and decentralized society. In the words of Craig Morris and Arne Jungjohann, "Germany's push for renewables stems from the people, not the government."[34]

This story, however, shades into triumphalism and disregards the energy problems that this movement from below itself created. Equally important, it overlooks a broader political economy in the Federal Republic whose internal dynamic generated powerful reasons—often reasons quite different from those of the Green Party—for favoring a transition off fossil fuels or nuclear power. This Green *Sonderweg* narrative juxtaposes its protagonists against a technocratic, closed-door style of governance hell-bent on consuming ever more energy at the expense of both the demos and the environment. But by rendering a monolithic image of the establishment against which the Green Party reacted, it overlooks elements of change that came from political and economic insiders, above all from the reform wing of the Social Democratic Party (SPD), unions, some domestic companies, and mainstream experts.

Rather, Germany's distinctive approach to energy emerged from a synthesis of outsiders and insiders, who saw the benefit of working through market devices to shape energy consumption and production, but who accepted that energy markets were inherently political. This was corporatism in matters of energy, which brought various and often competing stakeholders together during inflection points and crises to negotiate the energetic future of the nation. Instead of trying to "encase markets" and protect them from politics, as neoliberals did elsewhere, German policymakers embraced the fact that social actors and politics should determine the basic orientation of the energy system.[35] But in contrast to dirigiste systems, as in France, the Federal Republic deployed market mechanisms that it guided with a strong hand.

The Story in Brief

The anti-nuclear movement and the Green Party are only part of the story behind Germany's distinctive energy policy, which actually begins much earlier and was profoundly shaped by the country's turbulent energy history. In fact, since 1945 the Federal Republic has experienced some of the sharpest energy transitions of any industrial economy: no less than five of them. The first was the rise of oil, a pan-European phenomenon but one in which no country participated more powerfully than West Germany, once the hard coal capital of the continent. Next came nuclear power: a transition in which West Germany was also superlative in many respects. Starting its atomic program after the United States, France, and Britain, the country caught up through an incredible effort by state officials and scientific experts. Then in the 1970s reformers reacted against the breakneck speed of nuclear development and the 1973 oil shock to launch the country's third transition, to energy efficiency. At the same time, the Federal Republic became entangled with the Soviet Union in building a vast new infrastructure that propelled the rise of natural gas throughout Central Europe in a fourth transition that came to fruition in the 1990s. Finally, after 1989 the newly reunified Germany embarked on a transition to renewable power in an effort not only to tackle global warming, but to solve a welter of economic and political problems.

The first of these transitions is where West Germany's experience with energy began to diverge already from the United States, Britain, and France. The country experienced a more rapid shift to oil than any other industrial country (see Figure I.3). After 1950 foreign-owned oil flooded the country and wrecked the once powerful hard coal industry. This transition wiped out hundreds of thousands of jobs, destroyed an immense amount of capital, and forced domestic companies to move into more dynamic sectors.

This first postwar transition, moreover, led the nation into great dependency on a foreign-owned hydrocarbon supply chain over which the Federal Republic had little control because it lacked a multinational oil company. From an early stage, this supply chain proved susceptible to crises: closures of the Suez Canal in 1956 and in 1967, and coal crises in 1950, 1958, and 1966. Well before the pivotal oil shock of 1973, during the very Economic Miracle that is so often described as an era of stability, the nation suffered no less than five energy crises.

This turbulence turned energy into a politicized arena in West Germany much earlier than in North America, and shaped policy in several ways. First, as energy became politicized, West Germans began paying closer attention to how it was priced. Beginning with vitriolic debates in the 1950s, politicians,

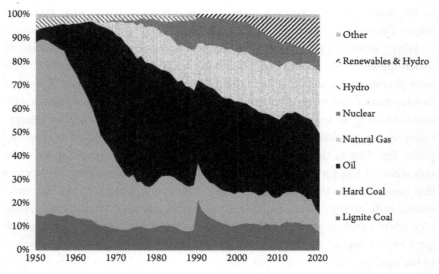

Figure I.3. The Federal Republic's mixture of primary energy consumption, 950–2020. Percentage of total energy consumed. *Source:* Data from AG Energiebilanzen e.V.

experts, and the public came to view the price of energy not only as something that the market could not determine, but as something that it *should not* determine. Miners, environmental activists, and others began calling on the state to guide the price of energy to achieve specific goals. These goals would change with time, and eventually most groups came to accept that energy was a political market, and many argued for a high energy price to stimulate conservation and encourage alternative energies.

Second, because of energy's politicization, West German economists began studying it as a coherent field of knowledge earlier than elsewhere. On the one hand, energy economics became a pillar of West Germany's closed-door technocratic governance, which would spark a backlash that demanded a more democratic energy system. On the other hand, this field of knowledge did not *dematerialize* the economy in theory, as did mainstream economics in the United States. Because energy was so crucial to their models, German economists grappled with the effects of energy crises more than their counterparts across the Atlantic. This eventually led them to advocate an active government agenda instead of leaving energy to the market, and to champion the new energy paradigm.

Third, the politicization of energy made security a major issue in West Germany well before 1973, as politicians, the coal industry, and nuclear advocates criticized the foreign-owned hydrocarbon supply chain already in the 1950s and 1960s, to be joined later by environmentalists and natural gas boosters. Conservation could gain more traction among political insiders than in the United States or Britain because it was presented as a solution to the long-standing problem of security. So too did natural gas from the Soviet Union.

West Germany's shift to oil, in other words, created a fertile landscape in which the seeds of a green energy transition might later take root, even if these were planted for reasons that had little to do with ecology or climate. But other developments contributed to this landscape, above all West Germany's self-understanding as an export nation. With an economy powered by energy-intensive exports, like steel and chemicals, after 1950 Bonn worked to keep fuel prices low. But as the global economy transformed and exposed these older industries to foreign competition, German leaders began seeking new sectors that could excel in the global marketplace, which it assisted with an "entrepreneurial state" that guided research and created new markets.[36] In the 1960s Bonn kick-started nuclear power because experts believed reactors would be the next great export engine. As the prospect for nuclear exports withered, some looked to the construction of natural gas infrastructure across Eurasia as a new engine for export contracts. But others fastened onto technologies that could either conserve energy or produce it from new sources, redirecting the entrepreneurial state toward renewables in order to stimulate exports, creating avenues for collaboration between business elites and environmentalists.

After 1980, in sum, many West Germans wanted to overhaul their nation's energy system. But calls for a transition were coming from two different directions. The Greens hoped to democratize the energy economy through a decentralized, citizen-run power grid. Meanwhile, insiders—the SPD, some unions, conventional experts, and key business groups—wanted to forge a new energy system in the name of security and sales abroad.

What brought these strands together were the languages of Ordoliberalism and Ecological Modernization. The former shared similarities with neoliberalism in calling for a strong state to set ground rules for competitive markets. But its early theorists never sought to extend market logic to all walks of life, and they fixated on economic concentration as the root of most problems. This philosophy profoundly influenced West German politics after 1950, and it provided a common language for parties—even the Greens, eventually—to justify their programs to a public proud of the social market economy. Over time, however, its formal ideas became hollowed out, and by the 1990s, Ordoliberalism meant different things to different groups. But this was its strength as a language of politics, which the Red-Green coalition would use to sell its energy agenda after 1998.

Substantively more important was Ecological Modernization, the bedrock of the new energy paradigm, which highlighted the social costs of production not included in market prices. After 1980, West German economists integrated ideas about the social cost of energy with new theories of technological change to create a novel justification for state-guided prices and investment. When global warming became a burning issue in the 1980s and 1990s, policymakers thus had a language that was well-suited for integrating economic and environmental goals, for building alliances across parties, and for justifying robust state action in an era defined by market fundamentalism.

In 1998 these two strands—insiders and outsiders: Greens on the one hand; Social Democrats, traditional experts, some unions, portions of the business community on the other—came together to pass groundbreaking legislation that envisioned a holistic transformation of Germany's energy system. And while the immediate spark for the Red-Green agenda came from anxieties about global warming, it arose, in fact, from ideas and from a political economy that had been in gestation for decades, and for reasons that initially had little to do with climate.

The German story that culminates after 2000 shows how the history of energy is essential for understanding the twentieth century. The rise and subsequent revolt against high energy, carbon-intensive society is entangled with the pivotal, conventional moments of postwar history; we cannot understand one without the other. The Economic Miracle; the emergence of consumer society; the rise and the faltering of Social Democracy; the integration of Europe; the Cold War and the collapse of Communism—all of these events have shaped, and in turn been shaped by, Germany's rapid succession of energy transitions since 1945. In the Federal Republic, the politicization of energy began well before the 1973 oil shock, the emergence of climate change as a global issue, or the rise of the Green Party. This timing helps explain the nation's idiosyncratic energy trajectory, which was made by a tense interaction between political insiders and outsiders. And it explains as well the intensity with which Germans approach matters of energy today.

But the story that follows also shows how energy transitions are multilayered historical processes shaped not only by technology and prices, but also by human agency and the ability of individuals and organizations to mobilize for or against particular fuels. Alternative paths abound throughout history: the essence of the discipline is excavating these alternatives to understand why some triumphed and others failed. Germany's energy path hinged on the interaction of global and domestic markets, political parties, experts, and grassroots mobilization, and it was determined by those actors who effectively built alliances, linked energy to other issues, deployed powerful visions of the future, exploited crises, and ultimately, overcame their opponents. As Hermann Scheer poignantly observed, transitions have spawned some of the most intense social conflicts in history, precisely because energy touches so many aspects of life. Germany's postwar history illustrates just this point: how energy has been a battlefield in the twentieth century, and will continue to be so in the twenty-first.

PART I

THE OLD ENERGY PARADIGM

1

Energy Price Wars and the Battle for the Social Market Economy

The 1950s

The old industrial heartland of Germany is a transformed landscape, many times over. The Ruhr, a hilly region criss-crossed by the Emscher, Lippe, Ruhr, and Rhine rivers, emerged as one of Europe's largest urban agglomerations in the nineteenth century. The storied names of firms such as Krupp and Thyssen, which drove German industry and which armed a nation that twice engulfed the world in war, hailed from the Ruhr. Their success rested on a mixture of technical know-how and patriarchal labor relations, but also on the energy found beneath the Ruhr. Coal fueled Europe's industrialization, and it was hard coal from this region that brought steam, heat, and electricity to the cities of Central Europe. The Ruhr was Germany's dirty beating heart, a coal civilization with its own rhythms, culture, and power centers. At its height during postwar reconstruction of the 1950s, the Ruhr's population peaked at 5.7 million, nearly five hundred thousand of whom worked in the mines.[1]

Today the last mines in the Ruhr have closed. The hard coal civilization that took a century to build has collapsed, leaving behind industrial relics and a culture of conservation, as grassroots organizations work to turn slag heaps into parks and mining sites into some of the largest museums in the world.[2]

Hard coal's collapse began in the 1950s, the very moment when it seemed Germans had overcome the dislocation of World War II. During this decade West Germany's first postwar energy transition began. Between 1950 and 1970 the nation's primary energy structure would change more rapidly than that of any other European country, as the new Federal Republic transformed from the continent's largest consumer of hard coal into its largest consumer of oil.[3] Yet, at the time, few realized what was to come. For in 1950 West Germany still belonged to an older energy system where horses and coal stoves were more

Energy and Power. Stephen G. Gross, Oxford University Press. © Oxford University Press 2023.
DOI: 10.1093/oso/9780197667712.003.0002

common than cars and refrigerators—a regime of consumption that paled in comparison to the voluminous energy used by the citizens of the United States.

The wrenching transition out of this older energy system had slowly started in the years before World War II, but it accelerated qualitatively in the 1950s and 1960s. The catalyst came from the pioneer of high-energy society, the United States. Marshall Plan aid that promoted oil and inexpensive coal from the mines of Appalachia promised unprecedentedly inexpensive energy and placed West Germany's hard coal producers under immense pressure.

But if America provided the spark, West Germany's own government, under the firm control of the Christian Democratic Union (CDU), accelerated this transition by pursuing a policy of low-priced energy and welcoming anything that would compete with the mines of the Ruhr. The popular economics minister Ludwig Erhard treated energy as a commodity like any other, to be produced as cheaply as possible. In the drive to advance exports and growth, contain inflation, and shatter the coal cartels that had thrown their support behind Adolf Hitler in the 1930s, Erhard's new low-priced energy regime led to a spectacular rise in energy consumption and carbon emissions, vaulting West Germany into the "Great Acceleration," that mid-twentieth-century moment in which nearly every metric of global consumption and pollution grew exponentially.[4]

This first energy transition, however, was deeply traumatic, imprinting on West German policymakers and public a deep anxiety over energy crises. For the Federal Republic was born in an energy crisis. While the oil shock of 1973 looms large in histories of the United States, energy shocks came much earlier to the Federal Republic, and they helped turn energy into a coherent field of political action sooner there than in America. Almost immediately after 1945, the Federal Republic's energy supply was exposed to the double-edged sword of postwar globalization, making foreign energy a tense issue that divided the nation's public because it entailed trade-offs and redistribution from one group to another.

In the 1950s these energy crises—first the Korean War in 1950–1951, then a far more disruptive shock in 1958—sparked intense debates. Should the state make cheap energy and efficiency the sole criteria of policy, favoring oil multinationals, exporters, and consumers, and placing the burden of any adjustment entirely onto coal companies and miners? Or should it pursue different priorities by embedding energy in a broader social framework and treating it as something other than a pure commodity? And to what extent should energy fall under supranational rather than national oversight? For these crises came as Western Europe was making its first drive toward integration through the European Coal and Steel Community (ECSC).

The stakes could not have been higher, because this discussion revealed the limits of the fabled social market economy. Scholars have too often portrayed

Germany's postwar development through a triumphal lens, as a market-driven recovery catalyzed by the application of free market principles. This narrative, however, distorts our understanding of postwar development, because many fonts of growth bear the fingerprints of an activist state policy, while the market price mechanism so praised by Erhard extended to only portions of the economy, and in many ways not at all to energy.[5]

Well before climate change became a social issue, before the modern environmental movement emerged, before the dramatic oil shocks of the 1970s, energy became politicized in the Federal Republic. Through crises, West Germans came to understand energy less as separate fuels than as a coherent category that warranted its own distinct policy. They increasingly saw price as a channel for distributional conflict as much as a means for balancing supply and demand, and energy prices became a political football as interest groups struggled to shape West Germany's economy. Unlikely coalitions formed that linked energy to first-order issues like exports and social stability. In the process, the debates and political maneuvering surrounding this first transition began to change how Germans approached energy, leading many to see energy as more than a mere commodity, as nothing less than a fundamental component of social and political life.

Born in an Energy Crisis

The end of World War II marked the most profound political rupture in modern German history. Total war led to the total defeat of Nazi Germany at the hands of the Soviet Union, the United States, and Great Britain. By the spring of 1945 Adolf Hitler, whose bid to forge a racial empire sparked a global war, had committed suicide. His lieutenants were dead or imprisoned. His war machine and death camps lay in ruins. His state was dissolved. His country was occupied by the victorious Allied powers and partitioned into four zones. Political collapse was mirrored in the physical destruction of Germany from a conflict the Third Reich had itself incited. The war left a "blasted landscape of broken cities and barren fields." Allied bombing and Soviet bombardment had reduced once thriving cities like Berlin, Cologne, Hamburg, and Dresden to smoldering heaps. Returning home, German soldiers were astonished by what little remained: "as far as one could see, rubble, rubble, and rubble as well as ruins which jutted ghostlike into the sky." The transportation system, rail network, water pipes, and electricity lines were decimated. A humanitarian disaster lay on the horizon as Europe descended into famine. Most Germans, like the many Europeans whose lives they had smashed, struggled to find nourishment and shelter, much less the energy to heat their homes or power their factories.[6]

After their victory the Allies worked to restore this shattered continent, and chief among their problems was energy. In 1945 Europe had precious little of it. For Germany's coal sector, twelve years of economic dirigisme under the Nazis, a decade of underinvestment, and two years of strategic bombing had created a "hopeless situation," as America's most detailed study of the Ruhr concluded: a state of "almost complete chaos."[7] The Third Reich's bid for global power led to the death or maiming of a generation of skilled miners. American and British bombers had decimated two-thirds of the coal industry's housing, along with the above-ground infrastructure that moved this energy around the country. Coal production in 1945 was a tenth its prewar level. In many areas, supply was so scarce that civilians felled trees for heating and combed the woods for anything they could burn to survive the winter, raising anxieties about deforestation and environmental degradation.[8]

As the United States and Britain broke with the Soviet Union and began to rebuild the Western zones of the former Third Reich, they poured resources into the mines of the Ruhr. This region was the prime energy source not only for Germany, but for the Europe that lay under American hegemony—so much so that the Soviet Union proposed partitioning the Ruhr to extract reparations, while the French wanted to internationalize it under Western control. By 1948 the Allies had surmounted the worst of the coal shortages, but only barely. Production remained lower than before the war. And the Ruhr's mines were old: over half dated to the 1870s and ran on unmechanized technology.[9]

In 1949, when the American, British, and French merged their occupation zones to create the Federal Republic of Germany, their continued control over this new state reflected worries about energy. With the memory of Nazism still alive, the Western Allies passed two decrees that curtailed the sovereignty of the young Bonn Republic, named for its sleepy capital on the Rhine. The first reserved the Allies' right to oversee West Germany's foreign policy and any changes to its constitution. The second was the "Ruhr Statute" that required the country's energetic hub to fall under the authority of an international supervisory body.[10]

Upon its birth, the Bonn Republic still had both feet planted in an older energy system defined more by limits than abundance. Across the Atlantic, the United States had forged a high-energy society in which easy access to an incredible array of services supplied by prodigious amounts of fossil fuels had become a normal feature of life. By the 1950s, gasoline-powered tractors had largely replaced animal power on farms. Four-fifths of households had their own refrigerator. Television sets powered by electricity flashed in the living rooms of over 40 million homes. Over 40 million cars—three-quarters of all the world's automobiles—roamed America's roads. German visitors to the United States marveled at the high standard of living, the huge consumer power of the average citizen, and the immensity of the nation's mineral wealth in oil, coal, and gas.[11]

For the average West German, such energetic luxury seemed part of a different world given the devastation they saw around them. Even before the war, Germans had lived in a society that was half as rich as America, and limited in wealth and power by energy constraints. When it marched into the Soviet Union, Hitler's army relied more on horses, carts, and human legs than gasoline-powered trucks. Three million beasts of burden served in the Wehrmacht. As they rode in jeeps through Western Europe in 1945, Americans were astonished by the vast number of draft animals still used by the Germans. Hitler's ambitions for global dominance were constrained by scarce energy, despite the huge resources he devoted to synthesizing petroleum from coal and capturing the oil of Soviet Baku. The Third Reich was a coal country just as Imperial Germany had been before World War I. In 1939 over 90 percent of all the energy consumed by Germans came from this black rock, the same as when an imperial monarch had ruled Berlin in 1914.[12]

It was this older energy-regime that the Allies rebuilt at first, and it would take much longer until the Federal Republic began to resemble the high-energy society of America. If 1945 marked a political rupture separating the Nazi past from a more democratic future, in the world of energy the great break was still to come. Through the 1950s most German farms still ran on the muscle power of animals. Although many Germans had access to electricity, household appliances remained a luxury: few people owned a washing machine, almost no one a dryer, and less than one in ten a refrigerator. In 1957 there were but a million televisions in the entire nation. Many, particularly in the Ruhr, heated their homes and cooked their food with coal, lit from kindling embers brought home from the workplace. Most people traveled by rail, foot, or bicycle. As late as 1948 one could see wood-powered vehicles cruising the empty highways built by the Third Reich, and only the elite could afford gasoline cars.[13] In 1951 Germany's per capita energy consumption was roughly the same as in 1913, lower even than during the 1920s or World War II. Average citizens consumed less than 40 percent of the energy of their counterparts in America. And like its imperial and fascist predecessors, the new republic remained grounded in coal: in 1950 this source provided 90 percent of West Germany's primary energy.[14]

Less than a year after the Federal Republic became a state and held its first national elections, this coal economy experienced the first of many energy crises. When half a world away the North Korean army invaded South Korea in June 1950, German coal mines that had only just recovered from war found themselves on shaky ground. As American rearmament kicked into gear, supported by the manufacturing centers of Western Europe, demand for coal exploded and West Germany descended into a gripping energy shortage. This was accompanied, driven in fact, by a dramatic spike in the price of raw materials around the world as the Korean War generated an explosion of demand for steel, iron, food, and textiles. A new round of inflation erupted, alarming a nation that twice in a generation had experienced traumatic currency reforms.[15]

For Ludwig Erhard, architect of West German reconstruction, the Korean War threatened his entire economic agenda. The son of a middle-class business owner, Erhard had been deeply disturbed by Germany's hyperinflation following World War I. After 1945 he had risen through the ranks of local administrators to become director of Germany's economic advisory council to the American authorities. His signature policy was the liberalization of West German prices in 1948. In a controversial decision, Erhard freed the price of consumer goods, clothing, and machinery at the very moment the Allied Occupation authorities reformed Germany's dysfunctional currency. In his mind, in order to function, a new currency must have market prices that would extricate Germans from the legacy of the Third Reich's rationing system. As he highlighted in public addresses and private correspondence, "there is no free market without free prices."[16] Supported by a vocal group of Ordoliberal economists, Erhard overcame resistance to price liberalization from his opponents, the Social Democratic Party (SPD), and even from his own party, the CDU. After an initial bout of inflation, by 1949 the economy rebounded, productivity soared, and commentators heralded Erhard's gamble as a remarkable success. Price liberalization and competition turned Erhard into a political star. The following year the CDU made his economic ideas the foundation of their political platform, winning the first national election and cementing their rule for over a decade.[17]

The Korean War struck at the heart of Erhard's reforms. "Price inflation," he argued in a radio address shortly after the North Korean invasion, "must of course not reappear on the political level, otherwise this will lead to the reintroduction of price freezes, the return of smuggling and the black market, the supply of the population through an allocation system; in short, we will experience again exactly what we have just fortunately put behind us."[18] Erhard's fears soon materialized in the field of energy. Coal producers were barely keeping pace with the growing demands of German industry. Korea tipped the scales, and by the winter of 1950–1951 coal bottlenecks turned into total blockages, forcing Bonn to impose power cuts and ration electricity. By Christmas the lights went out, as rolling blackouts spread across cities. Shop window displays went dark, the German rail network reduced traffic, industry saw its electricity supply cut by a quarter, households began hoarding coal, and black markets re-emerged. Industry, households, and local governments struggled to find the energy they needed. For people who had just spent five hard years rebuilding from the wreckage of war, the return of energy shortages came as a shock.[19]

Coal producers struggled to respond to the demand surge because in 1948 Erhard had not actually extended market pricing to all sectors. In fact, his reforms created a bifurcated price system that favored consumer over producer goods. Price controls remained in sectors ranging from housing to public transportation, and coal belonged to this controlled half of the economy. Because so many production processes relied on the Ruhr for energy, Erhard feared a rise in coal prices would ripple through the economy and spark inflation. Instead, Erhard pegged the domestic price of coal below its market value, while the International Authority of the Ruhr (IAR)—created by the Allies to channel German coal to the rest of Western Europe—controlled the export price.[20]

Unable to raise the price of their good, coal producers found it impossible to generate the investment they needed to expand. On the one hand, this was nothing new for the Ruhr because the state had effectively controlled coal prices since 1919. But in a period of rapid growth, this was causing problems. Investment in German mines lagged behind that of other Western European countries, and growth in the coal sector was lower than nearly every other German industry.[21]

As tensions mounted, a three-way conflict erupted that pitted Ruhr coal, Erhard, and the American authorities against one another. Coal producers saw Korea as an opportunity to revive their older tradition of corporatism. In the nineteenth century a heavily organized industrial order had crystallized in the Ruhr, composed of huge cartels and investment networks that connected the owners of mines, collieries, and steel factories, and defined by fraught labor relations. Coal insiders saw cartels as a source of stability for an inflexible sector and a way to regulate a chaotic market. After 1945, Ruhr producers hoped to revive their traditional authority to set prices through collaboration instead of through government dictate. The Korean War seemed to offer coal producers a chance to reassert the power to price their own good.[22]

Erhard and one faction of America's occupation authorities, however, wanted to shatter the Ruhr's industrial conglomerations. The economics minister criticized this industrial order for leading to "unsocial concentrations of market power" that jeopardized democracy.[23] American New Dealers, along with internationalists in the State Department, likewise believed that cartels had aided Hitler's rise to power, and that breaking them up was crucial to de-Nazifying Germany. In 1950 de-cartelization was a pillar of America's campaign to democratize West Germany.[24]

Beyond the cartel question, however, Erhard and the American authorities found little to agree on. Above all, they were divided over whether to directly control investment into coal through state levers. Allied authorities wanted to stimulate mining through state aid, but Erhard and his administrators steadfastly refused anything that smacked of dirigisme.[25] In March 1951 the conflict reached a tipping point when John J. McCloy, the US high commissioner for Germany, in a letter to West Germany's first chancellor, Konrad Adenauer, demanded that Bonn actively channel investment into coal and other strategic sectors. "Only a significant modification of the free market economy," McCloy mandated, "can meet the challenges of the new situation."[26] Under pressure from McCloy, Adenauer conceded that Erhard's liberalization had gone too far and blamed his economics minister for the nation's first energy crisis: "you [Erhard] have apparently failed to recognize the greatest danger to our entire economic well-being, namely, the coal question."[27]

The Korean War thus opened a rift between Adenauer and Erhard—the two most important West German leaders—over energy. It also represented a setback for Erhard's liberalization. McCloy's ultimatum forced the Federal Republic to ration coal as well as investment, and Bonn, which had already dismantled its postwar allocation system, had to rely on the coal cartels themselves for distribution. Over the next two years, coal producers got a taste of their former power as they distributed their goods and investment to meet the demands of a growing economy.[28]

National and International Tensions

While this corporatist allocation relieved shortages, the resolution pleased no one and created a rancorous atmosphere that would infect policy for the coming decade. Erhard came out of the Korean crisis dissatisfied because coal remained an island of concentration in the sea of his Social Market. But under pressure from McCloy, Erhard pushed through a massive, state-guided investment program that flowed to coal and other bottleneck industries. This Investment Aid Law of 1952 levied a contribution of 1 billion DM from sectors benefiting from the low price of energy, which the government channeled into primary goods sectors, coal receiving a quarter of the proceeds. After 1952 coal producers received depreciation allowances to encourage investment, and Parliament began subsidizing housing for miners. By 1956 coal was the most heavily supported sector in West Germany.[29]

The need to rely on industrial self-organization during the Korean War underscored for Erhard how dependent he was on the barons of the Ruhr. So after Korea he doubled down on his effort to impose competition in the energy market. As he argued to Parliament, we can only achieve "a fundamental solution to the problem of our future energy supply by promoting competitive forces in the energy market." Energy, in his mind, should be *forced* to join the social market economy, like any other sector.[30]

Under the guidance of Theobald Keyser, coal producers vehemently protested Erhard's agenda. Keyser hailed from Bochum, heart of Ruhr coal country. During World War II he had helped administer the rapacious Nazi mining operations in occupied Belgium, Romania, and Yugoslavia before winding up as a prisoner of war. But he survived with his reputation intact, and relentlessly climbed the ladder of state offices overseeing mining. To counter the agenda of Erhard and the Americans, Keyser and other coal producers entered into an unprecedented collaboration with what had once been their greatest foe, the miners, who formed a new, national union for energy industries, the Industrial Mining Union (*Industriegewerkschaft Bergbau—IGB*). Together they proposed reorganizing their sector into integrated conglomerates that would regulate the entire coal production chain, from excavation to sales and marketing.[31]

Bonn rejected this proposal. Instead, Erhard found an unlikely ally in the emerging European Coal and Steel Community (ECSC). In the spring of 1950 the French foreign minister Robert Schumann stunned Europe by announcing his vision of placing the coal and steel sectors of West Germany and his own country under the collective control of a higher authority. Schumann and his allies saw the nascent ECSC as a revolutionary step toward lasting peace, and one that would pave the way for a United States of Europe. For the Federal Republic, the ECSC provided a way to thread a needle through thorny domestic conflicts. Adenauer forcefully supported the project as a channel for his country to rejoin Western democracies. The Left, however, wanted to nationalize coal like so many of the Federal Republic's Eastern and Western neighbors had done. To win over the unions and Social Democrats, Adenauer thus struck a bargain, by supporting their demands to give laborers a voice running mining enterprises, or co-determination. In return for accepting the ECSC, in 1951 unions gained parity with owners on the domestic boards overseeing coal companies. The Ruhr barons grudgingly accepted this framework as the least offensive outcome, for they retained private ownership over their assets.

Yet the ECSC served energetic as well international and social goals, above all for the technocrat and former cognac merchant who led France into this organization, Jean Monnet. Monnet had designed France's postwar economic plan, which centered on the steel industry and relied on German coal. But as the Federal Republic rebuilt its economy, it began making greater claims on the black gold of the Ruhr. Only transnational control, Monnet believed, would ensure his country had the energy it needed for growth. He hoped the ECSC would preserve French access to German coal and keep energy costs low. With the Paris Treaty of 1951, the ECSC came into being, governed by Europe's first supranational executive bureaucracy, the High Authority, which could make binding decisions on trade and production if approved by the European Council that represented the member states. Monnet wanted the High Authority to regulate prices and smash the cartels that characterized coal and steel in Northern Europe. And he harbored ambitions of expanding the ECSC to encompass all energies.[32]

While Erhard resented the dirigiste outlook of Monnet, he welcomed anything that could help him impose competition in the Ruhr. In conjunction with the ECSC, after the Korean War, Erhard reorganized the coal sector in the name of competition, leaving producers with far more enterprises than they or the Mining Union wanted. The High Authority, meanwhile, acquired the power to fix the domestic price of German coal.

Ruhr mine owners and unionized miners stewed, for the bitter legacy of this first energy crisis went beyond questions of industrial organization and extended to the contentious issue of pricing. After Korea, Erhard desperately wanted a low coal price to contain inflation and spur growth.[33] And under the ESCS's new authority, after 1953 West Germany had the lowest priced coal in the ECSC. This infuriated Keyser and his allies, since they believed the ability to set their own prices was a "matter of existence" for the industry.[34]

Erhard added fuel to the fire by pushing low-priced energy across the board. In 1951 the economics minister opened the door to coal imports from North America, which could be produced far cheaper than in the Ruhr. For the first time since industrialization, Germany began importing coal to meet its energy needs. And in an effort to end the black market in energy, in 1951 Erhard dissolved the oil rationing bureau that had been operating since 1945 and freed the price of this new hydrocarbon. In the short term this caused oil prices to spike, but over the long term it stimulated a wave of investment in refineries.[35]

Coming out of the Korean War, in other words, West German coal began to face intense competition from foreign sources of energy. The Ruhr, however, did not take this sitting down. Under Keyser's leadership they founded a new organization, the Business Association for Ruhr Mining (*Unternehmensverband Ruhrbergbau*–UvRb), which would pursue a relentless campaign to claw back industry control over pricing. With support from the mining unions, Keyser and his colleagues protested the powers of the High Authority for the next six years. They tirelessly pointed out how their nearest competitors—large oil firms— could sell their products at whatever price they wanted and to whomever they pleased. And they insisted on the contradictions of Erhard's policies, namely, that under the ECSC's price regime "the foundations of the social market economy simply did not apply to coal."[36]

The Zenith of Coal and the Rise of Oil

The Federal Republic, in sum, had become a sovereign state in the midst of a crisis that turned energy into an intensely politicized sphere. But after the Korean War these tensions were submerged by West Germany's astounding Economic Miracle. In 1952 the huge global demand for capital goods sent West German industries humming. Over the next six years the Federal Republic experienced the most rapid economic expansion in its history, which pulled along the entire energy sector. The mid-1950s marked the high point of West German coal in terms of tonnage mined, workers employed, and capital deployed. Erhard's 1952 Investment Law unleashed a wave of investment in mining, with fifty pits either newly dug or newly expanded. Keyser and the UvRb guided this new capital in a coordinated program, building new worker housing, improving above-ground processing facilities, and mechanizing mining. Demand for energy kept the pits running near full capacity, and by 1957 West Germany was producing over 130 million tons (SKE) of hard coal annually, up from 110 million in 1950. For the first time since the 1930s, Ruhr mines entered the profit zone, just barely.[37]

The boom, however, was not without its challenges. The urgent need for coal, for one, gave leverage to the Industrial Mining Union. Its energetic vice president Heinrich Gutermuth, who had been conscripted into the Wehrmacht and interned in a Soviet POW camp through 1946, used this newfound influence to push through pay raises. By 1956 mining wages were among the highest in the country, while mechanization was taking much longer than anticipated. Ruhr mines also suffered from financial obstacles. The long period required to dig new mines or to expand old ones—fifteen to twenty years for the former, five to ten years for the latter—tied up capital for such a long time that investment into coal was risky.[38]

While King Coal navigated its growing pains, American authorities and West Germany's Economics Ministry began laying the foundation for a new energy system. As a legacy of its wealth in hard coal, West Germany remained an oil laggard after 1945, using much less petroleum than the rest of Europe: 4 percent of total energy consumption, in comparison with 18 percent in France and 35 percent in Italy.[39] After 1949 the occupation authorities worked to change this. Oil, after all, was a pillar of high-energy consumer society, which American leaders hoped to extend around the world to legitimize their struggle against Soviet Communism. With the descent into Cold War, Washington came to believe its hegemony rested as much on ensuring the flow of energy to its allies as it did on providing a stable monetary or trading system. The war, moreover, had driven home to American policymakers the strategic importance of petroleum, leading officials from the State Department to the Interior Ministry to see their nation's multinational oil corporations as tools for maintaining political control over this critical resource.[40]

Thus when the United States designed its Marshall Plan in 1948 to rebuild Europe, it laid the foundations for a new energy in West Germany, in part for its own security agenda, in part to revive a devastated continent. Over 10 percent of all aid from the European Reconstruction Program went toward petroleum products, more than any other single commodity. The Economic Cooperation Administration (ECA), which dispersed Marshall Plan funds, paid for over half of the crude sold in Europe between 1948 and 1951 by American corporations. The ECA, moreover, pressured petroleum companies to keep prices low. Other American aid went toward constructing refineries in West Germany itself, but on the condition that Bonn give American companies equal access to its market. Through the early 1950s American corporations received over 90 percent of Marshall Plan oil funds going to West Germany.[41]

While the world wars had dashed German ambitions to become an oil power, some domestic firms retained a presence in West Germany's oil market. Before 1914 the country's largest financial institution, the Deutsche Bank, had muscled into the petroleum concessions of the Ottoman Empire, future oil heartland of the world, only to lose its shares when Germany lost the First World War. Two decades later, the Third Reich advanced synthetic petroleum and exploited crude in its occupied territories through the chemical giant IG Farben and the state-led Koninental Öl AG. After 1945 the Allies broke apart these corporate titans, with some of the pieces falling to other German companies. But the domestic oil firms that survived the wreckage of war remained small, nationally bound, and drew their oil from West Germany's limited and pricey domestic reserves or bought it from international companies.[42]

So it was the foreign oil majors that would dominate West Germany's energy market. These were vast, complex, multinational companies that spanned the globe, that held immense stocks of capital, that possessed extensive petroleum concessions, and that could mobilize intricate distribution networks. There were seven of these firms, five based in the United States, one in Britain, and one in the Netherlands, and to contemporaries they seemed like extensions of American, British, and Dutch power. Three of these would use their global reach and lobbying power to help their subsidiaries penetrate West Germany's market: ESSO-Exxon, British Petroleum (BP), and Shell, with headquarters in the United States, Britain, and the Netherlands, respectively. With their German operations based in Hamburg, these corporations each amassed a capitalization that dwarfed Germany's domestic energy companies.[43] Their initial success stemmed partly from Allied policy, since they benefited not only from Marshall Plan aid but also from the dismemberment of the Nazi industrial complex. More importantly, these companies controlled complete processing chains. In contrast to German companies, the majors possessed rights to the most productive oil fields of the world, where they extracted crude at extraordinarily low costs, refined it, shipped it to Europe, and sold it through their sales network across West Germany.[44]

After 1950 Bonn worked with the United States and the majors to build a modern oil infrastructure. The Economics Ministry cultivated cooperative technical ventures between domestic and American firms. It provided a range of incentives to encourage refinery construction in West Germany, including tax rebates, low interest loans, priority access to foreign exchange, an exemption from the sales tax on petroleum products, and after 1953, a tariff on imported crude. With these incentives, the majors built new refineries around the exterior of the Federal Republic, on the estuaries of Germany's major rivers—Rotterdam along the Rhine, Hamburg along the Elbe, Bremen and Wilhelmshaven along the Weser. Between 1950 and 1955 German refining capacity quadrupled to 13.2 million tons a year. Consumption of oil products quadrupled too, and by 1957 oil accounted for 12 percent of total energy use. West Germans were acquiring the taste for a new type of energy.[45]

Energy Price Wars

Both oil and coal producers at first thrived in this booming economy. Yet West Germany's energy system had hardly stabilized before a conjuncture of events unleashed a bitter feud about pricing that would disrupt the political economy of German energy and lead many to question the very nature of the nation's energy market.

This shift began with a set of forecasts that led Europeans to fear their continent would soon face energy shortfalls. In 1955 the Organisation for European Economic Co-operation (OEEC)—Europe's leading body of economic experts—commissioned its first comprehensive energy forecast. When Harold Hartley, the British chemist who led the study, released his report in 1956, it caused a sensation. Hartley predicted Europe was entering a "new world" of rapidly rising energy consumption, driven by a growing population and changing consumer habits. The gold standard of energy use was the United States, whose mineral wealth and mechanized production meant its workers could deploy three times as much energy as those in Europe, and thus enjoyed an unparalleled level of productivity. Europe must match this to compete. But where would it get its energy, for according to Hartley the continent would soon outgrow its own sources of fossil fuels? Europe, he concluded, "must have sufficient quantities of energy at the lowest possible price in order to preserve our position in this world of competition and to improve our standard of living."[46]

In 1956 Hartley defined a Western European energy problem: the continent must close the gap with the pioneer of high-energy society to sustain a high standard of living. These anxieties struck a chord in West Germany, for in 1956 the country became a net importer of energy for the first time since industrialization. Coal producers predicted the economy would hit its "capacity limit" because energy was in such short supply. Domestic forecasters and the Economics Ministry confirmed the dawn of a new age of energy scarcity.[47] Few contemporaries anticipated the immense energetic transformations that lay ahead. Experts like Fritz Baade, director of the Institute for Global Economics at Kiel and a parliamentarian for the SPD, believed the future lay with hard coal. In a widely selling energy forecast, he argued this rock would remain "the prime foundation" of Europe's energy supply through 1975, perhaps even 2000.[48]

Coal producers reacted to these forecasts by ramping up investment. So too did the oil industry. Erhard, meanwhile, pushed through three major changes to meet the nation's anticipated need for more energy. First, he increased coal imports from the United States by reducing the tariff on American coal. North American coal, already cheaper at the source than Ruhr coal, was now held in check only by high freight rates, and coal imports from the United States rose 50 percent between 1955 and 1957.[49]

Second, Erhard seemed to change his tune on the question of pricing. With the return of full sovereignty to the Federal Republic in 1955, the ECSC agreed to end the pricing regime that had kept German coal inexpensive for the rest of Europe, and it signaled that Bonn should have the final say over whether to fix the price of coal. With encouragement from his liberal advisors, Erhard agreed to end state control over coal prices.[50] On April 1, 1956, West Germany freed its coal prices, with Erhard presenting this as "a decisive step towards integrating hard coal into the market."[51] His decision, however, met resistance from many quarters. The cabinet passed it by the narrowest of margins. Opponents warned that freeing coal prices would spark wage demands that would unleash an inflationary spiral. The SPD and the Industrial Mining Union disapproved and instead called for energy import quotas and a massive state investment project to modernize the energy sector as a whole.[52]

In reality, though, Erhard never wanted the coal price to be determined entirely by supply and demand. As he noted in discussions with the European Council, "even if one wants to grant a market price, one must still have the right to influence the market."[53] Though technically freeing coal prices, he still wanted to retain indirect control. In his mind, German welfare depended on exports. Driven by traditional industries like steel, chemicals, and increasingly cars, exports were becoming the motor of the economy, accounting for 19 percent of national income. For the economics minister, "a plentiful supply of the cheapest possible energy is becoming ever more important. No one can dispute that preserving our position on the global market requires not the least a low cost of energy."[54] To achieve his goal of a technically free but in fact monitored and low energy price, Erhard struck a gentleman's agreement with coal producers, who pledged not to raise prices without first consulting the Economics Ministry. And he reinforced this with a new round of subsidies to keep coal prices low. With new wage support for miners, new contributions to pensions, and new depreciation allowances, all of which cost the federal budget 200 million DM a year, Bonn effectively subsidized coal by 3.5 DM a ton.[55]

Erhard paired this with a third policy that removed the tariff on fuel oil imports. This made West Germany the least protected oil market in Europe: France managed petroleum imports through a strict licensing and quota system; Britain taxed fuel oil heavily, and would even ban oil from the Soviet Union. The move demoralized coal producers. On one level, petroleum posed little threat to coal. By the mid-twentieth century, oil had become above all a fuel for the automobile. Since the industry began thermal cracking in 1913 it had maximized gasoline as a refinery output because of its profitability. Gasoline for cars, however, was a novel market that did not detract from coal sales, particularly in Germany, where automobile ownership lagged behind that of Britain and France. But as demand for cars grew across Western Europe, the refining process that produced gasoline also churned out byproducts, like fuel oil. And this oil derivative *did* rival coal across a range of markets. Ships began using fuel oil, including the growing fleet of supertankers that brought crude to Europe. Railroad companies started burning it in their trains. Industry and agriculture increasing employed it for machinery. Households consumed it for heating. Power plants even began turning to it for electricity. Coal's one major market that remained untouched by fuel oil was coking for steelmaking; by the late 1950s every other one now faced a new rival.[56]

In closed-door discussions between Germany's leading oil and coal firms, Helmuth Burckhardt, chairman of Bergbau-AG Lothringen in Bochum, one of coal's most influential lobbyists, and an energy expert for the CDU, admitted that coal was simply "not flexible enough" to compete with fuel oil.[57] Since 1953, coal producers had consistently predicted tariff-free fuel oil would lead to a disastrous rise in oil consumption at the expense of coal.[58] Erhard paid these complaints no heed. As he explained to the European Council, his three policies functioned as a cohesive strategy to "introduce as quickly as possible an all-embracing competition between [West Germany's] individual energy suppliers. Toward this end we are trying to strengthen the competitive position of American coal vis-a-vis Ruhr coal, and to promote the consumption of oil products." He wanted to force Ruhr cartels to modernize in a fight for survival in a low-price energy regime.[59]

While 1956 altered the playing field between West Germany's two main energy sources, the dire effects predicted by Burckhardt did not at first materialize.

That summer and fall, Egypt's new pan-Arab leader nationalized and then closed the Suez Canal, throwing the energy markets of the Federal Republic into turmoil. Oil prices spiked by nearly 50 percent, dampening the changes Bonn had just implemented and leading the OEEC to call on coal producers to ramp up production still further.[60]

However, 1956 did spark an intense debate about how the Federal Republic should manage the price of energy, raising questions of efficiency, inflation, distribution, favoritism, and equity that would animate energy discussions for the rest of the century. In public lectures, in the press, and in discussions before Parliament, Ruhr elites vehemently criticized Erhard's pricing regime. Keyser and Burckhardt estimated the fixed price of coal had cost Ruhr producers billions since 1948, and that this was the root cause of coal's problems.[61] Subsidies, meanwhile, had not stimulated investment as intended: over the past several years total investment in West Germany rose, while investment in coal declined by 10 percent. Continuing these policies, Keyser and Burckhardt concluded, would only "lead to new gaps that would have to be filled with new subsidies."[62] Instead, they demanded "a true price" for coal.[63] Yet their idea of a "true" price differed from Erhard's: it should be the Ruhr cartel that determined prices, and it was only fair that it should cover the costs of production as well as the new investment demanded by the rest of the economy.[64]

Confident from the forecasts of 1956, in October 1957 the Ruhr cartel exercised its new legal power and unilaterally raised the price of coal. The price hike came just days after West Germany's third national election, in which the CDU promised stable energy prices for exporters. The decision outraged Erhard, who saw it as a violation of the gentleman's agreement and a "declaration of war" by the Ruhr.[65]

In denouncing the price hike, Erhard argued that questions of equity had no place in Bonn's energy framework, only competition. In a detailed discussion at the European Council in Luxemburg, Erhard claimed that the ideas of Keyser and Burckhardt in no way "conformed to the principles of a free economy." For nothing in a market economy gives producers "a legal and moral claim to recover, at all times, their costs." If coal cartels had their way, it would lead to a "virtual dictatorship" by a sector with high costs and would place inflationary pressure on the rest of the economy. Instead, Erhard reiterated his standard position, that energy prices must be flexible, low, and under constant competitive pressure.[66]

In the price debates, Erhard and the Economics Ministry portrayed oil as the binary opposite of coal, a sector governed by the market on a global scale, and one that through competition would modernize West Germany.[67] Oil producers were happy to play this role, and they mobilized the language of the Social Market in ways that coal was never able to. Where coal spoke of fairness, covering costs, and maintaining employment, oil spoke of consumerism, productivity, exports, and efficiency. In a joint letter addressed to Erhard and Adenauer, the directors of BP, Shell, and ESSO-Exxon synced their arguments to the tenets of the Social Market. Competition between providers, they argued, is "the prerequisite for efficiently supplying the Federal Republic with energy." Protectionist measures meant to favor one type of energy over another would "trigger even worse crises at a later date."[68] Oil succeeded, the directors of the majors claimed, because their firms merely followed trends that consumers themselves were generating for more flexible and cheaper energy. Consumers, "with their growing demand for comfort," were the true "pace setters" of the energy market, not producers. This was captured in the slogan "everyone should live better." Coined at West Germany's largest industrial exhibition in Dusseldorf, this motto was adopted by Erhard and his publicists during the 1950s to promote their vision of transplanting American consumerism into West Germany. The oil majors quickly latched on. As Ernst Falkenheim, general director of Shell's German subsidiary, argued, motorization and the growing use of oil "belong to those things that are embodied by the motto 'everyone should live better.'"[69]

The majors likewise aligned their demands with the export agenda of the Economics Ministry. In a statement that could have come from Erhard himself, their representatives concluded that "the meaning and objective of economic policy in our country" was not about choosing one energy over another, but rather about "raising productivity, that is to say, raising the output for each hour worked so that West German industry will be competitive on the global market."[70] The problem, however, as oil lobbyists reiterated on countless occasions, was that West Germany lay behind its competitors in nearly every metric of petroleum consumption. An investment study by ESSO-Exxon's economic research department claimed that using more oil was the only way West Germany could "close North America's lead." Playing on the Hartley report, it emphasized how America's high standard of living stemmed from its "unlimited use of petroleum." To catch up, West Germany needed a multi-billion DM investment drive to expand refineries, lay pipelines, build tankers, and construct processing facilities. Fortunately, ESSO-Exxon concluded, West Germany need not fund this project on its own; rather, oil firms could do so with their deep pockets.[71] Playing on a new iconography created by boosters in the 1950s, West German oil men portrayed their industry as "the raw material of progress," which would offer the world "nearly unlimited possibilities" from automobiles to air conditioners and plastics.[72]

Despite this publicity, however, the majors did not emerge from the price debates unscathed. For just as a new heroic iconography of oil was emerging, critics began looking more carefully into the global supply of petroleum. The first criticism came in 1955, when a report by the United Nations Economic Commission for Europe (UNECE) suggested the majors were keeping the price of crude oil artificially high for European consumers. Before 1955, little empirical public data had been published on the price of European petroleum products. The authors thus presented some of the first rough, public estimates of global crude prices, and in doing so they inflamed the fraught question of pricing by concluding that the global market for oil "could scarcely be regarded as unified nor prices as resulting from the free play of competitive forces."[73]

Those countries that harbored the majors, above all the United States, condemned the study, as did the majors themselves. Realizing it was treading on thin ice, UNECE muted its conclusions in cautious language. The press, however, sensationalized the report and accused the majors of unjust profits and pricing. In West Germany, shortly after it appeared, the SPD initiated a parliamentary inquiry into the price of gasoline, asking why it was higher in West Germany than in other European countries and suggesting the majors might be reaping unreasonable profits.[74]

The SPD's salvo marked the first major publicity attack on West German oil. Lacking detailed price information, the Economics Ministry turned to the majors themselves for data. The subsidiaries of ESSO-Exxon, Shell, and BP responded defensively with reports that contested the SPD's charges. They argued it was not the government's place to "review the gasoline prices at service stations for their adequacy." Such an action, BP noted, "represented discrimination against the oil industry," a remark that betrayed a willful ignorance of how Bonn monitored coal.[75] They argued that profits from gasoline sales were low, that West Germany's market for oil was highly competitive, and that the subsidiaries of ESSO-Exxon, BP, and Shell earned profits that were typical.[76]

But the press remained unconvinced. For the UNECE and SPD had raised the question of whether the price of oil was actually made by the market, like Erhard and the majors claimed. As the nation's leading newspaper pointed out, West Germans simply had little information about the price of petroleum products; the reports from ESSO-Exxon, Shell, and BP were some of the first studies with data on cost that reached an external audience. The task of judging whether gasoline was unjustifiably expensive was made even more difficult because petroleum products such as fuel oil and gasoline were co-produced, or made from the same refining process. A major could use profits it made from selling one byproduct, like gasoline, to offset the costs of another. And because Deutsche Shell, Deutsche BP, and Deutsche ESSO belonged to international conglomerates, they could easily shift capital and profits across borders.[77]

More generally, the price debates of the mid-1950s began to reveal the lie at the heart of the social market economy when it came to energy: namely, that there was no actual market price for coal or for oil, at least not the competitive one Erhard claimed was necessary for West Germany's export economy. As analysts in the 1960s would later point out, the price of crude—a price that shaped that of other petroleum products further down the production process—was not set by supply and demand, but rather by the power interplay between the majors, the producing countries in the Middle East, and the United States. Before the mid-1960s nearly all crude oil moving across international borders flowed *within* the international majors, from one subsidiary to another subsidiary of the very same conglomerate.[78] As Shell proudly pointed out, "the field of the great oil firms has always been the whole world. National borders are for them almost always of secondary importance as a factor of economic location."[79] ESSO-Exxon, for instance, sold crude oil extracted from its concessions in Kuwait and Saudi Arabia to Deutsche Esso in the Federal Republic. The price associated with this internal flow, however, was anything but market-determined. In order to pay taxes demanded by producing countries, the majors imputed a nominal price to the transfer of crude between their subsidiaries. The first such "posted price" came in 1950 for Iraqi crude, and others soon followed. During the 1950s this became an artificial number that had nothing to do with supply and demand and everything to do with the tax haul negotiated between oil corporations and foreign governments.[80]

The international majors not only set the world price of crude, they also used their global network to concentrate profits where they faced the least competition, namely, extracting oil from the ground.[81] When the SPD investigated the price of gasoline for excessive profits, in other words, they were looking in the wrong place. As ESSO-Exxon stated in its response, this line of questioning allowed them to "quietly leave out of consideration" any discussion of crude prices and profits. Yet this is precisely where the bulk of the majors' profits lay: they sold crude to their domestic subsidiaries as well as to independent refiners at prices far above the marginal cost of production. While the profit rate of the majors' subsidiaries in West Germany may have been normal, the international holding companies they belonged to reaped massive returns. And because ESSO-Exxon, BP, and Shell supported a unified production chain within the Federal Republic, they could easily shift prices to their advantage, raising them in fields where they enjoyed market power and lowering them where they faced competition.[82]

By the middle of the 1950s coal producers and even government officials began to note with trepidation the power of the oil majors. They claimed these conglomerates were keeping prices high for gasoline, where they enjoyed an "oligopoly situation," in order to sustain low prices in a "struggle to destroy" competitors in other arenas, above all fuel oil.[83] During the late 1950s it was Keyser and the UvRb who first put these pieces together, suggesting that the "monopoly price of gasoline has allowed fuel oil, which is a by-product, to be disposed of at a price below the actual cost of crude."[84]

Figure 1.1. Unused coal piles in the Ruhr in 1960. *Source:* Farrenkopf and Przigoda (eds), Glück auf, Nr. 86, pp 244. Original from Montanhistorisches Dokumentationszentrum (montan.dok), Deutschen Bergbau-Museum Bochum. Fotothek 024900007001.

Energy Crisis Returns

In the 1950s, in other words, West Germany's energy system was hardly the one portrayed by Erhard. Energy prices were low, as Erhard had hoped. Yet they were never actually determined by the free interplay of supply and demand. Instead, massive international oil corporations dominated the markets for gasoline and fuel oil. Against them were aligned an ailing phalanx of coal cartels tied to the national market, dependent on subsidies, and enjoying little latitude to control the price of their product.

In 1958 this market descended into crisis for the second time in a decade. This time the problem was too much rather than too little energy. The descent from the halcyon days of 1956 happened rapidly, catching nearly everyone unaware. In 1957 the growth of energy demand slackened.[85] Unsold coal piled up at the pits (see Figure 1.1). In February 1958 mines in Essen and Bochum sent 18,000 workers home without pay. More work halts and unpaid leave followed, and within a year the industry had cancelled 2.7 million work shifts. At the end of 1959 the first major pits closed. Large firms weathered the storm by relocating capital and workers to their more productive mines and closing individual shafts.

For small mines, however, the downturn became an existential crisis. As West Germany's leading periodical proclaimed, "the death of the mines" had begun.[86]

The causes of this crisis were multiple. Problems began with UvRb's decision to raise the price of coal in September 1957. This was a strategic blunder, coming just as the price of coal's nearest rival, fuel oil, began falling. Yet the crisis was more than a self-inflicted wound. For the economy as a whole slowed unexpectedly in 1957, and this cyclical downturn overlapped with a broader trend in which firms began using energy more efficiently. Two mild winters, finally, led households to burn less coal. While GDP grew 10 percent between 1957 and 1959, total energy consumption actually declined.[87]

All of this might have been a road bump had it not been for American coal and foreign fuel oil. As a result of Erhard's policies, by 1958 there were over 40 million tons of foreign coal contracts outstanding, mostly from power plants.[88] Oil products, meanwhile, continued to advance across the continent. While the rest of Western Europe tightly regulated petroleum imports to protect their domestic producers, West Germany did not. The majors responded to Europe's burgeoning gasoline market by pumping surplus fuel oil, a byproduct of gasoline, from other countries into the Federal Republic. There, fuel oil consumption more than quadrupled between 1954 and 1958, and its price plummeted more rapidly than anywhere else. By 1958 fuel oil became, for the first time ever, cheaper in thermal equivalent units than coal.[89]

In Bonn the crisis struck a nerve. CDU politicians feared the mine closures would lead to "radicalization in the workplace and a potential Bolshevization of Germany." Adenauer himself thought the Ruhr might descend into a "witch's cauldron" of social upheaval if the coal issue was not resolved.[90]

But politicians disagreed on how to respond. Erhard and the Economics Ministry believed the problems were cyclical and wanted to advance their low-priced framework.[91] As Ludger Westrick, Erhard's state secretary in the Economics Ministry, put it, "only when coal remains economically priced in comparison to oil can coal mining make a claim to be the most important energy supplier. If it is not in the position to do that, it will have to face the consequences."[92] Erhard thought the situation was really "not that dramatic," the canceled mining shifts economically insignificant. Thus throughout 1958 the architect of the Economic Miracle dismissed any active response and insisted that competition must have full reign.[93]

At the European level, the ECSC floundered through the crisis and burned up its limited political capital. Initially the High Authority believed the coal surplus would be temporary, and it delayed declaring a state of "crisis," which would have activated its wide-reaching powers. By the spring of 1959, a full year after the troubles started, the High Authority finally declared coal to be in crisis, hoping to collectively restrict overseas coal imports and finance the stockpiling of this precious rock. But the European Council, which represented the six nation-states in deliberations with the High Authority, vetoed this proposal. France and Italy, with little domestic coal, withheld their consent. A leading Italian politician remarked that he could not recall when coal-rich Northern Europeans had shared their energetic abundance with his country. But Germany, too, rejected intervention from the High Authority. The CDU feared European meddling in a sector over which their state had only recently regained sovereign control.[94]

The members of the ECSC thus charted their own national courses through the crisis. Toward the end of 1958 Erhard realized the situation was worsening, but if anything, he saw an opportunity to drive through the harsher side of his liberalization agenda. Stagnation in coal, he hoped, would "act as a brake on the union's wage demands in all types of employment," an argument Adenauer warned him not to repeat in public.[95] The Economics Ministry used the crisis to shut down West Germany's least efficient mines, and Erhard suggested that coal must pass through an "adjustment" by reducing output and transferring workers to other sectors and other regions.[96] In his mind, the only real question was which people "should be taken out of the labor force."[97]

Arrayed against Erhard was an unlikely alliance of Chancellor Adenauer, the Industrial Mining Union, coal producers, and domestic energy companies. They saw the crisis as a structural one and demanded that Erhard temper his competitive framework, prioritize goals other than low prices, and prevent what they feared would become a social catastrophe. Gutermuth of the IGB was the first to recognize that coal faced something other than a mere a cyclical downturn. "Structural transformations" in the economy and "far-reaching changes" in the energy system were leading him to question whether Germans still lived "in the era of iron and coal," or whether they were on the cusp of entering a profoundly different sort of energy society. At the union's annual conference in June 1958 he roundly criticized Bonn for failing to realize how West Germany's different energies were becoming part of a single market. He feared the nation was entering a new era in which oil would become an existential rival to coal.[98]

This structural change was so pressing that it led Gutermuth and other union leaders to rethink industrial relations as a whole. The relationship between unions and mine owners had never been smooth. The sheer depth of the crisis, however, forced the Industrial Mining Union and coal owners to begin advocating branch solidarity instead of class solidarity. In July 1958, Gutermuth joined with UvRb leader Keyser to present a unified front to Bonn. They called for stronger planning mechanisms for energy as a whole, not just coal, by establishing a "Council of Economic Energy Advisors."[99]

The crisis, in other words, opened the space for collaboration between mine owners and mine workers to save their ailing sector. Gutermuth, Burckhardt, and Keyser now went around Erhard to the chancellor directly. In Adenauer they found a politician more sensitive to the social needs of their energy. Since early 1958 Adenauer had been warning Erhard not to underestimate the Ruhr's problems.[100] Adenauer found Erhard's market policies, which required coal to shoulder the full burden of adjustment, far too harsh.[101] As he pointed out in a closed-door cabinet meeting at the height of the crisis, "mining is a special type of profession. Miners have a true sense of community. The notion, for a miner, that he could migrate into a different career represents but a meager comfort."[102]

By 1958, moreover, Adenauer began to worry that the power of international oil companies might be a greater threat to the economy than the coal cartel. The majors, so Adenauer believed, were delivering fuel oil in huge quantities at low prices with the explicit aim of "squeezing German coal out of the market." The problem was that "no one can prevent these oil suppliers, which monopolize the oil market, from then raising their prices after they have managed to reorder West Germany's economy from coal to fuel oil." Bonn, however, had to preserve coal for social reasons. If the federal government did not reassure miners of their jobs, the region would become a political "flashpoint."[103]

In August 1958 Gutermuth and Keyser convened their first meeting with Adenauer. Gutermuth played up the danger of radicalization and Keyser the threat of layoffs to demand that Bonn limit coal imports, cancel contracts for American coal, and establish an Advisory Council for Energy affairs. Adenauer responded favorably to these ideas, but the Economics Ministry opposed nearly all of them, most viscerally the last one. Erhard again stalled in implementing any substantial policy. By December 1958 Burckhardt, the new director of UvRb, announced the likelihood of 100,000 layoffs unless Bonn changed course.[104]

The crisis continued to mount, and in January 1959 Gutermuth organized the largest public demonstration since the founding of the Federal Republic; 80,000 miners marched through Bochum, an iconic coal town, to show how the miner's "living standards were being sacrificed on the altar of liberal market principles," and how Erhard's liberal price regime was creating an "economic Stalingrad on the Rhine and the Ruhr."[105] Erhard finally admitted that the price competition unleashed by his policies might not be functioning as intended, and was partly to blame for the crisis since it had "led to unnatural dumping prices."[106] Yet instead of adopting the proposals of the Industrial Mining Union or UvRb,[107] Erhard encouraged energy producers to solve the challenge themselves through a joint coal-oil cartel.[108] Coal producers, meanwhile, began buying out contracts for American coal through a new German Coal Mining Emergency Organization. Yet neither measure stopped the crisis. The cartel soon collapsed as foreign companies from France, Italy, and the Soviet Union dumped fuel oil into Germany.[109] Coal kept piling up at the pitheads, and by late 1959, Ruhr producers had laid off 60,000 people.

The climax came in September, when Gutermuth and IGB organized a second protest, this time in the capital of Bonn itself, where 60,000 miners marched silently through the streets waving black flags in a call for federal assistance (see Figure 1.2). Erhard finally changed his stance and moved toward Adenauer's perspective. Though never admitting the crisis was an economic problem, he now accepted it as a "human problem" that the state must address. On public radio he explained his about-face, noting that "it is one thing if each year hundreds of thousands of people in the Federal Republic willingly leave their workplace. But it is another thing when thousands of people are forced because of structural shifts to move into other professions."[110] Following the march in Bonn, the cabinet passed a draft bill for a tax on fuel oil. Westrick and Erhard accepted it, but only because it was better than "letting the Ruhr go up in political flames."[111] At the same time, the cabinet approved a new fund to support miners. With help from the ECSC, Bonn paid out 150 million DM for lost wages, and the government began financing a massive new retraining

Figure 1.2. Miners' demonstrations in Bonn, September 1959. Over 60,000 came from cities throughout the Ruhr. *Source:* Süddeutsche Zeitung Photo / Alamy Stock Photo. Image ID: 2A2B50H.

program to helped miners find jobs in other industries. For the moment, the crisis seemed to have passed.[112]

Conclusion

In some respects, this was a remarkable resolution. The new policies helped West Germany manage the *effects* of this energy crisis with considerable success, above all the threat of mass unemployment. The measures of 1959 represented the largest state social expenditure in German history until then. And although this "social" component of the social market economy was forced on Erhard, the Economics Ministry did move toward Adenauer's more accommodating position by cushioning the adjustment for an energy source that was declining under the weight of global competition.

For decline is what began to happen to hard coal after 1958, as West Germany's first energy transition accelerated. The year 1958 was an inauspicious watershed for the pioneering institution of European integration, which was founded on an energy weighed down by problems, and which emerged from the coal crisis with its reputation on the wane. The Council's resounding rebuff curbed the supranational aspirations of the budding European Community, and forced the High Authority to recognize its dependence on the member states. Across Western Europe, it was domestic policies that brought coal through this first crisis.

In West Germany, insofar as the policies of 1959 addressed the *causes* of coal's demise, they were a failure. Erhard's decision to open his nation's energy system to competition from foreign sources would have major consequences over the coming decades. The fuel oil tax is telling. Because the Economics Ministry received a flood of petitions protesting the measure from coal-poor states like Bavaria and from the oil industry, the final version was watered down from what Keyser and Burckhardt wanted.[113] It did little to level the playing field between coal and oil, a fact strikingly revealed by the stance of the majors. In public, ESSO-Exxon, BP, and Shell protested the tax. In private, they predicted it would do nothing to stem the "relentlessly rising pressure of the growing supply" of oil on the global market. In 1960 Herbert Fischer-Menshausen, director of ESSO-Exxon, told Adenauer that low oil prices were here to stay and that Bonn had best build its energy policy around this hard fact of life.[114]

The energy policies of the 1950s helped catapult West Germany into a new high-energy society, as the Economic Miracle raised living standards and gave Germans the ability to buy a host of new products, from cars to refrigerators. This was made possible by a specific and conscious policy of low-priced energy that aimed to spur growth, contain inflation, and promote exports. But the unintended consequence of this was an explosion of carbon emissions (see Figure

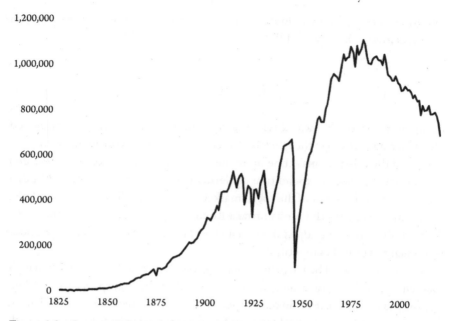

Figure 1.3. German carbon emissions, 1825–2019 (thousands of tonnes). *Source:* Data from Hannah Ritchie and Max Roser, "Germany: CO_2 Country Profile," *Our World in Data*.

1.3). For before it began to actually diminish hard coal production, this first postwar energy transition raised the consumption of energy across the board. With the exception of the five years of postwar recovery after 1945, the 1950s saw the most rapid and sustained increase in carbon output in the nation's entire history—an emissions growth rate of 4.7 percent a year—as West Germany joined the Great Acceleration as the world's third largest source of carbon after the United States and the Soviet Union.

More broadly, the Korean energy crisis, the price debates, and the coal shock of 1958–1959 altered how West Germans would approach energy. These crises turned energy into an intensely politicized topic and a coherent field of action that numbered among the most pressing national priorities, in contrast to the United States. The price wars, moreover, disabused policymakers of the notion that energy could be treated like other commodities. Many in the Federal Republic began arguing in the 1950s that there was no real free market for energy. Union leader Gutermuth went the furthest, calling the entire notion of the social market a "fiction" when it came to energy. In the heady days of the Economic Miracle, Germans had forgotten that "in the end, all economics is planned. The large conglomerates must plan, as must the individual small and medium firms." The only question was whether planning was effective or not, and in the field of energy West Germany had failed, in his opinion. Gutermuth wanted a national Energy-Economic Council. Many others agreed.[115] As one immediate result of the crisis, Bonn organized the first formal postwar parliamentary inquiry into energy, inaugurating a practice of corporatist collaboration between state officials, politicians, industry insiders, and academic experts to guide energy policymaking.[116]

The ramifications of this disenchantment with free energy markets were powerful, for the price wars began to change how West Germans framed the very question of energy. Many, including influential political insiders like Chancellor Adenauer himself, began to see energy as too important, too critical to life, to be left to the market, opening the door to the possibility that policy should address goals beyond cost. Specifically, German leaders were coming to see the international majors as much a cause of instability as a force for modernization. Some began to fear their new nation might lose control over its own energy system, that energy now lay outside Bonn's sovereign control. Adenauer, and a bevy of other ministers, worried that West Germany's "economy could be battered by price manipulation from the large oil conglomerates," with their immense wealth and global reach.[117] After 1958 mine owners and union leaders exploited these concerns to suggest a new concept: a "security premium" for energy. As Gutermuth suggested, "For such security one must pay a certain price. The Federal Republic already spends billions for military security. It is not presumptuous to provide resources for the security of our supply of energy, for the security of the mining industry, and for the people who create and work in that industry."[118]

After 1959 many thus still hoped that with new policies—the right policies—coal's "future prospects [would look] much more favorable."[119] But would they? For the crisis raised more questions than it answered. How far and how fast would this energy transition proceed? How long would coal remain king? And what did the future hold as West Germans began consuming energy in quantities that far surpassed the expectations of the 1950s?

2

The Coupling Paradigm

Conceptualizing West Germany's First Postwar Energy Transition

An adequate supply of energy is one of the essential conditions for economic advancement.[1]

The first postwar energy transition had grave consequences for the social fabric of West Germany, destroying thousands of jobs in the Ruhr while creating new ones in Hamburg and Bavaria, leaving behind abandoned mines and slag heaps while raising new refineries across the nation. But the effects of this transition spread beyond miners in the Ruhr or oil executives in Hamburg, to reshape the way Germans thought about their economy, raising pressing questions that animated public debate and academic research. Why was this energy transition taking place? What role should energy play in a modern economy? What goals should energy policy aspire to achieve? What should be left to the market and what should fall to the state? What did the future have in store for West Germany at it created a new infrastructure rooted in foreign petroleum instead of domestic coal?

In the 1950s, answers to these questions came primarily from Ordoliberals, a diverse group of intellectuals who profoundly shaped economic discourse in the new Federal Republic. After World War II, Ordoliberals and their patron, Ludwig Erhard, crafted a language of economics that set the tone for policy for the coming decades and that reverberates to this day in Germany's current *Energiewende*.[2]

While the rest of Europe was nationalizing its energy sector, Ordoliberals were calling for competition and private control. But they had little to say about energy in particular, and they eschewed any pretensions of forecasting the future, since they believed markets were inherently unpredictable. By 1960, however, policymakers increasingly sought guidance about energy, above all what the future might bring—for the coal crisis of 1958 overlapped with a new mentality in which social scientists began seeing the future as a sphere of political action that needed to be managed, rationalized, and brought under control. As a wave of technological, economic, and social change inundated Europe, an entirely new form of inquiry arose to meet the newly perceived need of "knowing the future," which Ordoliberalism could not fulfill.[3]

"Future studies," as it was called, found an ardent patron in West Germany's political elite and by the 1960s politics was becoming increasingly future-oriented and geared toward information analysis and prognosis.[4] As the style of politics shifted, so too did the goals. After West Germany recovered from war and managed to safeguard the basic freedoms crushed by the Nazis, new aspirations gained urgency, above all the quest for long-run economic growth.[5]

To meet these emerging political needs, a new discipline of energy economics took over where Ordoliberalism left off. Based at the Energy Economic Institute (*Energiewirtschaftliches Institut—EWI*) in Cologne and the German Economic Research Institute in Berlin (*Deutsches Institut für Wirtschaftsforschung—DIW*), these economists advised the government on energy, beginning with the first formal parliamentary inquiry into coal. And while they spoke a similar language to that of Ordoliberals by emphasizing competition and markets, they added new instruments to West Germany's policy toolkit, above all forecasting, which they adapted from methods first used by the Organisation for European Economic Co-operation (OEEC). Under their stewardship, the goal of energy policy began to depart some from the Ordoliberal agenda, and with time their ideas hardened into a paradigm: that economic growth in a modern economy was inextricably and inherently coupled with growth in the supply of energy.

In crafting this coupling paradigm, German energy economists differed from neoclassical theorists in the United States. Scholars often portray the 1950s and 1960s as the moment when economists began paying *less* attention to energy and raw materials as "a practical limit to economic possibility." Neoclassical economics, particularly in the Anglo-American world, became a science of money or prices, that paid little attention to the "material forces and resources of nature and human labour."[6] Mainstream American models described the economy as increasingly disconnected from material inputs—dematerialized even—as producers could switch from scarce to abundant resources at the change of a price.[7]

German energy economists, by contrast, did not dematerialize the economy as did their counterparts in the United States. The new experts departed from Ordoliberals as well as from Americans by portraying energy as a uniquely important input. As expectations for growth expanded, so too did forecasts for the consumption of ever more energy, while the models of these new experts naturalized oil's success as the new foundation of West Germany's economy. But in the process, their assumptions about how different energy types interacted in the market grew increasingly simplified and rigid, and they put forth *one* future trajectory as the conventional wisdom—the massively rising consumption of petroleum—to the detriment of alternative future scenarios.

This divergence from Anglo-American theory gave West German experts a different set of priorities than their peers in the United States, where the pursuit of growth trumped all else. Because development remained so tied to energy in German models, and because the nation's prime energy was increasingly foreign oil, security came to rival growth as a goal. This understanding of energy security, moreover, was deeply shaped by West Germany's position as a middling state in a broader transatlantic community. Having lost its bid for both an empire and a multinational oil company in the world wars, the Federal Republic had little ability to control energy reserves beyond its own borders. The costs of security thus seemed high while the tools to achieve it were limited. German theorists consequently portrayed the trade-off between security and growth in stark terms, conceiving of energy policy as a dilemma. And by the 1970s, they had developed their own distinctive way of understanding energy's place in the economy, which would have great ramifications for the nation for the rest of the century.

The Language of Ordoliberalism

Ordoliberalism—an ordered or structured liberalism—was the founding economic doctrine of the Federal Republic of West Germany. Ordoliberal thinkers bequeathed to the new state a language for economic matters that its political leaders and public have spoken through since the 1950s. An eclectic group of thinkers who ranged across disciplines, from economics to law and sociology, Ordoliberals began crafting a new vision of political economy in the 1930s in response to the Great Depression and the Third Reich.[8] The economic collapse of 1929 had laid bare for them the inadequacy of laissez-faire capitalism, in which a minimalist state did little to generate order in the market, and instead left matters entirely to firms and individuals. Without a mechanism to ensure a competitive playing field, so Ordoliberals argued, oligopolies and cartels inevitably formed, distorting the market and subverting the state through their lobbies. For Walter Eucken, a founding father of Ordoliberalism, the trend toward economic concentration was "a ubiquitous phenomenon of socio-economic life." A conservative nationalist who disliked the Weimar Republic, Eucken also became a critic of the Third Reich after 1933 and developed contacts with the resistance. From his position at the University of Freiburg, which he held from 1927 until his death in 1950, he developed a novel approach to law and economics. For him, the fundamental issue was the question of concentrated power—above all, the private power of cartels and oligopolies—and how to limit it, for such power would both cripple the market and infringe on the liberty of individuals.[9] According to Eucken and Franz Böhm, the leading legal theorist at Freiburg, such concentration of economic power had led to the collapse of the Weimar Republic in 1933. How to keep organizations small and local, and consequently free, was Ordoliberals' overriding priority.[10]

Before 1940, some Ordoliberals had agreed with the Nazis on the need to restrict democracy. But over time, most came to see Nazism itself as the greater threat to a liberal economy: National Socialism was a reaction to the Depression, but one that was worse than the laissez-faire that preceded 1929. With the outbreak of war, for Ordoliberals the Third Reich came to embody collectivism, their bête noir, for using planning to destroy not only the price mechanism and competition, but also economic and political liberty. After 1945, more than anything else, Ordoliberals wanted to distinguish the new Federal Republic from the Third Reich by preserving personal freedom as well as markets.[11]

This idiosyncratic analysis of the Depression and National Socialism informed the vision that Ordoliberals laid out for postwar Germany. What Wilhelm Röpke first called the "third way" in 1937, and Alfred Müller-Armack renamed the "social market" a decade later, was a political economy that differed from both laissez-faire capitalism and planning.[12] Röpke, a student of economics and sociology, had fled the Third Reich to Istanbul and then to Geneva, where he helped found the Mont Pèlerin Society in a quest to reimagine classic liberalism. Müller-Armack, who initially worked with the Nazi regime only to disengage from them in the late 1930s, joined the CDU after the war. For these two, the guiding principle for the state must be to actively create the institutional foundation for a competitive market order, and energetically prevent this foundation from degenerating into large concentrations of power. As Eucken put it, "the fundamental principle not only calls for abstinence from certain economic acts.... Nor is it enough simply to prohibit cartels, for instance. The principle is not primarily negative in nature. There is, rather, a need for positive economic policy aimed at developing the marketing structure of unrestricted competition."[13] Competition, in other words, was crucial for achieving the ultimate goal of the social market economy: defending the "basic values and principles of a free personality" that the Nazis had dismantled.[14]

This Ordoliberal animosity toward concentration grew from a conservative critique of modern society, which they believed had taken a wrong turn in the nineteenth century by becoming infatuated with everything large. Röpke—the leading public figure of the Ordoliberal group—advocated this critique most forcefully, but others also vilified the propensity toward big-ness in all types of organization, public or private. Röpke described his Third Path as one of "moderation and proportion," which would "free our society from its intoxication with big numbers, from the cult of the colossal, from centralization, from hyper-organization and standardization, from the pseudo-ideal of the 'bigger the better,' from the worship of the mass man and from addiction to the gigantic."[15] In fact, he called for a radical decentralization of economic life, "decentralization in the widest sense of the term," from promoting local tradesmen and disaggregating large corporations to replacing urban sprawl with towns of fifty thousand people.[16] Yet this was a profoundly conservative philosophy that valorized Germany's past in order to avoid descending into a "Godless capitalism." In Röpke's opinion, the nineteenth-century cult of the colossal had given rise to Europe's industrial proletariat, which threatened Europe's Christian foundation.[17]

If competition and decentralization were one instrument in the Ordoliberals' toolkit, prices determined purely by supply and demand were a second. Ordoliberals abhorred collectivism not only because they thought planning would result in dictatorship and coercion,[18] but also because collectivists' "passionate but irrational advocacy" of price controls reflected an "inability to find a method of accounting suited to the different degrees of scarcity of various products." Prices determined by the state, Ordoliberals believed, would inevitably lead to lower productivity and output than under a free market.[19] The state must make "the price mechanism workable. This is the strategic position from which one dominates the whole field."[20] When combined with competition, market prices would prevent concentration of power and advance individual and consumer freedom. For "in a free price system, costs are necessarily borne by consumers, it is the *consumers* who decide what and how much shall be produced. Hence it is the consumers who decide how the factors of production themselves are to be used."[21]

Competition, market prices, and a strong legal framework to curtail economic power and promote consumer freedom—these were the hallmarks of Ordoliberalism. In the late 1940s these were radical ideas, Germans having lived for over a decade with state-controlled prices, wages, and investment.[22] This doctrine might have remained a minor theory had it not been for the popularity of economics minister Ludwig Erhard. Though Erhard was a pragmatic thinker independent of any school of thought, he firmly believed in Ordoliberal principles. His decision to end price controls after the war was deeply informed by his Ordoliberal advisers, who had been calling for this since 1942.[23]

Erhard's 1948 price liberalization became the "founding myth" of the West German "Economic Miracle," and Ordoliberals benefited immensely from their association with these policies.[24] These reforms, along with Erhard and his publicity machine, propelled Ordoliberal ideas into the public limelight. Röpke became a celebrity economist on both sides of the Atlantic. Müller-Armack gained widespread popularity as the man who coined the term "social market economy," later directing the Economics Ministry's policy department. Eucken authored what would become most influential economics textbook in the Federal Republic.[25] By the 1950s, when West German politicians increasingly rooted their legitimacy in the success of the economy, Ordoliberals came to enjoy great prestige, and their ideas profoundly influenced discussions of energy in the Federal Republic's early years.[26]

Ordoliberals believed that energy was just like any other part of the economy. The only two spheres they deemed special in any respect were agricultural and labor. Energy, by contrast, should be a purely competitive market in which consumers would choose their fuel through the interplay of free prices and on the basis of affordability. For Röpke, production was, at its core, "nothing else than a perpetual transaction with Nature by which we seek to exchange on the most advantageous terms our efforts against the produced commodities." The upshot: energy, as nature's raw material par excellence, should enter the economic system as freely and as inexpensively as possible.[27]

At a time when the rest of Western Europe had nationalized much of the energy sector and firmly integrated it into the planned economy, the Ordoliberal approach set Germany apart. In Britain, the legacy of the Great Depression had left policymakers of the Left-Liberal tradition with a keen sense that public enterprise could be more efficient than private. For energy, "competition was not just inferior: it could be positively harmful, and hence, had to be made illegal." In France, the state was heavily involved in energy after World War II, and the leading school of economists believed they could make state-owned energy firms function as efficiently as private ones. In both countries, these sentiments and policies resulted in vertically integrated, state-owned monopolies providing electricity, mining coal, and refining gas, and state-supported domestic firms working the oil sector.[28]

This Ordoliberal approach found its clearest expression in the coal debates of the 1950s. As this energy fell under the authority of European institutions, Röpke and others railed against the European Coal and Steel Community (ECSC) for its dirigisme. Müller-Armack thought the ECSC's price controls robbed coal producers of their flexibility, while Ordoliberals more generally demanded competitive markets for all energy.[29] They believed that coal stood in the midst of a structural transition that was eroding its monopoly position in Germany's energy supply, and that this development should be encouraged, not resisted. Ruhr coal, after all, epitomized the colossal: a sector dominated by interconnected corporate groups that should be shattered by promoting competition from other energies like oil. If consumers preferred fuel oil or diesel, Ordoliberals argued, coal must simply "adjust." Bonn's only remit was providing mine owners and workers some "breathing space" to organize their sector's inevitable "contraction."[30]

Futurology and Forecasting

By the 1950s Ordoliberalism had fundamentally shaped the language through which West Germans discussed energy. Politicians and interest groups wanting to change the energy system would speak through the conceptual apparatus bestowed by Röpke, Müller-Armack, and Eucken of competition and market-based pricing. Ordoliberals, moreover, offered a powerful critique of centralization and scale, alongside a paean to small, decentralized networks that would endure into the twenty-first century.

Soon, however, the demands policymakers placed on their economic advisers began to change as the future itself became an anxiety-ridden minefield. With the needs of postwar reconstruction met, a host of new challenges and uncertainties pushed their way into the vision of politicians on both sides of the Atlantic. In West Germany this period saw ecological degradation become a widespread concern, with nitrate runoff polluting the ground water and oil spills marring the rivers. Globally, the threat of unchecked population growth and atomic war appeared to endanger humanity's very existence. New technologies, meanwhile, from television to the refrigerator to the automobile, were dramatically altering traditional ways of life and creating novel and unpredictable consumer lifestyles.[31]

As Jenny Andersson has illustrated, in the 1960s the future "came to be understood as posing distinct challenges to the functioning of societies, both East and West, and as a sphere in need of intervention." These anxieties in turn brought a greater urgency to forecast and plan for unknown eventualities.[32] Nowhere was this more apparent than in the field of energy. At the European level, energy saw some of the earliest and most rigorous forecasts, since European experts feared their continent's supply of fuel might not meet its burgeoning need. Fritz Baade, director of the Kiel Institute for the World Economy and author of West Germany's first popular energy forecast, captured these anxieties when he cautioned that the world was "living through the most dynamic [period] in history," driven by the "almost limitless" demand for energy generated by an unprecedentedly "prosperous humanity."[33]

In 1954 the OEEC—a transnational forum of civil servants and economists—began worrying that Europe's rapid growth might lead to energy shortages that could throttle development. The OEEC tasked Louis Armand, a French engineer who would become the founding father of Euratom, with convening a report on energy. Armand completed his study in 1955, warning that Europe's demand for energy was escalating rapidly and that the continent was becoming dependent on foreign oil for the first time in history.

Armand believed that "the history of energy resources was marked by sudden changes, veritable revolutions that constitute important stages in the advance of civilization."[34] Europe stood on the brink of such an energy revolution, and his urgency spurred the OEEC to organize the first detailed energy forecast, to be coordinated by British chemist and Oxford professor Harold Hartley. Published in 1956, at a time before gross national product (GNP) had become *the* measure to compare societies, the Hartley report argued that energy consumption per capita was the best gauge for quantifying the material progress of a nation. By this index, Europe lagged far behind the United States, which, by consuming three times as much energy per person, was enjoying higher productivity, higher standards of living, and higher competitiveness. But it was not only for reasons of development that Europe must pay attention to energy. Given the rapid growth of European oil imports, Hartley was concerned about security: the threat of an oil cutoff, and whether Europe could actually acquire the energy it would need in the future.[35]

Hartley's report was the first of many forecasts organized by the OEEC and its successor, the Organisation for Economic Co-operation and Development (OECD), part of their effort to pierce the veil of uncertainty hanging over Western Europe as the continent became a net importer of energy. The next one, by Austin Robinson of Cambridge University, published after the 1958 coal crisis, used methods that were virtually identical to those of Hartley. Together the two reports codified a set of best practices by developing two techniques for forecasting energy consumption. They first used a global estimate that predicted GNP growth, from whence they derived total energy consumption based on historical data for the energy intensity of an economy. The second method divided the economy into sectors and subsectors—industry, transportation, households—estimating the final usable energy each would need, converting this figure into primary energy by accounting for waste in the supply chain, and then summing the subtotals to reach a complete figure for the total energy consumed in the economy at a future date.[36]

Crucially, the assumptions used by OEEC experts revealed how they believed Europe's energy supply to be in flux, and how they thought the outcome of this fluctuation depended on price competition between energies. The reports by Armand, Hartley, and Robison all started from the premise that one type of fuel could easily be substituted for another. Energy, in their view, was price elastic: if the price of one energy rose too high, consumers or industries could and would switch to a cheaper form. As Robinson tellingly noted, "while the consumption of energy in total increases in a reasonably predictable relation to gross national product, the consumption of individual forms of energy must depend on the *relative prices*, as well as the technical advantages and conveniences of particular fuels, and thus ultimately on their costs of production."[37] Put differently, "demand will change in response to relatively small and almost unpredictable variations of relative prices."[38]

Hartley and Robison, in other words, believed that energy was malleable as an input in the production process, and that the outcome of current Europe's energy transition hinged on the price relationship between oil and coal. They had an openness to their approach, and they believed the "pattern of demand for primary sources of energy [was] over a sufficiently long period *very flexible*."[39] But as critics pointed out, Hartley and Robinson relied on intuition to incorporate this flexibility into their models. And intuition could err. While the reports anticipated that Europe's energy transition would continue, the speed with which it came to pass would prove utterly surprising.[40]

West Germany's First Forecasts

The OEEC's energy reports circulated widely and served as a model in West Germany, which began turning to forecasting in the late 1950s and 1960s to better understand the economy (see Figure 2.1). Interest in prognostication stemmed from the growing influence of economists as political advisors: by the 1960s, practitioners of this social science came to outnumber lawyers in high-level bureaucratic positions in the federal government. At the same time, Germany's economics profession began returning to its empirical roots. Where Ordoliberals protested predictions and saw the economy as an "open system" whose future could never be anticipated, a new generation of West Germans began adopting a data-centered approach that prioritized measurement. And where Ordoliberals had little to offer that could satisfy politicians' desire to know the future, a new sort of economist embraced forecasting with conviction.[41] As the SPD's parliamentary leaders put it, capturing these new sentiments, prognoses could provide "a view of the future, one that has a greater plausibility than another view. That is, after all, infinitely better than nothing. Better a vague sketch of unknown territory than just waving around a stick in the fog."[42]

In 1956, the same year as Hartley's report, Chancellor Konrad Adenauer (CDU) asked the Economics Ministry to conduct the country's first energy forecast. Using a crude extrapolation from past trends, the ministry predicted West German coal consumption would rise rapidly from 120 million to 165 million tons of coal by 1965.[43]

As the utter failure of this first prognosis became apparent after the 1958 crisis, both the CDU and the SPD agreed that they must study coal to better understand its future.[44] Following the lead of Fritz Burgbacher, director of Rheinische Energie AG and the CDU's energy czar, the nation's leading political

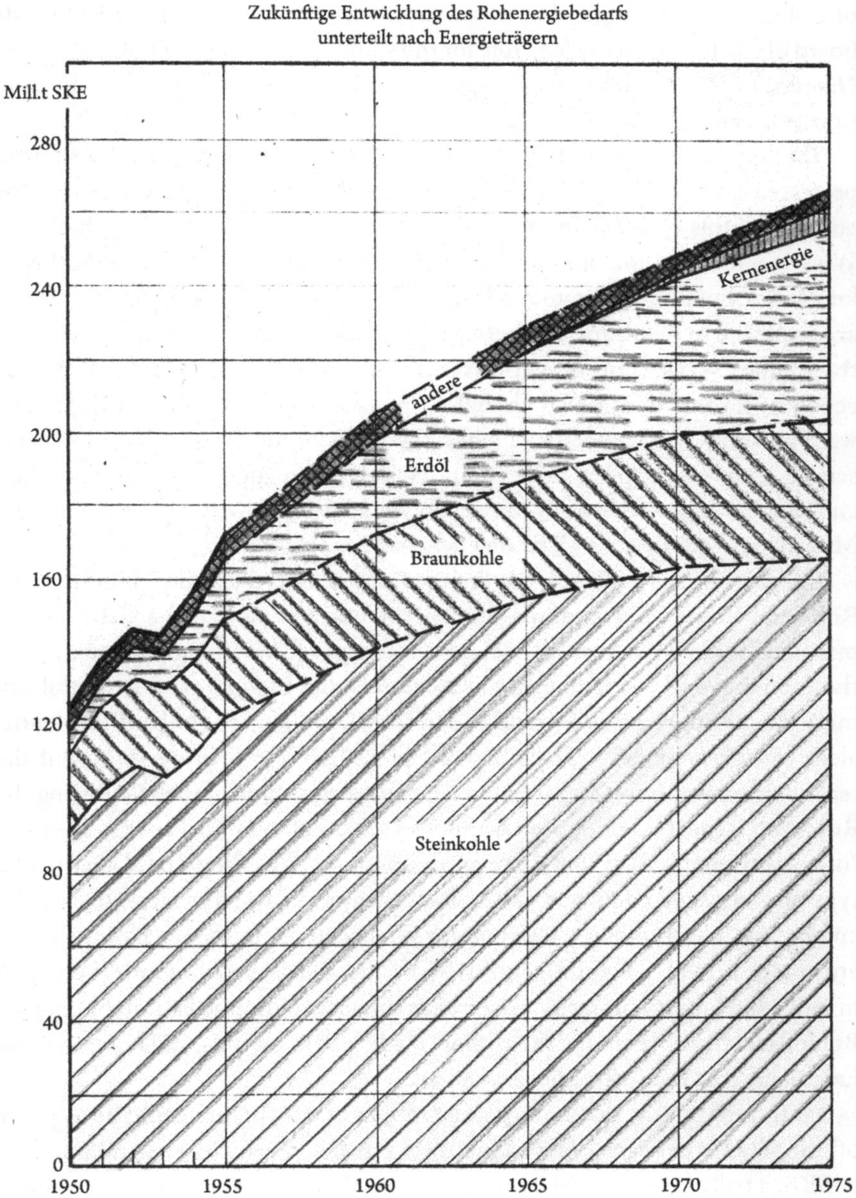

Zukünftige Entwicklung des Rohenergiebedarfs
unterteilt nach Energieträgern

Figure 2.1. Prediction of primary energy needs by energy type, 1956. Source: BBA, 26, Report from Wirtschaftsvereinigung Bergbau, November 1956. Image Title: "Zukünftige Entwicklung des Rohenergiebedarfs unterteilt nach Energieträgern."

party expanded the study to cover energy as a whole, arguing that the crisis illustrated how individual fuels could no longer be treated in isolation.[45] The government commissioned outside experts to conduct a forecast for all energies, with the special goal of targeting coal's problems in the conclusion. The EWI of Cologne, where Burgbacher was an honorary professor and president of the board of trustees, won the lead role for the study—known as the Energy Inquiry (*Enquete*)—solidifying this institution's reputation as the nation's leading authority on energy.[46]

The nuts and bolts of the Energy Inquiry were built by economists who professed far more confidence in prognosticating than did Ordoliberals. The epitome of this type of expert was Theodor Wessels. Born in the Netherlands, Wessels wrote his doctorate on Leon Walras, a pioneer of equilibrium theory known for building intricate models of the market. Wessels spent most of his academic life in the Rhineland between Cologne and Bonn. He had avoided active military service during World War II by doing regional studies of the Dutch economy for the Third Reich, though he remained critical of the Nazis' turn toward autarchy. After 1945 he climbed the ranks of the University of Cologne, serving as its rector in the early 1950s. He reveled in applying theory to policy, and after 1948 he served on the academic advisory board to the Economics Ministry and advised the CDU on energy issues.[47]

Under Wessels's guidance, the Energy Inquiry used the same methods as the Robinson Report and assumed that energy was a competitive arena in which price determined the type of fuel consumers used. Using a global analysis of the economy and estimating from past trends, the Inquiry predicted total primary energy consumption would rise from 212 million tons of hard coal equivalent (*Steinkohleeinheit—SKE*) in 1960 to 320 million SKE by 1975. But the real question was how this growth would affect each type of fuel. Following the Robinson Report, the Inquiry assumed a portion of total final energy used was "non-substitutable," i.e., that some processes or services always used a particular type of energy and could not switch. For other production processes the type of energy was "substitutable": firms or consumers could, in theory, choose an energy based on the "price differential." To predict the specific mixture of energies in 1975, the Inquiry projected past trends forward for non-substitutable energy. But for substitutable energy the authors made judgment calls, and it was in these judgments that forecasting became an art.

Most importantly, the Inquiry claimed that by 1975 there could be a sphere of competition between coal and fuel oil that reached as high as 156 million tons of SKE, a truly monumental amount. If price alone determined the outcome of this competition, fuel oil would wipe coal off the map. For where the Inquiry predicted the price of oil would decline, it expected coal costs to rise because of wage hikes. In its formal forecast that it presented to the government, however, the Inquiry did not apply this pure price logic. Because fuel oil was a byproduct of gasoline, because refineries prioritized gasoline over fuel oil, and because the rest of Europe would be competing for fuel oil, the Inquiry concluded that the outcome of the competition between these two energies would be determined not by *price* but by the *quantity* of fuel oil available. And this quantity, it concluded, would be limited.[48]

Thus, instead of an utter collapse, the Inquiry predicted West Germany's domestic coal would decline from 126 million to 100 million SKE by 1975 if existing policies remained in place. Domestic hard coal, on the one hand, still had a future. On the other hand, the Inquiry made it politically possible to imagine reducing West Germany's coal output below 125–140 million SKE—the magical number that had been a political taboo to question as late as 1959.[49] Even the Inquiry's best-case scenario now forecast that coal must decline; that this was an acceptable and unavoidable development. As the executive report to the chancellor concluded, "the declining importance of coal in a global context has long been apparent, and in principle should be seen as a natural process."[50] The Federal Republic, in their view, was participating in an unstoppable but gradual global energy transition off coal.

The Inquiry won acclaim as the most sophisticated energy forecast in West Germany. Nevertheless, it came under criticism from many quarters, including the banking sector, nuclear advocates, and of course coal boosters.[51] But the most damning criticism came from the oil industry itself, the energy the Inquiry seemed to favor the most.[52] To support its conclusion that the struggle between oil and coal would be decided not by the price but by the quantity of fuel oil available, the Inquiry had made assumptions about the refining process that the oil industry found to be gravely mistaken. As the director of Deutsche Esso noted in response to the study, the ratio of fuel oil to gasoline in European refineries was not actually fixed, as the forecast suggested. Instead, the oil industry had shown incredible "adaptability" in changing this ratio in the past, and would do so again in the future. Events elsewhere, moreover, would affect Europe's oil supply in a way the Inquiry had entirely overlooked. Demand for gasoline in the Middle East was rising, for instance. But these countries had little need for the fuel oil they co-produced in their own refineries, and this excess would wind up in Europe. Esso's director expressed supreme confidence that his industry "could meet all the requirements of consumers in practice without limit." That fuel oil might be limited in any way by 1975 was, in his mind, a totally "erroneous idea."[53]

Growth, Dematerialization, and the Rise of a New Energy Paradigm

The Inquiry aspired to bring more certainty to the ambiguity surrounding West Germany's changing energy structure. Would hard coal survive, or would oil sweep all others before it? By the 1960s, these were questions politicians desperately sought advice on because the economy was changing and expanding faster than ever. As it did, a new policy goal was emerging in Western Europe and North America: the pursuit of quantitative economic growth, what J. R. McNeill has called "the most important idea of the twentieth century."[54]

The growth paradigm grew, in part, out of a postwar international movement to create a standard framework for measuring the income of a nation, or GNP. While many economists looked upon such measures with skepticism, by the 1950s national accounting seemed to offer solutions to a range of challenges. As this measurement tool spread under the tutelage of OEEC/OECD officials and American experts, European leaders wanted not only to measure their nation's income, but to expand it. They came to see GNP growth as a panacea for almost any problem: it could resolve conflicts between labor and capital, ease budget disputes, even increase independence from American aid.[55] In the context of a hardening Cold War competition with the Eastern Bloc, by the 1960s countries across North America and Western Europe began entrenching this new growth paradigm in their policy agenda. In 1959 the United Nations reinterpreted its economic objectives to embrace growth. In 1960 the Robinson Report framed energy as a question of stimulating economic growth. In 1961 the OECD declared a decade of growth. As a leading British economist put it, "The cold war will last a very long time. Only by outgrowing the enemy can we keep on winning it."[56]

The rise of GNP and growth spurred a fundamental rethinking about the very concept of the economy. Where statisticians had previously compared the wealth of a nation by measuring material stocks—the output of coal, or the accumulation of gold—national accounting permitted a new level of abstraction. The economy became a process of monetary circulation. And where previous measurements suggested a limit because they were rooted in physical processes—a nation could, after all, exhaust its coal reserves, as the British economist Stanley Jevons poignantly warned in 1865—GNP had no obvious limit. In this new framework, monetary circulation could "grow without any problem of physical or territorial limits."[57]

In the United States, during the 1950s and 1960s neoclassical economics enshrined in its models this idea that growth was becoming dematerialized and unconstrained from energy or resources. Heterodox thinkers had once placed energy at the core of their analysis, like Lewis Mumford, who in *Technics and Civilization* argued that energy flows made economic systems: "the prime fact of all economic activity, from that of the lower organisms up to the most advanced human cultures, is the conversion of the sun's energies." But postwar mainstream American economists saw these earlier theorists as cranks.[58] Within the profession, only a handful grappled with resource questions. Among them was Kenneth Boulding, one-time president of the American Economic Association, who in the 1960s called on his profession to move beyond its obsession with growth and focus instead on resources, throughput, and the preservation of natural capital.[59]

In the late 1970s and 1980s Boulding's ideas would become foundational for ecological economics. But ecological economists were far from the center of prestige, and before the 1970s resource issues in American economics were confined to the sub-disciplines of agricultural and fishery economics. John Kenneth Galbraith, a leading American Keynesian, could remark in 1958 that there "hangs near total silence" over the resource question.[60] Instead, at mid-century Anglo-American neoclassical economists focused almost exclusively on growth, and their models belied a radiant optimism in technology, innovation, and the price system to overcome limitations. Roy Harrod, the British economist who had pioneered growth theory, dropped land and resources from his analysis altogether because, "in our particular context it appears that its influence may be quantitatively unimportant."[61] In 1953 the Paley Report, the first postwar American research effort to grapple with resource scarcity, argued that the acquisition of energy posed no threat to growth in the medium term. Instead, it highlighted how the value of all raw materials as a percentage of GNP had declined by half since 1900.[62]

The findings from this growing body of Anglo-American research suggested that advanced economies were becoming less dependent on natural resources like energy as they developed technologically. In 1960, one of the earliest conferences that applied neoclassical growth theory to resources concluded that the nineteenth-century classical economists' concern with scarcity was "no longer found relevant; for its founders wrote prior to man's invention of the method of invention and his subsequent innovation and institutionalization of the art of innovation," which was now divorcing growth from material inputs. The idea that resource prices would rise and throttle growth, the great fear of classical economists like Thomas Malthus and David Ricardo, must be relegated to "our stock of folklore." According to American neoclassicals, modern technology had created "chains of substitutes" for natural resources: "one can always be made to take the place of its scarce neighbor."[63] Even Resources for the Future (RFF), a think tank that continued the Paley Commission's work, advanced this optimism in dematerialization. One of its most influential publications, the 1963 report *Scarcity and Growth*, claimed that market pricing and innovation could overcome nearly all barriers to growth, because any increase in the "scarcity of particular resources fosters discovery or development of alternative resources not only equal in economic quality but often superior to those replaced." Pushing this logic further, the report concluded that technological progress in the United States had become "automatic and self-reproductive," to the point where "the process of growth thus generates antidotes to a general increase of resource scarcity."[64]

The pinnacle of such thinking came in the growth model crafted by the doyen of neoclassical economics at MIT, Robert Solow, which laid the foundation for Anglo-American economic analysis until the 1970s. This abstract framework—dubbed the "kingdom of Solowia"—included only the variables of population and capital, and focused exclusively on savings and investment. In the short and medium term, growth stemmed from the accumulation of capital; over the long term, it came from technological advance. Neither resources like energy nor patterns of consumption had any place Solow's theory. This emphasis on labor and capital underscored just how much the American profession found energy to be peripheral.[65]

Across the Atlantic, the growth paradigm came to West Germany in the 1960s, though it was never as dematerialized as in the United States. Economic expansion had certainly been a priority during the 1950s, but Ordoliberals believed expansion should be moderate and never unending. As Röpke protested in *The Social Crisis of our Times*, exalting rapid growth—in the economy or in anything else—"reduces qualitative greatness to mere quantity, to nothing but numbers." Having been cut off from the international debate over statistics between 1933 and 1945, Germany was slower in setting up a system of national accounts than other Western states, but by 1955 Bonn completed its first such estimate.[66] As the Nazi period receded into the past, the Ordoliberal concern that policy should prioritize individual freedom and the limitation of power was replaced by growth as the overriding priority. In 1961, for the first time ever, Bonn used the concept of growth in a formal government declaration. That same year the Federal Statistical Agency finally began publishing a more detailed set of figures on West German GNP. In 1963, the slowing of the "Economic Miracle" led Erhard to establish the Council of Economic Experts to advise the government on growth. In 1967, in the wake of a disruptive recession, Bonn passed the Law Promoting Stabilization and Growth in the Economy.[67] By the late 1960s the CDU, which had governed since the founding of the Republic, could state that "it is the declared goal of our economic policy to stimulate economic growth and increase prosperity in general."[68]

The rise of the growth paradigm led experts to rethink the role of energy in West Germany's economy. For during the 1960s West Germany not only adopted growth as a prime priority, that growth helped turn the country into an oil nation. By 1966, far earlier than the Energy Inquiry had predicted, oil surpassed coal to become the largest source of energy in the Federal Republic. Oil rose from 20 percent of West Germany's total energy supply to 55 percent in just twelve years, a tempo rarely seen in any other energy transition in history.[69] If the Inquiry had underestimated oil's rise, its hard coal predictions proved even more inaccurate, as the collapse in mining happened nearly twice as rapidly as expected. Oil's price, meanwhile, continued its swift decline after the Inquiry, falling steadily at a time when the overall cost of living in West Germany was rising. The sheer speed of this transition took nearly all informed observers by surprise, even the oil majors.[70]

Oil's rise changed how German economists modeled energy. Where earlier frameworks had assumed a large portion of energy consumption fell into a sphere that was substitutable, by the mid-1960s this assumption fell to the wayside. For one, because fuel oil use was growing so rapidly—its success had not been curtailed in the manner predicted by the Inquiry—the sphere of substitutability now theoretically seemed much narrower: oil had won big. Second, new studies conducted by Wessels and others began to claim it was actually difficult to substitute energy with other inputs. In an extensive report commissioned by the Economics Ministry in 1963, Wessels's EWI concluded that "fundamentally, and especially in the sphere of production, the demand for energy has a very low elasticity."[71] Wessels's team showed that in theory firms could, over the long term, replace energy with other inputs such as labor or capital. But such substitution, or even simply economizing on the use of energy, would require the adoption of entirely "new technology," something Wessels was reluctant to include in his model. Whereas Ordoliberals believed the fluid application of new technology was the essence of a market economy, Wessels saw technology as fixed in the medium term, and in the long run, something that changed quite slowly.[72]

These new assumptions facilitated forecasting, but they also led to a narrower understanding of technological change, in which firms passed any rise in energy costs on to the consumer instead of innovating. According to Wessels's models, in other words, the economy was *not* dematerializing, but was remaining as dependent as ever upon energy. If anything, energy became even more important in West Germany's economic profession during the 1960s, in contrast to American growth theorists. And where Ordoliberals saw energy as simply just another *normal* good, Wessels and others came to see energy as *special* because of its irreplaceability. As he pointed out on numerous occasions, all production processes in a modern economy now required vast amounts of this seemingly unique input, a fact illustrated by the rapid rise in the value of energy consumed in the Federal Republic, from 8.8 billion DM in 1950 to over 36 billion DM in 1963.[73] How could one heat a room, drive a car, build a house, forge iron, process chemicals, or live a modern existence without energy? In the words of another leading expert, energy had acquired a "meta-economic significance." It was a "foundational element of the economy and thus of human existence, just like land ... or air and water."[74] The massive use of energy in the economy combined with its "low level of substitutability" made it fundamentally different from other goods, for which consumers could find alternatives if sufficiently motivated by price changes.[75]

If growth inherently depended on energy, for West Germans in the 1960s this seemed entirely unproblematic from the standpoint of resource scarcity. While the 1930s and 1940s had been plagued by resource fears, and the 1950s, too, saw scarcity concerns spike during the Korean War, once the vast new fields of the Middle East began pumping petroleum into Western Europe, global energy reserves seemed limitless. Already in 1958 Baade, in his estimate of global energy use, argued that the claims of oil scarcity were just the "dramatic stories of false prophets."[76] By the 1960s this became the normal assumption, and experts like Wessels remained confident that "for a long time, the supply of crude oil will be enough to guarantee sufficient production of fuel oil even with very high rates of GNP growth and the corresponding increase in energy demand."[77] If energy were to ever pose a problem for growth, it would come not from resource exhaustion but from geopolitics.

But the strongest claim made by Wessels was that energy actually *complemented* other production processes: the more energy a sector used, the more it produced; the more energy a country consumed, the more its economy expanded. Energy, in a word, was inherently coupled with growth. Following this logic to its fullest extent—that energy was essential, non-substitutable, and complementary to other production processes—Wessels argued that to maximize growth a country must hold its energy costs *as low as possible*, particularly for an export-oriented country like West Germany that had to compete on the global market.[78]

In the 1960s, in sum, Wessels adapted energy economics to meet the requirements of the growth paradigm, but in a way that differed from economists in the United States. Where Solow was advancing the notion of dematerialization, Wessels was showing how modern economies were as rooted in energy as ever before, though fortunately the world still seemed rich in reserves. His ideas proved so influential that they hardened into a new paradigm in which growth and energy consumption were so inextricably coupled, the former required the latter to such an extent, that prices and technological developments were relegated to the margins. In 1967 Hans K. Schneider—Wessels's successor as EWI director, and West Germany's leading energy expert in the 1970s—captured this new paradigm in an influential summary of the field:

> The overall substitution elasticity of energy as a whole is very low over the long term. Energy, considered as an aggregate amount, is a complementary factor of production or a complementary consumer good. The total demand for energy is therefore determined primarily by the size and composition of national production, the current state of technology, and consumption patterns, and only to a small extent by the

ratio of energy prices to other prices (for production factors or consumer goods).[79]

By the mid-1960s forecasters adopted this coupling paradigm wholesale. They radically reduced the space in their analysis where one energy could be substituted for another, and by the early 1970s they dropped pricing from their models entirely.[80] The new generation of forecasts, in other words, eliminated the intuitive estimates used by the OEEC and the Inquiry, but at the cost of drastically simplifying their assumptions. They now hung nearly everything on an economy's *historical* energy intensity and the question of how much its GNP would rise in the future. If the former changed or the latter were wrong, predictions could go disastrously askew.

This theoretical narrowing generated predictions of immense energy consumption in the future: onward and upward. And this mattered because West Germans were doing more forecasting with more confidence than ever before. After 1968 the Federal Republic averaged no less than five energy prognoses a year, and forecasters displayed a strident confidence in their ability to model the future. As Hans Joachim Burchard, director of Deutsche BP and a leader of industry modeling, put it, "our forecasting methods have today reached a very high level of perfection."[81] According to its seers, West Germany was bound for a future of ever more energy.

Competition and Security

The new paradigm entrenched the idea that West Germany had little scope to change its energy supply now that it was becoming an oil nation. It became a "natural law" that demand for energy would roughly double every decade; little else could substitute for it. New technologies, new price relationships, new ways of economizing on energy—none of this could alter what had become a hard fact of life, the coupling of energy with growth.[82]

For experts the question then became how the Federal Republic could meet the vast energy demand predicted for the future. How could it push down the costs of this vital factor of production? In answering these questions, West German economists mapped out a distinctive set of policy prescriptions that prioritized competition, privatization, and market pricing in energy. On the one hand, such ideas were shaped by West Germany's history. Where Britain, the United States, and the Netherlands had all forged massive corporations that could collaborate with the state to secure foreign oil before World War I, and France had done so before World War II, Germany's efforts at this had crumbled during its two bids for global dominance. In the context of Allied oversight after 1945, the new Federal Republic lacked the institutional ability to build a national champion of its own that could venture beyond Europe to secure petroleum concessions in regions that had oil, but that had once been European colonies or protectorates. Besides regulation, through the mid-1960s market competition thus seemed to be the only mechanism available to ensure the steady flow of oil.[83]

On the other hand, this push for competition as a regulative mechanism reflected the enduring influence of Ordoliberalism.[84] Wessels in particular argued that an expansionary energy policy, in which private enterprises willingly invested in new energy infrastructure, could only succeed when firms received the proper "profit signals." "It is therefore of great importance that, in all branches of the energy economy in which investment is desirable from a national economic standpoint, we arrive at a relation of prices and costs that does not push [energy] profits below the rate that is achievable in other spheres of investment."[85] In areas where energy companies did not compete directly with one another—such as the distribution of electricity—Bonn should make sure it did not "impair" whatever forms of rivalry might emerge. In other spheres, the state should do its utmost to break down barriers to competition.[86] Above all, Bonn should "*not* pursue a policy of artificially increasing the price energy," Wessels argued. Instead, it should promote competition, which "turns uncertainty into certainty; competition is the best creator of true information."[87] The priority must be low costs, which meant sustaining competition and doing nothing to slow oil's inundation of the economy.[88]

Nevertheless, the expectations of vast energy consumption in the future still gave German experts pause, for it raised a concern that set them apart from their peers in Western Europe and the United States: the fear of an energy supply cutoff. During the 1950s such concerns about energy security had spiked on both sides of the Atlantic. In the United States, during the Korean War, President Truman had tasked a commission to study the country's resource needs. Issued in 1953, several years after the United States became a net-importer of oil, the Paley report called for a comprehensive energy policy to manage the security implications associated with America's energy consumption. After 1953, Resources For the Future (RFF) continued Paley's work, while the National Petroleum Council conducted an array of emergency planning reports, so there was no shortage of energy security studies during the 1950s. Meanwhile, the Suez Canal closure led the Eisenhower administration to impose first informal, and then formal, quotas on oil imports to protect the domestic petroleum industry, justified by the rhetoric of security.[89]

After 1960, however, anxieties about the domestic economy lacking energy were replaced by concerns about managing abundance. Already in 1955, the Cabinet's Committee on Energy Supplies and Resources reversed Paley's concerns, concluding that America's energy supplies "will be adequate, not only for economic growth, but also for at least the first year or two of full mobilization."[90] Under the Kennedy and Johnson administrations, the government's closest advisers reached similar conclusions. During the 1960s American experts firmly believed in the power of technology and innovation to overcome any domestic energy shortfall. Security in this sense became a non-issue, or a cover for giving financial succor to the domestic oil industry.[91]

A telling example of this technological optimism was Victor McKelvey, the leading expert for the US Geological Survey (USGS), one of the most influential organizations conducting energy forecasts for the American government. During the 1960s McKelvey profoundly shaped how American experts approached questions of scarcity and security. At a time when pessimists claimed America had oil reserves of 150–200 billion barrels, McKelvey put the figure at nearly 600 billion. Under his guidance, the USGS became an apostle of "market-driven abundance." In his own words, "Resources of usable raw materials and energy may be increased to an unpredictable extent by the development and application of ingenuity." The upshot: the United States had little to fear in terms of security from any shortfall in energy.[92]

Security remained a concern elsewhere in Western Europe, but it never animated economists as much as in West Germany. Take France, which during the 1960s fully subordinated energy to the growth paradigm. After the 1958 coal crisis, Paris embraced the penetration of cheap oil wholeheartedly, doing less than any other Western European country to protect its domestic coal industry. To stimulate growth, the state set a purposefully low price for petroleum products, something French leaders could do in their country's regulated energy sector. Meanwhile, the state monopoly on electricity generation kept prices so low—often below cost—that it was chronically in debt and had to receive subsidies from the state. Marginalists—the leading school of energy economics in France—prioritized efficiency above all else, even security. They understood security through the lens of France's defeat in World War II, which they attributed to their nation's economic weakness. Improving the economy's performance was thus itself a source of security. Toward this end, they favored importing cheap Russian oil and Polish coal, even if it meant undermining domestic coal and increasing France's exposure to the Eastern Bloc. As Maurice Allais noted, France's leading Marginalist after the war, "all history shows that the inconveniences of a certain insecurity are relatively temporary, while a protectionist policy carries a permanent cost." Security considerations, he concluded, could not be allowed to raise energy prices by more than a small margin.[93] In any case, after 1960 French officials were aggressively supporting their own quasi-state companies in tapping the hydrocarbon reserves of their former empire in North Africa. For French experts, the new oil and gas fields of Algeria were a success, and "become as much a part of the popular picture of Europe's energy resources as the coal fields of the Ruhr and Lorraine."[94]

In the Federal Republic, by contrast, the question of security cast a shadow through the 1960s. Oil's share of total primary energy rose more quickly there than anywhere else in Western Europe.[95] This dependence on foreign energy was relatively new to Germany, which before its militarization in the 1930s had been nearly self-sufficient in terms of energy as a coal economy. But now the nation was becoming ever more reliant on fossil fuels from "politically unstable countries in the Middle East," to use the words of the Economics Ministry.[96]

Energy security thinking, in fact, experienced a "renaissance" in West Germany after 1960.[97] After the coal crisis of 1958 spawned the idea of a security premium, Helmuth Burckhardt, the "Coal Pastor" of the CDU, used this to justify a large, fixed quota for coal production each year, arguing that Germans must pay a cost for having a reliable source of energy. But the gray area was what, exactly, the trade-off should be. As a leading industry journal put it, "What is the cost of a secure energy supply?"[98]

Wessels himself grappled extensively with security issues, and framed matters as a dilemma. If a country like West Germany wanted more security, it must pay a price by producing more expensive domestic energy to replace cheap foreign oil.[99] Yet he also believed the Federal Republic could circumvent this dilemma, in fact it must do so, for the growing volume of energy imports meant that "the security of our energy supply cannot be achieved within the framework of a *national* economic policy."[100] Because the nation had lost its empire after World War I, it could not follow the French or British models of privileging their own companies in the search for oil overseas in its former colonies. As a geographically small state, meanwhile, Germany had few prospects of finding new fossil fuel reserves under its own soil. World War II, meanwhile, had discredited the national approach to self-sufficiency on the continent. Wessels noted how in the 1930s the Third Reich had applied the "entire range of economic policy instruments" to secure the food supply—a different sort of energy measured in calories. But after the onset of war, even "forced labor from foreign workers" and "politically coerced food imports from occupied regions" failed to bring Nazi Germany the nutritional autarchy its leaders so desired.[101]

The disaster of this national strategy led postwar German theorists to advocate "close international integration" and a division of labor among national economies as the true solution to energy security. By diversifying the foreign sources from whence it derived its imports, Germany could, hopefully, "eliminate any one-sided dependency on certain supply regions that are located in politically tense zones."[102] But the scale of any such zone of cooperation was contested. Despite the calamity of the Third Reich, some experts rehabilitated earlier concepts of geopolitical thought from the 1930s, like Siegfried Balke, the second minister for Nuclear power, who argued that the Federal Republic must forge a "large area (*Grossraum*)" through the European Community (EC) to safeguard its "life processes" like the supply of energy.[103]

Wessels, by contrast, believed the EC was too small to actually enhance anyone's energy security. Not only was it failing to harmonize oil or coal policy among its member states, most European countries imported energy like Germany, and had even less of an ability to increase domestic production at a reasonable cost. Instead, Wessels called for energy cooperation within a broader Atlantic Community, since Europe together with North and South America would possess a greater diversity of energy resources as well as economic structures. Such an "economic bloc" would not be as efficient as an integrated global economy, in which the division of labor could play out to its fullest extent. But any productivity losses with a transatlantic strategy of energy security would be much smaller than through a European or a national strategy. In Wessels's estimation, even if Cold War or Middle East tensions confined West Germany to the Atlantic, his nation could continue to do what it did best and specialize in the export of high-value manufacturing products in exchange for primary resources.[104]

Security, in other words, could be achieved by expanding the scale of the market in which West Germany participated. Because Europe's oil market was supplied by the majors like BP, ESSO, or Shell, this meant cooperating with these massive companies whose intricate global supply networks were flexible and allowed them to "adapt optimally to changes in the political constellation of the producer states." Through this logic, many German experts embraced the status quo of the 1960s, arguing that whatever advanced the sales capacity of these vast companies also advanced the security interests of West Germany and the EC. "A policy directed against the non-European oil corporations," as one expert put it, "would go against the vital interests of the member states."[105]

But by the late 1960s, the continuing anxiety of a sudden energy shortfall—exacerbated by a second Suez Canal closure in 1967—led others to move beyond Wessels's market-based approach and explore new security strategies. Wessels's successor at the EWI, Hans Karl Schneider, and Urs Dolinski, Hans-Joachim Ziesing, and Manfred Liebrucks at the DIW in Berlin thought the coupling paradigm had elevated security into a fundamental theoretical challenge. As Schneider put it, for "a highly developed industrial economy it is *impossible* to replace energy as such with other goods (including services). A shortfall in the delivery of energy *must* therefore, on account of the complementary nature of energy, lead on the one hand to a cutback in production and, on the other hand, to restrictions in consumption."[106]

Mitigating the effects of an energy shortage, according to these newer experts, required more than just diversifying the source of oil. Competition and free pricing alone could never balance security with growth because energy as a sector had particular needs that the market could not always meet. Schneider pointed out, for instance, that energy companies had to think in decades in order to carry out their complicated investment projects—whether for refineries, pipelines, or mines. Markets, by contrast, excelled at providing short-term marginal information about prices, but they could hardly be trusted to deliver the long-term information that energy firms and consumers needed to thrive. Schneider thus recommended a mixed strategy, what he called "an interventionist market-oriented approach, that combines market elements with procedurally planned public interventions." Procedural planning lay at the heart of his strategy, in which the government would work out a quantitative forecast of possible and *desirable* energy futures. Although he did not call for formal targets, Schneider thought forecasts could do more than just predict, and actually could lay out Bonn's priorities for the economy by including an order of magnitude for how much each energy type should contribute, taking security into consideration. The government, moreover, should require all energy companies to publicize their plans for expanding capacity or building infrastructure, so others could plan their own investments. These economists hoped such procedural planning would not only avoid the overexpansion that led to price collapses, as in 1958, but also the underinvestment that could turn bottlenecks into security issues.[107]

Beyond this, Schneider advocated specific state interventions to minimize security threats, like sustaining strategic energy reserves and financing multipurpose power plants that could switch from one fuel type to another. Most radically, he argued that Bonn must actively support the creation of several large, German-owned energy companies that could compete with the oil majors. Only the creation of massive German energy firms could bring a semblance of competitiveness back to a market increasingly dominated by foreign-owned companies. To meet its security needs, Bonn should provide start-up funds, expand its preferential treatment for domestic energy companies, and encourage them to combine into West Germany's own international major. In the name of security, in other words, by the late 1960s a subset of German experts began to break with the language of smallness and anti-concentration enshrined by Ordoliberalism, embracing large-scale enterprises as a necessary and desirable attribute of the energy sector.[108]

Conclusion

By 1970 an energy paradigm had crystallized among West Germany's most influential advisors. The work of the Energy Inquiry and of experts like Wessels firmly coupled economic growth with energy consumption in the minds of politicians and the public. Where one expanded, so must the other; where one collapsed, so too must the other; the more energy the better; the cheapest energy the best. While this new orthodoxy preserved much of West Germany's original economic language—of competition, markets, and free pricing—its proponents departed from Ordoliberalism by elevating forecasting as the policy-tool par excellence, gaining the ear of policymakers who hoped to peer through the looking glass separating present from future.

In crafting their narrative of the future, however, proponents of this coupling paradigm lost the fluidity of earlier frameworks, turning generous assumptions into absolute truths. In their quest to construct elegant models, forecasters paid less attention to how different energies might compete with one another, and by 1970 the relative price of an energy no longer mattered. Instead, experts assumed that present trends of rapidly rising oil and energy consumption would persist far into the future. Their paradigm thus portrayed West Germany's postwar energy transition to oil as a force of nature—driven by the vast deposits of crude in the Middle East—not of policy.

Their work turned energy into a coherent body of knowledge and elevated this new branch of economics into a handmaiden of state policy earlier than in the rest of Europe and North America. It also conceptualized energy as a special sector because of its irreplaceability in the workings of a modern economy. Unlike the neoclassical growth theorists of America, West Germany's experts remained deeply convinced about the material rootedness of modern economies in energy. And this elevated energy security as a major preoccupation, in contrast to the United States. Energy became a public good: Bonn must assume responsibility for ensuring not only a market order, but also a stable flow of this precious input. While one solution to this security dilemma endorsed the status quo, a second, newer solution created the intellectual premises for a new type of energy that would be domestic *and* expansionary. This would be nuclear power, which many hoped could solve the paradoxes raised by the energy-coupling paradigm. And where Ordoliberals castigated any semblance of economic concentration, new economists like Schneider began justifying agglomeration in the energy sector to balance the power of the majors, whom they began to see as much of a source of instability as stability.

Equally noteworthy is what the coupling paradigm left out: the costs to anything that was not production. Its advocates framed policy strictly as a *dilemma*. As late as 1975, leading experts still maintained that "energy policy has *two tasks*, which are in conflict with one another: attaining the security of a sufficient supply of energy at all times, and guaranteeing a supply of the lowest priced types of energy as possible."[109] The possibility that consuming energy might impose a different type of cost—on the environment, for instance, or on society itself—and that there might be other goals to consider had no place in this paradigm.

These ideas would profoundly shape federal policy during the 1970s, when forecasts for a relentless expansion in West Germany's appetite for energy became conventional wisdom. By 1972 leading experts claimed that the coming decade would rival the 1960s in the growth rate of energy consumption.[110] The coupling paradigm would be used to justify the expansion of hydrocarbon production and imports, as well as the headlong rush into nuclear power. But equally important, West Germany's particular concern with security—coming from the very energy establishment that advised the government—would open novel lines of research that pointed as much toward conservation as other solutions. That a group of experts purported to know the energy future could not have had greater implications for the fate of nation that, by the 1970s, was becoming a high-energy society.

3
Chains of Oil, 1956–1973

> *The process of substituting one energy source for another . . . will have far-reaching and profound implications that will effectively change the face of our countries. Substitution processes should, in any case, not be judged only by economic standards; it would hardly be acceptable if they were measured merely by the criteria of short-term profitability.*[1]

When a military coup toppled the Egyptian monarchy in 1952, West Germany's public hardly took notice. Given the urgency of reconstruction and the partition of the nation, the affairs of Egypt seemed part of another world. Four years later, this would change dramatically when the architect of the coup, Colonel Gamal Abdel Nasser, nationalized the Suez Canal. Since it first opened to traffic between the Red Sea and the Mediterranean in 1869, the canal had been operated by European investors under the auspices of the British state. For Pan-Arabists like Nasser, the canal was a hated symbol of foreign domination and its nationalization marked a step toward Egypt's liberation. For the British, nationalization was a geopolitical nightmare.

For West Germans, the change in ownership of a waterway two thousand miles away generated an array of anxieties. Each day nearly two million barrels of oil flowed in tankers through Suez or in the nearby pipeline networks to Europe. When Israel, Britain, and France deployed military force to reverse Nasser's nationalization, Egypt responded by sinking ships that blocked traffic through the canal. Its Arab allies blew up the pipeline stations that pumped crude from Iraq to the ports of Lebanon, while Saudi Arabia halted oil shipments to Britain and America.[2] In one swift move, the new president of Egypt throttled a vital link in an emerging global hydrocarbon supply chain, cutting off more than two-thirds of Western Europe's oil. For West Germans, 1956 revealed the "dangers of an exclusive dependence of Western industrialized countries, without their own basis of oil," on the Middle East.[3] A new era of energy insecurity was dawning.

Energy and Power. Stephen G. Gross, Oxford University Press. © Oxford University Press 2023.
DOI: 10.1093/oso/9780197667712.003.0004

The fate of the canal gripped West German policymakers because their nation stood in the midst of a profound transformation. Each year, the energy transition unleashed during the 1950s intensified as more petroleum flowed into the Federal Republic, fueling the rise of a new high-energy society. Today, the sites of oil and gas extraction, the tankers, refineries, pipelines, and filling stations that provide the Global North with its hydrocarbons constitute one of the world's most expensive supply chains. This web distributes oil and gas at such low cost that fossil fuels seem naturalized as the modern world's energy foundation. But this was not always so. It was during 1950s and 1960s that this "second crust" of the Earth truly emerged, forged by a combination of state policy, corporate investment, and the rise of a new set of products and processes that hinged on the internal combustion engine (ICE).[4]

Oil belonged to a novel development block, an interconnected system of "technology, infrastructure, energy sources and institutions" that shapes the direction of economic growth. While nineteenth-century Europe was defined by coal and the steam engine, after 1945 the ICE drove a new economic constellation that revolved around the gasoline-powered automobile, but that included everything from mass air travel to petrochemicals. By the 1960s this development block began pulling oil toward Western Europe at rates that defied the expectations of contemporaries. Highways and suburbs, refineries and filling stations, synthetic fertilizers and plastics transformed how West Germans lived their lives, marking a profound break with the past. By 1970 nearly limitless energy undergirded daily life, and the Federal Republic joined the United States as a high-energy society.[5]

Following the flow of oil through West Germany reveals developments that are often hidden from other vantage points. One can see the "routes of power"—political as well as energetic—along which different actors insert themselves to reap economic gain, achieve leverage over people or resources, and shape the evolution of states and societies.[6] To understand West Germany's evolution into a society of affluent consumers, to understand its shift away from Ordoliberalism toward a more technocratic governance in the 1970s, and to understand why its political elite came to take the material problems of fossil fuel use so seriously, we must look to the hydrocarbon supply chain that crystallized during the 1960s.

While this sprawling oil infrastructure brought cheap energy that fueled a decade of growth, it also generated intractable problems. For one, the geography of oil was lumpy. Known reserves were scattered across a "Golden Girdle" just north of the equator, such that the world's centers of consumption received their oil through supply chains that extended for thousands of miles. During the 1960s West Germany came to depend on the Middle East as well as on the international oil majors—corporations of vast size and wealth—that controlled critical nodes in this network. Informed by the coupling paradigm, all West German political parties came to see the petroleum states and the international majors as double-edged swords: fonts of growth, but also security threats if they failed to deliver the oil.[7]

Second, the new hydrocarbon network demolished the older, hard coal development block that had risen from the wreckage of World War II. As coal collapsed, it threatened to bring down with it an immense amount of capital. Collieries and power plants have life spans of a half-century or longer, and investments in these material artifacts take decades to recoup.[8] Under the onslaught of oil, the devaluation of capital embedded in West German coal had the potential to become a social catastrophe, raising the prospect of hundreds of thousands of layoffs and the dislocation of entire communities. The new hydrocarbon supply chain generated intense resistance from the social factions tied to hard coal, forcing the federal government to make difficult choices about how to manage the repercussions of capital flight on a massive scale.

A tipping point came between 1965 and 1967, when a conjuncture of events forced West Germany to reorient its energy policy. Across Western Europe, leaders worried about the stability of their nations' oil supply. But a collective European energy policy never materialized to manage the challenges generated by this new hydrocarbon supply chain. West Germans, like other states, faced the issue of security largely on their own. An alliance of the Social Democratic Party (SPD), mine owners, and mining unions began pushing the country away from its Ordoliberal-inspired energy policy to embrace a technocratic corporatism that favored new, gigantic, German-owned firms. This would change the market structure and energy agenda of the Federal Republic, as large conglomerates curried favor with the state by virtue of their size and their claims about security and stability. In the process, West Germany's first postwar energy transition began to reshape not only the nation's consumer lifestyle, but its political and economic institutions as well.

Waves and Chains of Oil

Western Europe cruised into the 1960s on a wave of oil. The swell came from the Middle East and it was unleashed by the policies of the United States. Resource worries spawned by World War II had led the American government to aggressively exploit oil outside its borders. After 1945 the United States formed an oil-for-security arrangement with Saudi Arabia. Two years later, the largest American consortium in the region—ARAMCO—expanded dramatically, with Washington blessing the move and waiving its anti-trust laws. At the same time, President Truman announced his intent to halt the spread of Communism through Turkey and Greece to the south, establishing a security umbrella around the oil-rich Persian Gulf. America's postwar tax code, meanwhile, generated massive incentives for investment in the Middle East, bringing down the tax rate of American oil companies operating there far below what industry paid at home.

American policy combined with the incredibly low cost of extracting crude in the Middle East to turn this region into the hub of the global petroleum industry. During the 1960s capital poured into Saudi Arabia and neighboring countries. Between 1958 and 1967 investment in Middle Eastern oil nearly tripled what it had been in the previous decade, and Saudi Arabia became the single largest destination for American investment. Most of this capital came from the seven oil majors—ESSO-Exxon, BP, Royal Dutch Shell, Chevron, Gulf, Texaco, and Mobil—companies of vast size and wealth with operations that spanned the globe. These companies, five American, one British, and one Anglo-Dutch, possessed nearly all the oil concessions in the Middle East, which became the crown jewel of their commercial empires.[9]

Petroleum-rich states like Saudi Arabia cooperated with the majors to expand production as rapidly as possible. Global oil output surged: most majors doubled production between 1957 and 1966; some quadrupled it. Between 1950 and 1960 global crude oil production doubled to 1 billion tons annually. By 1968 it doubled again. During these postwar years, oil production reached a tipping point, oil and gas overtaking coal as the most important form of global energy, driven by the incredible output from the Middle East (see Figure 3.1). The effect on Western Europe would be revolutionary.[10]

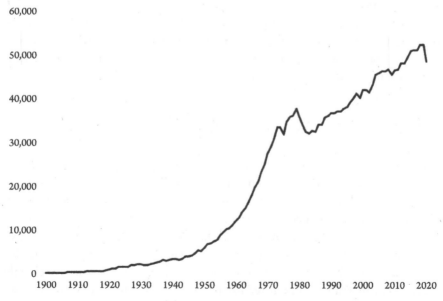

Figure 3.1. Global oil production (in TWh), 1900–2020. *Source:* Data from *Our World in Data.*

To bring this oil to Europe, the majors built infrastructure that fixed the directional flow of energy. After 1945 these companies laid thousands of miles of pipelines linking the fields of Saudi Arabia, Iraq, and North Africa to the Mediterranean. The Trans-Arabian Pipeline (TAP), built by ESSO-Exxon, Chevron, and Texaco in 1950, carried crude over 1,000 miles from Saudi Arabia to Lebanon. In 1952 the Iraq Petroleum Company (IPC)—a consortium of Shell, BP, ESSO-Exxon, and Chevron—opened a new pipeline from Kirkuk to Syria. More lines soon followed, connecting the inland regions of Libya and Algeria to the coast.[11] The majors and other corporations also invested in enormous ships dedicated to oil. Between 1955 and 1968 the global oil tanker fleet quadrupled in tonnage, the 40-thousand-ton ships of 1960 surpassed by behemoths approaching 200,000 tons a decade later. By 1970 nearly half the world's cargo ships were dedicated to oil products. And under pressure from the majors, Europe's leading ports like Rotterdam and Hamburg revamped their facilities to accommodate the huge new tankers.[12]

This new era of energy abundance, however, created problems for the industry as well as advantages. For one, the promise of untold riches brought new companies to the game of exploration, who began jostling for Middle Eastern oil. The majors responded in 1959 and 1960 by lowering the posted price at which they sold oil. But these price cuts incited remonstrations from the producer countries, who coalesced into a fractious alliance that strove to assert more control over the resources of its member states. By 1960 a new counterweight to the majors emerged that hoped to change the very nature of the oil market: the Organization of Petroleum Export Countries (OPEC). Lastly, the huge volume of Middle Eastern oil threatened American oil firms that depended on the more expensive crude found back home in the United States. These companies lobbied Washington for protection, their efforts culminating in 1959 in a quota system for foreign oil, which fractured the global oil market in two—America and the rest of the world. The majors with their Middle Eastern reserves, now limited in their ability to sell in the United States, began aggressively marketing their oil in Europe.[13]

By the 1960s oil was thus reaching Europe in quantities that were unthinkable the previous decade. But oil was not only pushed there by the discoveries of the Middle East, it was also pulled by the rise of a high-energy society with its centerpiece, the ICE. Since its invention in the 1880s, the ICE promised to revolutionize the way people traveled, worked, and even lived. With the spread of the Model-T in the United States in the 1920s, the gasoline-powered car became the paragon for a new lifestyle of personal mobility, adventure, leisure, and recreation.[14]

The ICE, though, needed more than cars to thrive. To bring this technology within reach of the middle class required a novel infrastructure with interconnected elements from roads, factories, and repair shops to filling stations and a reliable supply of oil. The car thus spread across the world unevenly. Though German innovators like Rudolf Diesel and Karl Benz had pioneered the earliest engines, their nation was slow in adopting the automobile on a mass basis. The United States became a car-nation in the 1920s. But across the Atlantic, the Weimar Republic taxed automobiles and invested in public transportation. Only under Adolf Hitler did the tertiary components needed to support the internal combustion engine slowly begin to emerge. The Third Reich built the first installment of the Autobahn; with funds plundered from trade unions it assumed the financial risk of creating a domestic, mass-production car company named Volkswagen, while its armament policies spurred the rise of a petrochemical industry.[15]

World War II slowed the spread of the ICE in Europe, but its diffusion accelerated after 1945.[16] During the late 1950s and 1960s car ownership in the Federal Republic skyrocketed as the mass production of smaller models by Ford and Volkswagen, with its iconic beetle, brought the cost of a car within reach of millions of consumers. In 1956 West Germany became the world's second largest car market after America. In 1961 Volkswagen celebrated its five millionth car; in 1965 its ten millionth. Between 1950 and 1970 the number of German automobiles rose from under 500,000 to over 15 million: 1 car for every four people in the country (see Figure 3.2) The expansion was so fast that experts worried about a "traffic emergency," as car use outpaced road capacity. In the early 1950s the highway system had ranked low among Bonn's priorities, but after 1958 the government began spending massively on roads, funding highway construction with gasoline taxes and exempting road building from the yearly hassle of budgetary negotiation. By the 1960s regional planners dreamed of replacing the "pedestrian city" with the "motor city." City governments devoted more money to streets, and between 1960 and 1970 annual public funding for road building tripled.[17] In the worlds of Wolfgang Sachs, "the wheeled society was launched."[18]

For Germans the car became a symbol of the postwar "good life": of material affluence and mobility in which growth and accumulation became the focus of not only state policy, but social life more generally. It offered a stark contrast to the wartime and postwar years of scarcity. "More and more, better and better," proclaimed a common Volkswagen ad. Or as the Association of German Automobile Clubs pronounced in its 1965 manifesto, the car "grants us the possibility to make life nice, broader, and freer."[19] In fact, the car was the cutting edge of a modern consumer lifestyle that was emerging, fueled by a doubling of incomes during the 1960s and one with an unquenchable thirst for energy. The emergence of a European consumer society had started slowly in the 1950s. But only in the 1960s did West Germans achieve the disposable income to begin

Figure 3.2. New Volkswagen Beetles rolling off the assembly line in Wolfsburg, 1956. *Source:* Keystone Features, Hulton Archive / Getty Images. Nr. 3432698.

speaking of "our consumer society." Durable goods like televisions, refrigerators, washing machines, and electric ovens became the norm for middle-class households during this decade, as the Federal Republic began to mass produce and mass consume these energy-intensive products. In 1970 the nation boasted ten times as many supermarkets, reliant on electricity-powered refrigeration, as in 1960. And as West Germans became wealthier, they worked less, freeing up time for leisure and for travel. The cost of European flights plummeted and oil-intensive air traffic surged to unprecedented levels. Tourist agencies blossomed and West Germans began vacationing, at home by car, and abroad by air to the Mediterranean, North America, and even Africa and Asia.[20]

If West Germans were not quite wrapped in a cocoon of energetic luxury as were middle-class white Americans, by 1970 the majority had nevertheless experienced a profound transformation in daily life since 1950, when horses and coal-fired stoves were more common than cars or refrigerators. This revolutionized the way Germans consumed energy, catapulting oil ahead of coal. In 1960 the average German consumed in a single day the amount of petroleum that had lasted an entire week in 1950. From there petroleum consumption only accelerated, and the nation absorbed the biggest swells of the oil wave coming to

Europe. Where France added 49 million tons of oil to its market between 1957 and 1967, the Federal Republic added 115 million tons. Total consumption of the black liquid grew nearly sixfold during this pivotal decade, as West Germany became a nation that ran on oil.[21]

Oil Infrastructure in West Germany

To consume such incredible volumes of petroleum required a new infrastructure. Infrastructure needed capital, and it was the majors who doled out the funds to link the Federal Republic with the emerging global petroleum supply chain. As Emil Kratzmüller, chairman of Deutsche ESSO-Exxon's managing board, reflected, "the fact that in the past year oil consumption has reached about twenty times that of 1950 makes clear which economic problems the oil industry had to solve.... All of that would have been impossible without heavy investment in oil exploration, process, transport, and distribution." To quench West Germany's thirst for oil, Kratzmüller's industry sunk 12 billion DM in infrastructural funds into the Federal Republic. Given the capital-intensive nature of oil, each employee had 300,000 DM invested in him, ten times that of the average German worker.[22]

Most of this capital went toward refineries. In the 1950s the majors had clustered this infrastructure in the estuaries of Western Europe's rivers. Now the rapid growth of inland cities like Frankfurt and Cologne, combined with improvements in transporting crude over land, meant firms could locate refineries where petroleum products were consumed, and tailor them to the specific needs of the locale. Refineries crept inland from the coastal cities of Rotterdam and Hamburg, first to the Ruhr, then to the Rhineland, and even to Bavaria. Between 1955 and 1970 West German refining capacity rose ninefold. The goal: to locate a refinery within 150 kilometers of every point in the Federal Republic, something the industry achieved by the end of the 1960s.[23]

Inland refineries required pipelines, and the construction of a network linking the coasts to the interior was the second focal point of investment. Before 1960 ships, rail, and trucks transported oil from the coasts through a network that was "fractured in multiple ways," and at great cost.[24] But in 1956, under the financial leadership of ESSO-Exxon, BP, and Shell, this began to change. West Germany's first pipeline connected the port of Wilhelmshaven to the ESSO-Exxon refinery near Cologne.[25] A year and a half later the second line opened, connecting Rotterdam, the largest refining center in Europe, to Cologne and the Ruhr. After this the race to the south was on. In 1962 the majors built their first direct link to the Mediterranean by connecting Laverna near Marseilles to Karlruhe in Southwest Germany. In 1963 the industry added a branch from Cologne to Frankfurt. A year later they opened a line that extended from Karlsruhe to Ingolstadt in Bavaria. In 1966 a different, Central European Line that started in Genoa and flowed through the Alps began pumping to Ingolstadt. Finally, in 1967 a consortium of ENI, BP, ESSO-Exxon, and Shell finished the Transalpine Pipeline (TAP) from Trieste, Italy, to Bavaria. With a cross section of 1 meter and a carrying capacity of 54 million tons, TAP became Europe's largest and most expensive pipeline yet. In 1955 West Germany had not a single pipeline; a decade later these metal tubes carried over 40 percent of oil traffic across the country.[26]

Ingolstadt, the new oil hub of Bavaria, illustrates the lengths to which oil companies went to expand their infrastructure, as well as the narrative power these projects gave the majors. Located on the upper reaches of the Danube, the city had been a sleepy university town before World War II, best known as the setting for Mary Shelley's *Frankenstein*. After researching over 150 Bavarian townships, ESSO-Exxon chose Ingolstadt because within a 100-kilometer radius lay Munich, Nuremburg, Augsburg, and Regensburg.[27] According to industry estimates, Ingolstadt's refinery operations saved Bavarian consumers 200 million DM a year on energy costs. The vice president of ESSO-Exxon's holding company used this new complex to illustrate how the majors were "better and more reliable than any other institution." As a result of ESSO-Exxon's investment, he opined, "undoubtedly will Ingolstadt grow into a city over the next twenty years, while the neighboring farm communities will also participate in this growing prosperity based on this inexpensive energy source, oil."[28]

With projects like Ingolstadt, the majors expanded the narrative they had crafted in the 1950s that oil served the social market economy better than any other energy. ESSO-Exxon, Shell, and BP portrayed their industry as one that operated under extreme competitive pressure, that had to innovate to survive, that drove down energy prices, and that fueled exports. They now could claim that their infrastructural programs advanced regional development, as the localization of refineries and cheap oil began to reverse the prewar trend in which production clustered near coal areas.[29]

Bavaria, once a relatively agrarian state, now industrialized rapidly on oil. In the 1950s Bavaria still imported coal from the distant Ruhr at a cost 40 percent higher than in the country's industrial heartland. Transporting oil products from the coasts to Bavaria was also expensive: in the 1950s fuel oil cost 50 percent more in Ingolstadt than in Hamburg. By the 1960s the localization of refining eradicated these energy price differentials and broke the limits that geography had imposed on Bavarian development. By 1970 three-quarters of Bavaria's final energy came in fluid form, while the growth rate of total energy consumption there far surpassed the national average. As a consequence, southern Germany formed a pro-oil voting bloc in Parliament that resisted any effort to dam the river of oil.[30]

But not everything was rosy, for these infrastructural projects created an alarming trend of overexpansion. Oil firms found it difficult to judge future refining needs. While excess capacity would eventually be absorbed by rising demand for oil, under-expansion meant losing market share to rivals. Pipelines, moreover, had to surpass a threshold of flow before they could operate economically. As German BP explained, the "operating costs of such lines require a certain minimum throughput in order to be economical. The supply of just one inland refinery through a trunk line is thus basically out of the question." Building a pipeline network thus led companies to cluster refineries along the route. These dynamics drove oil firms to expand their infrastructure to the limit, and by the early 1960s contemporaries, including state officials, worried that "the growth of refinery capacity in the Federal Republic is growing too quickly, and perhaps too copiously as well."[31]

Suez Canal and the Security Premium

By the mid-1960s West Germany was embedded in a new global hydrocarbon supply chain. A barrel of crude destined for the Federal Republic typically came from the ground in Saudi Arabia, Iran, or Iraq; after 1964 from Libya. A workforce that lived in company towns and that bordered on slave labor, often migrant in nature, operated the oil wells under terrible conditions for little pay. This crude was owned by consortia of foreign conglomerates, which determined the quantity extracted through opaque, long-term contracts that prevented the abundance of oil from undercutting their own markets, and that were disconnected from the pressures of supply and demand.[32]

From the wellhead in these arid regions, oil flowed by pipeline through Syria, Jordan, and Lebanon or by ship through the Suez Canal to the Mediterranean. It was then carried across this azure sea by tankers of enormous size that ferried crude to Marseilles, Trieste, Genoa, Rotterdam, or Hamburg. From these ports, oil was loaded into pipelines that rushed the black liquid to refining centers in the Ruhr, Baden-Württemberg, or Bavaria that resembled rectilinear, silver spider webs connecting towers of concrete and metal. These refineries—operated by the majors or by domestic energy firms who purchased oil from the majors—transformed crude into usable end forms under the watchful eyes of well-paid German technicians. The new army of German motorists then bought their gasoline at filling stations owned by the majors, large German energy corporations, or a dwindling number of independently operated firms.

This supply chain brought incredible growth to West Germany, but at the same time it spawned new risks, locking the nation into dependence on distant regions. Germans experienced the first intimations of these hazards in

1956 when the decline of the British Empire sparked Nasser to block the Suez Canal. With the canal closed to oil, work-arounds proved hard to find. Rerouting tankers around the Cape of Good Hope nearly doubled the trip length through Suez. Sourcing oil from the Americas raised its own logistical challenges. High-priced crude was the result: Germany's Economics Ministry estimated the Suez crisis would raise the cost of petroleum products 60 to 100 percent. Bonn, moreover, was woefully unprepared for an oil interruption, having no emergency guidelines, no agencies to facilitate rationing, and no idea how much oil its companies actually held in storage.[33]

The Suez crisis unleashed a heated debate about whether state dirigisme or market measures could best resolve the crisis. The OEEC quickly laid plans to ration oil in Europe on the basis of need, not price. Many Germans, including high-ranking cabinet officials, sympathized with this approach, while numerous complaints sent to Economics Minister Ludwig Erhard reflected a widespread interest in re-establishing price controls that had only been abolished in the 1940s.[34]

Erhard, by contrast, saw Suez as a test of the market and refused to countenance such controls. Under Alfred Müller-Armack, a leading Ordoliberal and director of the policy department, the Economics Ministry resisted rationing. Instead, the OEEC should restrict itself to working through the "versatile system of corporate relationships" and markets that coordinated Europe's oil supplies.[35]

In fact, West Germany overcame Suez through neither dirigisme nor market allocation. Instead, from the peak of the crisis in November through early 1957, it relied on the oligopoly of the majors to manage its oil. Under authorization from the Eisenhower administration, ESSO-Exxon, Texaco, Mobil, and other American companies formed the Middle East Emergency Committee (MEEC), which collaborated with the oil majors' subsidiaries in Europe. Through the MEEC, the majors redirected shipping from the Mediterranean to the Atlantic and the long journey around the Cape, and brought massive oil shipments from America's East Coast to Western Europe.[36]

As these challenges unfolded, German policymakers worked closely with the majors to supply their own market with whatever oil they could find. In mid-November, Economics Ministry officials met with corporate oil leaders, who proposed a coordinated reduction of West Germany's oil supply. The cabinet subcommittee for economics accepted these reductions on the condition that "the public not be made aware of the exact extent of these cutback measures."[37] Erhard subsequently met with oil companies to convey Bonn's approval of this regulation of supply. As one of Erhard's close advisors explained, Bonn aimed "to ensure the highest possible degree of freedom and support to these oil firms, so they can cooperate with their parent companies or other enterprises around the world to obtain for Germany the largest possible portion of the globe's remaining oil supply."[38]

By ceding control over its oil to a handful of international corporations, Bonn avoided price controls but still indirectly managed supply, through closed-door negotiation between state officials and the majors. Erhard admitted as much when responding to Parliament about Suez. "West Germany's supply of oil products depends critically on the efforts of the large oil companies. . . . I have done everything I could to support the endeavors of these oil firms."[39]

Suez thus left a contested legacy that would shape energy policy through the 1970s. The OEEC emerged confident in its ability to manage future interruptions to Europe's oil supply through planning. In the Federal Republic, however, Erhard and his allies used the relatively minor economic impact of the crisis to claim victory for the liberal market.[40]

After Suez, West Germany's oil sector would soon become even more liberal. In 1957 West Germany joined France, Italy, Belgium, the Netherlands, and Luxemburg in launching a new phase of European integration. In March of that year, the Treaty of Rome created, alongside the ECSC, two new institutions of European unity: the European Atomic Energy Community (Euratom) and the European Economic Community (EEC). While the EEC aspired to collective policies in everything from agriculture to transportation, at its heart lay the quest for a common market that would achieve the free movement of goods, people, services, and capital across borders in the name of growth. Member states were given less than a decade to reduce their subsidies and internal tariffs, including those on petroleum. Not all countries followed EEC guidelines—France made the opening of its oil sector conditional on the development of a common energy policy, which never materialized. In West Germany, by contrast, the Economics Ministry used the EEC's framework to reform the nation's oil tax code in 1963 and slash the tariff on foreign crude, which it had established after the war to promote domestic exploration. Erhard, put simply, bet the wave of cheap Middle Eastern oil would grow. After 1963 West Germany boasted the most open oil market in Europe.[41]

But Erhard's zeal for liberal markets overlooked the fact that oil was hardly a market, and more an organized oligopoly. And many were skeptical of Erhard's narrative. After Suez, West Germany's coal companies, mining unions, and the SPD—the lead opposition party to the CDU—all claimed their nation must be willing to pay a security premium for energy, because it "belongs, alongside food, to the vital necessities and irreplaceable goods" of a modern economy.[42] They portrayed Suez as the tip of the security iceberg. Such disruptions would become common as West Germany's reliance on Middle Eastern oil grew. As the Energy and Mining Union (IGBE) warned, who could guarantee that "with the labile political situation in the oil states of the East, from which the Federal Republic gets sixty percent of its crude imports, situations similar to the Suez Crisis won't emerge and under these new circumstances last even longer."[43] Specifically, they criticized Bonn's confidence in the majors. Erhard's assumptions, the IGBE went on, "that its oil supply is best secured through the international oil firms against disruptions in its Middle East diplomacy, and that the Arab states are forced to consider the Federal Republic because they are dependent on its market, have turned out to be a total mistake. On the contrary, the oil supply of the Federal Republic from third parties is now burdened with a threefold political risk": disputes between America and Middle Eastern countries; between the majors and Middle Eastern countries; and between the Federal Republic itself and Middle Eastern countries. The IGBE, mine owners, and the SPD instead called on the state to support *domestic* energy in the name of security.[44]

By 1964 these concerns began gaining traction in the Economics Ministry, once the bastion of Erhard's liberal policies. As West Germany's new oil infrastructure took root, officials worried that too much of it was owned by foreign companies. The ministry feared the usual suspects like ESSO-Exxon and BP, but after 1959 other majors began buying up West Germany's energy infrastructure too. Where German companies had owned nearly half the nation's refining capacity in 1950, by the 1960s that figure had dropped to a quarter. International consortia controlled virtually the entire pipeline network. Shell and ESSO-Exxon owned the largest tanker fleets flying the German flag. Foreign companies controlled nearly 60 percent of German gasoline stations, more if Mobil's stake in Germany's largest distribution conglomerate were counted. By late 1964 the ministry concluded that, "the situation could not be more serious. . . . The dependence of Germany's economy on these firms is large and threatens each year to grow larger."[45] Officials, moreover, pointed out how West Germany was the only country in EEC that "surrenders its energy market to international enterprises." France, Britain, the Netherlands, and Italy all had major oil companies of their own, and far more state oversight of energy. By contrast, no Germany enterprise owned oil concessions of any significance; all relied on the majors or the nascent spot market in Rotterdam.[46]

Capital Flight

Throughout the 1960s, then, West Germany's entanglement with the hydrocarbon supply chain generated anxieties. Yet these concerns were not only for security; domestic stability loomed large as well. As the Economics Ministry warned with only some hyperbole, "the expansionary needs of the great international firms alone are determining the future sales potential of Germany's coal industry."[47] The capital invested in oil threatened to destroy the old capital sunk in the coal sector, creating the potential for social dislocation on a massive scale.

For the burgeoning constellation of refineries transformed the crude arriving in Europe not only into gasoline, but also into fuel oil. And this threatened many of coal's traditional sales markets: warmth for homes, fuel for shipping, power for industry, even the raw inputs burned by power plants as they electrified the continent.

After 1958 the growth rate of fuel oil was incredible. Households burned just 3 million tons (SKE) in 1957; by 1965 they were using 28 million tons. In industry these figures rose from 5 to 24 million tons. Fuel oil for electricity generation exploded too, from almost nothing in 1957 to account for 12 percent of all electricity by 1972. As a percentage of final energy in the Federal Republic, fuel oil rose from 6 percent in 1957 to 30 percent in 1965, while coal fell from 66 to 34 percent. What began as a byproduct was becoming in its own right a massive market for the oil companies of Europe.[48]

The oil industry, of course, portrayed this as the organic result of oil's inherent qualities. "The natural advantages of fuel oil are so powerful," so Walter Bauer of German Shell claimed, "that the substitution process is stemming from the consumer, and the oil industry is merely meeting their requirements."[49] While Bauer may have touched a grain of truth, others demurred. Many believed the vast price scissors that opened up after 1958 between fuel oil and coal was driving this transition, for the new infrastructure built by the majors brought the price of fuel oil down. After 1958 coal prices stayed even, while those of fuel oil declined by over 30 percent on average, more in southern Germany. By 1967 a ton of fuel oil cost 20 DM less than a ton of coal (in energy equivalent terms).[50]

Bonn's own policies exposed the nation to the oil wave far more than other countries. Where France and Britain regulated oil through licensing or by heavily taxing substitutes for coal, with the 1963 tax revision West Germany had only a minor levy on fuel oil that hardly affected its consumers. During periods of energy abundance German oil prices fell lower than those of its neighbors, as excess oil that could not enter restricted markets like France wound up in the Federal Republic, sold by the majors to independent refiners at low cost. Fritz Burgbacher, the CDU's leading energy expert, captured this dynamic in a speech to Parliament:

> The great oil companies of England, and above all of the USA, can easily absorb red numbers in our market. For one, because they enjoy higher prices in every other country. And second, because they can say to themselves, and I would say the same were I in their place: if the Federal Republic has the most liberal energy policy among the great free countries of the Western World, then we can go there with our surplus, because we can at least get better prices there than if we dumped our excess into the sea.[51]

After 1963 falling fuel oil prices sparked capital to flee from the Ruhr, and hard coal began to look like a sector with no future. Coal companies paid lower dividends than other industries, reflecting their falling profitability. During a period when investment in the economy was booming, coal investment shriveled, from 1.1 billion DM in 1959 to less than 300 million a year by the end of the decade. Mining companies assumed debt simply to manage operations. The structure of the industry, meanwhile, prevented companies from shifting gears, since mining required capital to be fixed in time and place over long periods. Only after operating for 25 years—nearly 35 years after the moment of initial investment—would a mine achieve a normal profit. And because mines required a disproportionately large share of capital to be sunk into immovable assets that were specific to the task of mining, it was hard for investors to repurpose this equipment.[52] Experts estimated that reducing output by 35 million tons—a figure that seemed utterly likely—would destroy 2.75 billion DM worth of capital.[53]

Companies with roots in other sectors eagerly sought to get out of coal. But who would buy aging collieries that were unprofitable? Who would even buy the surface land that had been damaged by decades of subterranean mining?[54] Almost no one, it soon became clear.

As in 1958, the European Communities offered little help to the continent's ailing coal industry. In 1960 the ECSC's High Authority proposed a common energy policy to balance the desire for low-priced fuel with the need to stave off social collapse in mining regions. Just three years later, however, it accepted that low-priced foreign oil was here to stay, signaling its willingness to let global oil costs set the price of European energy. As compensation to coal producers, the High Authority suggested a European subsidy scheme. Coal-scarce states like Italy and France, however, balked at the proposal, and in 1963 the Council of Ministers—representatives of the EEC's six nation-states—rejected this stab at a common energy framework. The entire process dealt another blow to the High Authority's aspirations to build a supranational institutional architecture for Western Europe.[55]

By then even the CDU, West Germany's largest pro-market party, admitted the Federal state must help. After Erhard became chancellor in 1963, the CDU fought off demands from mine owners and IGBE to raise the price of energy as a way of sustaining coal production. They countered with plans to shrink the industry with state aid.[56] Bonn passed a law, Promoting the Rationalization of Hard Coal Mining, which set up an agency that paid firms to physically close down their mines: 12.5 DM of federal funds for each ton of coal they chose not to produce. The government, in other words, would cushion the destruction of fixed capital by opening the public purse. Over the next two years 31 large mines and 20 smaller ones shut down at a public cost of nearly 350 million DM.[57]

Closing inefficient shafts boosted mining productivity. But the CDU's opponents lamented the sad irony of giving public money to private corporations to *not* produce. IGBE called Erhard's new agency the "death club." The SPD castigated it as the most primitive form of socialization, "namely, the socialization of loss and not much more."[58] Most problematically, private companies controlled the allocation of these funds, and they closed those mines that were the least *profitable*, not the ones that were least productive. This led to distressingly irrational outcomes. The Graf Bismarck was one of West Germany's largest and most efficient mines, owned by Deutsche Erdöl AG (DEA), one of the nation's biggest zebras—those domestic companies that operated both oil and coal facilities. Even though DEA had just invested 70 million DM into this mine—the "Crack of Gelsenkirchen"—it earned lower profits than its oil holdings. In the face of 15,000 protesters in 1966, DEA closed Graf Bismarck and laid off nearly 4,000 employees (see Figure 3.3).[59]

Erhard accompanied his program with flanking measures that encouraged private companies to solve Germany's energy problems on their own, and the first Electricity Law of 1965 gave tax breaks for utilities building new plants that burned coal. But these initiatives failed to change the fundamental fact of fuel oil's plummeting price.[60]

Figure 3.3. Miners protest the anticipated closure of the Graf Bismarck mine near Gelsenkirchen, February 1966. *Source:* Picture Alliance / Getty Images. Nr. 1137740388.

Turning Points

Down but not out, hard coal producers struggled as West Germany clung to the political economy designed by Erhard that privileged low-priced energy. But in 1965 a concatenation of events unfolded that would topple these policies and create the space for a new approach to energy.

That year, in an election that saw the CDU win a lackluster victory, the West German economy did the unthinkable: it began to slow. The central bank, eager to avoid inflation from rising wages and public spending, restricted its monetary policy. Industrial production, consumer spending, and employment levels stalled in 1966 and began falling in 1967. Chancellor Erhard, in a controversial speech, declared that the postwar era was over and called on Germans to work longer and tighten their belts. The media, however, blamed Bonn. The "Erhard Recession" had begun.[61]

The slowdown had an immediate impact on West Germany's traditional energy. Despite rationalization and mine closures, coal costs were not coming down. Wage rates, pension payments, and interest rates on ever-larger debt burdens worked against mine owners. The recession, combined with a strong push from falling oil prices, thrust coal into its second crisis in less than a decade. By 1966 the volume of coal sold sank below 120 million tons for the first time since 1948. Stockpiles of hard coal piled up, and coal firms found their scarce capital tied up in illiquid assets that were impossible to monetize. Mine owners responded by canceling thousands of workers' shifts. Others exploited Erhard's "rationalization" program to close more mines, including DEA, which shut the Graf Bismarck that September (see Figure 3.4). Since 1960 Ruhr coal had already shed 120,000 mining jobs. Now the pace accelerated. During the next two years over 135,000 people would leave the Ruhr. As Theobald Keyser, former director of the Ruhr's Association of Mining Firms, remarked, "A deep depression and resignation has seized our colliery officials and miners. The earlier achievements of mining and miners during reconstruction have been forgotten. Job insecurity, worries about the future, and the unconvincing 'gratitude of the Fatherland' weigh on all people in the coal district."[62]

Some saw the crisis as the dawn of a new era that would transform the Ruhr in a positive way. But for most, their region seemed to lie at death's door. Miners lamented the end of a lifestyle. The media pushed the topos of a "dying region."[63] Most of this anger was channeled toward Ludwig Erhard. "The situation of the coal industry became untenable," lamented a mid-level manager forced into early retirement by the crisis:

> So many pits were closed, especially the small ones in the south of the region disappeared from the scene. The huge piles of unused coal (soon

Figure 3.4. Demolition of the Graf Bismarck mine in July 1968. *Source:* Farrenkopf and Prizigoda, *Glück auf*, Nr. 107, pp. 263. Originals from Montan.dok, Fotothek 024901743001 through 024901743002, and 024901744001 through 024901744005.

there was hardly any land for new heaps), the disappearance of well-known pit names, the slashing of the workforce gave a clear picture of the coal crisis. It was obvious that Erhard's liberal policies, which had helped rebuild West Germany's economy and brought an enormous upswing, severely damaged the coal industry.[64]

Mine owners and the IGBE fastened onto Bonn's *policies* as the root of the problem, downplaying the new hydrocarbon supply chain. As Keyser asked the unions: "Why do we not defend ourselves from the onrushing oil wave with the same or similar measures as the other great industrial states? While other states shield themselves, why must fuel oil expansion in the Federal Republic proceed at a pace like that of no other country of the world?"[65]

As the situation deteriorated, by May 1967, 90 percent of IGBE's members voted to go on strike. The old myth of the Ruhr as a center of upheaval reared its head again.[66] The "existential anxiety" generated by this second crisis, so the SPD's parliamentary leader Helmut Schmidt warned, was spawning a "breeding ground for political radicalization, to both the left and the right." This region of six million people, now the "sick man of the Federal Republic," became the cardinal danger to the nation's politics in Schmidt's opinion. Bonn responded with more social spending and a second electricity law.[67] But most commentators agreed that ad hoc policies would fail, and elites across the political and economic spectrum scrambled to devise plans to restructure their energy system. In the words of DEA's directors, "the process of substituting fuel oil for coal ... must be channeled in an orderly fashion to prevent the destruction of national wealth invested in the mines."[68]

As the crisis rolled on, West Germany's political landscape became more volatile. At the regional level, the radical right Democratic National Party gained momentum. Federally, growing coal subsidies turned the budget deficits of the previous years into a fiscal emergency. In September 1966, when Erhard presented a draft budget to raise taxes, his liberal coalition partners walked out, precipitating the first minority government in West German history. The CDU faced a vote of no confidence, and the sense of political turbulence reached heights not seen since the 1940s. After a month of negotiation, the CDU and SPD agreed to a Grand Coalition that brought Social Democrats back to power for the first time since the Weimar Republic. While the CDU retained the chancellorship, the SPD gained the Economics Ministry, which it staffed with a charismatic, Keynesian economics professor from Hamburg University, Karl Schiller.[69]

The SPD's entry into government opened doors in the sphere of energy, because this party advocated planning of a sort. For energy experts, the Grand Coalition could not have come sooner. By the mid-1960s pressure from foreign oil companies was reaching a crescendo, making the price of oil products "buckle."[70] Now even German-owned oil companies also stood on the precipice of disaster, with independent refiners worrying that the price cuts would lead to the "destruction of the free German oil market."[71] DEA lamented that "it can no longer be called normal and healthy when an entire economic sector cannot get an appropriate return on its capital."[72]

The dynamic that had rocked Ruhr coal, in other words, was now reshaping the nation's domestic oil industry. The majors were selling refined oil in the Federal Republic at such low costs because they were reaping large profits extracting crude in the Middle East. To compete with these behemoths, Germany's domestic oil companies began taking on debt or forming special relationships with firms in other sectors to tap new sources of capital.[73] The biggest blow came in 1966 when DEA, West Germany's largest domestic oil company, opened talks with America's Texaco. Founded in 1899 as a drilling company, DEA had grown into Germany's largest producer of domestic oil and a major refiner, but it also had substantial coal holdings. Headquartered in the oil city of Hamburg, it boasted a network of 5,000 gasoline stations that attracted the eye of Texaco—a participant in ARAMCO in search of an outlet for its huge surpluses. In April 1966, Texaco made a bid for 51 percent of DEA, sparking a backlash in the media and in Parliament.[74]

The cabinet deliberated intensely over the Texaco deal. Allowing the purchase might set a bad precedent. The Economics Ministry worried, "there will no longer be a German oil industry, since the entire capital stock will have passed into foreign, above all American, hands."[75] But given its fiscal crisis, Bonn could not afford to buy a controlling stake in DEA, as some proposed. So that summer Texaco's buyout moved forward, and within three years DEA was renamed Deutsche Texaco AG. Bonn spun the deal as a way to secure cheap oil. Yet the affair put the Economics Ministry on edge, and it began considering legislation to restrict the majors' access to Germany's market.[76]

On the heels of the Texaco buyout, anxieties about the nation's energy system rose still further in 1967, when border clashes between Israel and Syria degenerated into a regional conflict. A repeat of the Suez crisis of 1956, many feared, was finally coming to pass as foreign events again impinged on West Germany's energy system. On June 6, Arab oil producers called for a halt on all petroleum exports "to any nation assisting Israel in its aggression against Arab countries."[77] Saudi Arabia, Iraq, Kuwait, Algeria, the Gulf States, and Libya stopped oil shipments to Europe, while Egypt, Lebanon, and Syria blocked pipelines flowing to the Mediterranean and shipping through Suez. Altogether, nearly 65 percent of Western Europe's oil supply was now at risk.[78]

For West Germans, this second Suez crisis initially seemed worse than the first. The United States redirected oil to Europe, and declared a state of oil emergency, a prerequisite for the Justice Department to waive anti-collusion laws and allow the majors to collaborate. In June moderates like Saudi Arabia and Abu Dhabi resumed oil exports to all OECD members except the United States and Britain. But Libya—since 1966 the single largest source of West Germany's oil—showed no signs of lifting the embargo. With crude from Libya and Iraq still stalled, the Federal Republic stood to lose almost half its oil supply. As the

Economics Ministry put it, "German firms own no crude oil fields in the Middle East, while the tanker tonnage owned by German firms is negligible. . . . The Federal Republic's oil supply depends *almost exclusively* on the decisions of the international oil firms."[79]

Ministry officials and Germany's own companies warned Bonn to tread cautiously and to avoid alienating either the majors or the oil-producing states. As in 1956, Bonn did not ration petroleum, but worked behind the scenes with the majors to ensure the flow of oil. But just as these tensions were reaching a crescendo, the crisis passed. By July the Arab embargo broke down as Kuwait, Bahrain, Qatar, and later Iraq followed Saudi Arabia and began exporting to Europe again.[80] Nevertheless, the crisis scared policymakers, underscoring how little agency their nation exercised in the new hydrocarbon supply chain. The country was stuck inside a tense triangle of the majors, producer countries, and the United States. Something had to change.

A New Energy Order?

The conjuncture of a second coal crisis, the Texaco buyout, and a second Suez crisis forced German policymakers to grapple with the hard fact that they governed an economy that ran on foreign energy and that was dominated by foreign companies.

Much of Western Europe faced a similar quandary, and this created pressure for a common European energy policy. Already after the first Suez crisis, supranational experts called for a European framework that would stockpile petroleum reserves, diversify Europe's sources of oil, allocate petroleum equitably during a shortage, and facilitate cooperation between states and the majors. The OEEC's Oil Committee issued such recommendations in 1960, while the EEC suggested more far-reaching strategies, calling for a common market in petroleum and joint financing for exploration. The only concrete achievement, however, came in 1968 when the EEC required members to stockpile 65 days-worth of petroleum, a policy first implemented by West Germany several years prior.[81] That year the European Commission—a new institution created from the fusion of the administrative organs of the ECSC, the EEC, and Euratom—issued a major directive that made low energy prices the cornerstone of its agenda. Its report, however, reflected the unwillingness of member states to finance a collective energy agenda, and its broader agenda was torpedoed by disagreements between member states—those with an oil major versus those without one; those with coal versus those without it. With the High Authority's failure to lead through the coal crises still fresh, the directive fell back on the lowest common denominator: markets. The Commission rejected the premise that energy affordability and security were in tension, and instead claimed that markets could solve any crisis. Given the abundance of hydrocarbons around the world, competition among energy providers "compels them to become technically progressive, stimulates the natural processes of substitution, and brings with it a differentiation of supply." The Commission hoped that markets and competition, along with a sanguine optimism that the global surplus of oil would persist, should carry Europe through another Suez.[82]

European countries, consequently, addressed energy security on their own. Here France became a model for Bonn, though one whose history and colonial legacy gave it a head start over West Germany. After suffering its own oil scare during World War I, France had leveraged its colonial mandates in the Middle East to access the oil concessions of the former Ottoman Empire. In 1919 Paris acquired Germany's share of the Turkish Petroleum Company, with rights to Iranian and Iraqi oil, which it gave to a newly created Compagnie Française des Pétroles (CFP) that was 35 percent state-owned. This created an eighth oil major, though the weakest one, which became the foundation of Paris's quest for oil autonomy. Thirty years later, discoveries in French-controlled Algeria seemed to bring this vision to fulfillment. Paris initially asked the EEC to grant trade preferences to Algerian oil. With virtually no overseas reserves themselves, West German companies clamored to get a piece of the pie, while some CDU experts suggested accepting French requests for EEC protection in exchange for European aid for German coal. But Italy, the oil majors, and the United States opposed such preferences.[83] So France built yet another national champion to exploit Algerian oil, ELF-ERAP, and it required its refineries to purchase Algerian crude. Even after Algeria won independence, the Treaty of Evian cemented this petroleum bond with France, and by the end of the decade French companies were producing more crude oil than France could consume.[84]

In many respects, West Germans responded to the conjuncture of the mid-1960s by trying to remake their energy system in the image of their neighbors like France. Both the CDU and the SPD now moved away from Ordoliberal pretensions of limiting economic concentration to do the opposite, namely, using state levers to create domestic corporate behemoths in the field of energy. Large German-owned firms, they hoped, would mitigate the social challenges posed by the new hydrocarbon supply chain. As the Economics Ministry put it in 1968, a "new order" in energy was needed.[85]

The initial impetus came not from the state, but from West Germany's energy union and energy corporations. In 1965 IGBE began demanding more planning in the field of energy, by concentrating mines into a single enterprise. By uniting all forty-two mining companies into one enterprise, they hoped not only to streamline production, but to carry out "as smooth and ordered reduction in capacity as possible," that focused on social needs instead of profits.[86] IGBE couched its demands in the language of security, criticizing the fact that nearly all of Germany's oil flowed through the majors. In 1966 it proposed a single mining organization, and the following year the union began demanding that Bonn integrate "all energy sources"—not just coal—"into a common energy program."[87]

Before it was purchased by Texaco, meanwhile, DEA also began calling for a radical change: "an energy policy that includes plans and measures that altogether can be understood as an 'active oil policy,' as exemplified by the governments of other countries like France."[88] The Economics Ministry was reaching similar conclusions. In an internal memo it argued that Bonn must do more to ensure that "the towering market position of the international consortia not be abused. This requires public control of the large oil companies, or a commitment by the government to create a counterweight to the international oil enterprises in the energy market. This is the most important challenge."[89] With Texaco's takeover, the ministry began exploring "a merger of the remaining national enterprises."[90]

That the SPD controlled the Economics Ministry in the new coalition allowed these ideas to flourish. Minister Karl Schiller had read the internal memo and was himself critical of Erhard's liberalism.[91] Since 1964 he and other SPD luminaries like the mayor of Berlin, Willy Brandt, had been trying to "develop a public energy firm into an effective instrument of energy policy." Upon becoming minister, Schiller brought together unions, mine owners, and the local government of North-Rhine Westphalia (NRW) to discuss a completely new solution for West German coal.[92] By 1967 an elite consensus formed among that the nation's energy must be reorganized along more nationalist lines and into increasingly gigantic entities.[93]

So began the reorganization, first, of West Germany's coal industry. While IGBE provided the concept of a single firm for the industry, corporate leaders supplied the practical details for how this might unfold.[94] To facilitate the movement of sunk capital out of coal, Helmuth Burckhardt, director of the Ruhr's powerful mine-owners association, suggested creating a bad bank financed by the federal government to buy out unproductive mines. His plan had failed to gain traction with Erhard. But in 1967 the president of one of the largest zebras revived a modified version in which owners would lease their mining assets over twenty years to a consortium owned by Bonn, NRW, and a privately funded rescue company. In return, the owners would receive a small annual payment for their leased assets. Bonn would underwrite the deal through a debt guarantee of 7.2 billion DM.[95]

Though at first wary of these ideas, Schiller warmed up as negotiations unfolded. By the fall he accepted a combination of IGBE and the coal owners' plans, but with a twist. Reorganization must be overseen by a government commissioner with the legal power to coerce recalcitrant mining companies to sign on. The commissioner, moreover, would determine which mines to close. Most importantly, the commissioner could determine the tempo of closures based not on profit, but on how "quickly new jobs can be found for those people" who were laid off.[96] The owners vehemently opposed the idea of a commissioner, but by 1967 they had little leverage. Schiller pushed his advantage and demanded better financial terms for the government. The 7.2 billion DM loan asked by the owners was too high. With such a sum, reflected Schiller, "the state could buy the entire coal sector and convert it into common property."[97]

Under immense pressure from Parliament, Schiller, IGBE, and the coal firms finally reached an agreement in 1968. The commissioner determined the "optimal size" of any new coal enterprise to be so large that it could only be fulfilled by merging all mines in the Ruhr into a "total firm" to be named Ruhrkohle A.G. (RAG). Alongside the state, the shareholders included major coal consumers like the public utilities. Mining firms handed over their assets to RAG along with their liabilities, in return for 2.1 billion DM. The commissioner received "extraordinarily broad" powers with "almost unlimited jurisdiction." He would vet mine closings and tie them to a broader plan of social support for workers.[98] Schiller paired this agreement with a massive development program to revitalize the Ruhr, of 17 billion DM from the European Communities, NRW, and Bonn. By 1969, 94 percent of Ruhr coal production was in RAG's hands: 52 mines, 29 coking factories, 20 colliery power plants, and 183,000 workers. Overall, the new enterprise produced three-quarters of West Germany's hard coal.[99]

While corporations disliked the details, they jumped at the chance to extricate themselves from a dying industry. Zebras like VEBA dove into more profitable sectors like chemistry, oil processing, and electricity.[100] The deal was one of the single largest net transfers of wealth in West German history, in which nearly five billion DM of fixed assets changed hands as companies rushed out of coal. RAG became, if not a zombie enterprise, then one on life support. Through the 1970s subsidies to RAG surpassed the amount the enterprise spent on investment. The state, in other words, was covering the cost of maintaining the capital sunk into this older energy.[101]

The trend toward concentration, though, was not limited to coal. For the debates of 1967 spurred SPD politicians to call for the formation of a mega-German oil company, to pair with RAG against the majors. In justifying RAG before Parliament, for instance, Helmut Schmidt underscored that Bonn could no longer approach coal and oil separately, but must develop an "overall energy plan" based on large enterprises. Gesturing to the national champions of France and Italy, he lamented how "a country like the Federal Republic, structured around exports and intertwined with the global economy, must not allow itself to become completely dependent on the huge international companies in this important sector."[102]

In the summer of 1969 the CDU-SPD government collapsed and the SPD emerged as the leader of a new coalition with the FDP, proclaiming it would "create modern Germany." In the sphere of energy, "modern" now meant large-scale and technocratic.[103] Even before 1969, Schiller's Economics Ministry had begun working to build a fully integrated oil company. It first looked to Gelsenberg Benzin AG (GBAG), unique among German companies in actually possessing the rights to foreign crude, in Libya.[104] Bonn hoped to buy GBAG shares and sell them to VEBA—the largest domestic energy firm by sales volume, and the largest domestic refiner—to forge a counterweight to the majors. The asking price was too high and the deal fell through. But the seeds were planted for using state levers to merge two of the nation's largest domestic energy companies.[105]

With its newfound power, after 1969 the SPD began pursuing a different tactic of integrating the nation's domestic petroleum companies into a consortium. Following the Texaco deal, the Economics Ministry had already encouraged eight of the largest German oil firms to form Deutsche Erdoelversorgungsgesellschaft mbH, or Deminex, and agreed to support their exploration in Libya, Egypt, and the North Sea with federal funds. Yet the capital endowment of this first Deminex was miniscule, and the federal subsidies never fully materialized.[106] In 1969 the SPD rebuilt DEMINEX on a more robust financial foundation, and the government "partially assumed the risk of oil exploration." Bonn organized a Start-Up Program that would lend 575 million DM in low-cost, forgivable loans between 1970 and 1974 for exploration.[107]

By 1970 it seemed more urgent than ever to prepare for another energy supply shock, as global demand for petroleum began to outpace supply. For German officials, the first intimation of a novel situation came from a shocking announcement that year that the United States might not be able to provide an oil buffer to Western Europe in the event of a new crisis. At the same time, OPEC began demanding more control over the price of oil and more ownership over the consortia operating in their countries. First Libya, and then other OPEC countries, imposed tax hikes and price increases on the majors, driving the price of crude to new heights for the first time since the 1950s. As the chief of Deminex warned, "our children will have to pay twice as much for each ton of crude oil as we do today."[108] In 1972 Schmidt, following Schiller's resignation now the combined economics minister and finance minister, captured these anxieties when he warned that "restructuring" of the global oil market was "heightening the danger of a major confrontation between the producer countries and the oil majors, and could have a major backlash on the consumer countries."[109] Fearing Germany could no longer rely on America, he and others worried that companies like ESSO-Exxon would prioritize their own nation's markets over those of the Federal Republic.[110]

Deminex was to be the solution, by expanding West German influence along the new hydrocarbon supply chain, by acquiring concessions under German ownership, and by cooperating with the producer countries. As one of West Germany's leading energy officials explained:

> We want to give domestic oil companies their own oil foundation and thus give the Federal Republic the opportunity to be present in the global oil trade and play an active role. Having their own source of oil was in the past and is still today the foundation for the preeminent competitive position of the large international oil firms. From experience, the direct participation of domestic oil companies provides governments their best chance to influence the course of events during supply interruptions.[111]

With state financial support, Deminex opened negotiations with the National Iranian Oil Company (NIOC) to purchase oil for an emergency reserve, and to obtain direct ownership in Iranian fields. It likewise opened negotiations for a share of BP's concession in Abu Dhabi. For the first time since World War II, West Germany began trying to circumvent the majors and get to the source of crude itself.[112]

Conclusion: The 1973 Energy Program

The efforts of West German leaders to assert control over their nation's energy system and gain autonomy in the new hydrocarbon supply chain crystalized in 1973. In June, Iraq unilaterally raised the price of its crude. In September, other OPEC states prepared to follow suit. Geopolitically, Anwar Sadat, Nassar's successor as president of Egypt, began threatening Israel with war if it did not return to its 1967 borders.[113]

These foreign developments now shaped West Germany's domestic politics in a way that was inconceivable during the 1950s. For by 1972 the country's first postwar energy transition was complete. Whereas West Germany's prime energy source had been domestic hard coal as late as 1958, now over half of its energy came from oil, nearly all of which was imported. Cars, roads, suburbs, air travel, and the consumption of mass energy had become a natural part of West German life.

Against the backdrop of these tangled developments, Bonn presented West Germany's first formal energy program to the public on September 26. The program presented a vision of how different energies should fit together in pursuit of Bonn's twin goals of achieving a secure yet inexpensive supply of energy. With forecasts prepared from leading experts, the 1973 program predicted a truly massive rise in energy consumption—nearly doubling by 1985—while trying to manage problems that came with extreme dependence on foreign oil. For the abstract risks of the 1960s now seemed to be "real dangers," as petroleum became more concentrated in the hands of a few powerful producer states, as America became less able to help Europe, and as energy prices rose.

The program thus formalized in writing many of the ideas that had been brewing since 1960. West Germany required a new political economy in which the state took a "high degree of responsibility" for the nation's energy system, by managing infrastructural investment needs that would "snowball" in the future; by establishing an "in-depth exchange of information and opinions" with the oil industry; and by actively building up German-owned companies that could be a "link to the producer countries." Energy was increasingly seen as more than a mere commodity, but as a public good that must be safeguarded with care.[114]

The program epitomized how West Germany's political economy had changed as a result of its first postwar energy transition. Under Ludwig Erhard an Ordoliberal approach had reigned through the early 1960s, in which the smashing of economic conglomerations was supposed to bring Germans the lowest possible price for energy. But the rise of a new oil infrastructure that locked in the directional flow of oil from the Middle East to Central Europe, itself partly a result of Erhard's own agenda, disrupted his liberal vision by exposing West Germany to new risks. After Erhard's fall, West Germany under the SPD shifted to a politics of "bigness" and a technocratic corporatism to harness the oil wave while managing the problems that came in its wake.

After 1966 the size of a corporation and its market power became an asset rather than a liability. To a certain extent, this was a product of the sheer complexity of the new hydrocarbon supply chain. Only capital-rich firms that spanned the globe could afford the concessions, pipelines, freight capacity, and refining and storage centers needed to extract and move oil over long distances and process it. The physical nature of this infrastructure, put simply, created an incentive to go big. This period from the late 1950s through 1973 created the very energy network and corporate conglomerates against which the environmental movement of the 1970s and 1980s, and advocates of the *Energiewende* in the 1990s, would revolt against so viscerally. Nevertheless, West Germany remained the odd one out among larger European countries in having such little national-corporate influence over the global hydrocarbon supply chain. Britain through BP, the Netherlands through Royal Dutch Shell, and France through CFP and ELF-ERAP all controlled crude beyond the border of Europe. Even Italy was moving in this direction. By 1970 German hopes for their own major remained just that, a hope.

But it was not just the material nature of oil that led to economic concentration: under the SPD, West Germany actively promoted domestic conglomerates to balance the power of foreign corporations. Bonn cooperated with energy titans like RAG, Deminex, and VEBA—and also, if need be, the majors like ESSO-Exxon or BP—to manage West Germany's exposure to the vicissitudes of Middle East geopolitics, and to mitigate the social effects of the oil transition.[115] RAG marked the pinnacle of this new approach. With its sheer size and the powers vested in the state coal commissioner, it had "practically no other parallel in the industrial sector of the German economy." Through RAG, West Germany skirted social disruption during its first postwar energy transition because the state facilitated the flow of capital out of the old energy system by assuming risk as well as financial losses. Private capital was not destroyed, it moved.[116]

As a result of this turn toward technocratic corporatism, however, the most important decisions about West Germany's energy system were now made by a clique of bureaucrats, corporate directors, and union leaders that was increasingly walled off from the public. This closed-door method of policymaking was, after all, how the Federal Republic had weathered the energy security crises of 1956 and 1967. Given the belief that interruptions would only become more frequent, why would politicians discard a proven method?

By 1973 the powers of the government-corporate energy axis had fewer checks than ever. And this showed, for the new energy program minimized many concerns that were becoming more pressing for the citizens of West Germany, above all ecology. Tellingly, it focused almost exclusively on supply and security, relegating any thought of how energy was consumed or how it affected the environment to a short, vague discussion of efficiency and ecology.[117] What the program did prioritize, though, besides oil, was nuclear power. If oil emerged victorious from West Germany's first postwar energy transition, so many contemporaries thought, nuclear power would triumph in its second.

4

The Entrepreneurial State

The Nuclear Transition of the 1950s and 1960s

The final goal is not a technical one; it is rather a social one, technology being the means to an end.[1]

Humankind is currently dissipating, at an accelerating rate, the chemical fuel reserves that the sun has built up over millions of years in the forests and in the animal-life of the oceans, the remnants of which have given rise to oil. At the same time, the world urgently needs these carbons for the purposes of chemistry—from plastics to synthetic rubber, nylon, and aspirin. Chemistry and energy consumption are thus in a race to deplete the most important reserves of the world. It is clear that without nuclear power, in a few decades this competition would lead to catastrophe, because then the only way out would be the creation of power plants that directly use the rays of the sun or the tides of the ocean, and this would require extremely expensive equipment.[2]

On April 1, 1969, Siemens and AEG, two of West Germany's most storied companies, announced the merger of their nuclear power subsidiaries into a single enterprise: Kraftwerk Union (KWU). With one stroke, the deal all but eradicated competition in the nation's atomic market. Besides KWU, only a handful of foreign firms possessed the depth of capital and the scale of production to construct the gigantic nuclear facilities that electricity companies hoped to bring online. "Building the manufacturing infrastructure for such huge units," a KWU director lamented, "demands a higher capacity than either of our firms alone is able to provide."[3] But the surprising merger also grew from the hope of competing in what promised to be a global market of billions, as electricity demand in Western Europe and the developing world exploded.[4] No less than half the new firm's business, the board projected, would come from abroad. As a leading periodical pointed out, "the global market is the only proper scale for Kraftwerk Union." Bonn gave the merger its seal of approval, and by 1970 KWU became one of Europe's largest companies, building reactors and power plant turbines, winning competitions to construct the first nuclear plant in Holland as well as the coveted contract for the largest atomic reactor in the world, on the Rhine River in the heart of West Germany. More orders from South Africa and Columbia came rolling in, while the firm opened discussions in Austria, Switzerland, Spain, and Turkey. West German nuclear technology seemed poised to set the new global standard.[5]

A motor for exports, a field of incredible size and concentration, a sector firmly supported by the state and science, a technology that ensured a position at forefront of economic development—this was the image of nuclear power crafted by West German elites during the 1950s and 1960s. Unlike the Federal Republic's adoption of oil, the nation's turn toward nuclear energy grew from a highly coordinated effort by state officials, party leaders, scientific experts, and, to varying degrees, business magnates. If oil was a wave unleashed by American policy that West German politicians reacted to and welcomed, nuclear power belonged to a utopian and technocratic campaign to modernize the Federal Republic.

Nuclear power in West Germany, in other words, was an energy transition crafted by elites. That a state known for its Ordoliberalism could build an entirely new energy system comes as a paradox. But in fact, states have long been engines of innovation and market creation without which entrepreneurs could never have developed some of the most important technologies we now take for granted, from biomedical products to the internet.[6] Germany is no exception. The German state—or states—has a long history of stimulating technology, from founding universities and technical institutes in the nineteenth century to using state banks to stimulate strategic industries.[7] After World War II this tradition continued in a new guise, coalescing around nuclear power, the technology that many believed held the key to unlocking the developmental potential of humankind. By the 1950s the Federal Republic began functioning as an "entrepreneurial state," stimulating, organizing, financing, and assuming the risk of nuclear power in an effort that far surpassed its interest in any other energy type until the renewables of today.[8]

State officials, however, promoted the atom not because they thought West Germany needed it for electricity. Indeed, before the late 1960s those organizations most closely tied to electricity saw little use for reactors. Instead, nuclear advocates in the state cultivated a coalition with scientists, the electrotechnical industry, energy experts, and the SPD, who all hoped nuclear power would sustain West Germany's standing among the world's leading industrial and export countries. If French elites pursued atomic power for national prestige, and American elites to maintain their edge in the Cold War, West German elites did so because they believed this new energy held the key to the country's "overall economic development."[9] But how to organize and finance this novel industry—through state guidance, private initiative, or international control—remained hotly debated questions.

By the 1970s the huge construction plans for reactors promised to change West Germany's economy and even the nature of its political system. The sheer cost of atomic research, the scale of capital required to build reactors, and the dynamics of the electricity market meant this energy transition was characterized by even more concentration and in-transparency than West Germany's move to oil. And this brought incredible risks, as the elites making policy assumed a cavalier attitude toward decisions that affected the lives of millions of their citizens, raising fundamental questions about how West Germany should actually use its energy, and how policy should be made and vetted in a democracy.

Spitting the Atom: War or Peace?

At his refuge in Cambridge, Massachusetts during the 1940s, the émigré German-Jewish philosopher Ernst Bloch was moved by Europe's descent into fascism to pen *The Principle of Hope*. Reflecting on the fear spawned by Nazi Germany, Bloch countered that humans were inherently utopian: that dreams of a brighter future would motivate people to change the world for the better. After the Nazis, humankind now faced the "question of learning hope." Bloch's three-volume magnum opus reflected on the development of humankind, and the energy that led down this road of hope was atomic power, "the colossal discovery of our age." While the pace of invention had slowed to "that of a mail coach" in most fields, the atom was the exception. Atomic power would yield "energy of cosmic proportions," creating:

> fertile land out of the desert, and spring out of ice. A few hundred pounds of uranium and thorium would be enough to make the Sahara and the Gobi desert disappear, and to transform Siberia and Northern Canada, Greenland and the Antarctic into a Riviera.[10]

Bloch's words captured the imagination of people across the political spectrum for a new world supported by the atom, which was first split in a self-sustained chain reaction in 1942, and which three years later was shown to have devastating powers when atomic bombs decimated the cities of Hiroshima and Nagasaki. If only mankind could tame the "dread secret and the fearful engines of atomic might" that the American military had unveiled over the skies of Japan, a new world awaited. In 1953, in his Atoms for Peace Speech, President Eisenhower presented his dream of a peaceful international order based on cooperation around the atom, which would transform everything from medicine to agriculture. In 1955 Louis Armand, France's leading advisor to the OEEC, spoke of an energy that would lead to an "economic revolution."[11] In West Germany, before 1957 the possibilities for the atom seemed so limitless that it became a vessel into which elites poured their hopes for every type of advance, from medicine, chemistry, or industry to international order. Even redemption: rejoining the "West" through science, many Germans hoped, would restore their new republic from the crimes of its Nazi past.[12]

As Bonn began building its own nuclear infrastructure, however, these inchoate dreams coalesced into a specific vision for what the atomic age could offer this fledgling republic. Work on atomic energy began remarkably soon after the end of World War II. In 1949 West Germany, still occupied by the Allies, began reserving Marshall Plan funds for atomic research. In 1950 it began prospecting for uranium. In 1951 the German Research Council included atomic energy in its agenda. And in 1952 the Allies indicated they would permit Bonn to deploy research reactors. The man driving this campaign was Werner Heisenberg: Nobel Prize winner, founder of quantum mechanics, theoretician of nuclear fission, and a leading player in the Third Reich's nuclear program. A figure of profound prestige who enjoyed contacts among both Germans and Americans, Heisenberg pushed the pace of atomic development. As West German independence loomed on the horizon, he began calling for a national atomic commission. And he served as a go-between for Bonn and the Allies, helping to design nuclear protocols once Allied oversight ended.[13]

But from the very beginning, Heisenberg disagreed with both Ludwig Erhard and Konrad Adenauer of the CDU over whether the state should actively craft a research agenda. In contrast to Erhard's Economics Ministry, a bastion of Ordoliberalism, Heisenberg believed the task of creating a nuclear program would be so difficult that Bonn could not leave it to business alone. Without aggressive state support, he feared, they must, "for the time being, forgo German participation in [nuclear] development entirely." Adenauer, however, postponed an atomic commission until after signing the international treaties that would give West Germany its full independence, fearing a nuclear program would complicate the thorny question of rearmament, sovereignty, and integration into the North Atlantic Treaty Organization (NATO).[14]

Once the Federal Republic gained full sovereignty in 1955 and joined NATO, the context changed fundamentally. And change came in the nick of time, for the first International Conference on Atomic Energy in Geneva that August made an "extraordinary impression" on German participants, "dispiriting" them by underscoring just how far their rivals had already advanced.[15] Industry lobbyists alongside leading scientists now joined Heisenberg in arguing that West Germany needed national institutions to coordinate research between science, business, and the state, as well as major federal funding if the country were not to fall further behind the United States, Britain, and France. That October, Adenauer gained cabinet approval to form a Ministry for Atomic Affairs, which he placed under the guidance of Franz-Josef Strauss, leader of Bavaria's Christian Socialist Union (CSU) and a polarizing figure whose opponents placed him "to the right of Genghis Khan." The cabinet also formed an Atomic Commission of dignitaries to advise the new ministry and craft a legal framework for nuclear development. Formed in January 1956, the commission was led by Strauss and included leading scientists, industrialists, and bankers. Its vision for atomic energy: nothing less than "a national challenge in the true sense of the word. It is not about military or political power, it is not about prestige; for us it is about asserting and securing for the German people their place among the industrial nations of the world."[16]

By 1956 West Germany had in place the institutions that could develop nuclear technology. But to what ends—peaceful energy or the bomb? Over the next year and half, it became apparent that military considerations motivated not only the chancellor but also Strauss. While West Germany had renounced nuclear weapons upon joining NATO, this did not close the debate on the new country's arsenal. Strauss, for instance, linked his appointment as atomic minister with Adenauer's promise to put a CSU member in a defense field. Adenauer chose Karlsruhe as the site of West Germany's first nuclear research center in part because of its military defensibility. Between 1956 and 1957 nuclear weapons came to dominate national politics. When a suggestion by US Admiral Arthur Radford was leaked that the United States might cut its conventional forces in Europe and replace them with a nuclear deterrent, Germany's press and political elite erupted into turmoil. By 1957, an election year, the question of atomic armament by Adenauer's own admission, "overshadowed . . . all of politics."[17] Should West Germany's military be allowed to deploy nuclear missiles under NATO? Should it develop nuclear weapons of its own? And what should become of its nascent atomic energy program?[18]

The scientific community and the SPD worried intensely about these military aspirations. At a NATO meeting in late 1956, Strauss—now minister of defense—expressed a strong desire to acquire nuclear weapons as part of a broader geopolitical reorientation. Adenauer agreed behind closed doors and moved forward with plans to equip the army with NATO-controlled nuclear-launching systems.[19] The following April he seemed to confirm the fears of his skeptics when he remarked at a press conference that tactical nuclear weapons, "are, in principle, nothing other than a further development of artillery." To not arm West Germany would be a "gift" to the Soviets. His comments unleashed a storm in the media, and validated the public's existing understanding of nuclear issues, namely, as an object of military strategy not energy policy.[20]

Although this undercurrent of angst framed nearly all discussions of atomic energy, the scientific community was working to detach this new technology from its association with the bomb. The intervention of eighteen leading scientists helped do so. As the military overtones of West Germany's nuclear program became more apparent, in late 1956 Carl Friedrich von Weizsäcker—son of the former state secretary of Hitler's Foreign Office, leading physicist, and member of Heisenberg's nuclear research team in the Third Reich—met in Heisenberg's house to draft a confidential letter of appeal to the new minister for atomic affairs, Siegfried Balke. Heisenberg, Weizsäcker, and over a dozen premier scientists signed the letter, which rejected nuclear armament as the "wrong path" for the Federal Republic, one that could lead to the "total destruction of Germany" (see Figure 4.1). They threatened to refuse to participate in any effort related to atomic armaments. And while the scientists initially kept their complaints confidential, Adenauer's April press conference sparked Weizsäcker and Heisenberg to go public, stoking a political firestorm. Refuting the chancellor was a rare thing in the conservative political culture of the early Federal Republic, but it earned the respect of the public. After other notables added their voice to the clamor against nuclear militarization, Adenauer issued a communiqué clarifying that "the Federal Republic will not produce its own nuclear weapons."[21]

Figure 4.1. Signatories of the Göttingen Manifesto Meeting in Bonn in April 1957. Otto Hahn (Left), Walther Gerlach (Center), Carl Friedrich von Weizsäcker (Right). *Source:* Ullstein Bild / Getty Images. Nr. 541080905.

Nuclear Power, Economic Development, and Exports

The Göttingen Manifesto helped shift public opinion, but it took more than a letter to decouple the peaceful use of nuclear energy from atomic weapons. Only with the discussions surrounding Euratom, and the dawning realization that a global competition for nuclear energy was underway, did a non-military vision for nuclear energy triumph in West Germany.

After a groundbreaking push to integrate Western Europe through the European Coal and Steel Community (ECSC), momentum ebbed after a string of setbacks. But in 1955 the architects of European unification hoped to revive their dream with a new initiative. They presented their vision that summer, when Jean Monnet of France and Paul-Henri Spaak of Belgium suggested that atomic energy could become a symbol of modernity that would capture the popular imagination and create enthusiasm for European cooperation. For European countries to succeed in atomic energy, Monnet and Spaak believed they must become more like America, with a vast market and the resources of an entire continent that could sustain this expensive new technology. In April 1956 Spaak homed in on creating a common market and cooperating in atomic energy as the twin paths toward integration. That fall Monnet convened a committee of "Wise Men" to study nuclear technology, and after traveling to the United States, the chairs—Armand of France, Franz Etzel (CDU) of West Germany, and Francesco Giordani of Italy—issued their widely read *Target for Euratom*, calling for an unprecedented investment program that would produce 15,000 MW of nuclear electricity by 1967.[22]

Euratom advocates sold their agenda by claiming that nuclear power would achieve two distinct goals. On the one hand, the Wise Men portrayed nuclear power as necessary to meet the European energy gap predicted by OEEC experts. The Wise Men argued that Europe's "future growth is threatened by the power supply situation. America produces the power it needs, cheaply. Western Europe, where new power is twice as expensive, must import more and more of it, dearly. Western Europe has become the one great industrial region of the world that does not produce the energy necessary for its development." Nuclear power, they claimed, would let Europe close the energy gap with the United States.[23]

On the other hand, Monnet saw Euratom as a way to quicken the pace of political integration. As a virgin economic sector with few vested interests, atomic energy, Monnet hoped, would provide an easier path toward unification than establishing a common market for all goods and services.[24] Monnet soon drew others on board, including Chancellor Adenauer and the SPD, which hoped

Euratom could prevent the proliferation of atomic weapons. By 1957 this conviction in the political potential of atomic energy became so entrenched among key European players that Euratom formed one half of the larger deal that inaugurated the second, more lasting wave of European integration brought by the Treaties of Rome. German industry, which had initially resisted atomic cooperation with France, accepted Euratom as the quid pro quo for France moving ahead with the European Economic Community.[25]

Euratom thus formally came into existence in January 1958, less than a year after the Göttingen Manifesto. But the institution soon encountered insurmountable problems as cooperation fractured over the contentious issue of who should actually own fissile material used in any joint nuclear venture. Armand and Monnet wanted Euratom to control all uranium; German industry resisted what they feared would turn into a French monopoly. And the entire notion of European control of fissile material ran counter to the nuclear agenda of West Germany's science and chemical community, which hoped to develop their own line of heavy water reactors that used natural uranium—not enriched uranium produced in France or the United States—generated from a domestic supply chain of West Germany's own companies.[26] These disputes watered down the final treaty: technically Euratom gained the authority to own all fissile materials, but its monopoly was riddled with exceptions. By the early 1960s Euratom devolved into a joint research program, that by its own admission "supplements national programmes; it does not replace them."[27]

Nevertheless, Euratom changed how West German elites thought about nuclear power. The three-year-long debate exposed them to the rapid progress made by the United States, Britain, and France, leaving them with a palpable sense of falling behind. It also focused attention on what nuclear power could provide their country, honing down the multifarious visions of the early 1950s into a single theme that was *not* about electricity, medical advances, or atomic weapons. As Siegfried Balke, West Germany's second atomic minister after Strauss, put it, nuclear reactors must no longer be understood as an "arcanum that could solve all the economic and political malaises of the world. The peaceful use of atomic energy must be taken out of the sphere of the sensational and integrated into the logical development of technology and economy."[28]

The Euratom debates did just that, elevating economic development as the raison d'être for the entire undertaking. Cultivating the engineers and scientists needed to bring this technology to fruition; amassing the infrastructural capacity to handle uranium; building the cooperation between industry, science, and the state to orchestrate the monumentally expensive undertaking that a commercial reactor represented: all of this was necessary for Europe to keep its edge in an age when science and human capital were becoming ever more critical to development. The Wise Men, for instance, claimed that without directing its energies toward nuclear power, "Europe will rapidly become an underdeveloped territory."[29] The preamble to the final Euratom agreement pushed this discourse still further, describing "nuclear energy as an indispensable resource for the development and revival of the economy."[30]

In West Germany, this infatuation with economic development assumed a particular form: nuclear energy as an engine of exports, a high-tech global sector in which their nation could excel. For by 1958 the notion of an impending energy or electricity gap had vanished, as the nation descended into its first coal crisis, one of surplus energy not scarcity. If nuclear power could be justified, it was in the name of a new technology that would generate exports, not electricity.[31] In public pronouncements, Strauss hammered home the importance of this new technology for an export nation like the Federal Republic. In the introductory edition of *Atomwirtschaft*, the lead forum for West Germany's nuclear community, he pointed out how "the battle for the global market has already commenced.... For us it is not so much a question of political power as it is a matter of existence for a people living in small space." The Atomic Commission spoke of being "outclassed" and losing international "competitiveness" if West Germany failed to keep up in the "Atomic Economy."[32]

The strongest advocate of this export vision was the Economics Ministry, where officials lost their reticence to reactors when they realized the potential to sell this technology in emerging markets like India and Brazil. For the ministry, the advances on display at Caldor Hall in Britain or Shippingport, Pennsylvania, illustrated how the Federal Republic, if it did not join the nuclear race, would be "excluded from the global market in the promising field of nuclear industry and suffer, moreover, from a diminution of its technological reputation in other spheres of industry."[33] Underpinning this vision was the belief that West Germany's living standard ultimately depended on "the export of technical achievements," that the country's industry and science must stay at the cutting edge of technology by cultivating experience, human capital, and infrastructural capacity in what were considered to be the sectors of the future.[34] As Rudolf Kriele, atomic minister after Balke, argued to the chancellor's office, "if [Germany] wants to retain the rank and reputation of a modern, high performance industrial state it must quickly put itself in the position to export small and medium-sized reactors into the vast deserts of the developing countries."[35]

Nuclear Power through Private Enterprise?

If a broad nuclear consensus was emerging in the late 1950s, the concrete details of how any nuclear program should actually proceed remained undecided. Most importantly, how much should the state guide development? For while

Ordoliberalism prevailed in the CDU, other countries were engaging in massive state support for their atomic industry, raising fears that the Federal Republic would continue to fall behind if it clung to the free market.

A range of experts now followed Heisenberg to argue that the very nature of nuclear power necessitated an interventionist state. In *Atomwirtschaft* scientists and legal scholars came out in favor of a state-run program. As a leading jurist pointed out, "more than any other field, atomic energy is dominated by the state, as it were, from the nature of things: in terms of mining privileges provided by the state, international state treaties, the dangerous nature of materials and production processes involved, and in the current or potential significance for the conduct of war." The very need to handle radioactive material, to overcome the "mis-investments" that were bound to result from a project of this magnitude, and to ensure that knowledge generated from research spread throughout different sectors made joint state-private enterprises seem the way of the future.[36] Both Britain and France were moving in this direction, as the former struck forth on a costly ten-year program under its publicly owned Central Electricity Authority, while the latter channeled vast sums to nuclear research through two state-owned organizations.[37]

The SPD, too, pushed for a state-run nuclear program. Since the early 1950s it had been calling for public ownership of all energy and raw materials. The atom fit perfectly into this program, representing the element that could most improve human life, "free people from their worries, and create prosperity for all," as the party explained in its 1959 Bad Godesberg program. As a leading parliamentary delegate for the SPD put it, energy firms "act according to the considerations of private enterprise. They cannot take on investments that in reality must be taken on by the entire economy."[38] Even Strauss sympathized with this view, arguing that "the public hand and private enterprise should together finance the technological development" of nuclear power through a "nonprofit" limited liability company. And West Germany organized its first large research center in Karlsruhe along these lines, with Bonn putting up 30 percent of the initial shares, Baden-Württemberg 20 percent, and a consortium of enterprises the remaining 50 percent.[39]

This statist approach, however, did not at first penetrate the power centers that controlled funding in the federal government, nor did it at first appeal to industry. The Karlsruhe Institute aside, officials in the Economics Ministry believed nuclear power must be firmly embedded in the social market. But so too did Germany's second atomic minister, Siegfried Balke. A chemical engineer from Munich, Balke reached the higher echelons of industry in his career, and thought nuclear power offered incredible potential for West Germany's export industries. He called at first for "as little state as possible." In his opinion, "this new technology [could] be meaningfully integrated into our existing economic system. There is no need for public action in this field, because all of the challenges and problems that might arise in the future can be regulated on a private sector basis." Pure atomic research should be state-funded, but experimental power plants and the equipment and construction that made these possible should be left to private companies.[40]

The notion, moreover, that nuclear energy would fit particularly well with the social market—an economy that prized competition, decentralization, and small scale—was widespread during the 1950s. Strauss thought the low cost of transporting uranium—so little of it by weight was actually needed to generate power in comparison with other fuels—meant that with nuclear power "the problem of location plays virtually no role, and this opens entirely new perspectives. Power plants can be built practically everywhere." Industry need not cluster near the mines of the Ruhr, or in the riparian ports where the great oil tankers docked.[41] Leo Brandt of the SPD, and cofounder of West Germany's nuclear research center in Jülich, agreed, enthusing that with nuclear energy "decentralized power plants will once again have a chance . . . it hearkens back to the era of a steam engine in every factory." That West Germany had no huge state enterprise guiding its nuclear program, so many argued, was a good thing.[42]

The CDU designed the Atomic Law with private enterprise in mind. Passed only on its fifth iteration—Ruhr anthracite producers having killed a version in the midst of the first coal crisis—the Atomic Law went into effect in 1960 and provided the legal basis for West Germany's nuclear program. The final version gave the state strict oversight over the handling and distribution of uranium, and made use of fissile material subject to its approval, but ownership remained in the hands of companies, institutes, or utilities. More importantly, the law vastly curtailed the insurance risks associated with constructing a nuclear facility, liberating companies from a potentially unbearable burden. As even advocates of the atom admitted, the sheer lack of experience with this technology made the risks assumed by producers utterly "incalculable" from the standpoint of those providing the insurance.[43] An accident, in other words, could render the entire program un-insurable and thus financially impossible. But the show must go on, as a delegate from the Atomic Commission remarked, "the risk of catastrophe should not smother the development of nuclear power."[44] To solve this impasse the law capped the total liable damage at 500 million DM—a figure taken from American law—while requiring the actual reactor operators to cover no more than 15 million, the state assuming the rest. In effect, the law shielded private enterprise from risks that would have been prohibitory by assuming public responsibility for them.[45]

This private enterprise outlook extended, in part, to financing, which the Economics Ministry thought was the key to the entire nuclear endeavor. In the 1950s the Federal Republic suffered from tight credit, making it impossible to raise the capital for nuclear projects through private markets. Before the 1960s financing came largely from firms' retained earnings or from state-like institutions such as the European Recovery Program and the Reconstruction Credit Organization (*Kredit-Anstalt für Wiederaufbau*), a government-owned development bank. Nuclear power was no exception, but rather an example of how constrained the Federal Republic was in embarking on major infrastructural projects. As the Economics Ministry noted in 1958:

> the incomparably high risk that comes with investment into nuclear power, given today's current stage of development, limits private entrepreneurial activity since the risk exceeds the financial power of private enterprise and paralyzes private initiative.[46]

How to overcome this financial risk remained contentious. The Atomic Commission wanted to compare reactors with coal-fired plants, and have the state assume the expected difference between the up-front cost of these two sources of electricity. State aid would come in two forms: either through massively subsidized loans or by offering tax-free status to funds that firms dedicated to nuclear reactors.[47]

The Economics Ministry feared these plans would place most of the cost and "practically the entire risk" onto the state. And this went too far. Instead, the ministry put forth a program that conformed more to market principles, that limited the state's expenditures to mainly research and development, but that still tried to bridge the gap between how much risk the private sector could assume, and how much risk a world-class nuclear program actually entailed. Put simply, the private sector should bear most of the risk of any financial loss, while the state should help bring this down to acceptable levels.[48]

These debates informed West Germany's first atomic plan in 1957–1958, the Eltville Program, which aimed to build five nuclear plants by the early 1960s. While West Germany was already building research reactors based on foreign models, Eltville aimed to help national industry craft larger plants of their own design.[49] But Eltville was largely a failure. Only three projects began under its framework, and federal funding never materialized on the scale called for by the Atomic Commission.[50] In reality, Bonn followed the Economic Ministry's advice of leaving much of the financing and initiative to private enterprise. More generally, the program had no clear vision: the planned reactors were too large to complete without more state aid, but too small to function as a power generator that could feed electricity into the grid. Just two years in, Balke changed his mind and began calling instead for small, German designed, multipurpose reactors of 25–50 MW, arguing that these would be the easiest to export. The Economics Ministry countered that industry would never gain the experience it needed with such small plants, and demanded that domestic firms cooperate with American companies to build large, foreign-made models to learn to operate reactors on a commercial scale.[51]

If Eltville suffered from competing agendas, the larger challenge plaguing West Germany's first program was that industry simply had little interest in sinking capital into atomic energy. As nuclear development in the United States and Great Britain encountered cost problems, the utopian aspirations spawned by Eisenhower and the Geneva Convention wore off. The predictions of an energy gap, meanwhile, were proving wildly inaccurate as a wave of oil began flowing from the Middle East. In West Germany the electricity industry saw the entire nuclear endeavor as a burden they wanted to foist onto others. For the capital needs of these utility companies were already immense in the 1950s and early 1960s, as they struggled to cover the cost of building generation and transmission infrastructure to keep pace with their nation's burgeoning appetite for electricity. No less than two-thirds of all investment in West Germany's energy economy was going to its power lines and electricity plants; utilities had little left over for a technology that seemed decades away.[52]

Key players in the electricity sector, moreover, actively opposed nuclear power at first. Rheinisch-Westfälisches Elektrizitätswerke (RWE), West Germany's largest utility and one of its richest companies, declined to enter the nuclear race at first. Like other utilities, much of its capital was going elsewhere: building hydro plants in Bavaria or power plants in the Ruhr. Uniquely among West German utilities, this conglomerate had acquired ownership of massive lignite coal deposits. Nuclear power thus ran counter to its own interests, since its control over cheap, dirty, brown coal gave it an edge over its rival. RWE's director, Heinrich Schöller, downplayed nuclear utopianism and asked the nation to "take a breather" on this technology in the depths of the 1958 coal crisis. Would not the eight billion DM that Bonn expected to spend on nuclear power, he suggested, be better spent on reviving coal? With that sum, the entire coal sector could modernize and provide far more electricity than nuclear power. RWE's internal estimates, furthermore, claimed that nuclear-generated electricity would be four to five times as expensive as coal-fired electricity, while even the most generous external forecast expected it to remain uncompetitive until the mid-1960s.[53]

The chemical industry, an early advocate of nuclear power, also grew frustrated with this technology. As a sector that could potentially supply inputs for reactors, like heavy water as a moderator or enriched uranium, it stood to benefit from the atom. Yet by the late 1950s it was clear that the United States could undercut many of the supply chains that German chemical firms hoped to forge. Euratom had paved the way for bilateral agreements with the United States, which offered to sell, rent, or lease enriched uranium to European companies at strikingly low prices.[54] The US Atomic Energy Commission (AEC) saw these bilateral agreements as a channel to help American firms penetrate the European market, announcing major price discounts on enriched uranium, while the US Export-Import bank offered cheap credit to European institutions that imported uranium. Despite complaints about American dumping, German chemical companies' plea for a subsidy fell on deaf ears in the Economics Ministry.[55]

Among private enterprise, then, the electro-technical industry was left to spearhead nuclear energy. Companies like AEG and Siemens, after all, would be the ones building reactors. To compete on global markets, moreover, the directors of these firms believed their companies had to master this new technology, lest they fall behind the Americans, British, or French. Brown, Boveri, and Cie (BBC)—a Swiss-owned electro-technical firm based in Baden—joined the machine tool manufacturer Krupp in constructing the research reactor for Jülich.[56] Siemens partnered with America's Westinghouse to build the research reactor in Karlsruhe, using heavy water and natural uranium in the hope this technique would help Germany carve out an export niche. The entire affair, West Germany's leading newspaper noted, "is not only about solving scientific problems . . . and certainly not only about the future energy supply of Germany, rather it is in fact about German exports."[57] In 1961 Siemens's archrival, AEG, won the contract for a large reactor in Kahl. Ironically, RWE bought and paid for Kahl in cooperation with the local utility, Bayernwerk, for defensive reasons. It wanted experience with reactors should this technology ever become competitive.[58]

The Entrepreneurial State

Kahl went online in January 1962 in Bavaria, a region that suffered from a high coal price and thus seemed to offer the space for nuclear electricity to thrive. But in 1962 Kahl was West Germany's only successful, large reactor feeding electricity into the grid; the other projects were either devoted to research or suffered from cost over-runs and delays. The Federal Republic was hardly closing the chasm with the global leaders, and in fact new gaps seemed to be opening before their very eyes. Wilhelm Alexander Menne, member of the Atomic Commission and a vice president of the influential Association of German Industry, captured the sense of inertia in 1961, when he reported that West Germany was still ten years behind because the state had failed to fund this new technology. In his mind, lack of state aid bedeviled research and development across the board, something he and others hoped to change by reinventing Bonn's technology policy. He went on to reflect:

In the three most modern fields of industrial development we are not at all represented in the manufacture of aircraft or rockets, and in the atomic industry we are still at the beginning. In the long run, deficiencies in these modern developments will not be without repercussions on the rest of our industrial sectors.... If we in Germany want to keep up with the high-level advances in other countries, we must achieve similar accomplishments in at least one of these fields of modern technology, namely, atomic energy.[59]

Experts from across the political spectrum increasingly agreed. Balke feared engineers would emigrate if West Germany did not change course. In an about-face, the CDU began worrying that without more state aid their nation would fall behind. Between 1958 and 1963 the United States, Britain, and France far outspent Bonn on nuclear research. And now the United States seemed poised to deliver turnkey reactors to Europe.[60] The atomic community concluded that their country had little space for error if it wanted to uphold the competitiveness of German industry. If it failed now, the nation would "sink to the level of a nuclear undeveloped country. If one considers that reactor technology is almost universally the strongest engine of general technological advance, the consequences of backwardness in nuclear energy production for our future economic potential are utterly incalculable."[61]

Balke now walked back his conviction that nuclear energy must be led by private enterprise. Already at the end of 1958, after frustrating delays and another atomic conference in Geneva that again illustrated how far behind his nation was, he lashed out to the finance minister. "To break the ice," he argued, "it is necessary that we finally build an actual power reactor.... If private enterprise continues to delay, these projects should be implemented by the public hand." By 1961 Balke demanded that Bonn triple funding for nuclear power and cover the risk of constructing a commercially viable reactor. Statist ideas were winning the day.[62]

At the same time, Balke and the atomic community began calling for a different sort of reactor: West Germany must go big. "The ultimate goal," Karl Winnacker of Höchst plainly put it, "is the construction of massive power plants."[63] Developments in the United States were already pointing in this direction: in 1959 General Electric completed a 200-MW plant near Chicago and the rest of the world took notice. Germans believed they needed a similarly large reactor to gain experience for commercialization, to push the boundary of their knowledge, and to ensure their nation remained competitive in future export markets.[64]

This was the context behind RWE's new plan to cooperate again with Bayernwerk in building a massive reactor in Gundremmingen that might, given the right mix of subsidies, be competitive with coal. But the Atomic Commission and the Atomic Ministry were the real motivating force behind Gundremmingen (see Figure 4.2). RWE, in fact, was reluctant to assume the full risk of this 250-MW facility. As even Heinrich Mandel, RWE's main atomic enthusiast, noted in the negotiations, "our free market economic system will be overburdened if one expects the utility companies, during a moment in which atomic energy is neither needed nor is in the strict economic sense competitive, to crank out atomic power plants on an assembly line."[65] In his mind, the project hinged on whether electricity from nuclear could compete with coal power plants. Mandel wanted Bonn to provide RWE with tax-free status for its reserve funds. The Economics Ministry rejected this, but the enthusiasm within the Atomic Ministry for Gundremmingen led Bonn to find other ways to subsidize the project. In the end, of the total 340 million DM cost, RWE and Bayernwerk together covered just 100 million, with the rest coming from Euratom, the ERP, Bonn, or America's Export-Import Bank.[66]

Gundremmingen was a sweet deal for RWE. In their eagerness to get the plant built, the Atomic Ministry and others accepted a tardy safety report that did not even include drawings of the reactor. Construction began in 1962, and in 1966 the plant, the largest in the Federal Republic and one of the largest in Europe, began feeding electricity into the Bavarian grid. RWE gave the subcontract again to AEG—providing the electro-technical company with the largest contract in its history to that date—and again chose the American light water model, dampening the hopes of Germans who were pining for their own heavy water technology.[67]

Figure 4.2. Gundremmingen nuclear power plant. *Source:* Süddeutsche Zeitung Photo / Alamy Stock Photo. Image ID: RMB46N

After 1963 federal money began flowing more freely as the CDU allocated new funds to the Atomic Ministry, now reformed into the Ministry for Scientific Research. With help from regional governments, Bonn took over financing the main research centers in Karlsruhe and Jülich. It passed West Germany's second atomic program, the Spitzingsee Agenda, committing to 5.3 billion DM over the next four years for experimental reactors, human capital development, basic research, and Karlsruhe.[68] Work began on two large new commercial reactors in Lingen (250 MW) and Obrigheim (350 MW), both of which were financed on the same model as Gundremmingen, with Bonn, local government, and Euratom together supplying most of the funding. The state, moreover, agreed to cover 90 percent of any losses on these commercial plants, up to 100 million DM, removing nearly all the risk.[69] This was the moment, in other words, when the state became the creditor to the atomic economy, taking on risk that had once deterred private industry and crafting an agenda that set the course of development for the coming decade.

Big Nuclear Takes Off

Although West Germany entered the nuclear race well after the United States, the hope that atomic technology might become an export engine kept interest alive just long enough to nudge Bonn, private industry, the utilities, and both major political parties toward a more statist political economy of technological development.

But in the mid-1960s the leaders of West Germany's atomic community still felt their nation was behind its rivals, above all the United States. This was the decade when America's aggressive postwar technology policy began yielding formidable results. World War II had launched this new approach, when the government vastly expanded its capacity for research through the Manhattan Project, and when a new cohort of state officials emerged to advance technology in strategic directions. In the 1950s the Lawrence Livermore, Los Alamos, and Sandia labs grew into the largest research complexes in the world, while the American military began massively funding other research organizations like Bell Labs. After the Soviets launched Sputnik in 1957, the United States commenced a new phase of state-sponsored technological development through the Defense Advanced Research Projects Agency (DARPA), whose "mission-oriented" approach aimed to solve technological problems with a long time horizon. By the 1960s this massive, state-sponsored complex put America at the cutting edge not only of nuclear power, but nearly every technology that seemed to matter, from computers and semi-conductors to solar power.[70]

Europeans took notice. In West Germany, experts and politicians feared a research and development chasm now separated the two sides of the Atlantic. In 1966 Gerhard Stoltenberg, the CDU's new minister of scientific research, conceded there was "still backwardness in German development" in the three technologies he believed were transforming the world: civilian nuclear power, space research, and electronics.[71]

> In all three spheres sustained intervention by states has fundamentally influenced the course of events. To a great extent the immense—often worrying—scientific and technological predominance of the global powers, which is already turning into commercial dominance in key markets, is due to these interventions.

Europe, he feared, "in many areas has been forced into a peripheral role."[72] By 1966 the United States had achieved a breakthrough in the commercial use of nuclear power that was overwhelming to West Germans. In that year alone the United States commissioned 20,000 MW of nuclear capacity, raising its total to 40,000 MW, roughly the same capacity of *all* electricity production in the entire Federal Republic.[73]

West German scholars and politicians argued that their nation must mimic the United States. At the annual meeting of Germany's Technical and Scientific Association in 1965, participants pointed out how the Social Market, which asked Bonn to leave technology to the market, was proving "completely unsatisfactory." The problem was a cleft between theory and praxis insofar as the global economy, in which German firms operated, in no way conformed to a competitive market. The Association pointed to the computer as a new field called into existence by American military spending. The Federal Republic must emulate this through a new ministry devoted purely to technology. Balke, ousted as atomic minister in 1962 but still active in these debates, called for a closer relationship between state, industry, and science. The former must no longer just channel funds to universities and leave the research agenda to scholars, it must actively guide research in a direction that "promotes the maximum benefit for the economy of the state—and thus for the wellbeing of its citizens."[74] The state must bridge the gap separating basic research and the application of technology, above all in those industries where advances required high risk and long-term funding. For nuclear power, as a CDU expert argued, this meant "forcing the pace of constructing a commercial facility."[75] The SPD, which had long criticized the CDU's technology policy, began demanding the unification of all research and development into a single ministry that would think in decades, not years.[76]

This new, long-term statist approach and these anxieties about a technological gap overlapped with other developments that would unleash an atomic building spree. For one, this was the moment when the new coupling paradigm went mainstream, which held that economies could only expand by consuming larger quantities of energy. This paradigm's influence was particularly strong in discussions of electricity, the consumption of which was growing more rapidly than other energies, driven not only by industry but also by the spread of power-hungry appliances like vacuum cleaners and refrigerators. Between 1950 and 1965 electricity consumed by German households grew almost sixfold, by German industry nearly fivefold.[77] And it seemed this tempo would not relent but in fact would increase. Euratom and West Germany's most respected forecasters all predicted electricity demand would more than double between 1965 and 1980. The country, it seemed, would require vast new investments, reviving anxieties about an energy gap from the 1950s.[78]

Second, the SPD, an organization that fully embraced the coupling paradigm, joined the government in coalition with the CDU in January 1966. Since reforming their platform in 1959, Social Democrats argued that the Federal Republic needed growth to generate the public funding that could pay for social reforms. And growth, so the party argued, required inexpensive energy. More than this, they hoped that nuclear power could provide cheap electricity to regions far from the Ruhr, and "for the first time in the history of industrialization," equalize energy as an input and even out the lumpy economic geography of the Federal Republic.[79] The 1966 address to the nation, coauthored with the CDU, highlighted nuclear power and the technology gap, advocating government "promotion of research in key spheres of technological development."[80] No less than Willy Brandt—an SPD luminary, foreign minister in the CDU-SPD coalition, and after 1969 the first Social Democratic chancellor—believed "the unhindered civilian use of the atom is a vital interest of the Federal Republic of Germany." In his mind, "the future of the Federal Republic as a modern industrial state depends on it."[81]

Lastly, consolidation in the electricity market during the 1960s created the conditions that allowed this vision of massive, state-supported nuclear power plants to flourish. Since it became law in 1935, the national electricity framework encouraged the formation of large networks through mergers and the annexation of territory. What had been a welter of diverse and disconnected electricity networks was gradually evolving into large zones dominated by a single utility. Where Germany boasted over 16,000 utilities in 1934, less than 400 were still operating by the late 1960s. The twelve biggest produced two-thirds of nation's power. Two numbered among West Germany's largest firms by capital and sales—RWE and VEBA. Power plants, too, grew in size to capitalize on the economies of scale made possible by the consolidation of territory. In 1964 300-MW plants were considered top of the line. Seven years later the first 600-MW plant joined the grid, and by 1975 900-MW plants were coming online.[82]

This trend toward bigger utilities, controlling larger swaths of territory, powered by ever more massive electricity plants, opened a window for nuclear power. While the initial investment for a commercial reactor was high, utilities in the United States were showing that large plants could spread this fixed cost over a bigger volume of electricity to yield prices low enough to be competitive. By the late 1960s German experts argued that such economies of scale could be achieved with nuclear plants of 600 MW or larger. But only the very largest utilities—at most five or six companies in the entire Federal Republic—could sustain projects of this magnitude.[83]

Nuclear advocates in the new Grand Coalition drew these strands together— the fear of a technology gap, the implacability of the growth paradigm, the trend toward ever larger utility companies—to overcome the lingering Ordoliberal penchant within the CDU and push for West Germany's third and largest nuclear program. They were helped by the recession of 1966–1967, which Minister of Science and Technology Stoltenberg used to sell his plans as a way of investing in the technological future of German industry.[84] With support from the SPD, Stoltenberg rolled out a new federal program in 1967 that would spend 6.2 billion DM on nuclear research and construction over the next four years. The financial commitment caused the Economics Ministry to flinch, but the Finance Ministry accepted predictions about a looming breakthrough in nuclear power and accepted the plan.[85] Most funding went toward research institutes in Karlsruhe and Jülich, but a variety of financial perks—duty-free uranium imports, free land leasing, generous depreciation allowances on investments, guaranteed low-interest or interest-free loans—made reactor-building more attractive than ever.[86]

With this new program, Stoltenberg brought state officials together with the major builders and the operators of nuclear facilities. The former, Siemens and AEG, were enthusiastic about more reactors. Among the latter, however, RWE remained unconvinced about the need for nuclear electricity despite its Gundremmingen plant. In 1966 it had just opened the largest brown coal plant in the world at Frimmersdorf, deepening its interest in this pollutant energy. Stoltenberg, however, asked the utility giant to send him their prerequisites for embarking on a larger reactor, and RWE responded with an incredible list of conditions. Most importantly, robust economic growth and continued rise in electricity demand were an absolute necessity, as well as a streamlined authorization process. In their quest to go big, Bonn's nuclear acolytes acceded to RWE's demands, and in a decision that would set the stage for social upheaval in the 1970s, the federal government agreed to shelter the licensing process from public view and deal with safety questions at a later stage.[87]

At the same time, however, Stoltenberg and his allies tried to force RWE's hand by helping its rivals enter the nuclear game. In 1967 Bonn helped orchestrate contracts by Preussen-Elektra, a subsidiary of VEBA and RWE's main competitor, for two massive reactors in Stade and Würgassen, of roughly 650 MW each. After a vicious struggle for the contracts, AEG won the bid for Stade, and Siemens for Würgassen. RWE responded by shedding its nuclear reluctance and beginning construction on the largest nuclear plant in the world, in Biblis near Frankfurt.[88] With a capacity of 1,200 MW, RWE's new plant would be the first to cross the 1,000 MW threshold and was considered a monumental achievement. As a later technology minister put it, for the political establishment the endeavor represented the best of West German industry, "a cutting-edge technology for the coming decades."[89]

Stade, Würgassen, and Biblis opened the floodgates. In the following decade West Germany would begin construction on no less than fifteen more commercial reactors, all of which were light water reactors using enriched uranium. This breakthrough had profound spinoff effects on the political economy of West Germany. The sheer size of the new reactors not only left smaller utilities in the dust, allowing RWE and VEBA to dominate the industry, it also led to further concentration among supplier firms. Most tellingly, Biblis brought the nation's two major electro-technical companies together for a joint venture in the field of turbine and reactor construction.

During the competition for Gundremmingen, Stade, and Würgassen, both Siemens and AEG began to realize that cooperation could be profitable. As early as 1967 Bonn had given its explicit approval for collaboration, believing it was in West Germany's interests to have a larger, more competitive, high-tech firm at a time when American multinationals were on the move across Europe.[90] Concrete negotiations began in 1968, and in April 1969 West Germany's two largest, and only, domestic reactor-building companies merged their nuclear operations into Kraftwerk Union AG (KWU). The deal fit perfectly with the SPD's new approach to economic organization, which prized size in the name of modernization and global competitiveness. As Willy Brandt noted in October 1969, in his first address to the nation as chancellor after his party's electoral victory that fall, "economic concentration is indeed necessary."[91]

KWU's successful bid to build the turbines and other components for Biblis only seemed to confirm the decision to form a massive enterprise. The new firm boasted 1 billion DM of contracts during its first business year, with the expectation of another 1.5 billion to come, making it the world's fourth largest reactor firm.[92] Global opportunities awaited, given that electricity demand was forecast to grow 7 percent a year in Western Europe and even more rapidly in developing markets like India, Brazil, and Argentina. As Klaus Bartelt, KWU's director, rhetorically asked, "Is there any other industry with such an assured rate of growth?"[93] In 1968 Argentina had ordered Latin America's first nuclear reactor from Siemens. It seemed that the dream which had initially kept nuclear power alive—of an export engine—was finally being realized. So argued West Germany's press, who hyped the Argentine contract as the deal of the century, even though the financing was supported by the German state. SPD leaders went further, predicting that in a decade and half the electro-technical industry would be exporting 70 billion DM worth of reactors, roughly the size of the current federal budget. KWU's successful bid for Holland's first nuclear reactor in 1969 confirmed the hype. The new firm, so it seemed, was an export "prodigy."[94]

Resource Fears Already?

But as the nuclear race took off, many wondered if the world actually contained enough high-grade uranium. And would the atom actually provide more security than oil, given that the Federal Republic imported virtually all of its refined uranium from the United States?

Initially, advocates had harbored little doubt that atomic power would revolutionize the relationship between energy and the material constraints of the earth. In 1956 *Atomwirtschaft* pointed out that, although reactors were similar to other energy systems in relying on resources from the ground, the nature of this element differed qualitatively, since the energy unleashed from uranium dwarfed what fossil fuels could yield. By one estimate, a single kilogram of yellow powder could provide as much energy as three million tons of coal.[95] And while experts quickly determined that West Germany was poor in uranium, and Western Europe hardly richer, they believed deposits in Europe's colonies and dominions could cover the continent's needs. Taken to its extreme, this thinking led the most avid atomic apostles to declare that humankind would be liberated from material limits. As the president of Euratom put it:

> Mankind no longer risks being faced with a power shortage. There is no further need, therefore, to be thrifty with natural resources—quite the contrary—apart from the quantities intended for chemical uses, we should take the fullest advantage of investments which have already been made by using up the reserves before they lose their value as a result of competition from cheaper nuclear energy.[96]

The real question was how much enrichment would cost, and where it should be done: in the United States or in Europe. Through the early 1960s America's processing facilities easily provided West Germany with enough enriched uranium at a low enough cost that resource constraints hardly seemed an issue.[97] But with the expansion of the industry in the 1960s, and the creation of market devices for nuclear energy that allowed for better forecasts, experts across Europe came to see the United States' monopoly on enriched uranium as a problem. With the vast expansion in global electricity demand, they began warning that in the capitalist world, deposits of top-grade fuel—uranium with a high percentage of U235, which would be further enriched for use in light water reactors—were not nearly as abundant as once thought. By the mid-1960s leading West German geologists claimed proven reserves of high-grade uranium were just a third of previous estimates. Low-grade natural uranium, meanwhile, which has a high proportion of U238 and is not easily fissible on its own, was proving even more expensive than expected to transform into usable form. By the late 1960s the OECD began publishing global resource estimates that predicted a looming uranium shortfall and warning, along with other experts, of rising costs.[98]

In the late 1960s West Germany's atomic community consequently began investing in the idea of enriching their own uranium. The Third Atomic Program devoted more money to this task, while Stoltenberg opened discussions with the Netherlands and Britain for a new, joint enrichment venture.[99]

But the more momentous response to these resource anxieties was the vision for an entirely new type of reactor that could, in essence, generate its own fuel: the fast breeder. At the end of World War II, American military planners had themselves started worrying about the scarcity of high-grade uranium. Scientists around Enrico Fermi responded by suggesting a new type of chain reaction that could turn natural low-grade uranium into plutonium. Thus was born the fast breeder, which would combine a small, initial amount of plutonium with low-grade uranium to produce energy, plus even more plutonium than had gone into the reaction in the first place. Put differently, the process would "breed" more nuclear fuel as it generated energy, all while relying on the more common, naturally occurring U238.[100]

While breeder experiments in America and Britain encountered problems, by the mid-1960s "breeder fever" captured the imagination of Western Europe. In West Germany, the nuclear community hoped their program would culminate in the fast breeder as a way to attain resource independence.[101] But this chain reaction remained an academic problem until Wolf Häfele, the director of Karlsruhe's theoretical research department, returned from a year at America's Oak Ridge National Laboratory as an advocate of this new technology. As a project swinger and a proponent of "Big Science," Häfele began weekly seminars devoted to the breeder, playing up fears about a technology gap and uranium scarcity to solicit funds. In 1963 Häfele's group began working out the details for an actual breeder reactor.[102] That same year, experts across Western Europe began pushing this technology as the long-run solution to their shortage fears. West Germany's delegate to Euratom, for instance, argued that the breeder meant nothing less than "energy for an incalculable length of time and independence for the European Community from all fossil fuel imports—which would help temper the price policies of the oil states."[103] By the mid-1960s "breeder fever" reached the highest levels of the government, forming the centerpiece of a parliamentary debate about the nation's technological development. Here the CDU's lead protagonist called the breeder a "scientifically based fact," a claim that went uncontested among the parties.[104] Karlsruhe began work on a sodium-cooled experimental breeder, collaborating with an industrial consortium led by Siemens, while AEG began experimenting with steam-coolant. The following year the third atomic plan placed the fast breeder at the center of its research agenda.[105]

As with previous reactor models, West Germany's state assumed the financial risk. But the fast breeder was a much greater gamble than the light water reactor because of its potential for accidents. According to German estimates, to be commercially viable the amount of plutonium produced by the breeder was still much too low.[106] More problematic was the risk of a catastrophe. To support the unique chain reaction, the fissile core had to be more concentrated than in standard reactors, creating the need for a more powerful coolant. Sodium was the major contender. But sodium interacts violently with water and air, meaning any minor leak had the potential to unleash a full meltdown. More worrisome, the breeder reaction would continue if coolant were lost, leading in the worst-case scenario to a nuclear explosion. America's first experimental breeder in Idaho—nicknamed Zinn's Infernal Pile after its designer—went online in 1951 but suffered a partial core meltdown just four years later. A second breeder, the Enrico Fermi Plant near Detroit, started generating electricity in 1963, only to experience its own partial core meltdown in 1966.[107]

West German researchers and state officials were fully aware of these setbacks, but advocated the breeder into the 1980s.[108] For Häfele and other enthusiasts, these risks were the cost of pushing the boundary of technology. As he argued in a widely cited lecture, "The Ancient Egyptians built pyramids, the Middle Ages magnificent cathedrals, and the Early Modern Era the great palaces. Today it seems atomic cities and rocket stations most compellingly capture the desires and the abilities of modern industrial states."[109]

Conclusion

By the early 1970s West Germany boasted one of the leading peaceful nuclear programs in the world. The research coming from Karlsruhe and Jülich was first class, the reactors along the nation's river system were some of the most sophisticated in the world, and the companies building them were poised to become export powerhouses. In September 1973 Bonn's first energy program predicted that the Federal Republic would soon generate over 20,000 MW of electricity from reactors, more than Britain and France combined. Credit for this remarkably quick energy transition—or at least the beginnings of one—went to the state, which assumed the entrepreneurial risk and financed huge portions of this atomic program at a time when industry was reluctant to venture into this technology. Considering that many utilities, including RWE, were themselves partially owned by local municipalities, the nation's entire nuclear agenda seemed to be an affair of the state, even if the state was fragmented and often functioned like a private corporation. Even the most liberal of states, if sufficiently motivated and possessing the right narrative, financial firepower, and ties to industry, can orchestrate an energy transition from the top down.[110]

More than any other energy, it was nuclear power in which the Federal Republic departed most intensely from the Ordoliberalism that informed the architects of West Germany's Economic Miracle. While some, particularly scientists like Heisenberg, thought a statist approach was needed from the beginning, not until the 1960s did a cross-party coalition emerge that would push ahead with massive state investments into nuclear technology. This entrepreneurial state thought nuclear energy would solve the challenges of rising electricity consumption and a looming technology gap with the United States while upholding West Germany's claim to be a global export leader. Between 1958 and 1972 nuclear power received twice as much state support as West Germany's traditional energy, coal, which once employed over half a million people. By the early 1970s the new SPD-led government had so thoroughly embraced the concept of state-led technological development that it came to see research, above all nuclear research, as "part of politics as a whole," where Bonn must, "by tactically setting the focal points [of research], orient scientific policy to society's needs."[111]

But the price tag of this entrepreneurial state would be high, and the bill would come due the following decade. The effort to force the pace of nuclear development brought together an alliance of scientists, managers in the electro-technical industries, directors of the largest utility companies, and state bureaucrats who conducted technology policy through closed-door decision-making that claimed to be above politics, and that was hardly open to democratic review. From the beginning, in fact, this alliance was built on the original sin of secrecy: the Atomic Commission's constitutive session embraced a gentleman's agreement to keep discussions free from public scrutiny so its elite members could have free reign to voice their ideas.[112] This fostered a cavalier attitude toward all things nuclear, which fetishized the speed of development and which saw atomic energy as a national project. Menne of Höchst best captured this attitude, when he described the nascent industry as one of "superlatives":

The highest intensity of research, extreme demands on technical capacity, a furious pace of development, extraordinary capital expenditures, exceptional risk—these are the essential characteristics of this unique and new creation of our technological era.[113]

Over time this technocratic approach, and the nuclear program more generally, helped reshape West Germany's political-economic landscape, contributing to economic concentration and the formation of massive enterprises. To the list of huge energy corporations like RAG, Deminex, or oil majors like Esso, were now added KWU and the electricity giants RWE and VEBA. The latter two grew immense as local grids were consolidated through mergers. Injecting nuclear power into the mix accelerated the trend toward bigness, since only the largest utilities could afford the massive reactors coming online in the 1970s.

While this closed-door decision-making by a small group of elites and the ensuing economic concentration accelerated the development of nuclear technology, it opened the door to problematic policies, or policies that favored particular social groups or a particular vision for a high-energy society over other groups and visions. The utility companies, after all, had predicated their dive into nuclear power on the conviction that electricity use would continue its incredible rise. But what if these assumptions proved wrong? Would the investments by RWE or KWU hold up in a climate of stagnating energy consumption? The decision to locate and build nuclear plants, meanwhile, was determined almost exclusively by the utilities with almost no public oversight—something Bonn accepted to get RWE into the nuclear race.[114] But what if the local or national publics resisted these decisions or questioned whether nuclear power actually served the interests of the nation? Lastly, the rush by Bonn to commercialize nuclear technology privileged speed in the quest to close the technological gap with America. But what if the government's investments failed to pan out? And what if safety precautions were flaunted in the drive for commercialization? All of these concerns would come to haunt the federal government, and above all the SPD, as West Germany struggled to navigate the changing social, economic, energetic, and environmental landscape of the 1970s.

5

Shaking the Coupling Paradigm

The 1973 Oil Shock and Its Aftermath

The dramatic developments in our energy situation over the past weeks have galvanized even lay people, and helped them experience the fateful importance that sufficient energy has for our civilization, indeed for our life. In the end, this is not merely an energy crisis. Rather, it is a worldwide political crisis that is manifesting itself at a neuralgic point—our energy supply.[1]

"For the first time since the end of the war, tomorrow and the following Sundays before Christmas our country will be transformed into a pedestrian zone.... The young generation will experience, for the first time, what scarcity can mean."[2] So Willy Brandt, West Germany's Social Democratic chancellor, remarked on televisions across the nation in November 1973, when he announced the prohibition of all car travel for the coming Sundays. Looking back, what Germans remembered most about the oil shock that year were not long gas lines or price controls, as in the United States, but empty highways and vacant city streets. The absence of Sunday car travel seemed to herald the possibility of a different world (see Figure 5.1)—a world which, in the words of Brandt, might actually consider "what type of economic development benefits our society, and what merely burdens it."[3]

The chancellor's announcement came in response to an unprecedented rise in the price of oil that fall. Since 1971 the Organization of Petroleum Exporting Countries (OPEC) had watched the value of its revenues deteriorate as the dollar—the currency in which crude was traded globally—depreciated relentlessly. After negotiations with the oil majors collapsed in October 1973, OPEC unilaterally raised the posted price of crude by 70 percent. Ten days earlier, Egyptian and Syrian forces had invaded the Golan Heights and the Sinai Peninsula to retake territories lost in the Arab-Israeli War of 1967. After the United States began airlifting arms to Israel, Saudi Arabia led the Organization

Figure 5.1. Celebrating car-free Sundays on the Autobahn, November 25, 1973.
Source: Ullstein Bild / Getty Images. Nr. 1174154981.

of Arab Petroleum Exporting Countries (OAPEC) in declaring a boycott of petroleum shipments to the United States. The oil shock of 1973 had arrived.[4]

For oil-producing states, 1973 was an effort to claw back sovereign rights over the resources in their own soil. For West Germans, however, 1973 was a disaster arising from affairs that lay outside the control of their politicians and experts. In one sense, the crisis was just the latest in a sequence of shocks that had come to seem inherent in the new hydrocarbon supply chain. By 1973 West Germans had already experienced three energy scares in a generation: in 1950, 1956, and 1967. But in another sense, 1973 seemed a different order of magnitude: more perilous and packing more potential to break the foundation of West Germany's growth-oriented society. When OAPEC announced its embargo, West Germany's media reacted with hysteria, proclaiming a new "era of scarcity," wondering whether the "entire complicated machinery of Western economies" would shut down. Would the "lights go out in Europe?" Would the oil crisis lead to an "economic collapse of catastrophic proportions"?[5]

If Chancellor Brandt raised the prospect of a more energy-conscious future, the immediate impact of the oil shock worked in the opposite direction, reinforcing the coupling paradigm, which held that growth would stall without energy. In a speech before Parliament, Hans Friderichs, liberal minister of economics, captured these anxieties when he reflected that "in our national economy it is not just a bit more or a bit less comfort that depends on energy. It's not just hot water, watching television, or driving a car, but ultimately every job. In other words, to put it bluntly: the lives of the people of this country."[6]

What if the source of over half the nation's energy became prohibitively expensive? This was one of the burning questions of the 1970s, a decade characterized by transformation and crisis. The rise of new industries in East Asia, the reappearance of inflation, the growth of inequality, and the emergence of domestic terrorism during this decade seemed to mark the end of Europe's postwar era. With it went an entire "order of industrial life" that had emerged since the 1950s—of confidence, progress, planning, and quantitative growth. But what would replace it? Given the deep uncertainty gripping Western societies—characterized by contemporaries as ungovernable democracies based on an uncontrollable capitalism—policymakers across the Atlantic at first answered the energy question with what they knew best. They doubled down on the coupling paradigm and claimed that abundant energy was more essential than ever to keep the postwar boom alive.[7]

In the Federal Republic this response created a slew of paradoxes. After 1973 Bonn tried to stabilize the nation's energy supply through neo-mercantilism. It extended West German influence along the global hydrocarbon supply chain through huge corporate formations, moving further away from Ordoliberal governance. Bonn tried to emulate other Western European countries in building an oil major that could secure petroleum outside the borders of Europe. Through its aspiring oil major VEBA, and other large companies, German industry would pursue infrastructural projects in Iran, Saudi Arabia, Siberia, the North Sea, and Brazil in an effort to smooth the flow of energy to the Federal Republic. But in pursuing neo-mercantilism, Bonn weakened the very multilateralism and European cooperation their politicians so highly prized.

Meanwhile, West Germany poured money into new technologies to replace petroleum with other energies. These supply-side initiatives generated a utopian vision, crafted by an unlikely alliance between apostles of the atom and representatives of coal, and accelerated the tendency to tackle energy challenges through centralized, inaccessible organizations. In the process, the SPD further changed the structure of the energy market, overriding the courts to authorize some of the largest mergers in West German history.

Nearly all of these supply-side programs, however, failed to meet expectations, and the Federal Republic suffered a more severe reduction of oil during the crisis than the rest of Western Europe. Nevertheless, West German policymakers and the public came through the 1970s confident that their nation had mastered the crisis more effectively than their neighbors. This paradox hinges on how oil was priced and sold in West Germany, whose relatively liberal energy market spurred companies and consumers to begin conserving and switching energies more rapidly than elsewhere. In the process, the oil shock opened the space to reimagine the place of energy in a modern economy, and helped place the Federal Republic on a new trajectory, the endpoint of which the country is only beginning to realize today.

Energy Planning and the Failure of Multilateralism

For Western Europeans the oil shock hardly came as a surprise; it was no energy "Pearl Harbor," as in the United States. Suez crises in 1956 and 1967 had revealed the vulnerability of the new hydrocarbon supply chain. Across Western Europe, confidence in global oil markets waned after 1970 as experts warned that oil shortages loomed on the horizon, fueled by America's announcement that Europe could no longer rely on its spare capacity in the event of a crisis.[8]

Under the SPD, the Federal Republic met these anxieties by diving deeper into planning. The Economics Ministry had traditionally been a bastion of Ordoliberal market enthusiasm, but by the late 1960s leading economists began to question whether energy prices actually reflected scarcity and abundance, or simply the power of large international corporations like ESSO-Exxon or Shell. New officials in the Economics Ministry proved receptive to these qualms, above all Ulf Lantzke. A lawyer by training, the ambitious Lantzke had started his career in the ECSC, the pioneering institution of integration that was inspired by Jean Monnet's dirigisme. In 1968 Lantzke took over and expanded the Economic Ministry's division for energy and raw materials, and would go on to become a director of the International Energy Agency. In 1970 he urged Bonn to stockpile oil products. The following year, he and other ministry officials began expanding the scope for energy crisis planning to include instruments not used since the postwar scarcity of the late 1940s, drafting laws that would permit Bonn to ration oil products, regulate energy companies, and impose driving bans.[9]

On the eve of the national election in 1972, the ruling SPD-FDP coalition tasked Lantzke with collecting these strategies into a comprehensive energy agenda. Working with the country's leading forecasters, Lantzke helped craft West Germany's first Energy Program: the closest the Federal Republic would come to planning in the sphere of energy. While competition remained a guiding principle, the program argued that the state must "assume a high degree of responsibility" for the country's energy supply given the importance of this input to social and economic life.[10] Unveiled in September 1973, a month before the energetic life of Europe was torn apart, the program represented the apotheosis of the coupling paradigm in aiming to secure abundant and inexpensive energy. It predicted astounding energy growth rates going forward, and suggested the nation would need to build one hundred new power plants by 1985. To cover these enormous needs, the program called on the state to open a closer "dialogue" with energy suppliers, to provide forecasts to help firms coordinate investments, and to cooperate closely with companies in large infrastructural projects. Lantzke's brainchild, put simply, focused almost entirely on expanding the supply of energy.[11]

The nation was just embarking on this novel agenda when the oil shock erupted. The impact of OPEC's price hike and OAPEC's embargo quickly surpassed expectations. By January 1974 oil imports had fallen 13 percent, the import of refined petroleum products by 32 percent, and the price of crude had tripled.[12] Seen through the lens of the coupling paradigm, the shock seemed to threaten the nation's economic foundations. Helmut Schmidt, finance minister after the 1972 election, believed the crisis would "compel us to change our way of life and the pattern of our thinking." Chancellor Brandt went further, stoking fears of zero-growth, unemployment, and inflation: problems West Germans had not experienced since the 1940s. "Not everyone will be able to keep their job," he cautioned in November.[13] For Brandt, 1973 marked a rupture. "After twenty-five years of building up the Federal Republic," he warned, "before us now lies a new era" in which energy would become a coveted and contested product.[14] The Council of Economic Experts—the nation's most influential body of advisers—supported this dire interpretation. It noted how the economy had hitherto only ever been limited by the availability of labor or capital. But now West Germany was entering an entirely "new situation"—dramatic words from authors known for their caution—in which production might, for the first time, stall because of the energy supply.[15]

One faction in the SPD saw this ominous situation as an opportunity to radically alter West Germany's economy. The party's left wing and its avant-garde younger delegates demanded price controls and the nationalization of foreign oil companies.[16] And the initial legislation responding to the shock gestured in this direction. On November 9, Parliament passed the Law to Secure the Energy Supply, giving Bonn wide-ranging authority to regulate the production, distribution, and consumption of energy and activating many of the tools drafted by Lantzke. Bonn imposed speed limits, banned Sunday driving, established a clearing office for petroleum, and even empowered the state to set price ceilings on oil, natural gas, coal, and electricity. By December the nation's roads were vacant on the weekend, suggesting that profound change might soon come to a society that had only recently embraced the car as the centerpiece of its lifestyle (see Figure 5.2).[17]

Centrist Social Democrats, however, shied away from the more expansive regulatory powers permitted by the new law. Brandt believed his party must balance market ideas with dirigiste ones if it were to thrive. But more important for energy policy was the SPD's second in command when the shock hit,

Figure 5.2. Car-free Sundays for a nation still new to mass-car society, November 25, 1973. *Source:* DPA Picture Alliance / Alamy Stock Photo. Image ID: D9WJGB

Helmut Schmidt. Following revelations about an East German spy among his advisors, Brandt resigned as chancellor in early 1974, handing the reins of power to this conservative Social Democrat and avid believer in the coupling paradigm. Hailing from a working-class part of Hamburg, Schmidt had joined the SPD after the war to lead its student league. He was catapulted into the party limelight during the energy debates of 1958, when he confronted Chancellor Adenauer over the push for nuclear weapons. Yet Schmidt had always been an independent politician, splitting with his party over military affairs and frequently berating his colleagues. In economic matters he governed from the center. Having studied economics and worked in Hamburg with Karl Schiller—the Keynesian mastermind behind the coal consolidation of 1968—Schmidt believed in private enterprise, disavowed full state regulation, and wanted to prevent the SPD from returning to its Marxian roots. Since 1972 Schmidt had worked with Lantzke and saw a growing energy supply as crucial for growth.[18] And throughout the shock he collaborated closely with Economics Minister Hans Friderichs, a member of the liberal FDP who, while willing to compromise Ordoliberal dogma, believed markets must remain the main "system of order" for West Germany.[19]

Instead of regulating the economy, Brandt, Schmidt, and Friderichs initially hoped to keep oil flowing through West Germany's long-standing axiom of multilateral cooperation. Most politicians saw the crisis at first as an *international* one that should be handled through international institutions. Throughout November, Brandt hammered this home to Parliament. In a speech that began by appealing to European solidarity, he emphasized:

> The Federal Republic gets its oil and uranium predominantly from far away, from overseas. Those pipelines that are important for our supply begin in Rotterdam, Trieste, Marseilles or the Soviet Union. The major energy corporations are all based outside our country. . . . Hence the fundamental need for our economy to strive for the most free and unhindered access to both the global market as well as to the single internal market of the European Community (EC).

No consumer nation besides the United States could sustain its energy supply by itself, so the oil problem must be addressed multilaterally. Specifically, he mused, "[t]he energy question will show what the EC is really worth."[20] Schmidt, too, as an anglophile and Atlanticist, prioritized multilateralism and hoped to avoid the beggar-thy-neighbor policies that had plunged the world into depression during the 1930s. He, along with West Germany's foreign policy establishment, feared that without robust cooperation mechanisms the world might degenerate into rival continental consumer blocs that would unleash a price run beyond what OPEC had already instigated.[21]

So when US secretary of state Henry Kissinger proposed an energy action group that December, West German officials responded enthusiastically. "Given the energy potential of the United States," the Economics Ministry put it, "Europe cannot politically afford to reject this outstretched hand."[22] At the Washington Energy Conference in early 1974, Kissinger presented his plan for a counter-cartel of consumer nations. Western Europeans, however, hesitated to join anything directed explicitly against OPEC, for fear of further alienating oil-producing states. France in particular balked at Kissinger's proposal. But after securing concessions from Washington, the EC with the exception of France moved ahead. The result was a major new international organization under the auspices of the OECD: the International Energy Agency (IEA).[23]

But the IEA came into being far too late to address the oil shock itself. More generally, Bonn's hope of solving the crisis through multilateralism failed, above all at the European level. In 1973 the EC was, in fact, still a relatively weak cluster of institutions. Only in 1968 had the three executive organs of the ECSC, Euratom, and the European Economic Community fused into one. Only later in the 1970s did the EC begin to play a more significant role in the domestic economies of its members, after a series of court cases consolidated European law. Just as the oil shock struck, meanwhile, EC was expanding to include Britain, Denmark, and Ireland, bringing more voices to the fray of policymaking. Energy, meanwhile, a field that had launched integration in the first place with the ECSC, remained underdeveloped. Brussels had only passed its first real guidelines for oil in 1968, while the Council of Ministers—the EC's most influential organ— hardly discussed energy before 1972. The member states, lastly, had wildly diverging attitudes toward both oil majors and oil-exporting countries. Britain and the Netherlands, respective hosts of BP and Royal Dutch Shell, resisted calls to force oil companies to divulge information on their prices, while France and Britain, with their long-standing ties to the Middle East, wanted to approach OPEC states on their own rather than through a common framework, as other member states wanted to.

On the eve of the oil crisis, the Commission tried to formulate a comprehensive energy strategy, recommending the EC harmonize oil taxes, share supplies, coordinate state intervention, and jointly finance new exploration. Brussels feared a supply shock would unleash a scramble to buy oil, unnecessarily driving up prices. But nothing came of this push, and the competitive over-bidding the Commission had feared erupted at the end of 1973. There was no European-wide oil-sharing scheme, as Brussels wanted, nor any uniform reduction of oil consumption. Instead, Britain, Spain, and Belgium stopped exporting refined petroleum in order to boost their own economies. France signed a bilateral oil agreement with the Saudis in late November 1973, and Britain quickly followed suit.[24] The wildly diverging pricing policies in Europe, meanwhile, prevented a coordinated intervention to the crisis. Earlier that year Brussels had called its first energy conference since 1970. The meeting lasted deep into the night and resolved nothing beyond underscoring the "profound difference of opinion" on how to price energy. Though most members controlled oil imports through licensing and regulated the price of crude, the Federal Republic did not, and it tirelessly defended its liberal position at every EC energy negotiation, thwarting the search for common ground.[25]

Oil, lastly, was so closely tied to the Arab-Israeli conflict that it proved impossible for the EC to stake out a common position. When OAPEC declared its embargo, it divided EC states into friendly, neutral, and hostile categories depending on their support for Israel. The Netherlands was the only European state to be labeled hostile and suffer an embargo, and requested a declaration of solidarity. Britain and France refused.[26] And while West German officials sympathized with the Dutch, they too felt growing pressure from OAPEC to avoid anti-Arab declarations. In the estimation of contemporaries, the EC's collective energy policy through the shock was a "shambles."[27]

Mercantilist Response Take I: Hydrocarbons

As the hope for multilateralism withered, the Federal government followed a different, more mercantilist path to navigate the new world of oil, which found its clearest expression in the First Revision to the Energy Program in 1974. Bonn hoped to manage foreign oil dependency by ramping up the supply-side policies initiated in the 1960s: large domestic corporations; greater influence over the global hydrocarbon network; bilateral deals; more nuclear power; a more entrepreneurial state that could commercialize new sources of energy.

The highest priority was creating a German oil major—going big in the field of petroleum exploration—and in 1973 its energy masterminds now looked to Italy for inspiration. For over fifty years, seven massive corporations, based in the United States, Britain, and the Netherlands, had dominated the global production and trade of petroleum. By the 1960s these "seven sisters," alongside an eighth French company, controlled more than 80 percent of petroleum produced outside the United States and the Soviet Union, as well as the supply chain that brought oil to Europe. Governments in Washington, London, Amsterdam, and Paris believed these companies would be potent allies in any energy crisis, as they had been during the Suez Canal closures. Italy had been late to the game of oil, missing out on key concessions that gave British, Dutch, and American companies access to the energetic wealth of the Middle East. But Rome had been working for decades to rectify this. Following World War I, Italy mimicked France in building a state-owned, domestic oil champion, Agip. After briefly flourishing under Benito Mussolini, Agip struggled because it lacked access to the petroleum wealth of the Middle East. But in the 1950s a wartime resistance leader and ambitious businessman, Enrico Mattei, began a campaign to forge a new major, in the process creating a model for countries without colonial influence in the Middle East. In 1953 Mattei gathered Agip and other Italian state oil entities into Ente Nazionale Idrocarburi (ENI) to challenge the "seven sisters," a moniker he allegedly coined, by securing rights to Soviet oil and by cutting a deal with Iran that was unprecedented in how it favored this oil-producing country. In 1962 Mattei died in a suspicious plane crash, but his efforts suggested a way for other aspiring newcomers to strike novel bargains with oil-producing countries.[28]

West Germans took notice of Mattei, in particular Rudolf von Bennigsen-Foerder, the scion of a noble family from Hannover who directed one of the nation's largest domestic energy companies. Bennigsen-Foerder had traveled to Iran and Saudi Arabia, where he became convinced that West Germany needed direct access to crude. Without a clearer commitment to nuclear energy, he mused before the oil shock, "politically insecure oil will continue to be the foundation of Western energy supply.... West Germany as well needs a national energy company." His company, VEBA, would be the starting point. Founded in the 1920s as a consortium of coal and electricity companies, VEBA had been owned by the Prussian state and expanded into oil processing during the Third Reich's armament drive. After the war its managers were arrested, and state control passed to the new Federal Republic. Headquartered in Düsseldorf, the company expanded into petrochemicals to become the nation's largest corporation by sales volume. After shedding its coal holdings into RAG with the reorganization of Ruhr mining, VEBA came to specialize in refining, electricity, petrochemicals, and glass. After the CDU partially privatized the company in 1965, VEBA became for a time the world's second largest publicly traded company, and was run as a for-profit business with little interference from Bonn. But the state still owned 40 percent, and this, in the mind of energy experts, gave VEBA a "special status" as the potential nucleus of a national energy champion.[29]

Bennigsen-Foerder quickly found like-minded experts in the Economics Ministry, and on the eve of the oil shock the economics minister promised VEBA's CEO that the state would support its search for crude in the Middle East. Schmidt initially demurred, but with the oil shock he changed his tune and became convinced that it was an advantage to have a state-backed oil firm.[30] With Friderichs at the Economics Ministry, they envisioned transforming a group of domestic companies into a conglomerate that could act like a single unit in the global oil market, a peer to other large oil companies with the requisite expertise to bargain with producer nations, to exchange technical know-how for crude, to prospect for deposits, and to better represent West Germany in the EC or in negotiations with other consuming nations.[31] As Schmidt put it, the majors had "played an important and benevolent role both in the past and in the recent crisis." Why not build a German one?[32]

With oil tensions high, the Energy Program proposed creating West Germany's own major by merging VEBA and Gelsenberg AG (GBAG). VEBA was one of the only German companies with substantial refining capacity, but it relied entirely on the oil majors or traders in Rotterdam for its crude. And as crude prices rose, VEBA began to hemorrhage money. GBAG, by contrast, was the only German firm with any noteworthy holdings of crude outside Europe, in Libya. The merger seemed like a match made in heaven, and in fact the Economics Ministry had been wanting this since 1968. But it took the crisis to turn a dream into reality. That fall Bonn opened discussions with the electricity giant RWE, owner of a large stake in GBAG, to buy its shares and transfer them to VEBA, with Bonn forking over 641 million DM in taxpayer funds to support the deal. When the Federal Cartel Office threatened to stop things, Friderichs overrode its protests about economic concentration, couching the fusion as a special measure that advanced the security of the nation. A year later the deal was finalized, and the Federal Republic now seemed to possess what its officials had been pining for since 1967: an integrated energy company that could forge its own cooperative relationships with the oil states of the Middle East. For Bennigsen-Foerder, this was West Germany's "last chance" to join the multinational giants in the race for oil.[33]

At the same time, Bonn enhanced Deminex, the loose prospecting consortia it had reformed in 1969, using this group to "strengthen its own crude oil basis." When Deminex's subsidies ran out in January 1974, Bonn expanded the program, first to 600 million DM then later 800 million.[34] The state cultivated a "tighter entrepreneurial decision-making process" in the consortium by reducing the number of participants from eight to four, with VEBA holding two-thirds of the enterprise's capital. And Bonn urged all of them—the largest domestic refiners in West Germany—to consolidate their exploration programs and personnel into the new enterprise.[35]

In contrast to its rhetoric about multilateralism, Bonn now actively worked with VEBA and Deminex to try stabilizing access to the hydrocarbon supply chain through bilateral deals supported with generous loans. Earlier in 1973 VEBA had opened discussions with Iran about its crude. After the oil shock, Friderichs visited Tehran to lobby for an "oil for investment" contract, in which German companies would build a refinery on the Persian Gulf in return for crude to VEBA-GBAG, which hoped to circumvent the majors to get Middle East oil directly. Friderichs returned the following May to try to solidify what might become one of the largest export triumphs in German history.[36] The Iranian deal was the first of these attempts. VEBA followed in 1974 with another effort to buy 3 million tons of oil from the Soviet Union, on the model of Mattei's ENI. A third attempt unfolded when VEBA representatives, alongside a flood of other German industrialists, descended on the capital of Saudi Arabia offering steel mills, truck plants, and chemical factories for crude.[37]

Deminex, meanwhile, accelerated its push to buy concessions anywhere it could. Since 1971 oil-producing states had started favoring "government-to-government" agreements, and they saw Deminex as an untarnished "quasi-national" organization with which to do business.[38] While the consortium had been exploring concessions throughout the Middle East before 1973, these only began to yield results after the shock. The consortium's 800 million DM subsidy was the single largest budgetary line in Germany's new energy program. And by the end of the 1974 the consortium was pursuing seventeen projects in twelve different countries, ranging from the Niger Delta to the Caribbean, from the Persian Gulf to the North Sea, with a total concession area under contract larger than West Germany.[39]

But oil was not the only hydrocarbon over which West Germany tried expanding its influence. Natural gas, a relatively new source for industrial and household heating, offered another path out of 1973. Because so many West Germans relied on fuel oil to warm their homes, switching households to gas could reduce the nation's dependence on the Middle East, if only the infrastructure to pump this fuel to the Federal Republic could be built. The oil shock dramatically accelerated efforts already underway to build a pipeline network that would bring the energetic wealth of the Soviet Union to the cities of West Germany. After 1973 West Germany's steel and engineering companies would collaborate with the USSR to tap the gas reserves of remote Western Siberia.[40]

The Mercantilist Response Take II: Nuclear and Coal

If oil and gas were the pillars of Bonn's mercantilist response, they were the affair of a small coterie of elites, and the efforts to forge a new oil major remained relatively hidden from public view. The circumstances surrounding nuclear power could hardly have been more different. For the atom captured the national imaginary, as the energy of the future or the technology that might bring about the death of West German democracy.

Energy officials fell firmly into the former camp, and more than any other group, nuclear advocates vigorously defended the coupling paradigm. Since 1970 the industry had been warning that without more nuclear power the Federal Republic would suffer brownouts and shortages. But this rhetoric reached a fevered pitched with the 1973 shock. As the editor of *Atomwirtschaft*, the industry's leading trade journal, put it:

> The energy crisis has brought to light what energy experts have long said but which has found little resonance in the public: it won't work without nuclear energy. Without nuclear power plants our electricity supply, along with that of all other industrialized countries, cannot be secured over the next two decades.[41]

After 1973 Hans Matthöfer, minister of research and technology, emerged as the leading advocate of these views inside the government. Born of a working-class family that had suffered unemployment, Matthöfer became a stalwart SPD member after the war. Having studied economics at the University of Wisconsin, he returned to West Germany to become a union leader. Matthöfer came of age in the coal town of Bochum, and he ardently hoped to retain the black rock of his *Heimat* as the economy's foundation. But he gradually became a proponent of nuclear power through his work with the atomic institutes in Jülich and

Karlsruhe. After ascending to become minister of technology with support from Schmidt, in a series of widely read interviews in 1976 Matthhöfer laid out the quintessential argument for nuclear power in the wake of the oil shock. Energy and growth were, he argued, "directly and closely linked"; a dearth of the former would endanger "hundreds of thousands of jobs." If West Germany "wants more growth, then we need more energy."[42] As global fossil fuels dwindled or became concentrated in geopolitically fraught regions, only nuclear power could provide the energy needed for growth. Unlike petroleum, uranium could be sourced from many different locations. Through new technologies, nuclear power could even free the nation from its dependence on foreign oil.[43] As Matthöfer bluntly put it, a Federal Republic without the atom would be a country of "zero growth," a nation defined by "mass unemployment and economic crises."[44]

This formed the core of the nuclear argument: a domestic, secure, abundant source of power for an economy that needed energy to thrive; the nation had "no realistic alternative." Layered atop this were other justifications, above all about the atom's environmental benefits. Compared to its nearest competitor in generating electricity, hard or brown coal, reactors emitted no carbon, no methane, and no sulfur dioxide, which by the 1970s was becoming the scourge of urban residents. Bonn's main concern about nuclear power was not environmental degradation but reactor safety. Officials, however, were confident that proper safety precautions, the standardization of reactors, and the steady advance of technology would reduce any risk to acceptable levels.[45] Thirty years of work, Matthöfer reassured the citizens of the republic, had shown that nuclear power can "certainly be mastered. The inherent dangers of this technology are understood."[46]

Armed with these arguments, Bonn accelerated its nuclear program after the oil shock. Its revised energy program raised the target for atomic electricity by 1985, signaling the state's commitment. Bonn promised a more "speedy" and "streamlined" licensing system for utilities.[47] The Ministries of Interior and Technology, meanwhile, doubled down on large power plants, supporting industry's call to build huge standardized reactors through a more "rational" and "frictionless" construction process.[48] By the mid-1970s the Federal Republic had more commercial reactors on order than any other country besides the United States, and boasted the world's largest per capita investment in nuclear energy.[49]

But the nuclear drive did not stop at West Germany's borders. A fundamental motive behind the program in the first place was the quest for exports, and this gained urgency after 1973. For the oil shock intertwined with a slowdown in productivity growth, falling profits, the rise of East Asian export competitors, and monetary instability to unleash a new era of turbulent capitalism. Energy prices, already rising, surged after 1973, unleashing a vicious global recession as firms saw their costs spike and their profits tank. Looking around the world in 1974, Schmidt feared this downturn would ricochet back to the Federal Republic, as the collapse in global demand put Germany's export-dependent economy at risk.[50]

Schmidt fought this looming recession with international cooperation through a new transnational institution of industrial nations, the Group of Seven (G-7) summit. But others in Bonn pushed nuclear exports as a way to sustain employment during this period of uncertainty. Matthöfer and his allies played on long-standing tropes to justify anything that might stimulate exports. "As a country poor in resources and energy we depend on exporting modern and complex technologies." Nuclear technology, he believed, has a "leading position among German exports" and could generate tens of thousands of jobs if supported by the state.[51] His arguments resonated across party lines, while the nuclear industry did everything it could to bolster them. As KWU, the domestic reactor-building conglomerate, put it, "The Federal Republic's economic destiny is not only the compulsion to export," after all, "but to concentrate on the export of technically advanced industrial goods."[52]

This was the context behind KWU's drive for a monumental export deal with Brazil, the first of its kind that would provide a developing nation with the facilities for a full nuclear cycle. Preliminary negotiations had started in 1969 after Brandt visited Brasilia. The oil shock raised the stakes for Brazil, as its leaders had to manage the growing financial burden of importing fossil fuels.[53] But the military dictatorship there had a second goal in mind, seeing nuclear collaboration with Western Europe as a way of achieving greater autonomy from the United States. In exchange for a lucrative contract and raw uranium, Brasilia wanted help building its own reactor industry from the ground up. KWU could provide its experience with largest reactor in the world at Biblis, along with "sweeteners" that American firms were forbidden to provide, like assistance with enriching uranium.[54]

West German industry presented the deal as a way to secure an alternative source of uranium outside the United States, hitherto its prime supplier. But more importantly, the agreement promised new jobs back home. In June 1975 Bonn and Brasilia signed an industrial agreement financed by West Germany's state-run development bank. KWU agreed to deliver two of the largest reactors in the world, with Brazil securing an option to buy six more. Each would cost 1.5 billion DM and employ over six thousand skilled German workers a year. When added to KWU's domestic contracts, the deal let the nuclear industry boast of 250,000 "crisis-proof" jobs during a global recession. All told, the price tag was expected to reach 12 billion DM.[55] The press heralded this as the "greatest export deal in German history."[56]

The Brazil affair, moreover, revealed just how willing Bonn was to flout multilateralism when energy was in play. For the negotiations opened a rift with the United States, which feared Brazil would use German technology to acquire the bomb. President Jimmy Carter and other American dignitaries lambasted the deal as a "serious danger" to global peace, while domestic critics in Germany worried it would place a "huge burden on German-American relations." But the government refused to back down, and exports trumped good standing with the United States.[57]

This was just the tip of the technocratic nuclear iceberg; beneath the surface lay even more incredible visions of reforming the entire energy sector to run on the atom. Since the 1960s the nation's research centers had been working on two new reactor lines: the fast breeder, which promised to generate more plutonium than it consumed; and a thorium reactor that produced such intense heat it could unlock a panoply of new industrial processes. The oil price spike brought both into the realm of the possible, with nuclear boosters arguing that these technologies would liberate Europe from its fossil fuel security dilemma and fears of resource exhaustion. Wolf Häfele, a student of Carl Friedrich von Weizsäcker, became the poster boy for this techno-nuclear utopianism. After pioneering West Germany's breeder program in the 1960s, Häfele moved to Austria, where he led the International Institute for Applied Systems Analysis (IIASA). The oil shock let Häfele ramp up publicity for what he called the "breeder power economy."[58] By his estimation, global energy demand would rise tenfold over the next half century as population growth and development in the Global South altered the world's energy calculus. With no policy shift, he believed, oil reserves would be gone by 2010; conservation might delay the inevitable for just a few decades. The result was an energy paradox, first popularized by the Club of Rome's report on the *Limits to Growth* in 1972, but one that became more pressing after 1973: how to support growth in a world of finite fossil fuels. In Häfele's opinion, "these unpleasant choices cannot be avoided by turning off a few lights, or by exhortations to travel less, or by harnessing wind and garbage power. For large-scale energy supplies, there are no near-term alternatives to fossil and nuclear fuels."[59]

West Germany could solve this paradox by rebuilding its energy system around new reactor technology. The breeder, Häfele argued, could extract one hundred times more energy from uranium than conventional light water reactors, and would provide electricity for "several hundred thousand, if not [a] million, years."[60] But the more radical vision was the thorium reactor, which promised to replace the entire fossil fuel network by extending nuclear energy into entirely new applications. Nearly three-quarters of the demand for final energy was really the demand for heat, either by households or industry. If heat from high-temperature reactions could be transported over long distances to cities, or factories relocated next to atomic reactors, nuclear energy could not only generate electricity but also warm homes and power industry. Häfele and others hoped to repurpose gas and oil pipelines to transport heat generated by industrial processes into the homes of European citizens, allowing "central nuclear power stations to play the role of natural gas fields far away," with reactors even floating in the ocean. The intense heat from high temperature reactors—905 degrees Celsius—moreover, could open up entirely new industrial applications.[61] At his most extreme, Häfele mused about using this heat to split water molecules into liquid hydrogen that could replace the oil powering Europe's cars, airplanes, and trains. With dedication, he believed, Europe go all nuclear in just sixty years.[62]

This was big science writ large, with utopianism at its core and a long-term agenda to remake society. But in the short term it was connected to a more down-to-earth set of technologies, for the oil shock also turned the liquification and gasification of coal into a potential reality. West Germany had some of the most extensive experience with these processes dating back to the Third Reich, which had led the world in coal. Some companies had continued experimenting with these processes after the war. But the oil shock presented a new opportunity, and soon energy companies like Ruhrgas, the coal conglomerate RAG, the nuclear giant KWU, the steel firm Mannesmann, and the electricity titan RWE began lobbying Bonn to fund coal gasification and liquification.[63] Mining unions also became interested. Despite the reorganization of mining in 1968, the industry continued to hemorrhage jobs, leaving union leaders grasping for ways to sustain their workforce.[64] The oil shock, they hoped, heralded a "renaissance" for hard coal. In 1974 the piles of unused coal, a normal sight in the Ruhr, began to melt "like snow in the sun" as coal consumption by steel producers ticked upward for the first time in years and government subsidies for coal-fired electricity generation grew. This new technical vision added to the industry's hopes that coal's time in purgatory was nearing an end.[65]

To justify synthetic coal-oil and coal-gas, this alliance of domestic companies and unions turned to energy security, a language they had crafted in the 1960s and that now found great traction. Success would create "double security," they claimed: more jobs at home and less dependence on OPEC abroad. Such arguments resonated with the SPD, and immediately after OAPEC's embargo Chancellor Brandt began formally calling for coal-enhancement technologies.[66]

As in the 1930s, technological hopes for coal gasification flourished against the backdrop of global trade tensions. But now the hope was to wed coal to the atom, which promised to remove the main roadblock to this chemical process: the need for heat.[67] Thus the oil crisis brought together not only domestic energy firms with unions, but also two energies that otherwise might have been rivals. Coal and atomic energy needed one another other to break beyond the electricity market. That both were the target of the environmental movement only strengthened their bonds. Indeed, this alliance lay at the heart of Matthöfer's research agenda, which aimed to replace nothing less than the very hydrocarbon supply chain built in the previous decades.[68]

To execute this vision, the entrepreneurial state would have to expand dramatically. Since the SPD joined the government in 1966, Bonn had more proactively shaped technological development. But after 1973 it extended its "Big Science" approach to other domains, for its leaders believed the market alone could not stimulate the technological advance the country needed. Energy was so far from a market that the scarcity indicators needed to push innovation in the right direction hardly existed: oil prices were determined by an oligopoly; coal prices by state subsidies; electricity prices by regional monopolies. As Matthöfer put it, "Market forces are not enough to give the necessary impulse to basic innovation, for the development of key technologies"—the growth engine on which Germany's economy relied. "That is why we need government-funded research."[69]

Energy research thus grew dramatically after 1973. In November, Brandt demanded more research funding for energy. Following the oil shock, the Fourth Atomic Research Program, in gestation already, formally began, devoting 6.1 billion DM to nuclear power, more than what Bonn spent on defense research. In January 1974 the Ministry for Research and Technology (*Bundesministerium für Forschung und Technology—BMFT*) inaugurated an entirely new project when the cabinet funded the first Basic Energy Research Program, which explicitly aimed to reduce oil imports by building the coal-nuclear matrix. Conservation found little place in the SPD's broader agenda. The party instead aligned with traditional nuclear experts in forecasting nothing but upward sweeping curves of energy consumption (see Figure 5.3).[70]

Mercantilism Failed

The hopes of state officials, experts, Social Democrats, and the country's largest firms to alter the nation's energy system through supply-side mercantilist policies, however, yielded preciously few successes. Few of these long-term term aspirations—for more influence along the hydrocarbon supply chain, for the next generation of reactors, for liquified coal—materialized. Most failed, in some cases spectacularly.

The first to go was the quest for an oil major. The very experience of 1973 disabused oil analysts of the prevailing dogma that states could use majors as a tool of policy. Prime Minister Edward Heath of Britain had called on British

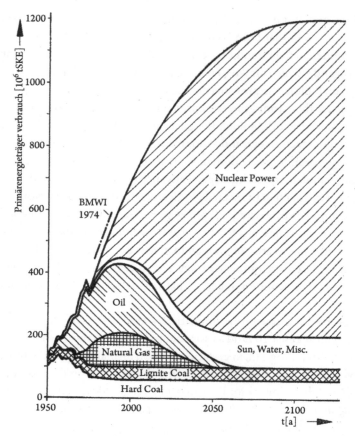

Figure 5.3. Prediction of future energy consumption. Produced by experts at the Jülich Nuclear Research Institute in 1976 and reproduced in the SPD 1977 Energy Convention. *Source:* SPD, *Forum SPD. Energie: Ein Leitfaden zur Diskussion* (Bonn: May 1977), 52.

Petroleum (BP) to favor his nation with oil. The government, after all, owned half of this massive petroleum company. But BP's leaders demurred, leaving Heath ashamed at his impotence.[71] France likewise demanded that CFP, partially owned by the state, prioritize the national market. CFP too refused. Exxon went one step further, and was encouraged by the Saudis to hand over information on where the US Navy purchased its oil. The sheer power of the majors to flaunt the whims of their home governments spurred contemporaries to see them as "a kind of temporary world government," and to conclude that "anyone looking for confirmation of the view that it paid a country to have an oil company based within its own jurisdiction would have found scant support for such a hypothesis."[72]

West Germany's efforts to forge an oil major ended with a fizzle. After 1974 Deminex boasted some minor successes, above all its Thistle concession in the North Sea. In May 1978 the first tanker with Germany's own North Sea crude landed in Wilhelmshaven. But as a chairman of VEBA remarked, it would take three or four Thistles to make any difference.[73] During an era when oil exploration was moving away from the low-cost Middle Eastern wells to non-OPEC fields in costly off-shore regions, Deminex never acquired the financial firepower to compete. Bonn invested over 1 billion dollars in the consortium through low-cost loans. But this paled in comparison to the capital deployed by the likes of Shell or BP, who mobilized 30 to 40 billion dollars for exploration every year. Thistle strained Deminex's finances, but it brought the consortium's total production to a mere 2.4 million tons of crude a year, less than 2 percent of West Germany's need.[74]

VEBA's ambitions collapsed even earlier. By the mid-1970s it was crushed between the high price of crude on wholesale markets and the falling prices it began facing in the market for final petroleum products. The oil fields it acquired through GBAG were in Libya, and still subject to high taxes and thus high prices. And in any case, GBAG's supplies were limited. VEBA consequently still bought most crude through long-term contracts from the majors at high prices. But by the mid-1970s the rise in crude prices suffocated European demand for oil. To the astonishment of contemporaries, the continent consumed less petroleum in 1977 than in 1973.[75] The result was refinery overcapacity, to which refiners responded by dumping petroleum at rock-bottom prices on the spot market. Nearly all of this was snatched up by independent companies for resale in the unprotected, liberal market of West Germany. There the profit margin between crude bought in OPEC and the final product sold to consumers—like gasoline or fuel oil—fell to nothing, or even less than nothing. Deutsche Esso reported losing 15 DM a ton; Deutsche Shell 18 DM. VEBA, with no non-OPEC source of oil, bested even these in its losses, losing 38 DM a ton in 1975.[76]

VEBA, consequently, began trying to get out of oil to concentrate on chemistry and electricity. It found a willing partner in British Petroleum, announcing a major deal in 1978 in which it sold GBAG alongside refinery capacity to Deutsche BP for 800 million DM and a long-term contract to buy crude at competitive prices. Now with no substantial source of oil outside Germany, VEBA would never achieve the relative success of France's CFP and ERAP, or Italy's ENI in exercising some influence over the global hydrocarbon supply chain. But VEBA's oil division was just one part of a diverse enterprise that included some of West Germany's largest chemical plants and its second largest utility. Going forward, this German conglomerate would remain a pillar of the old paradigm by concentrating on these other spheres of energy. Bonn, admitting its hopes for an integrated German major were dashed, opted instead for closer collaboration with BP. In 1978 the economics minister again overruled the Cartel Office to advance the sale of GBAG to BP, anchoring this British oil major into the Federal

Republic in the name of "the long-term security of our energy supply." And so the foreign oil majors would continue to dominate West Germany's market.[77]

Meanwhile, the campaign to rebuild West Germany's energy system on a nuclear foundation ran aground on the twin rocks of rising costs and technocratic arrogance. Already in the early 1970s the safety requirements for new reactors began to rise as regional governments responded to local complaints. In 1972, for instance, a crack in a reactor containment system let loose a flood of radioactive water that led to a building freeze. The federal nature of West Germany's safety system, meanwhile, in which the states, Bonn, and the courts all had a say, added complexity to an already lengthy approval process. As a result of tighter safety measures and mounting legal complications, reactor building times doubled. Costs rose even more quickly as KWU, the nation's reactor-building giant, had to increase its cash reserves to cover the potential cost of future safety breaches. Adding to this were the monetary aftershocks of 1973: as nominal long-term interest rates spiked, so too did the cost of borrowing funds to build these capital-intensive behemoths.[78]

By the mid-1970s, meanwhile, the export market for reactors began to wither. KWU had difficulty financing many of its foreign contracts. More generally, the global recession stalled demand for electricity and thus for new reactors. In the mid-1970s KWU had a dozen orders for reactors outside West Germany, six alone in Iran. The firm lost all of the latter following the Islamic Revolution of 1978–1979. Among the others, only a handful ever materialized.[79] Even the Brazil contract ran into insurmountable obstacles: the building site was far from local industry and entailed high costs; local firms had close ties to the government, and little incentive to cut costs; and the Brazilian government looked dangerously insolvent. The enrichment technology that West German firms promised, moreover, had never been commercially tested and was itself incredibly energy-intensive. By 1978 the project looked "technically complicated, economically absurd, and financially dangerous." Just one reactor was ever built.[80]

The most important cause for the stalled hopes of atomic apostles, however, was the domestic reaction to this massive nuclear program, sparked by the insulated manner of decision-making about energy affairs. Even before the oil shock, key atomic decisions were mostly shielded from outside review, including the decision of where to site reactors. But after 1974 Bonn created new channels to bring big industry, big science, and the state together to advance the coal-nuclear matrix, like the Expert Committee for Energy Research and Technology. This link with industry served as the "coordinating body for all energy-related advisory councils" and was staffed by visionaries like Häfele alongside leading representatives from VEBA, RWE, and other energy conglomerates. Citizens had no input.[81]

This insulation from criticism fostered a cavalier attitude toward nuclear energy in the Technology Ministry and the major research institutes. Here Häfele is exemplary. In 1974 he authored a widely read article that tried to naturalize the dangers of atomic energy by embedding them into "the normal conditions of life," which in his mind were replete with risks. One must weigh the chance of a nuclear accident, for instance, against the certainty of the sulfur dioxide generated by fossil fuels. Through such comparisons, he believed people would realize that the benefits of the atom far outweighed the drawbacks. The "practically infinite amounts of energy" offered by the breeder and high-temperature reactor, he concluded, represented "an unusual kind of benefit which exceeds any so far experienced," and justified nearly any risk.[82]

More generally, insiders believed nuclear power was beyond the grasp of the public and that its technical nature "overstrained" the average intellect.[83] Bonn's nuclear publicity campaign was tellingly called "Educating the General Public on the Necessity of Building Nuclear Power Plants," with an aim to "correct" the "misleading" information provided by the media.[84] "Such a complex matter like building and operating nuclear power plants," a leading atomic apostle reflected, "in many contexts cannot be assessed by citizens based on their own knowledge. It thus falls to the judgment of experts."[85] Those who questioned this process were merely suffering an "emotional" reaction to a new technology.[86]

This approach, however, would spark an intense social backlash that eventually slowed down reactor construction through the courts. KWU received its last partial reactor authorization in June 1974. By 1977 it had ten reactors worth 20 billion DM of contracts on hold because of a court-mandated building freeze. Each month it was losing 5 million DM.[87]

Of the mercantilist responses to 1973, the least touted would show the most enduring success: natural gas. But the nature of this infrastructure meant it would be a decade before this response to 1973 began to take physical shape, following a massive new commercial deal to tap the resources of Siberia with German technology.

The Real Legacy of 1973: Pricing and Decoupling

In fact, none of the mercantilist initiatives helped during the immediate energy crunch of the 1970s. Instead, West Germany's actual path through the oil shock took a much different route. And the lessons that officials, politicians, and energy economists drew from this, and from the subsequent and unintended developments in energy markets, would pave the way for a truly novel approach to energy.

At the height of the oil crunch, the Federal Republic used the same quasi-oligopoly quasi-market mechanisms that had brought it through the Suez crises. To get the energy their country needed, federal officials combined a heavy reliance on the foreign oil majors with the only institution that approximated a market for petroleum in Western Europe, the Rotterdam spot market. The main problem facing policymakers during that winter was not the quantity of oil—there never really was an absolute shortage—but the price. OAPEC's embargo did little to curb global oil production, but it did unleash a bidding war on spot markets. Fearing the majors would prioritize their home countries, the consuming nations began paying unheard of prices, with Iranian, Libyan, and Nigerian national companies reporting bids of $20 a barrel from European companies, up from just $3 that summer. At home, German consumers responded to alarmist rhetoric by tanking up cars and hoarding fuel oil in their cellars. The result was a shocking rise in the price of final oil products in November and December 1973, despite a relatively minor drop in quantity.[88]

Economics Minister Friderichs responded by following his predecessors and courting the majors. "We must ensure," he proclaimed to Parliament, "that the interest of the international corporations in particular in Germany's market are protected; because we have no desire to see the quantities [of oil] flow out into other countries." Bonn formed a system in which the Economics Ministry, the Cartel Office, and the largest oil companies consulted with each other to manage the distribution and pricing of oil products.[89] In West Germany, this network held off releasing emergency reserves, and instead had the economy conform to shortages by encouraging conservation and by raising prices to curtail consumption.[90] German officials, meanwhile, worked out an informal agreement with the majors, who channeled non-embargoed Iranian oil to West Germany, but who also kept an unofficial ceiling on the price of key petroleum products at the height of the crisis. After the initial spike in October, from early November through late January the price of fuel oil, one of the most important consumer fuels, stayed steady for companies that refined their oil *inside* West Germany—meaning the subsidiaries of the majors like Deutsche Esso or Deutsche Shell, and large domestic companies like VEBA.[91]

But the majors did not supply *all* of West Germany's oil products. They owned over two-thirds of West Germany's refining capacity, which they furnished with their own crude. The remainder came through the Rotterdam spot market, known as the "garbage dump of Europe" or the "market of last resort."[92] In reality, Rotterdam was not a physical market in the Netherlands, but a network of several hundred firms spread across the Low Countries, London, Paris, Milan, Hamburg, New York, and the Ruhr who traded the oil products that serviced Western Europe. The prices published in Platt's *Oilgram Price Service* index gave the market its cohesion. Here in "Rotterdam" the majors and independent companies sold their surplus petroleum products to small importers. During supply gluts the price fell precipitously; during shortages they rose immensely. Between December 1973 and January 1974, even *after* the initial OPEC price spike, Rotterdam prices surged to incredible heights.[93]

Rotterdam's huge price spike created an unusual dynamic. Once prices there crested the fixed ceiling at which French or British companies could sell petroleum back home, these firms no longer bought on the spot market. But in the Federal Republic—the only large EC member state that did not cap the price of imported oil products—independent companies faced no limit on what they could charge consumers, and consequently they could buy Rotterdam petroleum throughout the crisis, outbid firms from other nations, and sell for a profit in West Germany. These high prices sucked in oil from not only Rotterdam but even France, leading West Germany's neighbors to level charges of "supply hogging," precisely the beggar-thy-neighbor policy Schmidt and Brandt publicly excoriated.[94] In reality, however, this was a conscious strategy. As Friderichs explained, "We are not dogmatic in this question, but we cannot permit a corset that cuts us off from urgently needed crude oil and oil products. Our EC partners depend neither on the Rotterdam oil product market nor on the import of oil products more generally [as do we]."[95]

Reliance on Rotterdam, however, created a massive "cleavage in the price of oil" in West Germany, between companies with access to the majors' crude through long-term contracts, and those who relied on the spot market.[96] During the crisis West Germany's oil market fragmented, with prices offered by independent refiners nearly double those of the majors. And while the latter made a killing selling their crude in Rotterdam, the *Mittelstand* of Germany's oil industry teetered on the edge of bankruptcy.[97]

Despite these pitfalls, the strategy of working with the majors and the spot market brought West German through the worst of the shock. For those who purchased petroleum from independent companies operating in Rotterdam, the price spike took its toll and precipitated a formal investigation into oil profits. But from December 1973 through March 1974—the height of the crisis, when the main goal was acquiring the energy to run the economy—the Federal Republic's economy sputtered but kept going. Economic growth remained positive through the first quarter of 1974 while it declined in the United States, Great Britain, and the OECD as a whole. During those four months the supply of total *energy* fell less in the Federal Republic than in other large Western economies: 1.9 percent compared with 3.3 percent in France, 7.5 percent in the United States, 16 percent in the United Kingdom, and 6 percent throughout the OECD.[98] This was due, at least in part, to West Germany's high oil prices, which drew in petroleum imports from across the continent.[99]

Just as important, the exceptionally high marginal price of petroleum from Rotterdam induced German industry to substitute coal for fuel oil much more than their neighbors. The Federation of German Industry, working as an interlocuter between state and manufacturers, helped the latter save over 400,000 tons of fuel oil a month by stepping up the consumption of coal. The switch was even more intense for chemical companies, some of the largest petroleum consumers, as they replaced almost a third of their oil consumption with coal and natural gas, and as they stepped up energy conservation through recovery systems and power-heat coupling technologies. All told, in 1974 industry used 13.5 percent more coal per unit of GDP than in 1973, while using 10 and 17 percent less of heavy and light fuel oil because of the price swings. Where coal consumption during the four crisis months fell in Britain by 44 percent due to the miner's strike, and in France by 0.2 percent, it rose in West Germany by 7.8 percent, evening out the loss of West Germany's crude imports.[100]

Taken together, price rises and substitution clarify a paradox. Even though West Germany's *oil* supply fell more than its neighbors during the crisis, its total *energy* supply declined less. And this dynamic permitted Bonn to conclude, with glasses that were only slightly rose-tinted, that "overall, the Federal Republic weathered this challenge better than most other industrial countries."[101]

West Germany, put simply, survived the short-term crisis by sustaining the flow of energy, writ large. The crisis, moreover, demonstrated that economies had more room to reduce oil use than experts had ever thought possible, suggesting the possibility of a new long-term strategy: conservation. After 1973 all Western nations pursued conservation or at least paid it lip service—even the United States, which actually pushed conservation on their European allies through the OECD.[102]

Initially, West German delegates to the OECD resisted these calls, fearing conservation would slow growth and raise unemployment. Out of eighty sections, the nation's first energy program devoted just one vaguely sketched set of paragraphs to reducing energy use.[103] But the aftershocks of 1973 yielded astonishing results. Not only did West Germany's economy substitute coal and gas for oil at the height of the embargo, it did the unthinkable. Between 1973 and 1977 West Germany's economy seemed, tentatively at least, to decouple, and to do so far more than its neighbors. Taking inflation into account, real GDP in West Germany grew by 19 percent between 1974 and 1977 while the total energy consumed by the economy declined by 0.27 percent. West Germany's energy intensity—the amount of energy required to produce a unit of real GDP—fell by 16.5 percent. Put differently, its economy became 19 percent *more* energy efficient. By comparison, France and Britain moved in the opposite direction, becoming 0.6 and 11 percent *less* energy efficient, respectively, during these same five years.[104]

West Germany's economy, in other words, was proving more flexible with respect to energy than expected. Premonitions of this appeared during the crisis itself but became more convincing with time, putting into question the coupling paradigm that had informed energy policy for over a decade. Tellingly, by the late 1970s the forecasts on which Bonn had based its first two programs seemed outdated at best, at worst disastrously wrong (see Figure 5.3). Whereas the 1973 program predicted West Germany would consume 406 million tons of SKE by 1975, the country in reality consumed just 347. By 1977 Matthöfer, one of the strongest advocates of the coupling paradigm, admitted that the past four years had shown how "the established correlation between economic growth and the growth of energy consumption is, in fact, a variable one."[105] In his estimation, the "errant prognoses" that had guided policy vastly underestimated the ability for a "forceful energy saving policy" to reshape the consumption of fossil fuels.[106]

The cause of this decoupling remained heavily debated. Nevertheless, most experts and officials believed it had to do with pricing. Without intending to, Bonn had created the conditions for experts to re-evaluate long-held beliefs about the relationship between energy, growth, and prices. The Federal Republic, after all, was the only large economy that did not control the marginal price of petroleum—and these spiked higher in West Germany in 1973–1974 than almost elsewhere.[107] West Germany's council of economic experts concluded that, contrary to what the coupling paradigm held, energy consumption was proving very responsive to price changes. The government concurred, recognizing that the "cost consciousness of industry and private consumers has grown significantly in response to the strong price rises. The Federal Republic thus expects that higher energy prices will lead to a fundamental intensification in efforts to use energy more rationally, especially in industry." The nation's energy leaders, in sum, were gradually coming to accept not only that it was impossible to return to the low-priced energy of the 1960s, but that doing so might not even be desirable.[108]

Conclusion

"That is the irrefutable insight: that in the past growth has come not only from better technological and organizational proficiency, but also—in fact not insignificantly—from the over-exploitation of our limited resources and our environment." So Hans Matthöfer, a leading policymaker in the Federal Republic, remarked in 1976. For him, the oil shock had changed the guiding question of energy policy. No longer must Bonn strive solely to expand the energy supply. Because 1973 had demonstrated the potential for decoupling, one must now ask, "How we can slowly switch to qualitative growth, that consumes fewer raw materials and less energy?" Such a transition would not be easy, Matthöfer believed. And he still questioned whether the economy could truly expand without more energy. Nevertheless, that the minister of technology—once a stalwart advocate of the coupling paradigm—would frame energy policy as the quest for a new type of growth illustrates just how much had changed.[109]

Indeed, the experience of the oil shock opened the space for the Federal Republic to change its approach to energy in three fundamental ways, moving policymakers to depart from the supply-oriented policies that had once held priority.

First, the crisis revealed the limits of the technocratic agenda that prioritized massive projects executed by large companies with state support. To be sure, the alliance between coal and nuclear forged during the 1970s remained a political force. This unlikely grouping of mining unions, nuclear scientists, conservative experts, and domestic firms would support the SPD into the 1980s.[110] But this alliance never actually gave West Germany the autonomy its leaders wanted. As the Technology Ministry admitted in 1977, despite its best efforts the country could still do little to counteract the hazards of the global oil market.[111] More generally, the neo-mercantilist plans to expand West Germany's energy supply failed. AEG pulled out of KWU as profits in the nuclear industry tanked. Deminex remained a dwarf among giants. VEBA spiraled into the red. Whereas the United States, Britain, the Netherlands, and even France and Italy all had international oil companies which exercised some control over the global hydrocarbon supply chain, West Germany did not, following GBAG's sale to BP. This, among other developments, created the space for a new approach to energy to flourish.

By the late 1970s even political insiders could admit that large-scale, state-supported, technocratic, supply-side energy projects were ineffective at best. At worst, they would prove costly and unnecessary if energy consumption continued to stagnate. For those outside the halls of power, the arrogance and impenetrability of the energy technocracy was itself cause for grave concern.

Second, the crisis shook the energy-coupling paradigm to its core. Government officials and experts—the very establishment against which the nascent anti-nuclear movement would direct its ire—began rethinking their approach to energy as they came to realize decoupling was possible. While some insiders would cling to the conventional strategies of the 1960s, others began calling for a new agenda to manage the security risks associated with foreign oil. The coal-nuclear alliance represented one strand, and they greeted high energy prices as a necessary cover for moving off oil. Coal miners and owners had, after all, been calling for higher energy prices since the 1950s. But another strand would come to see *conservation* as the better strategy. They would build on the "change of mentality" sparked by 1973 to move conservation into the center of discussion.[112] The roots of decoupling in the Federal Republic began, in other words, as much a response to energy security as they were a reaction to environmental challenges.

Third, the oil shock suggested that pricing itself could be a powerful tool of policy. The coupling paradigm had placed little weight on the ability of prices to change anything related to energy. But West Germany navigated 1973 by using its high marginal energy price to absorb oil from the Rotterdam spot market and to switch to different fuels. This experience profoundly illustrated how the economy could respond to large shifts in the price of energy.[113]

In the United States, the oil shock rocked the legitimacy of the government and created the space for a new economic philosophy to gain momentum, which called for less state intervention, deregulation of energy, and free market pricing. After the 1970s the Federal Republic would take a different turn. For both sides of the political spectrum believed West Germany's path through 1973 vindicated Bonn's economic governance. From the opposition, CDU representatives proclaimed the nation's energy pricing to be an "exemplary implementation" of the "functional capabilities of the social market economy."[114] The SPD-led government hailed 1973 as "a healthy reminder of the regulative capabilities of the social market economy."[115] The oil shock, in other words, reaffirmed the need for a flexible energy price. But for Germans this never meant removing the state's hand from the scales of pricing altogether. Instead, 1973 suggested to insiders that they could use the state to guide the price of energy to achieve specific goals. If an OPEC embargo could induce people to economize on energy, why not use the levers of the state to accomplish a similar result, but for different reasons? The 1973 shock did not discredit West Germany's state; in sum, it led officials to experiment with a managed energy price in the name of security and conservation.

But while the shock spurred West Germany's establishment to rethink their nation's energy system, transformation would also come from elsewhere. For the crisis overlapped with a powerful impetus for change coalescing among outsiders, one that had been growing since the 1960s and that crystallized around resistance to the state's flagship energy program, nuclear power. As the anti-nuclear and environmental movements gathered steam, they would transform West German politics and offer their own vision of a greener, more democratic energy system.

PART II

THE NEW ENERGY PARADIGM

6

Green Energy and the Remaking of West German Politics in the 1970s

> *The state is failing. It is no longer in a position to secure for us and our children a humane future. Through insane centralization, through its huge, unmanageable systems the state is no longer capable of solving problems like unemployment, a secure energy supply, or other pressing duties of this period.*[1]

During the late 1970s West Germany experienced the closest thing in its short history to a civil war, fueled by questions of democracy, the environment, and energy. In the fall of 1976 Gerhard Stoltenberg, the Christian Democratic (CDU) governor of Schleswig-Holstein, authorized construction of a nuclear plant near the small town of Brokdorf. When the work began, several thousand women and men moved to occupy the site, only to be driven out by mounted police, water cannons, and dogs. Two weeks later they rallied, and over twenty thousand people converged on this agrarian dairy region. Ignoring appeals to avoid provocation, a hard core of demonstrators laid siege to the site, sparking a violent confrontation with the police in which they exchanged rocks and Molotov cocktails for tear gas and high-pressure water. The largest police deployment in the history of the Federal Republic until then—augmented by border guards and riot police—cleared the area after injuring dozens. In early 1977 protestors returned to Brokdorf and expanded their efforts to other reactors (see Figure 6.1). That summer German activists joined in the largest demonstration in France against a fast-breeder reactor near Lyon. In September demonstrations climaxed, when over 50,000 people protesting West Germany's own fast breeder in Kalkar were greeted by tanks, helicopters, and thousands of police officers.[2]

The nation watched in horror, media outlets portraying the confrontations as a "civil war" between state and citizens. Protesters saw Stoltenberg's crackdown as evidence of a police state, and atomic power as the epitome of an energy

Figure 6.1. Protesters stand off against police at the Grohnde Reactor, 1977. *Source:* Picture Alliance / Getty Images. Nr. 1065790918.

system that was ruining the environment. Police, nuclear advocates, and CDU officials depicted the demonstrators as provocateurs hell-bent on ruining the civil foundations of the Federal Republic.[3]

Energy, specifically Bonn's ambitious nuclear agenda, sparked this turmoil by providing a cause that rallied an unlikely alliance of environmentalists, left-wing activists, disaffected scientists, Social Democrats, Protestant parishioners, and conservationists. Yet at the heart of these protests lay something more than energy. At stake was the nation's political system and the very environment in which its citizens lived. Was West German democracy degenerating into a centralized state that valued growth above all else? Were the security demands associated with energy creating a totalitarian system reminiscent of the Third Reich? Were the nation's air, water, and natural resources on the cusp of an irreversible decline? All of these questions motivated the protestors, who claimed to share a common "experience of powerlessness in the face of an overwhelming concentration of economic and political interests." They were fighting the entrepreneurial state that had been pushing a nuclear transition since the 1950s. But they hoped for something grander than just halting reactor construction. Their state, they believed, had become unmoored from democracy and controlled by a technocratic clique. And this had to change. "We do not just want to eliminate abuses and prevent undesirable developments," a protest leader intoned. "Rather, we have set ourselves the goal of creating a fairer, freer, and more humane social order."[4] They wanted a new politics in which "ecological principles [would be] given precedence over so-called 'objective economic constraints.'"[5]

In 1980 these demonstrations cohered into the Green Party, an organization that would transform Germany's political scene. The Federal Republic was the *only* large industrial country where an environmental party gained prominence before the 1990s—what one historian has called Germany's new *Sonderweg*.[6] The Greens became a political force for reasons that ranged from the particularities of the Federal Republic's voting system to changing social structures and the fraying of Europe's traditional party system.[7]

But the Greens also broke into West German politics because they mobilized ideas that resonated with different groups. They drew on mounting fears of ecological degradation and technology gone awry, and they skillfully linked this angst to their nation's history of National Socialism. In the process, they imported ideas from American ecologists to suggest the possibility of a different future, where society would follow a soft path of decentralized technology powered by a local, more democratic energy system that prioritized conservation over consumption.

Crucially, this outsider movement forced the nation's conventional parties to reorient, dragging the political center of gravity toward a greater concern for the environment. While America, Britain, and France also debated the merits of a new energy policy in the 1970s, only in the Federal Republic did a political force emerge to represent these views at the national level. Above all, these new ideas threatened the SPD—the lead party in Bonn's governing coalition—which feared losing voters to the anti-nuclear movement. As their base of blue-collar workers eroded under the pressure of deindustrialization, Social Democrats had to grapple with existential questions. What did the party stand for? Could social reform be achieved without abundant energy and nuclear power? Could ecological protection and jobs be reconciled?[8]

To counter the anti-nuclear and ecological movement, Social Democrats began a fiercely contested internal reform in which the party allied with new experts, adapted Green ideas for its own purposes, and placed ecological considerations at the heart of a new agenda. This realignment yielded policies that led West Germany to diverge from other large industrial countries in its approach to energy, unleashing an efficiency transition well before climate became a major issue. In the 1980s energy use rose in the United States, Great Britain, and France. The latter would become a bastion of atomic power that represented a distinctive energy model for other nations in Europe to emulate. West Germany, by contrast, carved out an alternative energy trajectory that, so it seemed, might achieve the impossible: decoupling energy use from growth. The seeds of Germany's later *Energiewende* would grow from this tangled interaction between outsider Greens and insider Social Democrats desperately seeking to refashion their party for a new era of politics.

Green Precursors

In 1958 the Austrian conservationist Günther Schwab published a popular fictional interview with the Devil, who was nearing the final stages in his satanic plot to destroy the natural foundations of life. Going through ten editions between 1958 and 1972, *Dance with the Devil* played on the legend of Faust to dramatize the findings of environmental science. In Schwab's telling, the Devil was trying to poison "everything that man needs for his existence" by employing nefarious sub-demons like the Stink Devil of air pollution, the Atomic Devil, and the Foul Water Demon of river pollution.[9] Schwab's novel epitomized a burgeoning, postwar genre of eco-apocalyptic literature in Central Europe that was at once reactionary and forward-looking. His tale highlighted the destructive potential of DDT four years before Rachel Carson's *Silent Spring*, while also warning that the accumulation of carbon might melt the ice caps. Others captured the fear of resource exhaustion that would bedevil the 1970s. "It is as if modern man had made a wager with the devil," suggested Anton Metternich in 1947, "that he would be able in a thousand years . . . to consume, use up, and squander everything that Nature created in uncounted millions of years."[10]

Eco-catastrophism was one of several tributaries feeding West Germany's environmental efflorescence in the 1970s. Such writing was motivated by the dawning recognition that economic growth came at a cost; for Germany, that the Economic Miracle which lifted the nation from the ashes of war brought not only wealth but despoliation. Schwab, after all, opened his dismal panorama with the devil of "Progress and Living Standards." As he and others pointed out, much of the problem lay with the production and distribution of energy.[11]

Before 1970 anxieties about pollution centered on the energetic hub of West Germany: the Ruhr and the Rhine. The former was the nation's coal heartland, known for the soot that perennially hung in its air. But in the 1950s and 1960s the Ruhr's pollution problems accelerated as coal smoke, blast-furnace gases, metal dusts, and other emissions became normal features of its ecosphere, drawing comparisons with other polluted cities of the world like Pittsburgh and Los Angeles.[12] Energy production was a leading cause. Dust from the mines and coking facilities that powered the new republic clouded the air. "Whoever sneezes in the Ruhr," so Germans grimly joked, "finds a coal briquet in his handkerchief." The region's terrible air quality sparked the nation's first systematic, cross-party effort at smog control. In 1960 the Ordoliberal and CDU official Alfred Müller-Armack called on the region to do more to keep its air clean. A year later Willy Brandt, the charismatic Social Democratic mayor of Berlin, shone a spotlight on the issue when he famously declared that the "the sky over the Ruhr must become blue again."[13]

Next door lay the country's largest river, Father Rhine—symbol of German identity; conduit for goods and people; and centerpiece of the country's budding nuclear program. As oil refineries crept inland along the waterways of Western Germany, riverine oil slicks became commonplace. In 1960 and 1961 water supply companies sounded oil alarms over forty times. The cause was not only spills or refinery accidents, but leaks from the hundreds of thousands of new underground fuel oil tanks that residents built as the price of this energy plummeted. Run-off from the petrochemical fertilizers that West German agriculture began using after 1945 only made matters worse. This toxic combination decimated the river's ecosystems. By the 1950s the Rhine's salmon catch was a fraction of prewar levels, and mass fish deaths happened with frightening regularity. In 1969 the nation's largest mass of swollen fish cadavers washed ashore along the Rhine itself, caused by pesticide from the nation's petrochemical giants. By 1970 the river, Europe's "largest cesspool," was biologically dead.[14]

Air and river pollution alarmed the public. But despite outpouring from writers like Schwab, Germans lagged behind the United States in developing a broad-based environmental movement. Through the 1950s ecological anxieties remained rooted in a closed network of conservationists, many of whom held reactionary views that connected ecological threats to moral decline. They hoped salvation would come by renouncing modernity. In Schwab's story, humanity was spared by appealing to God, who ended the devil's havoc only after the protagonists repented and subordinated themselves to the divine. And many still suffered from "avocado syndrome": green on the outside, surrounding a brown, authoritarian Nazi core. Schwab himself had joined the Austrian SA in 1930, and remained animated by Nazi notions of race after 1945, arguing that the devil's "Number One" department was the "population bomb," where a "wave" of "black" and "yellow" peoples would out-reproduce Europeans and extinguish the "white man."[15]

Nevertheless, conservationists like Schwab laid a foundation for more progressive ecologists to come, by raising awareness of environmental degradation and by crafting a powerful critique of technology. Referring to the myth of Prometheus, Reinhard Demoll, president of the German Fishing Association, called technology the "great seduction" that was outpacing human control and ethics. But he did not decry all technology, just that which was beyond humanity's scale: "dwarf-like is humanity—gigantic is its handiwork."[16]

Demoll tapped into a chorus on both sides of the Atlantic that condemned large-scale technology. In Europe this began as a conservative commentary, but after 1960 this critique migrated to the Left. This shift had started in the United States with the communitarian planning ideas of the New Deal, and gained momentum after 1945. By the 1960s these ideas were spreading to Europe through the work of Jacques Ellul—French resistance leader and sociologist—and Ernst Friedrich Schumacher. Ellul's 1964 tome, *The Technological Society*, connected criticisms of technology less to moral decay, as had Demoll or Schwab, than to politics. Ellul argued that modern technology was qualitatively different than in the past: highly centralized into massive systems that "encompass all human activities." And this was "inevitably anti-democratic," as centralized technology, operated by an aristocracy of technicians, would "cause the state to become totalitarian, to absorb the citizens' life completely."[17] These ideas were popularized by Schumacher, who had fled the Third Reich to Britain, where he worked as a development economist and later advised one of Europe's largest organizations, the National Coal Board. After seeing the destructive power that modern technology had on traditional life in Burma and India, Schumacher began calling for user-friendly technologies to meet the needs of local communities. In 1973 Schumacher put this passion for decentralized technology into his popular book, *Small Is Beautiful*, which criticized the seemingly "irresistible trend, dictated by modern technology, for units to become ever bigger." He called instead for "small-scale technology, relatively nonviolent technology, 'technology with a human face.'"[18]

If eco-catastrophism and concerns about technology prepared the ground for Germany's environmental movement, so too did the student mobilization of the 1960s. As a new generation came of age after World War II, they developed a different view of the Federal Republic than the generation that had experienced the Third Reich. Qualms about the state of the republic coalesced in 1966, when the CDU and the SPD formed a Grand Coalition. With the two largest parties and former opponents together controlling 90 percent of Parliament, no institution represented the aspirations of the younger generation. An extra-parliamentary opposition thus emerged, which claimed the state was becoming alienated from the people, and the parties themselves oligarchies that threatened the nation again with dictatorship.[19]

In 1967 the police shooting of twenty-six-year-old Benno Ohnesorg at a protest transformed this discontent into a public movement. This "first political murder in the Federal Republic," as many on the Left called it, sparked an emotional debate about the concentration of state power. Countercultural youth and university students filled cities across West Germany, protesting everything from the Vietnam War to gender inequality. But underneath these disparate issues lay a deeper agenda of atoning for the nation's Nazi past and guarding against the return of authoritarianism. Looking back, Joschka Fischer, a mercurial working-class youth from the Frankfurt alternative scene, reflected how "never again Auschwitz" became the rallying cry of 1968. The movement's leaders compared not only American bombing in Vietnam to the crimes of the

Third Reich, but also the growth of state power at home. Only participatory democracy from the bottom up, a demand that infused the Left, could solve the nation's social ills.[20]

The left-wing youth revolt was short-lived and the movement soon splintered. Nevertheless, 1968 profoundly influenced how West Germans would reimagine energy, growth, and politics, for it produced a powerful critique of centralized power that linked the current republic to the Third Reich, as well as an abiding enthusiasm for the Left. Following 1968 more youth swung leftward in West Germany than in the United States, France, or Britain.[21]

The 1968 youth movement devoted little attention to ecology. But after it fractured, one strand of it dove into local initiatives that spread ecological awareness beyond traditional conservationist circles. It began in 1969 when the energy giant VEBA bought land for a petrochemical complex near Duisburg. Supported by federal subsidies to revitalize the Ruhr, VEBA promised thousands of new jobs for a region that had been hemorrhaging them since the coal crisis of 1958. Yet the ecological impact would be grave. Duisburg's mayor feared sulfur dioxide would pollute his city's air, while spillage would threaten the water supply. These fears generated one of the first Citizen's Initiatives (CI) to gain national publicity as local residents, Duisburg's mayor, and the SPD's youth organization began a campaign to "Stop the VEBA-Moloch," delaying construction through demonstrations and the courts. After 1970 local CIs proliferated, giving Germans a new instrument to tackle ecological problems when the state failed to act.[22]

But the federal state was slowly taking action, spurred partly by the thriving environmental movement in the United States. In 1969, after the SPD won a dramatic electoral victory that ended the Grand Coalition, one of its first achievements with its new partner, the liberal FDP, was elevating the environment as a major policy issue. Willy Brandt ran for chancellor on an optimistic campaign of reform that paid attention to river and air pollution, which he hoped could harness the enthusiasm of young voters after 1968. In late 1969 the Ministry of the Interior, led by the FDP, brought noise abatement together with ecological issues into a new unit, Division U, for *Umweltschutz*. With this literal translation of "Environmental Protection" from English into German, the SPD-FDP coalition bestowed a name to an array of hitherto unconnected challenges, which it then formalized through its first environmental program in 1971. *Umweltschutz* flashed through the media, and CIs adopted the concept for their own national organization: the Federal Association of Citizens' Initiatives for Environmental Protection (*Bundesverband Bürgerinitiativen Umweltschutz—BBU*). A new sphere of political action was born.[23]

Accelerators and Sparks

By 1972 the tributaries to West Germany's Green movement were all in place: mounting ecological damage and rising public concern; a critique of state power and of large-scale technology on the Left; and a new form of grassroots organization. Two accelerators that erupted in 1972 and 1973 would bring everything together.

First, in 1972 the Club of Rome published a bombshell report that caught the attention of the world. An organization of scientists, business executives, and technocrats, the Club studied border-crossing issues that defied national resolution. Its report on the *Limits to Growth* suggested the world was reaching its economic capacity, and within a generation would descend into a full-blown ecological crisis as it ran out of resources as well as the sinks that processed the detritus of modern life. Warnings of resource exhaustion were nothing new. But when this highly regarded institution commissioned a computer simulation from the Massachusetts Institute of Technology, it gave the concept of ecological limits a new aura of credence.[24]

The *Limits to Growth* came under immediate criticism from neoclassical economists and even from many on the Left.[25] But in West Germany the study impressed politicians across the spectrum, spurring some to rethink the very nature of growth. Spearheading this was the ecological wing of the SPD under Erhard Eppler, a Protestant reformer from Baden-Württemburg (see Figure 6.2). Since 1968 Eppler had served as minister for economic cooperation, and his travels through drought-ridden West Africa had led him to think critically about the relationship between energy, the environment, and growth. In 1972 Eppler put his views into a coherent framework when West Germany's most influential union invited him to speak about what actually made workers' lives better. In his speech Eppler drew on the *Limits to Growth* to argue that "in the foreseeable future humanity will bump into limits, limits that just five years earlier we could hardly have dreamed of." If economies continued to grow as they had over the past century, life would become "unbearable." Western society needed a "fundamental change in our value system," to operate on a human scale, to replace GDP with a new "measure of progress," and to conserve energy instead of recklessly consuming it.[26]

Eppler borrowed from the United States the notion of "quality of life," proposing this as a new a rubric that could help society improve well-being without squandering the earth's resources. The concept caught fire after Chancellor Brandt adopted it as the SPD's slogan in 1972: "With Willy Brandt for Peace, Security, and a Better Quality of Life." Following another victorious election, Brandt reaffirmed the need for a new approach in his address to the

Figure 6.2. Erhard Eppler (left) and Chancellor Helmut Schmidt, 1979 SPD Party Conference. *Source:* Süddeutsche Zeitung Photo / Alamy Stock Photo. ID: RMJYWF.

nation: "it is more about the question of where, how, and for what the economy grows."[27]

The notion of ecological limits might have remained a slogan had it not been for a second accelerator, the 1973 oil shock. The dramatic confrontation between oil producers, oil majors, and consumer nations seemed to confirm the fears stoked by the Club of Rome, as West Germany's press and elites eagerly connected the oil crisis to resource scarcity rather than the power of OPEC. "In a few months," Eppler explained, "the energy crisis has achieved what warnings over many years have been unable to: the insight that our resources are limited." Liberals worried that the past twenty years of explosive growth were "a unicum in the history of humankind." As Carl Friedrich von Weizsäcker concluded, who had helped pen the Göttingen anti-nuclear weapon manifesto, 1973 illustrated how "our current system of energy consumption will eventually reach a limit that is determined by the laws of nature."[28]

The coupling paradigm thus met a new paradox: how to secure growth in a world of finite energy resources. And like any good paradox, this one defied easy resolution. Instead, it opened a rift in the SPD that pitted a younger, reform-oriented wing around Eppler against a conservative faction led by Helmut Schmidt. Both sides remained convinced that only a strong, centralized state could solve the paradox.[29] But where Eppler wanted to solve the paradox through energy conservation, Schmidt thought otherwise. Following Brandt's resignation in 1974, Schmidt became a new chancellor facing imminent recession, and he strove to restore the postwar growth consensus. At stake was nothing less than the SPD's reform program that had brought it electoral success in 1969 and 1972. Social reform required public funding, which required a growing economy. As Horst Ehmke put it, an ally of Schmidt, "I see growth today as a condition for progress. . . . The workers' movement, the party, and the unions were in fact always programmed on growth." And growth, according to Schmidt and Ehmke, required energy.[30]

In 1974 the SPD sided with Schmidt. At a time when expert advising was crucial to policy, there was still little empirical work supporting Eppler's idea that conserving energy would be anything less than catastrophic for the economy. For Schmidt's wing, nuclear power, more than any other supply-side policy, provided the silver bullet for their energy woes. This energy not only promised independence from OPEC, it could also solve the paradox posed by the *Limits to Growth*, since the energy from uranium seemed potentially limitless.[31]

The SPD's nuclear policies ignited the spark that would bring forth a new, ecological political movement. In 1973 West Germany commenced a massive atomic campaign that was concentrated in a handful of river watersheds. In this the country was not alone. The United States and Britain were diving headlong into nuclear power, but more importantly, so too were West Germany's immediate neighbors, Switzerland and France. The former needed to supplement its hydropower. The latter hoped atomic power would advance its aspirations to become a third force in a bipolar world. The transnational Upper Rhine—a "dreamland for nuclear engineers"—was to become the economic heart of Western Europe, powered by reactors that used these fast-flowing waters for the indispensable task of cooling the staggering heat generated by controlled fission. A race for the waters of the Upper Rhine commenced. In the Federal Republic, ground zero for this vision was Baden-Württemberg, a wealthy region with an agrarian way of life. On a stretch of the Rhine that ran through this state and beyond, no less than eight nuclear parks with seventeen reactors were planned by the Federal Republic, France, and Switzerland. Newspapers called it "the most colossal energy concentration on Earth." It was here, near the small town of Wyhl, that West Germany's anti-nuclear movement was born.[32]

The Rise of a Movement

Local concerns kick-started the revolt. The first stirrings of pushback came on the French side of the Rhine in 1971, when several thousand people, including many from Baden-Württemberg, protested the construction of a reactor in Fesselheim by the national utility, Electricité de France (EDF). This French demonstration failed, but not before inspiring farmers and activists in Southwest Germany to mobilize. For contemporary studies indicated that the necklace of reactors planned around the Rhine threatened this vital river network with thermal pollution, and the surrounding countryside with new weather patterns. The CDU government of Baden-Württemberg, however, paid these reports little heed, pressing ahead with the reactors and outraging the region's winegrowers. When local farmers and professionals attended hearings in Stuttgart, they were met with condescension from government officials. The entire process, according to one participant, was a "mockery" and revealed how the state acted like a "dictatorship" by making decisions with little input.[33]

In February 1975 workers arrived in Wyhl to begin construction. After a meeting of the local CI, 300 people, led by women and farmers, marched impromptu to the site, pitching tents and shutting down the excavators clearing the forest. Regional authorities responded by mischaracterizing the occupation as a militant struggle by disaffected Leftists, surprising the protesters with their aggressive rhetoric and using police to clear the location. Police arrests provoked a bitter reaction. Three days later several thousand locals, supported by people from outside Baden-Württemberg, returned to occupy the construction zone for another seven months.[34]

The demonstrations ignited a media frenzy. Activists from across the country flocked to Wyhl, culminating with an Easter gathering spearheaded by the new BBU. Among those speaking was Petra Kelly, a charismatic activist educated in America who had returned to join the SPD and work at the European Commission in Brussels. Wyhl turned into a "rolling experiment in participatory democracy" that attracted a cross-section of West German society. Members of the New Left from cities and university towns came. But so too did farmers, Protestant parishioners, scientists, professionals, and conservationists. After surveying the scene, Kelly concluded that nuclear power "would divide society," and she and others began working to spread this movement to other regions. Their chance came a year later in Schleswig-Holstein. In the town of Brokdorf on the Elbe and upriver at Grohnde, more violent protest and police retaliation erupted in 1976 and 1977. In France demonstrations also escalated when over 30,000 people protested the Superphénix fast-breeder reactor at Creys-Malville. There one person died and three were wounded, with French authorities blaming West German participants for the bloodshed.[35]

In these burgeoning confrontations, protestors moved beyond local concerns to embrace broader issues, above all democracy and the dangers associated with nuclear power. For evidence had been accumulating that challenged the view of the atom as a normal form of energy. In 1971 scientists from the Lawrence

Livermore Labs in California published a book suggesting the radioactivity levels labeled "safe" by regulators were far from it. These findings spread to the Federal Republic where, along with other damning evidence, it was compiled in a sensationalist book by SPD member Holger Strohm. Strohm recounted how radioactive material could show up in the milk of cows grazing near reactors and how dentists were finding strontium in the teeth of children downwind from nuclear parks. More reports followed about accidents in West Germany: a minor leak that killed two workers; a confidential report made public by the BBU about the potential danger of a meltdown.[36]

The nascent anti-atomic movement used these stories to portray nuclear power as a proxy for the ills of modern society. As Kelly, whose sister had died of cancer, put it:

> Splitting the atom; uncontrollable emission of radioactive toxins; the insanity of the nuclear, bacteriological and chemical weapons build-up; unrestrained economic growth spreading commercialisation to every aspect of our lives; overconsumption of goods and raw materials ... these are the conditions of modern industrial society, and these are the factors responsible for disease. In an epoch characterized by "gorging to excess," all sense of responsibility has disappeared. Cancer has become a fitting symbol for the disease of civilised society.[37]

But even more than radioactivity, atomic waste became the rallying point that enabled the movement to expand beyond local concerns and link nuclear energy with technology and state-power gone awry. Waste was a fundamental part of the nuclear energy supply chain, and since 1972 the Federal government had been exploring the potential for an integrated nuclear waste management center, replete with reprocessing plants and final storage. Officials had repeatedly said that the disposal of spent fissile material was on the cusp of being solved. But when protests erupted at Brockdorf and Grohnde the disposal question remained wide open.[38] Most spent fuel stayed at temporary facilities managed by industry, while Bonn assumed responsibility for final storage. Yet this division of labor was not working as planned. In 1974 radioactive material showed up in a community dump near the Obrigheim reactor, raising doubts about the security provided by either firms or the state. The accumulation of waste thus acted as a brake on further nuclear expansion, prompting the Federal government to plan a centralized disposal system. In late 1976 and early 1977 Bonn linked nuclear licensing to the development of a storage program, settling on Gorleben on the border with East Germany as the new site.[39]

But as critics pointed out, any integrated waste site would have to be monitored for tens of thousands of years, presenting an intergenerational problem unlike anything modern governments had ever faced. This became the prime question of the nuclear debate. For Ralf Dahrendorf, one of Germany's leading liberal theorists, "This is an issue on which our entire future depends, one way or the other . . . the danger that arises not so much from power plants themselves but from the disposal of nuclear waste is extraordinary."[40] Here the anti-nuclear movement found a weak link in Bonn's nuclear program, for West Germany's Atomic Law gave the courts wide latitude to assess questions of risk as well as the current state of technology, while the nation's decentralized judicial system proved receptive to arguments about the danger of long-term storage. In a sensational public trial, an administrative court revoked the license for the Wyhl nuclear reactor, citing the risk of a worst-case scenario explosion. Even more important, following legal appeals in 1977, the Schleswig-Holstein court ruled construction must stop in Borkdorf and Grohnde until the waste issue was adequately resolved, creating a de facto moratorium on new reactors.[41]

This new movement's legal success came in part because new counter-experts were willing to challenge the pro-nuclear scientists sponsored by state and industry. An early center of counter-expertise was the University of Bremen, location of an important conference series on energy.[42] But the main face of counter-expertise was in Freiburg, just twenty miles from Wyhl. In this university town Siegfried De Witt, a lawyer who spearheaded the court proceedings against the nearby reactor, created an Institute for Applied Ecology in 1977, otherwise known as the Eco-Institute. Staffed by scientifically trained anti-nuclear activists, many of whom who had participated in the Wyhl protests and court cases, the Eco-Institute aimed not only to fight nuclear power. It hoped to forge an entirely new energy future for the country.[43]

A Dawning Divergence

After 1977 the Federal Republic would diverge from its neighbor across the Rhine in matters of energy. The court moratorium was a blow from which German atomic apostles would never recover. France, too, experienced intense turmoil in 1977 as its own anti-nuclear movement reached a crescendo demonstrating against the Superphénix. Like West Germany, France harbored a growing environmental movement, captured by René Dumont's campaign for president in 1974 as an ecological candidate. And like Bonn, Paris inaugurated a massive nuclear campaign after the oil shock, which galvanized grassroots resistance around the nation.

But Paris held firm to the atom. After the Creys-Malville protests, officials cracked down against anti-nuclear groups, casting them as radical militants. A hard core of Leftist dissenters used the cause to do battle with the government, undermining public support for the anti-nuclear movement. The unitary nature of the French state, meanwhile, let officials make energy policy with little input from below. The central state gave generous aid to local officials who endorsed reactors while its national court system, which was tightly overseen by the Ministry of Justice, brooked less space for legal challenges. Paris, meanwhile, had a powerful ally in EDF, France's national electricity monopoly, which controlled the entire grid, employed over 120,000 people, and dwarfed even the largest West German utilities. This state-within-a-state mobilized expertise and immense financial firepower to push on with atomic power.

More importantly, the public and much of the political Left in France remained committed to the atom throughout the 1970s. Since the inception of their nuclear program, politicians like President Charles De Gaulle had linked the atom with French geopolitical grandeur in a way that was impossible for postwar West Germany to do under American supervision. De Gaulle believed his nation must have a nuclear strike force to be a third power in a bipolar Cold War world. France's civilian nuclear program was steeped in this military agenda, from its campaign to source uranium from former colonies to its investment in the Superphénix. This fast breeder promised not only to generate electricity but also to yield weapons-grade plutonium. La Hague on the British Channel, meanwhile, became the largest nuclear waste reprocessing center in the world. French leaders hoped the combination of post-imperial uranium mining plus reprocessing centers plus the breeder would give France energy independence from America's nuclear supremacy. State officials and EDF experts drilled this message home relentlessly, and through the 1970s public approval for nuclear autonomy actually rose. Even the Communist Left after 1977 supported France's nuclear program as a counter to American influence.

Protests thus dampened but never killed the French dream, and after 1980 the country became the hub of a transnational nuclear supply chain that serviced much of the EC. French companies with commanding stakes in the mines of Niger and Gabon shipped uranium by sea to power European reactors. Neighboring countries, including West Germany, sent their nuclear waste by rail or truck to La Hague for processing. And the Superphénix, supported by stakeholders from Belgium, the Netherlands, and the Federal Republic, went critical in 1985 as the largest fast breeder ever constructed.[44]

Green Visions

In West Germany, by contrast, a new force burst onto the political scene after 1977. Over the next three years, local protests transformed into a heterodox political party that wove together the strands of environmental anxiety, grassroots activism, and anti-statist, anti-technological, and anti-growth criticism that had been swirling around different milieus since the 1960s.

The new movement was young, educated, and rooted in the thriving tertiary and service side of the economy. For the very growth that came under criticism after 1973 had, in fact, created incredible social mobility over the preceding two decades. This mobility combined with the dramatic expansion of higher education to change West Germany's occupational and mental landscape. White-collar professions like law, education, and services grew dramatically, while employment in traditional industries like coal declined. And as people became better off, their values began to shift. Eppler had noticed this already in the early 1970s. By the end of the decade, surveys confirmed that Germans were prioritizing personal freedom, self-fulfillment, and post-material values over material goods, a shift famously termed the "Silent Revolution."[45]

In West Germany, the generation born after the war experienced this occupational and value transformation most profoundly, and they formed the core of the new movement. Two-thirds of people who would go on to join the Green Party in the early 1980s were under the age of thirty-five. Many were disaffected 1968ers seeking a new political outlet, or traditional liberals interested in local activism. Most came from middle-class families, were educated, and lived in college towns or cities. They belonged to professions like law, healthcare, and education, or they were underemployed and frustrated by their inability to find a career. In either case, they had little material connection to the heavy industries that were the nation's largest energy consumers and worst polluters.[46]

Uniting these disparate activists was the fear of an ecological apocalypse. Nearly every prominent intellectual who joined the anti-nuclear protests professed millenarian beliefs about environmental destruction, from philosopher and SPD-member Carl Amery, to CDU delegate Herbert Gruhl, to the founders of the Eco-Institute. Most believed that either technology and chemistry would "destroy the world" or that raw material shortages would lead to economic crisis.[47] Gruhl, a prominent conservative leader, spoke of a "physical" solution to the new "physical" conditions that population growth and rampant fossil fuel consumption had created, namely, "global catastrophe." In a 1975 best seller, he explained how collapse was the endgame to "a world historical situation that has never existed since humans have lived on this planet."[48] According to the Greens' self-professed philosopher, such existential angst was *the* motivating factor behind the movement. Politics in the most basic terms was "the art of the continued existence of humanity."[49] This movement took the *Limits to Growth* seriously: Armageddon was the likely result of the "growthmania" that gripped the West. For those on the Left of the movement, capitalism itself was destroying the natural foundation of life, or as Rudolf Bahro, an East German émigré, put it: "ecological crisis [would] bring about the end of capitalism."[50]

Although anti-capitalism was powerful, it did not pervade the entire anti-nuclear movement. What was universal was an extreme hostility toward traditional ideas of *quantitative* growth. Nearly everyone in the movement blamed the nation's political parties for propagating the "pure illusion" that unlimited growth was not only possible but something society should aspire to. As Carl Beddermann put it, founder of one of the first local Green electoral lists, "the current conception of growth is like the charlatan asserting he has invented a perpetual motion machine." From Bahro on the Left, to Beddermann in the middle, to Gruhl on the Right, the movement criticized economic thought for believing growth could continue indefinitely. As Beddermann explained, "in all natural systems we know there is growth only in a temporary and quantitatively limited form." The earth and its economy were no different. If society continued on its current course, "catastrophic collapse" would be the result.[51]

The growth paradigm became the movement's main target not only because they believed it was unsustainable, but because the very process of pursuing it seemed to be generating terrible social effects. Here the anti-nuclear movement wove together ideas from Ellul and Schumacher with their experiences at Wyle and Brokdorf to argue that perpetual growth created "tremendous complexity" that was reshaping the individual's relationship with the state, the environment, and the economy.[52] Specifically, they believed the technology required for limitless growth had become so complicated that it could not be overseen by any single entity, even the modern state. For Klaus Traube, a leading authority on atomic power who had defected from the nuclear construction firm AEG, "large-scale technology" was now utterly "intransparent." As a former atomic project leader, not even he had fully grasped the programs under his oversight. And this complexity, he warned, "means that the direction of development escapes all effort at control." By the late 1970s these ideas became widespread: less than a third of West Germans believed technology was a blessing, down from nearly three-quarters in the 1960s.[53]

But the problem did not stop at technology. For as technology metastasized, so too would the state in an effort to control it, the anti-nuclear movement argued, generating the seed for more dangerous developments. Big technology would lead to an expansive state that suffered from corruption and the concentration of power. Such anxieties became concrete in Wyhl, where revelations came out that members of the local government belonged to the regional utility's board of directors and were promoting reactors for personal gain. The state, in other words, was hardly neutral and better characterized as a "sleazocracy."[54] At the federal level, all major parties and industry shared a common desire for inexpensive energy and quantitative growth, and were keen to expand the size of energy corporations, such that it was difficult to discern where the interests of one left off and another began. In the words of Joschka Fischer, it was as though "all established powers have coalesced into a single party, the party of industry, growth, and technocracy."[55]

By 1980, moreover, the SPD-FDP coalition had been in government for almost a dozen years—half the lifetime of a college student—making West German politics susceptible again to the criticisms of 1968. As a participant in the earlier student movement, Fischer was among those who rehabilitated critiques of centralized state power. He and others argued that this "grotesque consensus" around growth and big technology extended not only from the Left to the Right, but beyond politics. As Schumacher, leader of the BBU, warned, the absence of an alternative vision for growth and energy "signaled a frightening and at the same time dangerous loss of state legitimacy, and reveals the alarming crisis symptoms of our representative democracy." This created a feeling of impotence on issues that Greens held most dear. For how could the greatest challenge of all, ecological collapse, be averted if the nation was governed by a centralized constellation of industry and state that was devoted to growth and unresponsive to the demos?[56]

In 1977 these anxieties acquired an evocative name when the audacious journalist Robert Jungk coined the term "Atomic State." A German Jew who had fled Nazi Germany in the 1930s, Jungk gained influence as a critic of nuclear power through popular books. By the 1970s many activists were beginning to call the Federal Republic authoritarian for its response to the nuclear protests. A best-selling author, Jungk pulled these criticisms together by comparing the Third Reich with the "Atomic State" he thought West Germany was becoming. Pointing to new surveillance techniques at reactors—from mobile command posts to undercover agents—he warned the nation was descending into a security frenzy. Monitoring fissile material for "tens, hundreds, thousands of years," would turn the nation into a garrison state. Comparing future to past, Jungk concluded:

> In time, laws designed to deal with a crisis—such as an outbreak of terrorism—will become the norm for what is in fact a permanent state of crisis. The watchword will be: Protect at all costs the source of energy. Atomic power continually under threat leads to a permanent state of siege. It brings about harsh new laws to "protect the people." ... In the end, it will justify everything. ...[57]

Given the backdrop of police power in Wyhl and Borkdorf, Jungk's hyperbolic claims had just enough resonance to gain traction. His ideas were taken up in internal SPD debates and repackaged in popular fiction.[58] And his picture of the slippery descent into a police state seemed all too real for a nation less than a generation removed from racial dictatorship. For by the 1970s West Germans were in the midst of a visceral debate about their Nazi heritage, which was reaching a crescendo with Brandt's genuflection at the Warsaw Ghetto Memorial in 1970, the murder of Israeli athletes at the 1972 Olympics, and the broadcasting of the popular "Holocaust" mini-series in 1978–1979.[59]

Jungk's idea of an atomic state captured core ideas of the anti-nuclear movement, which wanted the very opposite of the society he depicted: "We want simplification, decentralization, and deconcentration."[60] In an evocative analogy that would shape policy deliberations for the coming decades, the anti-nuclear movement imported ideas from America to argue that the nation must choose between a hard path and a soft one. In 1976 Amory Lovins, an American physicist and environmentalist, suggested a paradigm shift that could solve the energy paradox raised by the *Limits to Growth*. He described two energy trajectories that entailed radically different "technical and sociopolitical structure(s)," but that could achieve exactly the same level of welfare. The current, hard path depended on large, centralized, supply-oriented energy systems that were capital-intensive, fixed in time and space, and reliant on complicated technology. The alternative soft path, by contrast, would rely on small, decentralized, flexible, sustainable, and diverse energy sources that used easy-to-understand technology and that could be governed in a more democratic fashion.[61]

Lovins's essay became one of the most influential energy articles every published. And his ideas were popular in the Federal Republic.[62] The anti-nuclear movement adopted his dichotomy wholesale, using it to reorient the energy debate away from questions of price to those of systemic organization. They used the soft path to tie their disparate hopes together under a single rubric, for a decentralized, democratic society oriented around qualitative growth.[63] And energy was the pivot for this agenda, or as the Eco-Institute put it, the "central instrument of economic policy."[64]

More than anything else, those at the Eco-Institute wanted to dismantle the old energy paradigm by decoupling growth from energy consumption. Even more than renewable energy, *energy efficiency* became the most important policy the anti-nuclear movement would push through the 1980s. In a ground-breaking study from 1980, the Eco-Institute argued that by using existing energy-saving technologies more extensively, and by aligning the type of final energy used with the type produced, West Germany could achieve the same welfare with half the energy. Drawing on Lovins, they based these astonishing predictions on simple concepts, like the overlooked fact that heat accounted for nearly three-quarters of all energy used by final consumers. If consumers could avoid energy conversions—burning coal to generate electricity to in turn heat homes—and instead use heat from the initial stage of energy production, the nation could drastically reduce its energy footprint. As experts at the Eco-Institute put it,

"Anyone who heats his home with electricity behaves like someone setting fire to a forest to light a cigarette."[65]

What is more, the social benefits of transitioning to a softer system, the Eco-Institute believed, would rival the environmental ones. Focusing energy production in small units at the point of consumption would weaken the corporate energy conglomerates, who exercised great influence on politics. Retrofitting and servicing buildings with energy-efficient technologies promised an entirely new strata of local jobs in places where people already lived. The *right* energy transition, in other words, could lead to qualitative growth and a more democratic, more decentralized, and more egalitarian society.[66]

Turning Point: 1977

In contrast to Britain, France, and the United States, West Germany's anti-nuclear movement gained political influence because their ideas about state power and concentration resonated with a generation coming to terms with their parents' participation under the Third Reich. Their political ambitions, moreover, benefited from the Federal Republic's system of proportional representation, in which small groups faced a low barrier to enter Parliament, just 5 percent of the vote. Equally important, space on the Left was open in West Germany in a way it was not elsewhere. With a rival socialist state to the east, anti-communism had become a pillar of the country's political identity, which was juridically cemented when the high court outlawed the Communist Party in 1956. Where the 1968 movement in France and Italy found support in Communist parties, in West Germany there was no party left of the Social Democrats. As the SPD migrated to the center, a tide of Left-leaning, educated young men and women found themselves in the political wilderness.[67]

This began to change in 1977 when Carl Beddermann, a young lawyer who lived near a reactor site, founded the Environmental Protection Party. Beddermann had suffered harassment by police at Grohnde, and he hoped turning to local parliament would force politicians to take the new movement seriously. In preparation for regional elections in 1978, Beddermann's party merged with a neighboring Green List for Environmental Protection, gaining support from the Communist League and the BBU. In a surprising success, the new alliance received nearly as many votes as the FDP. But not enough to break into parliament, and Beddermann resigned. Nevertheless, his campaign heralded a new stage in the anti-nuclear movement, with more state elections on the horizon along with the 1979 election for the European Parliament. After a Green list crested 5 percent in Bremen in 1979, prominent leaders like Kelly and Gruhl came together in 1980 to found a national party, *Die Grünen*. In Kelly's famous metaphor, the Greens would have two legs: one kicking in parliament, the other rooted in extra-parliamentary mobilization. In March 1980 the Greens entered the Baden-Württemburg state parliament, and in 1982 they crossed the 5 percent threshold in West Berlin, Lower Saxony, Hamburg, and Hesse.[68]

Social Democrats looked on with trepidation as their Left flank threatened to buckle. Since the late nineteenth century, mass parties had governed Europe by cultivating long-term allegiances. Social Democrats, Christian Democrats, and Liberals all had "cradle-to-grave" participation that linked their programs to the job, class, or religion of their members. By the 1970s, however, this order was decaying and milieus were becoming less rigid following the Economic Miracle. For the SPD, cracks emerged already in 1974 when Brandt's replacement by Schmidt disappointed reformers. Kelly, Traube, Strohm, and others would leave this party of Schmidt.[69]

The civil turbulence of 1977 thus turned the SPD's divisions over energy into an existential rift. Schmidt refused to accommodate the anti-nuclear movement, finding the passion for ecology at best a "whim of bored, middle class women," at worst a distraction from the pressing issue of unemployment.[70] Brandt, by contrast, believed anti-nuclear demonstrators were the "SPD's lost children," and he began calling on his party to become green itself. Eppler's wing now reasserted itself. Eppler, who had resigned from the cabinet, reached out to the new movement and adopted many of their ideas, even installing solar panels on his roof and driving a diesel Passat.[71]

In early 1977 Bonn delayed a new revision to its Energy Program in hopes the SPD could resolve its divisions. But a series of party plenums opened as many doors as they closed. As the party's point man on energy, technology minister Hans Matthöfer entered these debates hoping to broker a compromise. In advance of its Energy Forum in Cologne that April, the SPD tasked Matthöfer with writing guidelines for the discussion. At this point, he still stood under the spell of the coupling paradigm. Drawing on research from traditional experts, he claimed that rising "energy consumption is, to a large extent, pre-programmed," and that quantitative growth was necessary for the SPD's reform agenda. He concluded that West Germany's energy needs would be "hardly conceivable" without the fast-breeder or high-temperature reactor.[72]

Matthöfer's guidelines, however, encountered stiff resistance within the SPD from some unlikely quarters. Three weeks before the Cologne forum, the union leadership declared their disagreement with the government over nuclear power. They would only support more reactors after the courts fully authorized a waste and reprocessing site. To a certain extent, safety motivated their call for more stringent building criteria. Yet the union leadership was more concerned about employment, fearing reactors might not generate as many jobs as new coal plants. So in early 1977 a traditional faction within the SPD temporarily broke with Schmidt and Matthöfer, calling for "only as much nuclear power as is absolutely necessary for the security of our energy supply."[73]

The relationship between energy and jobs thus emerged as a prime issue at Cologne. Eppler forcefully challenged the reigning paradigm, in which a certain employment level required "X percent economic growth and therefore Y percent energy growth." Taking a cue from the anti-nuclear movement, he argued that the SPD should promote "technological advances that could both save energy *and* create new jobs."[74] Here he stumbled onto a recipe that might unify the party: thinking of energy not as a *complement* to labor and capital, but rather as a *substitute*.[75] If economizing on energy actually produced jobs instead of destroying them, Eppler suggested, the SPD could promote "energy savings" as a strategy to both the CIs and the unions. As he put it for his colleagues at Cologne: "It could be that the alternative is not nuclear energy or unemployment, but rather whether we can manage the transition from a purely energy-wasting and job-saving type of technological progress to a form of progress that is, at least partially, energy savings *and* job-creating."[76]

At Cologne, Eppler found an ally in Klaus Michael Meyer-Abich, a physicist and philosopher who had established a Max-Planck-Institute in Starnberg with his mentor, Carl Friedrich von Weizsäcker. After moving to Essen, he organized an interdisciplinary group on energy that worked with Matthöfer's Technology Ministry. Meyer-Abich drew on Lovins to argue that energy demand was not demand for energy in and of itself, but rather demand for the *services* provided by energy. As he pointed out in Cologne, no one "demands oil. Oil stinks, oil is flammable and dangerous; no one actually wants oil." What people want are the heated rooms or the moving cars made possible by petroleum. Instead of trying to produce more energy, West Germany should use its existing supply more efficiently by matching final demand with the type of energy initially produced. Meyer-Abich reminded the SPD that until now Bonn devoted nearly all of its research to energy production. If this changed, he thought, the nation could decouple. Energy savings, in other words, could become a new energy source that was ecologically sound and domestic.[77]

After Cologne these criticisms snowballed. Eppler's wing gained momentum and the SPD rejected Matthöfer's guidelines.[78] But in the meantime, pro-nuclear interest groups organized a campaign that doubled down on fears of an energy gap. Kraftwerk-Union (KWU)—West Germany's largest nuclear firm—warned a moratorium would lead to layoffs and energy shortfalls. New studies reinforced KWU, suggesting the employment effects of a new coal and new nuclear plant were roughly equivalent.[79] In 1977 KWU sent a pro-nuclear petition signed by 35,000 people to Chancellor Schmidt. They followed this with a memorandum that claimed the reactor moratorium would destroy 170,000 jobs. The climax came in November when the leader of the powerful Mining and Energy Union organized pro-nuclear demonstrations in Dortmund.[80]

This lobbying struck a chord, and "the proximate fear of unemployment trumped the distant fear of radioactivity." In early November the unions reversed course after they accepted the arguments of KWU and the government's forecast of an energy gap. The SPD's fall congress saw a tight vote where forty percent of the party came out against nuclear power, far more than Schmidt anticipated. But the yeses were enough and the SPD announced its own U-turn, now in support of nuclear construction.[81]

On the surface, 1977 seemed to vindicate the coupling paradigm. But the debates laid the foundation for more lasting change by revealing widespread discomfort with business as usual. For one, the debates forced the government to modify its Second Revision to the Energy Program. For the first time ever, Bonn's program did not set a specific target for nuclear power, but instead conceded that the atom should only cover the "residual energy demand" that other sources could not meet. Just as important, it ranked energy saving as the highest priority. Bonn now began devoting a far larger share of research to non-nuclear energies, with the federal budget for efficiency research rising over tenfold compared to 1972.[82]

Most importantly, the program implemented practical policies to save energy immediately, accelerating the efficiency transition that was slowly beginning in the wake of the oil shock. Where the petroleum price spike had induced industry in West Germany's liberal, price-responsive economy to use energy more efficiently, the state now added its weight to this trend. After 1977 Bonn's top priority was making the nation's building stock more efficient. Households and small business accounted for nearly half of the nation's final energy demand, most of which went toward heating. Better thermal insulation, piping, heating pumps, and water storage systems, in other words, had great potential to reduce the nation's energy use.[83] In 1976 the government passed a conservation law that went into effect the following year to promote energy efficiency in new buildings through subsidies and investment assistance. The 1977 program made this the centerpiece of policy. Hitherto most houses were poorly insulated and had single-paned windows. But between 1978 and 1982 the new housing modernization program spent 4.35 billion DM to retro-fit nearly 2.5 million homes, 10 percent of the nation's total. Through subsidies that covered a quarter of the cost of any improvements, the law generated over 17 billion DM of investment to improve building efficiency.[84]

The program also reduced the massive volume of heat routinely lost generating electricity, by building power plants that could pump waste heat to warm nearby buildings, and by using industrial heat to generate electricity. After 1977 Bonn devoted 680 million DM in aid, as well as a 7.5 percent subsidy for new investments, into combined district-heating and heat-power co-generation plants. It made the regulated price for electricity, which had favored large consumers with cheaper rates, less degressive. And it combined this with levers to spread information about efficiency, like rebates on advising and requirements that appliance producers label how much electricity their products consumed. All of this was added to a new Keynesian "Future Investment Program" that devoted nearly another billion DM toward the rationalization of energy.[85]

Lastly, the 1977 debates spread Eppler's ideas through the party. Eppler found a new ally in the SPD's rising star, Volker Hauff, who was just thirty-eight when he took over the Technology Ministry in 1978. Hauff had initially supported the coupling paradigm, but in 1977 he began to change his tune. In part he saw an opportunity for the SPD to take the ecological initiative. But he was also responding to a flurry of new studies commissioned by Matthöfer in the preceding years. As the new technology minister, Hauff chaired many of these conferences, which often included Eco-Institute experts, and mounting evidence persuaded him that the assumptions informing energy policy were problematic. By 1978 he came to see "restructuring the energy supply system for the more rational and economical use of energy [as] *the* central political task." Nuclear power he ranked fourth.[86]

Hauff, moreover, had an eye for selling energy savings to a centrist audience, weaving questions of affordability, morality, modernization, and security together with the ecological claims made by the nascent Greens. West Germans were spending 20 billion DM more a year for petroleum than in 1972, Hauff noted. Consuming less oil through more efficient heating was thus the financially correct path. Consuming less oil was also the diplomatically correct path, for it would give "Third World" nations more room to develop and deflate the North-South conflicts.[87] It was the ecologically right path, since oil was now seen to be a "scarce and precious commodity" that strained the environment. Lastly, it was the right technological path, for creating a less oil-dependent economy promised to generate a "boost to innovation" across a range of fronts. The nation must pursue *both* hard and soft paths, Hauff argued, to transform its energy structure, modernize its economy, generate jobs in new sectors, and make the nation more competitive.[88]

Shocks, Security, and Energy Savings

Hauff called his synthesis the "Path off Oil" and it gave Eppler's wing of the SPD a novel strategy to reach out to the anti-nuclear movement while also appeasing conservative elements in their own party. But just as Hauff was formulating his ideas, a double energy shock erupted, reopening the SPD's ecological rift and underscoring yet again how the nation's energy system hinged on developments beyond the control of its leaders.

In January 1979 Mohammad Reza Pahlavi abdicated as the shah of Iran after months of demonstrations by activists and Islamicists. The global oil industry watched in anxiety as guerrilla troops brought the Ayatollah Khomeini to power in Tehran. The rise of a new Islamic Republic sent oil prices soaring and led to panic as buyers scrambled to find alternative crude. Seeing an opening, OPEC doubled its prices. In West Germany, which received nearly a fifth of its oil from Iran, crude prices doubled and gasoline prices surged by 60 percent. This second oil shock underscored, moreover, just how miserably West Germany's supply-side policies of 1973 had failed. The nation still relied as much on foreign oil as in 1972, and now it was again buffeted by sky-rocketing prices.[89]

Then in March, on the other side of the world, on the Susquehanna River in Pennsylvania, the accident that had only been imagined came within a hair of happening. When a valve on the Three Mile Island nuclear plant broke, water from the cooling system streamed out and the reactor overheated, nearly melting the core. It was ten days before the governor could assure locals the crisis had passed, and in the meantime panic spread.[90] Knowledge of Three Mile Island arrived in West Germany just as a small tractor parade of activists was making the trek from Gorleben to the capital of Lower Saxony. When the news broke, the protest surged and over 100,000 people marched to Hannover to protest the government's nuclear disposal proposal. They arrived just as public hearings over Gorleben were beginning, prompting commentators to remark that "the dear Lord is clearly on the side of the nuclear opponents."[91]

Facing these challenges, Schmidt retreated to traditional arguments about energy. At the European Nuclear Congress that year he demanded the revival of West Germany's atomic program, arguing that the atom was still the "foundation of modern industry with a large number of future-oriented jobs." If West Germany left the atomic race now, it would be impossible to return given the breakneck speed of development. Promising to massively expand funding for nuclear safety, Schmidt rammed through another revision to the energy program in 1981 over the heads of his own party. Allying with FDP leaders in his coalition, he entrenched nuclear power at the heart of this Third Revision and cleared a path to begin construction on new light water reactors. The program bypassed the problem of disposal by authorizing more onsite storage. Work had already resumed at Grohnde and Brokdorf. With the 1981 program, construction now began on a convoy of new reactors.[92]

This heavy-handed approach, however, alienated Schmidt from his party. Eppler's faction responded to the twin shocks by leveraging fears of energy insecurity in an entirely different direction. Hauff used the Iranian Revolution to play up the "potential for distributional conflicts over oil in the 1980s as the greatest danger to international peace," adding urgency to his call for energy savings. With the failure of Germany's supply-side energy policies, "rolling back" West Germany's use of oil was more important than ever, and energy savings emerged as the silver bullet: "the most important energy source of the future and the politically central task for the coming two decades."[93] Taking ideas directly from the anti-nuclear movement, Hauff argued the nation must now, "[l]eave the 'hard' path of expensive, inhumane, and socially retrograde large-scale technology, of centralized bureaucracy, technocracy, investment, and power and instead develop 'soft' technologies adapted to ecological and human conditions."[94]

Hauff's ideas caught fire at the SPD Congress in late 1979. After debating energy for fifteen hours, the SPD narrowly voted to keep the atomic path open, under intense pressure from Schmidt. But the rest of the program followed Hauff's agenda.[95] The Congress concluded that 1979 only made nuclear power seem more "problematic," the global oil market more "uncertain," and the importance of energy savings "more urgent." So urgent, in fact, the party itself formally called energy savings "the most important energy source for the near and medium-term future." They added another layer to their 1977 efficiency program, including a minimum level of insulation for buildings; a requirement that public offices be a model of energy efficiency; a depreciation allowance for investments that reduced energy used by households; and more transparency for renters about energy in their leases. The federal post office even began offering energy saving stamps to advertise the new agenda (See Figure 6.3).[96]

But the main hurdle to energy savings remained: the question of employment. This had torpedoed Eppler's ideas in 1973, and now it reared its head again as the Federal Republic entered the worst jobs recession in its history. Unemployment struck at the SPD's identity as a party of the working class, provoking a moment of deep reflection. In a public speech in 1981, former chancellor Brandt argued that Social Democrats must make room for other progressive groups lest the party lose the youth vote entirely. His remarks sparked a heated rebuttal from Schmidt's supporters, who argued the party was a product of industrial society and must remain first and foremost an advocate for workers.[97]

But times were changing, and by 1981 so too were the attitudes of the unions. In 1977 they had feared ecological reform would cost jobs. But after 1979 many unions, particularly smaller ones, realized the environmental movement was here to stay. Where many labor leaders had once seen a class divide between environmentalists and their own constituency, heightened nuclear fears after Three Mile Island and anxieties about pollution led many to find common ground with ecologists. Meanwhile, the consensus that ecological protection would cost money and siphon off investment needed for jobs was deteriorating under the weight of new evidence. According to studies, during the 1970s environmental protection and energy savings investments had created between

Figure 6.3. Federal energy saving stamp, 1979. *Source:* Borislav Marinic / Alamy Stock Photo. Image ID: 2BJXN17.

250,000 and 360,000 new jobs and were growing faster than the nation's traditional industries.[98]

From a political standpoint, the key was reimagining energy savings less as a cost than as an investment that could generate rewarding employment. As chair of the Commission for Environmental Questions and Ecology, Hauff began pushing this line in the SPD. An energy-savings transition would generate quality jobs like putting filters on coal plants, retrofitting houses, or building local pipe networks for heat.[99] At the party's Berlin congress in 1982 Eppler pulled these ideas together, arguing that the right sort of growth could forge an "alliance between the classical workers movement and the new social movements." Energy savings would be a lynchpin, for "it is simpler, cheaper, less dangerous, less intricate, above all more environmentally friendly and it creates more jobs to curtail energy consumption rather than exploiting new sources of energy. It is as simple as that."[100]

In the context of high unemployment, these ideas won converts. The German trade union association made savings their top energy priority. Matthöfer, now minister for finance, completed his exodus from Schmidt's camp and proposed a bold new fiscal program to implement these ideas. He wanted to stimulate energy-saving technology by taxing oil, and using the proceeds to kick-start a massive energy efficiency investment program to forge a new stratum of skilled, labor-intensive jobs for the nation's *Mittelstand*. The program, he hoped, would give Germans an edge in a new global export market of energy saving and environmental technologies.[101]

Conclusion

Matthöfer's oil tax ultimately floundered in the face of resistance from the conservative wing of the SPD, the FDP, the central bank, and traditional industries. Oil prices were too high, opponents argued, raising them further would crush any recovery. Market forces had worked well enough so far, they maintained, why tinker with them?

But before Matthöfer and Hauff could regroup, the SPD was rocked to its core. Mounting Cold War tensions, including a new arms race that would bring American medium-range nuclear missiles to West Germany, sundered the SPD-FDP coalition. Schmidt joined the FDP in pushing an aggressive stance toward the Soviet Union, alienating many in his own party. The fault lines followed those etched by the energy debates, with Eppler condemning Schmidt's nuclear agenda and endorsing what became the largest protest movement in the nation's history—the Peace marches of 1981 and 1982. These drew together young and old, liberals and Marxists, ecologists and feminists, Social Democrats and Greens, as hundreds of thousands of people demonstrated in opposition to the fact that their nation now had the largest concentration of nuclear weapons per square mile in the world. The deepening recession, meanwhile, in which unemployment approached two million, led many in the SPD to demand a huge deficit spending program. For the FDP this proved the last straw, leading them to join the CDU in a vote of no confidence on October 1, 1982.[102]

So ended the thirteen-year reign of the SPD. Out of power, the party began a soul-searching reform in which it placed energy savings and ecology at the heart of a new agenda. As older leaders like Schmidt lost influence, a younger generation assumed key positions and Hauff and Eppler's views spread. By 1982 a majority favored an exit from nuclear power, and by the mid-1980s the party embraced the soft pathway. Hauff's Program for the Ecological Modernization of the Economy in 1985, which he crafted in cooperation with anti-nuclear luminaries like Klaus Traube, gave "unconditional priority to energy savings," eschewed nuclear power, and demanded the "creation of an energy-economic infrastructure that mixed centralized, large plants with decentralized, small plants."[103] As the party declared in its new framework: "the old thesis, that economic growth can only go hand-in-hand with high energy consumption, lost its validity a long time ago."[104]

The visceral debate over nuclear energy and the rise of the Green movement profoundly changed West Germany's approach to energy. The country pivoted because the nation's lead party was forced to confront a new political rival by repackaging Green ideas for a centrist audience. But for the SPD energy reform need not involve the destruction of capitalism. Instead, Eppler, Hauff, and Matthöfer tied ecology to long-standing concerns like security, exports, and modernization. By highlighting how energy saving could generate new jobs in skilled sectors, help firms carve out new niches in the global market, and counter not only ecological degradation but also dependency on foreign oil, the reform wing of the SPD honed arguments that politicians would later use to drive through groundbreaking legislation.

This entire process, moreover, set West Germany on a different trajectory than the United States, Great Britain, and above all France. Energy consumption in these three countries continued its ascent after 1985. The latter became the model for an atomic economy as it built more reactors and supported the full nuclear fuel cycle. By the 1990s France boasted more reactors per square mile than any other country on earth, and these reactors generated 80 percent of the nation's electricity. The country's energy giant EDF had immense overcapacity, giving French consumers some of the lowest electricity prices in Europe.[105]

The Federal Republic, by contrast, began to decouple: total energy use fell even as the economy grew. The country's efficiency transition and its idiosyncratic path after 1980 resulted not only from the oil price shock, but also from policy. Experts estimated the SPD's efficiency programs reduced energy consumption by 3.5 percent. Over the coming decade, energy use per person among households and small businesses fell significantly even as the national housing stock expanded and average dwelling size grew. The vast demand for electricity that experts predicted never materialized; neither did the need for a nuclear economy. And so the Federal Republic passed the zenith of its atomic program. In 1981 Schmidt launched what proved to be the final round of reactor construction. The last light water reactor was commissioned in 1982, the last one completed in 1987.[106]

Most importantly, decoupling now seemed to be a realizable political goal, one that might benefit the economy instead of harming it. And not only Green activists and reform Social Democrats advocated this agenda. In the 1980s a new cohort of economists would emerge who supported these goals with an entirely new intellectual framework. In doing so, they diverged radically from experts in the United States in imagining both the commercial and the political possibility of a green energy transition.

7

Reinventing Energy Economics after the Oil Shock

The Rise of Ecological Modernization[*]

> One factor more than anything else was decisive for economic growth in the postwar period: the triumphant advance of petroleum. The economic and technical qualities of this energy were unique: it sprung forth from the ground abundantly and free of charge, so to speak; it could be transported everywhere through huge tankers and pipelines; and it could be stored with relative ease. . . . But this era of supplying the economy with cheap natural services in the form of abundant and inexpensive energy and other resources has come to an end.[1]

In the late 1970s and 1980s a new language of economics burst onto the scene in West Germany. Since the Federal Republic's inception, the creed of Ordoliberalism had informed the way most politicians spoke of their nation's economic challenges. In 1936 Wilhelm Röpke, an Ordoliberal who would attain international fame, had called for a "free price system" in which "it is the consumers who decide what and how much shall be produced. Hence, it is the *consumers* who decide how the factors of production themselves are to be used." After World War II, consumer freedom and market prices became a bedrock of West Germany's economic language.[2] Advocates of the energy coupling paradigm added new layers that departed from Ordoliberal orthodoxy, above all forecasting and the idea that energy should be understood as a special part of the economy. Nevertheless, most of the nation's leaders claimed that growth, employment, and social stability could best be achieved through free prices,

---

[*] A more detailed version of this chapter, with an international focus, appeared first in the *Journal of Modern History* in March, 2023. I am grateful to the JMH for permitting the publication of an adaptation here as Chapter 7.

competition, and consumer autonomy. Alfred Müller-Armack—Ordoliberal and state official—captured this sentiment with his catchphrase of the social market economy, and by the 1960s all political parties cast their goals through this framework. The social market economy was "an economic and social political program for everyone," the CDU proclaimed, in which "competition is the most effective instrument for steering the economy."[3] After overhauling their program in 1959, the Social Democrats agreed, affirming the power of "the free market, wherever there is true competition."[4]

The spiraling ecological degradation of West Germany's air and water, however, presented a paradox that neither Ordoliberalism nor the coupling paradigm could firmly grasp: how, if at all, could growth continue if it damaged the environmental and social fabric of the nation? Müller-Armack began to explore this question during the "Second Phase" of the social market economy in the 1960s, once the urgency of postwar reconstruction gave way to post-material concerns. But he was unusual among Ordoliberals, who exhibited a striking neglect of energy in their namesake journal and leading institutions.[5]

As the shortcomings of this older economic framework and the coupling paradigm became apparent, a new language arose to address the concerns of the anti-nuclear movement and a younger generation of West Germans coming of age. New experts began to speak of market failures, of the environmental cost of energy production and growth, of externalities to the very price system itself. Ironically, the currents feeding into this new language originated from neoclassical economists in North America, who were inspired by the last great pre-Keynesian economist, Arthur Cecil Pigou, to explore how markets failed to account for a range of environmental costs created by capitalism itself. While this work began in the 1960s, it accelerated dramatically with the Club of Rome's report on the *Limits to Growth* and the oil crisis, tremors that shook the economics discipline from its apathy toward ecology.

In the United States, however, the most creative findings of three subfields of economics—resource economics, environmental economics, and applied statistics—never coalesced into a coherent paradigm. Following Ronald Reagan's election in 1980, a new philosophy of market governance gained political favor and profoundly shaped energy policy there: neoliberalism. While American neoliberals bore similarities and sustained contacts with German Ordoliberals, they were more radical in seeking to extend market logic to all walks of life. In the United States, neoliberalism took resource economics to new heights of technological and market optimism. Conventional American neoclassical economists, meanwhile, accepted that markets could fail, but they remained fixated on static questions of efficiency, neglecting dynamic theories of technology and mechanisms that could reshape consumer behavior.[6]

Ideas generated in North America found a more receptive seedbed in the urban corridor of Europe stretching from Denmark down to Northern Italy, but above all in the Federal Republic. In the 1980s heterodox experts working in West Germany—the Swiss-German Hans Christoph Binswanger, Lutz Wicke, Hans Karl Schneider, Bertram Schefold, and Udo Ernst Simonis—pioneered a new energy paradigm of Ecological Modernization.[7] These economists broke with German Ordoliberalism. They came from an intellectual tradition that, in contrast to the American profession, was less beholden to abstract theory and deeply empiricist and historical. And they operated in a strikingly different political context than Reagan's America, one that included a powerful anti-nuclear movement, a reform wing in the governing party, and an ecological crisis that gripped the nation: the looming death of Germany's prized forests.

The oil shock led these experts first to question the coupling paradigm. But they quickly pushed their ideas further, weaving together new developments from American economics to justify an activist state that could redirect the economy to achieve specific environmental goals. In the process, they forged a novel way of thinking about energy that differed from Keynesian, neoclassical, Ordoliberal, and neoliberal economics. Like the followers of John Maynard Keynes, these German experts believed in the power of state-led investment. But they condemned Keynesians for exalting quantitative growth and paying little attention to how material flowed through an economy. Like Pigouvian neoclassicals, they believed markets could fail systemically. But where Americans homed in on efficiency, German experts aspired to dynamically restructure the economy and guide consumer behavior. Like Ordoliberals, they believed in competition and private initiative and they despised economic concentration, the *bête noir* of Röpke and Müller-Armack. But they believed the price system must be steered with a heavy state hand. And where neoliberals wanted to aggressively insulate markets from democratic politics, West Germany's new experts embraced the politicization of energy markets and prices.[8]

After 1980, in the United States and other parts of the world, neoliberalism would triumph over Keynesianism. There seemed to be few alternative visions to rival this new creed, few economists who opposed the "trust-in-markets revolution." But some did. In West Germany, Ecological Modernization would resonate across the political spectrum, providing a language that could bind together unexpected political alliances and enable the nation to diverge from the United States in matters of energy. Ultimately, this new paradigm would keep open the intellectual space for the Federal Republic to embark on a groundbreaking reform with the potential to launch a new, green energy revolution.[9]

New Impulses from North America

In late 1973 Robert M. Solow, professor at the Massachusetts Institute of Technology and one of America's leading neoclassical economists, gave the Richard T. Ely lecture to a packed hall of his peers at the profession's premier conference. He was taking stock of the mounting interest in resources, a topic where nearly every week it seemed a new paper appeared. After being "suckered" into reading the Club of Rome's report on the *Limits to Growth* and following the drama of oil prices in the news, Solow had directed his gaze to the exhaustibility of raw materials. Thinking himself clever enough to have rehabilitated a dusty set of questions, he found he was not alone: "It was a little like trotting down to the sea, minding your own business like any nice independent rat, and then looking around and suddenly discovering that you're a lemming."[10]

The twin events of the *Limits to Growth* and the oil shock worked like a magnet on economists, drawing them to the study of resources, which since the 1930s had been a neglected branch of the discipline. For Anglo-American neoclassical economists, one could read "the whole of the very extensive literature of the 1950s and 1960s," noted two prominent practitioners, "without ever realizing that the availability of natural resources ... might be a determinant of growth." In the 1970s this changed, as the profession turned the full weight of its analysis onto what seemed to be the most pressing paradox of the day: did the world have enough oil, coal, and natural gas to continue its incredible trajectory of energy-intensive growth?[11]

Like so much else in their discipline, the new resource models built by American economists were thoroughly neoclassical, focusing on the question of efficiency and exploring how societies allocate scarce resources among competing ends. By the 1950s and 1960s neoclassical economists used mathematical proofs to show how, in theory, competitive markets operating with free prices provide an *optimal* allocation of resources.[12]

As a subdiscipline of the profession, resource economics was heavily neoclassical, expanding the concept of equilibrium to look at how societies allocate resources between the present and the future. Do existing markets, its practitioners asked, efficiently balance the current uses of a raw material with future needs? Are resource markets in intergenerational equilibrium?[13] In 1931 Harold Hotelling had first answered these questions by suggesting the value of a natural resource in the ground grew at the real interest rate, and that this market-determined pace of depletion optimally balanced the needs of the present with those of the future.[14]

Hotelling's Rule informed the first forays into exhaustible resources in the 1970s by Solow and William Nordhaus of Yale. In 1973 Nordhaus constructed what became the seminal neoclassical model for understanding the long-term allocation of energy that redefined complex social issues into a question of maximizing the current value of a resource. And he built his model on several assumptions that would come under fire in Europe. For one, Nordhaus believed that energy prices have "for the most part been determined by market forces," and would continue to be so in the future. From the beginning, he saw energy as an economic market, not a political one.[15]

Second, because the nature of the problem was so long-term—"the duration of man's habitation on the planet"—Nordhaus introduced a new concept called the "backstop technology." When the price of an exhaustible resource rose high enough, he assumed a new technology would be activated that could provide a "virtually infinite resource base," thus giving an endpoint to his intertemporal model by assuring an ultimate solution. For Nordhaus, the breakthrough technology par excellence was the fast breeder, which he predicted would come to dominate the energy supply by 2020. His model, in other words, exuded optimism in technology and, in his opinion, illustrated "the inevitable transition from exhaustible fossil fuels to nuclear fuels." In defiance of the Club of Rome, he concluded that "we should not be haunted by the specter of the affluent society grinding to a halt for lack of energy resources."[16]

But third, while Nordhaus was optimistic about technology, he was less so about markets, believing they were partially flawed because of a lack of futures and insurance markets for commodities like oil. In reality, producers and consumers could neither buy nor sell petroleum for delivery at any point in the future, and bore risks associated with many issues, not just exhaustibility. So while Nordhaus was hopeful for a backstop technology, he feared markets might not get humanity there if left to themselves.[17] But the absence of futures or insurance markets was a rather tepid market failure to focus on, and it led American neoclassicals to propose a mild recipe for state action—mild because they believed governments had their own problems, which made it, in Solow's words, "far from clear that the political process can be relied on to be more future-oriented than your average corporation." Solow and Nordhaus thought governments should merely perform more and better forecasts, pay closer attention to price stability, and devote more funding to research.[18]

Elsewhere in the profession, American neoclassicals were applying their tools to the increasingly visible problem of pollution. Since the 1960s a new cohort of environmental economists had been trying to understand this particular type of market failure by grounding their work in the ideas of Arthur Pigou. In *The Economics of Welfare* (1920), Pigou had suggested that markets often suffer from externalities, which he defined as a cost or benefit generated by a transaction between two parties that affected a third party uninvolved in the transaction, but that were not included in the actual price of whatever was being exchanged. Put differently, an externality was the divergence between the private benefit or cost of a transaction and the overall social benefit or cost, and it represented one of the prime ways that markets failed.[19]

American environmental economists made Pigou's notion of market failures their guiding concepts, defining pollution as fundamentally "an economic problem." Here the work of Allen Kneese is emblematic. Kneese had built his reputation studying water systems in the United States and West Germany, and in 1977 he put his framework into a popular textbook, *Economics and the Environment*.[20] Like most Americans, Kneese believed that "real economic functions are more or less accurately simulated in market-type systems." If, that is, the practitioner was willing to include market failures. Through his empirical studies, Kneese became convinced that externalities were "not freakish random events . . ." but "a systematic part of the economic development process." This was so above all for the external costs associated with the detritus of industrial life. Too much of this flowed into the "common property": things that no single individual owned but that belonged to society as a whole. The social costs of this pollution were consequently not borne by the companies producing it. While Kneese believed pollution could be managed through market mechanisms, the government must first step in to act "as an agent for the public" and levy a price on the "destructive uses of the common property resource."[21] Kneese consequently called for an environmental tax to internalize the external costs of pollution. But he did not grapple with intertemporal issues and he largely ignored technological change, instead focusing on minimizing the costs of pollution abatement given the *existing* range of possibilities.[22]

At the same time, interest in energy was advancing along a third axis that was removed from high neoclassical theory. In 1971 America's Ford Foundation inaugurated a project to produce a coherent energy policy. The foundation commissioned the statisticians Edward A. Hudson and Dale W. Jorgenson of Data Resources Incorporated to develop a framework to understand how price changes shaped technological development.[23] Hudson and Jorgenson created a dynamic new model that changed the flow of goods through the economy as prices changed. As the price of one input rose, their model predicted that industries would shift to other inputs that were now relatively less expensive.[24]

The implications of their research were striking, for it suggested states themselves could manipulate the price of an energy to stimulate technological and structural transformation in the economy. Hudson and Jorgenson thus began pushing the notion of an energy tax for completely different reasons than Kneese: they wanted one not to achieve an abstract sense of efficiency, but to reshape production processes. And their work shifted attention away from the *supply* of energy to the *demand* for it.

Neoliberalism and the Reagan Revolution

During the 1970s American economists thus developed three distinct ways to understand energy: intertemporal resource economics; environmental economics; and price induced-technological change. But after 1980 these strands never converged into a coherent philosophy of governance. This stemmed partly from the formalism increasingly prized by the American profession. For neoclassicals, building theory became more important than empirical work, and younger economists became enthralled by "abstract matters that seem more and more remote from the real-world problems that overlap economics and other disciplines."[25]

This did not bode well for those who studied technology. In the 1950s there had been lively interest in price-induced innovation when John Hicks suggested that changes in the relative price of land, labor, or capital could act as a "spur to invention" by encouraging firms to economize on inputs that are more expensive.[26] But during the 1960s and 1970s interest evaporated in Hicks's ideas, for in the neoclassical world to say a factor is expensive is to also say it is productive, and why would firms economize on something productive? In general, neoclassical theory continued to treat technology as a "black box," and by the 1980s economists who studied energy let technology fall out of their analysis almost entirely.[27] Instead, they strove to find the most cost-efficient way of fighting pollution given existing technical choices. Nordhaus, who served on President Carter's Council of Economic Advisors, criticized price-induced technological development. His 1982 essay on grazing the global commons suggested pricing carbon through a tax *not* to stimulate technical change, but to minimize the cost of reducing emissions. His model was, in his own words, a "pure exercise in optimal economic growth": static, neoclassical analysis at its finest.[28]

If Americans' fascination with efficiency hindered the rise of a vigorous theory of state action, the neoliberal movement that began to sweep the nation only made matters worse. After a string of crises gripped the Atlantic economies in the 1970s—the oil shock, rising unemployment, exploding inflation—the state's ability to manage the economy lost credibility. So too did the ideals that had justified policy since John Maynard Keynes wrote the *General Theory of Employment, Interest, and Money* at the depth of the Great Depression. Keynesianism had grown from neoclassical analysis, and many leading postwar American neoclassicals followed Keynes in believing the state should aggressively deploy discretionary fiscal and monetary tools to boost consumption and investment during recessions, in order to return the economy to a path of quantitative growth.[29]

The malaise of the 1970s unleashed a bitter feud between old guard Keynesians and the rising creed of neoliberalism. Like Keynesians, many neoliberals drew on neoclassical analysis. But they believed theory basically corresponded to reality, and that any problems with markets came from states, unions, or other institutions interfering with the price system; never from markets themselves. Neoliberals wanted states to aggressively insulate markets from democratic politics, for they believed majoritarian political parties or particular interests would always strive to bend the economy in their favor. They sought to protect private capital, elevate efficiency and profit over other goals, and prevent politicians from making decisions that altered the free price system. More specifically, they believed that taxes hampered initiative, that private actors would anticipate state action and render it ineffective, that government officials could never have enough information to correct free prices, and that markets were fundamentally self-correcting. At the global level, neoliberals militantly defended the free flow of capital and goods across borders, and tried to dismantle obstacles that favored national companies. And where problems did arise, they were best solved with more markets. These core ideas grew out of a collective of thinkers that included Milton Friedman and Ronald Coase at the University of Chicago. Yet other variants professed even more positive views of the market, like the Austrian school inspired by Friedrich Hayek. For these philosophers, not even the neoclassical gurus of Chicago could capture the sublime power of the market, which they claimed was beyond human reason to predict or guide.[30]

American neoliberals differed from German Ordoliberals in various ways. Both valorized profits, competition, and market prices. But while the latter feared economic concentration would subvert the political order, the former were quite willing to tolerate oligopolies and massive enterprises. The University of Chicago developed a distinctive anti-trust doctrine, which held the size of a company to be irrelevant so long as it was efficient. Coase, for instance, argued that there were good reasons for firms to grow large, while Friedman thought worries about concentration were exaggerated. When confronted with the choice between regulation and private monopoly, the latter was the "least of the evils." Neoliberals, moreover, were more radical than early Ordoliberals in seeking to extend market logic to all spheres of life, from healthcare to the environment. Röpke and the grandfather Ordoliberal Walter Eucken, by contrast, believed a conservative, local, and family-oriented ethical life must underpin the market. Large enterprises and the technological change they wrought, moreover, threatened "overexploitation" or "loss" of the countryside, and for many Ordoliberals this warranted state protection.[31]

In the United States, Ronald Reagan's landslide electoral victory transformed neoliberalism into a nascent political order and accelerated its spread into energy economics. The former governor of California exuded optimism in technology and markets, called on the government to "get out of the way" and "set the oil industry loose," and touted the limitlessness of American innovation.[32] This enthusiasm found a scholarly foundation in the work of Julian Simon at the University of Illinois and later at the University of Maryland. A graduate of Chicago, Simon was inspired by Friedman and Hayek, but skeptical of neoclassical modeling. In a provocative essay in 1980 that challenged the *Limits to Growth*, Simon argued that resources were actually getting *more* plentiful because of innovation.[33] In his 1981 magnum opus—*The Ultimate Resource*—he turned this technological and market optimism into a doctrine of infinite substitutability. In Simon's hands, scarcity became purely relative, defined entirely by whether the cost of a resource rose over time. He claimed that technological progress "more than made up for the exhaustion of the more accessible" resources, and that the very pace of technological development was accelerating. Nothing less than the "total weight of the universe" was a limit when it came to resources.[34]

Simon believed, however, that only unfettered markets could achieve such infinite substitution. Shortages would create their own remedy, because advances "are not just luck . . . they happen in response to scarcity." States could never guide technological development because human ingenuity would always look beyond the "accustomed boundaries" of a system for replacements. Drawing on Hayek, Simon concluded that *existing* prices were the best indicator of the future, since forecasters and government officials had no skin in the game, in contrast to the "professionals who spend their lives studying commodities and who stake their wealth and income on being right about the future."[35]

At the same time, many American environmental economists were taking their own neoliberal turn. In 1960 Coase had published a now famous essay on the "Problem of Social Cost," which took aim at externalities. Though he did not question their existence, he argued they could be overcome with little recourse to the state. And he reframed externalities from a unidirectional into a reciprocal problem. Where factory pollution was killing fish, the question was not how to limit pollution, but whether "the value of the fish lost [is] greater or less than the value of the product which the contamination of the stream makes possible." In a perfect market, factory and fisherman could strike a deal where one compensated the other. Beyond defining property rights, for Coase state action was hardly necessary.[36]

If Coase showed that externalities need not be solved by government intervention, soon the very notion of an externality was coming under fire from neoliberals. Some argued that information problems plagued not just central planning, but states' ability to calculate the optimal level of an environmental tax. "We may not even know in which direction to modify the level of an externality-generating activity," an influential overview of the field concluded. These criticisms quickly snowballed, and neoliberals began arguing that externality itself was a "vacuous and unhelpful" concept, that "any general prescription of a tax to deal with external diseconomies is useless." Pollution, they argued, should be controlled by creating new markets where there had never been any before.[37] From here it was one more step to suggest trading in the rights to pollute, where states would manage emissions by creating brand new property rights. First, a state set the maximum level for a particular emission. Next, it assigned organizations the right to emit a share of that level. Finally, it let these organizations trade rights among themselves, so those that could reduce emissions more cheaply could sell their rights to others.[38]

In America, by the late 1980s and early 1990s emissions trading triumphed over alternative frameworks for solving environmental problems, including both regulations and ecological taxes. Emissions trading brought neoliberals together with conventional neoclassical environmental economists to offer a politically feasible path to tackle the most important environmental challenge of the day, acid rain. For the neoclassicals advising President George H. W. Bush on environmental problems, "*Cost effectiveness* [was] the primary focus of economists when evaluating public policies." Like Nordhaus, they kept faith that neoclassical models could accurately price externalities.[39]

After the end of the Cold War in 1989, a moment when it seemed free market capitalism might conquer the globe, advocacy by American neoclassicals economists for emissions trading overlapped with neoliberal calls to forge new markets. The centerpiece was an amendment to the Clean Air Act of 1990 that created a national emissions trading program to reduce sulfur dioxide generated by the electricity industry. Pointing to the "new spirit" of free markets emanating from the collapse of Communism, neoliberals portrayed emissions trading as nothing less than "environmental perestroika." And while neoliberals promoted trading because of its efficiency, they went further than neoclassicals by touting the ability of trading to create new markets.[40] For them, emissions trading bested regulations and ecological taxes precisely because it "allowed the *market* to value the property rights." Through the first decade of the twenty-first century, the United States would follow this model and become the leading advocate of using carbon trading to tackle global warming.[41]

Noteworthy, however, was what was absent from this agenda. American economists were supremely concerned about finding the most efficient abatement techniques for *producers*, by using market incentives to encourage firms to reduce waste or reprocess. But they had little to say about changing *consumer* behavior. Nor did conventional neoclassicals and neoliberals attend to how policy might direct technology in a particular direction. A hallmark of emissions trading was its very disregard for the type of technology deployed: techniques that cut pollution in the short term received equal weight as those that might sustain a long-term pathway off fossil fuels. For neoliberals like Simon who *did* focus on technology, meanwhile, the trajectory of innovation could patently *not* be steered by the state, but must be left to the wonders of the market.

Across the Atlantic

While American economists were offering solutions to their nation's energy woes that were either neoliberal in their market optimism or supremely focused on efficiency, West German energy economists took a different turn after 1980. There a new energy paradigm arose to undermine the old coupling paradigm that had informed policy since the 1950s, propelled by young economists sympathetic to the environmental movement and to Social Democracy.

The first cracks in the coupling paradigm, however, came from experts who had crafted the older framework in the first place, and from concerns about *security*, not scarcity. Two years before the 1973 oil shock, the Economics Ministry commissioned the nation's leading energy institutes to simulate how the economy might react to an international oil crisis. Lacking the necessary empirical data, the DIW in Berlin, together with the EWI in Cologne, polled industrial groups about the possibility of substituting one energy for another on the basis of discernible technological trends. Their results were surprising, for they challenged the notion that demand for oil and for energy were utterly inelastic, as the coupling paradigm maintained. The DIW and EWI estimated that during an emergency the nation could reduce its total energy consumption by 2 percent and its industries and consumers could find substitutes for 12 percent of their oil supply without having any economic side effect whatsoever.[42]

These were no small figures, and they suggested that the iron logic of the coupling paradigm may not be so firm after all. But at first these findings had little impact on policy because there was as yet no sustained empirical work on the relationship between price changes, energy consumption, and growth.[43] The oil crisis, however, delivered a shock to the profession as well as to the economy, and following the dramatic rise in petroleum prices in 1973, evidence mounted that gave economists a basis on which to question the axioms of the old paradigm, namely, that to grow an economy must have a cheap and growing supply of energy. For over twenty years the average price of motor fuel in West Germany had hardly budged, then after 1973 it exploded by nearly 50 percent in just four years.[44] But aside from a mild recession in 1974, the economy kept slowly expanding. Thus in 1977 when Berndt Lehbert at the Kiel Institute for the World Economy published the first econometric study on the oil shock, he could show that firms were proving quite able to switch from one energy to another when the relative price levels between them changed. In economists' language, fuel oil, gasoline, and coal seemed to have much higher price elasticities than previously thought.[45]

These cracks in the coupling paradigm may not have widened, however, without the hypertense atmosphere of the late 1970s. As the anti-nuclear protests reached a climax in 1977, and the governing SPD spiraled into a visceral debate over energy, politicians and the public began to realize just how little the coupling paradigm had to say about the tectonic changes erupting in energy markets, and how little Ordoliberalism had to say about crises. It was clear to everyone that the forecasts guiding policy had gone disastrously wrong (see Figure 7.1) None other than Hans Matthöfer, the SPD's minister of technology and one of the most influential energy decision-makers in the country, admitted that these "errant prognoses" illustrated just how much "the relationship between economic growth, energy demand, and jobs must be more closely and methodically analyzed than before."[46]

Beginning in 1977 the Ministry of Research and Technology commissioned a slew of energy studies that brought together the nation's leading economists. The results of this research blew a hole in the methods hitherto used by experts.

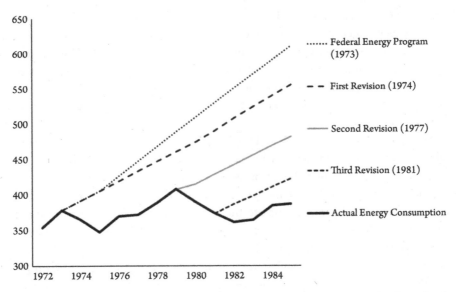

Figure 7.1. Comparison of actual and predicted total energy consumption by the Federal Republic's Energy Program and three revisions (1973, 1974, 1977, and 1981). *Source:* BSB, "Die Energiepolitik der Bundesregierung," DS 7/1057, October 3, 1973; BSB, "Erste Fortschreibung des Energieprogramms der Bundesregierung," DS 7/2713, October 30, 1974; BSB, "Zweite Fortschreibung des Energieprogramms der Bundesregierung," DS 8/1357, December 19, 1977; BSB, "Dritte Fortschreibung des Energieprogramms der Bundesregierung," DS 9/983; AEG Energiebilanzen e.V. "Primärenergieverbrauch–alte Bundesländer."

During the 1960s cheap oil and the steady rise in energy use had led forecasters to drop any pretense of including price changes in their models, which naturalized the demand for energy as purely dependent on the rate of growth in the overall economy. This led to astounding predictions for future energy consumption—no less than a doubling of the nation's current level by 2000.[47] But after 1977 new studies put price and elasticity back into the picture. They foregrounded Lehbert's idea that price changes could shape the demand for oil, and they began drawing directly on Hudson and Jorgenson's research to push Lehbert's findings further. Indeed, the American model of price-induced technological change became one of the most widely cited examples of how to update West Germany's own policy. In two widely read articles, Robert Dickler, an economic historian from the University of Pennsylvania working in Bremen, roundly criticized Bonn's forecasts for being static and grossly overestimating energy demand because they paid no heed to prices. Dickler concluded there was no looming energy gap in West Germany, rather a "research gap" with America where experts were developing better methods for mapping demand.[48]

After 1977 new empirical research showed not only that the price elasticity of various fuels was "quite large," but that the elasticity of energy with respect to GDP—how much energy was needed to produce a unit of growth—was far from fixed and had actually varied dramatically during the 1950s and 1960s.[49] Other data-based studies demonstrated that labor could substitute for energy, including research commissioned by the Interior Ministry that showed how investments into energy saving or environmental protection were creating far more jobs than they eliminated.[50] All of this shook the conventional wisdom. As an expert from the nation's leading energy institute in Cologne concluded in 1979, "the thesis of an historically observable close coupling between economic growth and energy consumption—much less the thesis of an energy elasticity of one—is simply not supported by the available data." Not only might the expansion of energy consumption *not* be essential for growth, but efforts to economize on energy might actually produce jobs instead of destroying them. The very "production structure" of the economy now seemed malleable, such that "energy price increases could be a market-based instrument for reducing energy consumption."[51]

Neoclassical Theory in Question

Attention to energy contributed to what was becoming a crisis of confidence among German-speaking economists. As in the United States, West Germany faced issues that challenged the profession: high inflation, rising unemployment, and growing state deficits. West German economics likewise became divided between Keynesians advising the SPD-government, old-school Ordoliberals, and a surging cohort of neoclassicals who advanced policies akin to, albeit less drastic than, neoliberals in America. By the early 1980s Bonn's Council of Economic Experts, once a bastion of consensus, was riven by discord. But beneath this division there lay even more fundamental doubts, that the very goals of Keynesianism, Ordoliberalism, and neoclassical economics were problematic. For the anti-nuclear movement had shone a spotlight on just how much West Germany's growth had come at the expense of the environment.[52]

Many economists thus began arguing that their profession must change to deal with a new era of ecological crises. By the end of the 1970s these reflections reached a crescendo, encapsulated by the NAWU report (*Neue Analysen für Wachstum und Umwelt*—New Analyses for Growth and the Environment)—an influential study organized by the Swiss-German economist Hans-Christoph Binswanger working at the University of St. Gallen. In this 1979 publication, Binswanger outlined the dilemma that now bedeviled economic policy:

> Continuous economic growth will inevitably create an environmental crisis. If the wheels of the economy were to run more slowly, however, then we would suffer the bitter consequences of industrial paralysis with its unemployment and state debt. ... What do we do if both views are right, if in fact only a growing economy is functional, but this growing economy destroys the environment? Under these conditions we only have the choice between two evils: the bankruptcy of the economy or the bankruptcy of the environment.[53]

Neither Keynesians nor neoclassicals could solve this paradox, Binswanger and his allies maintained, because they remained wedded to quantitative growth and left Nature out of their theories. Since the birth of neoclassical analysis, Binswanger pointed out, the economics profession had treated Nature as an "indestructible parameter" by subsuming it into the broader category of Capital. By the early twentieth century, Land and Nature vanished almost entirely from high theory, including the work of Keynes. Both Keynesians and neoclassicals paid little attention to how specific technologies or production processes affected the flow of materials from the environment through the economy then back to nature.[54] The quantitative growth advocated by both depended on an exponential increase in the "flow of goods through the economy" on the one hand, and the consumption of non-renewable resources on the other. Eventually, Binswanger maintained, these processes would overstrain the environment as well as the economy itself.[55]

Binswanger thus laid out a call to determine "how, and with which political or legal frameworks of control, would it be possible to move from the phase of exponential growth, in an orderly manner without economic crises, to a balance between economy and ecology." This, he believed, would require nothing less than a fundamental transformation in policy. And energy would be the pivot, because for the past two centuries it had both driven growth and caused the "largest proportion of our ecological damage."[56]

Energy economists in West Germany proved receptive to Binswanger's appeal not only because the oil shock suggested that structural change was possible, but also because they had come to see energy as a deeply politicized matter well before 1973. The state control of coal prices before 1956; the coal crises and bailouts of 1958 and 1966; the visible influence of massive conglomerates over the price of gasoline; the Suez Canal closures of 1956 and 1967; the state-led development of nuclear power: these experiences had disabused German energy economists of the notion that the sector they studied was shaped primarily by markets. West German energy economists, put differently, were less pure in their neoclassicism than Americans and less beholden to formal modeling. One of the nation's leading economists, Carl Christian von Weizsäcker, readily admitted that the market described in theory was "a nirvana" that diverged from reality as much as real existing socialism diverged from socialist utopias. Market processes would themselves not lead to an optimal use of resources. Instead, he thought that "in our current situation there are good reasons to hold the consumer price of petroleum products above their equilibrium value."[57] Or take Bertram Schefold of Frankfurt University, who coauthored with Binswanger an influential study on economics and the environment in 1983. Schefold was a follower of classical economics, which he thought was better at understanding political and institutional issues than neoclassical theory. Yet he was equally critical of Keynesians, whom he thought should ask what *type* of work ought to be done, not just how many jobs could be created.[58] Binswanger and Udo Ernst Simonis, an economics professor at Berlin's Technical University and director of the International Institute for Society and the Environment, thought the notion of competitive energy markets was a sleight of hand: following the business concentration in the 1960s, oligopolies ran the energy sectors of the world, and economic power motivated these companies far more than efficiency.[59] After the oil shock Hans Karl Schneider, director of the EWI, came to similar conclusions: that energy prices could get stuck far from their optimal levels for long periods of time. Equilibrium was hardly the norm, and likely a rarity. Exactly this had happened to the price of oil in the 1960s, he admitted, when it was too low by virtue of American policy and thus did not reflect the "real scarcity" of petroleum.[60]

This skepticism in neoclassical theory led West Germans to strikingly different conclusions about technology than Americans, and they greeted Nordhaus's backstop technology idea with disbelief. Schneider had participated in conferences with American resource economists, and he found their main conclusions "surprising": that the absolute scarcity of energy need not limit growth. The empirical work behind these claims was, in his mind, questionable and unable to undermine the central conclusions of the *Limits to Growth*.[61] Others were more critical. Binswanger complained that the neoclassical conception of technology was a "fantasy," wrongly assuming the entrepreneur had complete knowledge of the economy and all its raw materials, machinery, and labor. And if history illustrated anything, according to others, it was that technological advance rarely came through massive and discontinuous rupture like those Nordhaus built into his model.[62] For West Germans, Nordhaus could hardly have chosen a more ill-suited example of a breakthrough technology than the fast breeder. Many of the new experts sympathized with the anti-nuclear movement, and knew very well just how much mal-investment had flowed into West Germany's nuclear program since its inception. The breeder itself stood out as a red herring that produced nothing but cost overruns. By 1980 it was absorbing over a fifth of West Germany's energy research funding, surpassing the original cost estimates fivefold. In 1978 Parliament halted work over concerns about the hazards associated with a plutonium economy.[63]

Criticisms Catch Fire

After 1977 the anti-nuclear protests and the SPD party debates thrust this critique of neoclassical theory into the public limelight. For many of the ideas Binswanger and his allies formally analyzed were already on the minds of non-economists. Theory was not driving new ideas, but in many instances reacting to and trying to understand current political and environmental developments. That the energy sector was characterized by concentration; that markets were not handling resources well; that energy was not required for a high quality of life—all of these remonstrations were on the lips of public figures like Petra Kelly, Erhard Eppler, and Herbert Gruhl. And where neoliberals in America found support in a bevy of conservative think tanks, the opposite occurred in West Germany. There the anti-nuclear movement spawned new institutions to forge connections between ecologically minded social scientists and politicians: the Eco-Institute in Freiburg, the Fraunhofer Institute for Solar Energy Systems, the International Institute for Society and the Environment in Berlin, and later the Wuppertal Institute all publicized the work of these new experts and gave them a forum to network with policymakers.

Parliament, too, advanced these critiques when it launched a formal inquiry into nuclear power in 1979. The composition of the advisory committee reflected the growing influence of the new experts. Alongside architects of the coupling paradigm sat Günter Altner, cofounder of the Eco-Institute, and Klaus Michael Meyer-Abich, the SPD's leading apostle of energy savings. The inquiry embodied the deep political tensions over energy. All parties, the trade unions, as well as advocates and opponents of nuclear energy, had representatives on the commission, and to balance these rival interests its first major report in June 1980 adopted a radically new approach to forecasting. Instead of providing one projection of future energy consumption, as Federal programs had always done, the inquiry charted four different paths the nation might tread depending on the policies pursued, two of which represented the views of the new experts.[64] This very decision to provide scenarios removed the aura of scientific objectivity that had undergirded forecasting, since each path roughly corresponded to a political position, from the CDU to the Greens. This undercut the neoclassical idea that there was just *one* optimal equilibrium. For with close attention to detail, the nuclear inquiry demonstrated that each of the four paths could achieve the same level of welfare. Which one the nation followed depended on politics, not on some abstract notion of optimality.[65]

Political developments, however, were not the only thing propelling the critique of neoclassical economics into the mainstream; stimulus came from the environment itself. By 1980 Central Europe's forests seemed to be collapsing at an alarming rate. According to official estimates, a third of the woodlands in the Federal Republic were damaged, more than half in industrial areas like the Ruhr.[66] Experts had known for years that sulfur dioxide damaged trees, but now a new generation of German scientists hoped to spread their findings beyond the confines of academia. They found a willing abettor in the German media, which sensationalized this ecological threat after the nation's leading periodical, *Der Spiegel*, ran a cover story in 1981 with the headline, "Acid Rain over Germany: The Forest Is Dying" (see Figure 7.2). Its series made "Forest Death" a household concept by explicitly using *sterben*, the word for human or animal death, to evoke a bond between Germans and their environment. The imagery resonated not only with progressive ecologists but also with conservatives, since the forest was a symbol of German national identity dating to the nineteenth century. No less than Franz Josef Strauss, chairman of Bavaria's Christian Social Union, accepted this narrative of Forest Death, lamenting how "[w]e love our forests" and how Germans must "do whatever is possible to avert this threat."[67]

By 1982 a leading ecologist, Hermann Hatzfeldt, could claim that "it is no longer a question of whether the forest is dying or not. . . . the real question is rather: 'Why is it dying?'" Commentators located the answer in the nation's fossil fuel addiction, above all the sulfur dioxide that coal power plants and oil refineries had pumped into the air since the 1950s. Trees take time to die. That foresters were seeing tranches of completely dead ones in areas like the Black Forest that were far from industry suggested the cause dated back decades. Many

Figure 7.2. Der Spiegel cover image: "Saurer Regen über Deutschland: Der Wald stirbt [Acid Rain over Germany: The Forest Is Dying]" (1981). *Source: Der Spiegel,* "Saurer Regen über Deutschland," 47/1981.

argued that the beginning of tree sickness hinged on the spread of huge new refineries and power plants in the 1960s. And most agreed with Hatzfeldt that "ultimately, it is the dissipation of energy in the industrialized countries—West and East—that is causing the forests to die."[68]

Forest Death profoundly affected West Germany's energy economists not only because their object of study lay at the root of the problem, but also because it seemed to be a massive externality coming from the very Economic Miracle that had brought their nation prosperity. And commentators used economic concepts to make sense of the crisis. When Hatzfeldt noted that "preserving our forests is in the public interest," he portrayed the woods as a collective good that provided a range of services no one actually paid for. Refineries and coal plants were externalizing their costs onto the citizens of Central Europe to the estimated tune of 11 to 18 billion DM each year. Or at least this was how many saw it. The neoclassical notion that markets might achieve an optimal outcome, put simply, flew in the face of an ecological challenge that was gripping the nation.[69]

A New Paradigm

The new energy experts capitalized on Forest Death by opening many of their major essays with frightening images of a landscape losing its trees. When added to unemployment, high energy prices, Cold War tensions, and fears a of nuclear meltdown, Forest Death propelled a general sentiment of pessimism in the Federal Republic in the early 1980s, in stark contrast to Reagan's optimism in the United States. In this cauldron of anxieties, developments in energy economics began to fuse into a new paradigm.[70]

Concrete facts worked in favor of this fusion. Most importantly, the restructuring kick-started by the oil shock became more observable by the 1980s. Between 1973 and the early 1980s energy productivity—the amount of energy required to generate a unit of GDP—improved immensely in West Germany, by over 25 percent. By the mid-1980s energy as a proportion of overall costs in industry had not budged since 1973 despite the huge surge in oil prices, illustrating the ability of industry to use less fossil fuels.[71]

The new experts located the cause of this structural and technological shift in the energy price hikes of 1973 and 1979. Since Lehbert's pioneering study, more empirical work showed how the price elasticity of particular energies was astonishingly high for industry. New, more efficient methods of extracting aluminum from bauxite; new, more efficient turbine casings; new, more efficient methods of reusing waste heat at power plants; new hybrid heat reservoirs at factories: these and other techniques yielded efficiency gains that allowed German firms to produce more with less energy. The gains were even larger among consumers. Household energy use declined even as the size of the housing stock rose, a transition driven by new insulation, controlled ventilation, and better heat recovery systems in response to new policy incentives and the rise in energy prices.[72] As a survey of the field concluded: "the elasticities show that with suitable price signals, i.e. with appropriate price increases, the economic system can adapt over the long run. The question is whether the necessary permanent price increases are *politically* feasible."[73]

These findings were bolstered by historical research. In the United States using qualitative history to inform economic models was rare. West Germany's economics profession, by contrast, remained steeped in historical studies. When Rolf Peter Sieferle's *Subterranean Forest* appeared in 1982 as part of a collective energy project, it made waves. Sieferle's book depicted the Industrial Revolution in Europe as an energy transition driven by price-induced innovation: the breakthrough to coal only unfolding after population pressures had driven the price of wood to astronomical heights. His monograph, in other words, seemed to substantiate the idea that the resource challenges of the 1970s found their clearest parallel in the greatest technological revolution in European history.[74]

Arguments in favor of price-induced innovation gradually began to sway experts who had constructed the coupling paradigm in the first place. Here, no one was more important than Hans Karl Schneider, who participated in the Technology Ministry's energy conferences, who advised the Parliamentary Inquiry into Nuclear Power, and who after 1982 became a member of the Federal Council of Economic Experts. By 1980 this resource economist became a devotee of price-induced technological change, arguing that "the massive oil and energy price hikes of the recent past have opened up a huge potential for innovation." Augmenting this with energy taxes would unleash further "innovative potential that had been neglected in the past, the development of which will likely give powerful impulses to technical progress in energy use, energy conversion, and energy generation."[75]

These ideas crystallized into a new paradigm which claimed that demand for energy was highly malleable and responsive to price changes and to other incentives created by the state. Economists like Schneider now portrayed the production process as an interchangeable combination of technology, energy, capital, and labor, or TEKA (*Technologie, Energie, Kapital, und Arbeit*) that could shift depending on the needs of market, society, and state.[76] Binswanger had first suggested this in the NAWU Report. Policymakers had historically focused on investment that replaced labor with energy or capital—what he called Rationalization Type I. But now Binswanger proposed a different sort of investment—Rationalization Type II—that would instead replace energy with capital, new technology, or labor. But because environmental goods stood outside the existing price system, market processes could never generate Type II investment on their own; the government must do so.[77]

New ideas about the malleability of the energy economy found their clearest expression in a bevy of forecasts that followed the Parliamentary Inquiry into Nuclear Power in doing scenario analysis. Most believed energy costs would continue to rise, creating profitable opportunities for Type II investment. All of them, consequently, predicted vastly less energy use than the forecasts of the 1970s. Where Bonn's first energy program in 1973 saw the nation consuming 610 million SKE of primary energy by 1985, after 1980 no forecast even came close to this. The new experts argued that energy use could be brought down to 350 million SKE by the millennium, and even energy conglomerates like Shell Oil put the figure near 500 million SKE. Forecasts for West Germany's future energy consumption fell faster than those for any other Western European state or the United States, illustrating how quickly the new paradigm was spreading.[78]

A second layer to this paradigm came from environmental economics, where differences between American and West German economics were also deepening. Since the 1960s German scholars had been watching this subfield develop in the United States. But they also had their own tradition that dated to the government's first Environmental Program in 1971. Emblematic of West German environmental economics was Horst Siebert, who published one of the first German-language introductions to the field in 1978. Siebert's education was deeply transatlantic; he taught at Texas A&M University before doing research for Resources for the Future, where he worked alongside Kneese. And much of his approach echoed American environmental economics in being rooted in market failures and externalities.[79]

After 1980, however, West German environmental economists became more willing to accept political aims rather than chase the unicorn of optimal efficiency. Already in the 1970s Siebert began questioning some of his field's assumptions, including whether social welfare functions could be measured, or if one even "exists at all." In most cases he thought it would prove impossible to assess the exact costs of environmental damage, which undermined any hope of achieving an efficient allocation between production and pollution.[80] After 1980 qualms about the guiding assumptions of environmental economics grew, as others began to echo criticisms voiced by American neoliberals about the difficulty of quantifying externalities. As Schefold asked, using language reminiscent of Hayek, how could one actually calculate the marginal disutility of air pollution,

> which damages trees, therefore forests and ecological habitats with their survival value for rare species, their recreational value etc., which also damages lakes and rivers with analogous consequences, which damages buildings and works of art, which damages books and, last but not least, is a risk to human health? I should not believe in a figure given for marginal damage costs in this case, whoever the economist who presented it.[81]

But where such criticisms led American neoliberals to question externality as a concept, West Germans accepted that the exact value of environmental externalities would be inherently defined by politics. Siebert, for instance, argued that society must base environmental policy on a "fixed target" of pollution that it deemed acceptable, and this was a political decision. As his introduction put it, "the *political* system must balance actual environmental quality, which is a product of the technical-economic system, with the desired environmental quality." For Siebert the right amount of pollution was not some static level defined by a graph, but something the state would continuously balance through trade-offs with other goals.[82] Schefold similarly pointed out that preference formation among consumers was "to some extent of [a] political nature anyway," and that "critical decisions as to the degree of state intervention cannot be avoided because of the complexity of the sources of emissions and the number of people concerned."[83] West Germans thus not only retained externality as a concept, but elevated it because it allowed them to incorporate environmental goals into politics.

The work of Lutz Wicke exemplifies this political realism. A professor of environmental economics in Berlin, Wicke belonged to the CDU and had worked with the Federal Environmental Agency since 1978, becoming its scientific director in 1983 after the CDU gained power. A firm believer in the *Limits to Growth*, Wicke made his name with a guide to environmental economics. He opened this with an homage to Pigou, arguing that "the existence of external costs must be seen as one of the fundamental causes of ecological degradation in a market as well as a planned economy." While he pointed out the limits of quantifying externalities, he nevertheless thought the concept still had profound value for it could "depict the necessary challenge of balancing between environmental aims and other social goals."[84]

Crucially, German economists like Wicke incorporated price-induced technological change into their arguments in a way Americans did not. Already in the 1970s Siebert suggested that taxing emissions from energy production would give "the economic system an incentive" to develop new technology.[85] But with the TEKA conceptualization, ideas about guiding technology became more widespread. Wicke, for example, drew directly on the NAWU report to argue that "instead of labor-saving investments" the state must "push energy and resource-saving investments." In several popular books he fused the Pigouvian idea of market failures with price-induced technological change, using the former to justify the state actively pursuing the latter. He thought the future of environmental technology lay not in end-of-the-pipe equipment, like filters for coal plants, but in changing the very production process to minimize throughput and pollution in the first place.[86]

This new approach found its clearest expression in the field of renewable energy. The oil shock had raised interest in wind and solar power on both sides of the Atlantic. But the initial euphoria soon wore off as the hoped-for digression in the cost of solar and wind infrastructure never materialized. Where most experts had once criticized these technologies as too expensive, environmental economists like Wicke now argued that renewables were not failing, the market was. They pointed to the negative externalities caused by fossil fuels to justify a major state investment drive to advance renewable technology. In the 1980s new studies put a price tag on the social cost of energy.[87] Wicke published one of the first in 1986, *The Ecological Billions*, which priced environmental externalities at over 100 billion DM a year in West Germany.[88] Two years later Olav Hohmeyer completed a groundbreaking study on the social costs of energy at the Fraunhofer Institute for Systems Technology and Innovation Research. Using a firmly Pigouvian analysis, he estimated how much fossil fuels damaged the nation's flora and fauna, citizens' health, landscape, and even climate in the "seriously distorted market" of the Federal Republic. By his calculations, the overall difference between the positive externalities of renewable energies and the negative externalities of fossil fuels was the same order of magnitude as the real price gap between renewable and conventional electricity. If priced socially, in other words, solar and wind would already be cost competitive.[89]

Wicke and Hohmeyer's estimates did not go uncontested, and defenders of the old paradigm responded with their own counter-studies sponsored by utilities.[90] But this back and forth only affirmed externality as a legitimate concept that informed public policy, leading experts and politicians to wrangle not over whether the market erred, but how much it erred and what to do about it. Wicke, Binswanger, and Hohmeyer argued that if the state bridged the price gap between fossil fuels and renewables by including the full social cost in the market price, it could accelerate the introduction of solar and wind technology to the year 2000. The state itself, put simply, could orchestrate a price shift to kick-start a technological revolution.[91]

A Technocratic Price of Energy?

These new experts suggested many mechanisms for promoting technological development and internalizing the environmental costs of energy, including using the banking systems to subsidize Type II investment.[92] But their instrument of choice was the energy or ecological tax. These had been considered at the highest level in the Federal Republic since the early 1970s. But only when the new research became mainstream during the 1980s did this idea gain traction across the political spectrum. From the Left, Binswanger, Schefold, and Simonis argued that the "central instrument" should be a "targeted energy levy to make energy use more expensive." From the Right, Wicke argued that "energy prices in the long term must be raised to a high enough level that they are largely independent of fluctuations in supply and demand."[93]

This tax agenda differed fundamentally from Keynesianism, for it eschewed discretionary spending or monetary stimulus. Instead, these economists believed an energy tax would be an implacable force compelling behavioral changes among producers *and* consumers. Binswanger spoke of a tax that would "motivate consumers of every category... toward more energy-saving behavior." Wicke thought an eco-tax would unleash "structural change in the economy," by creating a boom in energy-saving technology. Simonis went further, calling on the tax to advance a "fundamental restructuring of production and technology," to promote a systematic "greening of economic, technology, energy, and transportation policy," and to "alter behavior to eliminate careless attitudes and wasteful uses of goods."[94]

Nor did they intend the tax to generate quantitative growth or consumption—goals dear to Keynesians. Instead, the nation must give up the "dogma" that GDP growth of 3–5 percent was necessary for social welfare, and look at qualitative measures that incorporated environmental preservation as a value in itself. Qualitative growth would pay more attention to the type of jobs in which people worked and the types of production processes that firms used, and use social indicators that captured metrics like health, nutrition, education, life expectancy, or child mortality. An energy tax could move society toward these ambitious goals, Simonis and Binswanger hoped, by reducing the throughput of energy and raw materials, and allowing the state to reduce taxes on other things highly valued by society. Against the backdrop of wrenching unemployment, this meant above all wage and salaried labor, the cost of which was burdened with pension payments and social insurance fees. A targeted eco-tax could discourage energy consumption while encouraging employment in sectors that were energy-saving and service-oriented, accelerating the nation's transition toward an economy that valued humans more than resources.[95]

The real beauty of the tax, they believed, was its ability to guide the economy while preserving individual initiative. In contrast to regulations, ecological taxes would "induce the individual agent to reduce his demand for nature without prescribing how to do so." And they would spur the "users of nature [to] *permanently* search for possibilities to lower emissions and energy use."[96] As Wicke summarized: "if the costs of pollution are high, then people will invest in environmental protection. And if environmental protection is profitable, then it will awaken self-interest."[97]

By the mid-1980s these ideas cohered into the new paradigm of ecological modernization. Proponents of ecological modernization hoped to build a full-scale and diverse social theory that paid attention to culture and institutions as well as the economy, and in fact political scientists numbered among its earliest promoters.[98] But its approach to the economy hinged on the ideas crafted by Binswanger, Wicke, Schefold, Siebert, and Simonis. Most important, ecological modernization moved beyond the zero-sum perception that environmental protection was at odds with economic development or employment, to optimistically argue that these could be harmonized. It said little about social struggles that were linked with ecological degradation. Instead, environmental problems were challenges for reform within the existing institutions of capitalism, which these modernizers believed could be steered in the right direction. As a regulatory philosophy, it moved beyond the neoclassical focus on efficiency to demand a "structure-oriented modernization of industrial society on an ecological basis," foregrounding the need to spur technological innovation and change consumer behavior.[99]

Ecological Modernizers believed their agenda retained the core pillars of the social market economy by placing self-interest in the service of environmental salvation. In doing so, their arguments about private initiative, profit, and competition would allow politicians to claim an eco-tax and other policy tools for the Ordoliberal tradition.

In reality, the architects of Ecological Modernization differed just as dramatically from Ordoliberals and neoliberals as they did from Keynesians. As experts, *they* should be the ones setting the goals society strove for, not consumers; they should be the ones determining the price of energy, not the market. For the original Ordoliberals like Eucken and Röpke this was anathema, since they believed technocrats would always err when setting prices.[100] But by the 1980s a new generation of Ordoliberals were becoming even more aggressive in valorizing markets and condemning state intervention, and they gravitated toward American neoliberalism in their willingness to extend market logic further than their predecessors. This younger cohort included economists like Lothar Wegehenkel, who studied at the University of Freiburg—where Eucken had launched Ordoliberalism—and who worked with the Walter Eucken Institute, and Fritz Holzwarth, who also matriculated at Freiburg, wrote for the journal *Ordo*, and later worked on water management for the Environmental Ministry. These experts rejected energy taxes as a manipulation of organic prices and a technocratic power grab that would lead to state overreach. They denounced externality as an ineffectual and unrealistic concept. They argued that government officials could never have the "comprehensive knowledge" of the economy to correctly predict the consequences of a tax.[101] They lambasted Ecological

Modernization as a "political price-formation process," in which, "ultimately, politics determines shortages instead of the market."[102] At best, such policies would become uncontrollable. At worst, they would require such a high degree of "state force" and "centralism" that they would "violate the very regulatory principles" of the Social Market.[103]

After 1980 these younger Ordoliberals took their cue from American neoliberals to offer an alternative to Ecological Modernization. They countered calls for an energy tax with their own demands for emissions trading. Such a policy would, they hoped, not only solve environmental challenges, it would generate "the opportunity for more markets."[104] These ideas were in turn roundly castigated by experts like Wicke and Binswanger. For Ecological Modernizers, emissions trading was not only *not* a silver bullet; its very premise was flawed. That one could easily measure and distribute property rights for environmental assets, or that there could be one clearly identifiable polluter and injured party were assumptions that hardly conformed to reality. Whenever an environmental good flowed into and out of a region, moreover, like the sulfur-contaminated air swirling over Central Europe, it would be impossible to exclude outsiders from influencing developments within a geographically bounded emissions trading system. Once trading began, licenses might concentrate in a single region, creating a sacrifice zone where pollution rose higher than ever. Most importantly, the new experts saw trading as the moral equivalent to "selling indulgences" in medieval Christianity, giving legal sanction to a social sin. As Simonis put it, cap-and-trade would "economize ecology," where the goal ought to be the reverse: "ecologizing the economy."[105]

Conclusion

Simonis's remark captured a core issue at stake during the 1980s and 1990s. How far would neoliberalism go? In most places, quite far. American economists would take a strong neoliberal turn. So too would many in Britain and at the European Commission in Brussels. But energy economists in the Federal Republic did more than just avoid the excesses of neoliberalism. They developed a new energy paradigm that they promoted through a well-funded network of institutions. Ecological Modernizers accepted that energy markets should be politicized and saw energy systems to be malleable and responsive to price changes. They suggested that a new sort of qualitative growth could continue while energy consumption declined, and they used the rampant evidence of market failures to call on the state to forcefully guide society over the long term toward new technology and to reshape consumer as well as producer behavior.

After 1980 American and West German economists thus increasingly disagreed about energy. Whereas the former turned to emissions trading with its promise of depoliticizing hard decisions and commodifying nature, the latter called for ecological taxation, directed investment into environmental technology, and strong, political price guidance for green energy. And the appeal of this new paradigm spanned the political spectrum in West Germany. For the CDU, Lutz Wicke outlined an Eco-Plan that played on the party's heritage by aspiring for a new "Economic Miracle" oriented around energy and the environment. In 1986 he called for a state-led investment program—paid for in part by a new tax on energy and pollution—to stimulate energy savings, develop new technology, and spark an employment drive in environmental protection. In 1989 he went further, calling for a global Marshall Plan whose centerpiece would be a "climate tax" on carbon dioxide and fossil fuels to start a revolution in renewables. For Wicke, "the oil crises of the 1970s [were] suitable as models for the design of global measures to reduce the consumption of fossil fuels." His goal was nothing less than the ecological equivalent to America's space program: if the world could put a man on the moon, "then why not rehabilitate the environment in ten years? The technological effects will be similar."[106]

The SPD, meanwhile, could draw on Binswanger, Simonis, and Schefold. These economists wanted to make quality of life, not quantitative growth, the central metric for guiding policy. The Federal Republic, they thought, could do this by turning energy from a passive into an active policy tool, and placing it at the heart of macroeconomic policy. Their instrument would be an energy tax that started small but rose each year. And they would supplement this by transforming the banking system into an agent of ecological modernization that promoted investment in energy-savings and renewables.

The Greens, lastly, could look to Peter Hennicke and others working with the Eco-Institute in Freiburg for an even more far-reaching energy reform program. Based at Osnabrück then later Darmstadt, Hennicke understood better than almost anyone the inner workings of the nation's energy system, consulting for the state of Hesse and the city of Bremen and closely advising the Greens on electricity policy. Like Wicke and Binswanger, Hennicke thought an energy tax was "sensible and necessary." But he lamented, "It would be naïve if we—like Pigou 70 years ago—tried to capture the many global catastrophes that are today foreseeable like the anthropogenic greenhouse effect, a war for oil, or nuclear devastation after a super meltdown, with the category of 'external effects' alone."[107]

Hennicke's writings revealed the limits of this new paradigm at the very moment it was gaining popularity. For one of the most important energy spheres of all, electricity, was not even a market, and thus insulated from the reform agenda of Ecological Modernization. The energy services that people actually cared about—food cooled by a refrigerator, clothes cleaned by a washing machine—were not a single product but an interconnected bundle of goods that required consumers to buy kilowatt hours from the utilities, appliances from retailers, and capital from banks. And this made it nearly impossible to guide electricity consumption, because there *were* no firms that could provide all of this—a final energy service—in one package. There were only utilities with the incentive to sell *more* electricity, retailers with the incentive to sell *more* equipment, and financial companies with the incentive to lend *more* capital. And this, for Hennicke, was the nub of the problem: that no social actor had an interest in ecological reform. One must first be created by overhauling the entire energy system.[108]

Hennicke and his Green allies thus argued it would take *more* than just altering prices to launch an energy revolution. If the nation were to shift off fossil fuels or nuclear power, it would have to fundamentally restructure its entire electricity system "from below," by creating a new agent with a vested interest in reform. This meant breaking the power of energy companies and placing the grid in the hands of local actors: people with an incentive to save energy, since they were the ones consuming it; people with an incentive to generate green power, since they would be the ones operating wind turbines or solar panels.[109] In the 1980s such a movement was, in fact, seeing the first stirrings of life, as grassroots groups began to challenge West Germany's powerful network of utilities. As the decade progressed, the nation would stand not only on the brink of political reunification, but also on the cusp of its most radical energy transformation since the oil wave of the 1950s.

8

Energetic Hopes in the Face of Chernobyl and Climate Change

The 1980s

The high living standard in the Federal Republic of Germany is based on a plentiful and low-priced supply of energy.[1]

A distinct upward adjustment in the price of energy must stand at the forefront of energy policy deliberations, since now as well as in the medium term there is hardly any other incentive to exploit the potential for rationalizing energy use and renewable energy sources.[2]

On the morning of April 28, 1986, an unusually long line formed at the entrance to the Forsmark-1 nuclear station in Sweden. In going through their routine screening to enter, nearly every worker was registering abnormal radiation levels. Confused, a technician sent the shoe of an employee to the lab to make sure the equipment was working. When the test came back with stunning measurements he telephoned his manager, who began to seal off the reactor. Soon, however, agencies elsewhere began reporting similar results, suggesting a reactor meltdown outside Sweden. By afternoon the Swedish National Defense Institute concluded that a radioactive cloud was moving into Scandinavia from the southeast, from what could only be the Soviet Union. The following day Moscow released a terse statement, buried beneath other news, about two people dying from a minor incident at the Chernobyl nuclear plant. European and American correspondents started hunting for information, and the details slowly leaked out as the radioactive cloud slipped south toward Central Europe. Radiation levels soared across the continent, leading politicians to demand details from Moscow. What, exactly, had happened?[3]

In West Germany, the government at first downplayed this news from the East. "We're 900 miles away from the site of this accident," counseled the interior minister, "It's entirely out of the question that the German people are in any danger." Politicians elsewhere reacted in similar ways, French officials even claiming that radiation stopped at the Rhine. But despite these pronouncements, the conservative government of Helmut Kohl struggled to acquire basic information about the catastrophe. Had thousands of people died, or just a handful? And how much radiation was actually in the atmosphere? Local authorities contradicted Bonn, issuing their own emergency orders to halt the sale of fruit and milk, and asking people to close their windows and bathe immediately after coming inside. May turned into the "month of the Geiger counter": strontium and cesium became household words; outdoor events were canceled; West Germans remained glued to their televisions and radios, awaiting the news about radiation levels.[4] All of this rekindled the passions of the 1970s. Demonstrations ripped across the country at power plants. Black-clad militants faced off against the police, demanding an end to nuclear power. Jutta Dittfurth, a leader of the Greens, captured these burning emotions, noting how "Chernobyl is everywhere.... Our life is no longer as it was. We will have to redefine what a normal life is."[5]

But in many ways the meltdown only accelerated a redefinition of normal that was underway even before 1986. In West Germany the sociologist Ulrich Beck put it best, in a popular book written before Chernobyl but published just after the catastrophe. The defining features of modern life were the novel risks generated by industrial civilization, which "escape perception," which endanger future generations, and which "boomerang" across borders beyond the control of nation-states.[6] In this, Chernobyl was hardly unique, for by the 1980s risks that defied direct perception were multiplying. Sulfur dioxide was damaging forests. Chlorofluorocarbons were tearing a hole in the ozone layer. Most ominously, carbon dioxide from the world's fossil fuel infrastructure was threatening to fundamentally alter the atmosphere.

Spreading awareness of these "hazardous side effects" of industrialization turned the 1980s into an "ecological decade" for West Germans, and environmental mobilization reached its zenith during these years.[7] Chernobyl sparked the creation of an Environmental Ministry. The nation's leading magazine ran cover stories about global warming. The chancellor declared climate change to be the "greatest environmental problem" of the day.[8] By 1989 West Germany seemed ripe for an energy transition to rival the one that had brought oil to the nation a generation ago. Hopes were high that the nation would embark on a new energy path and embrace a new energy paradigm. But in contrast to high politics, in the world of energy the decade ended with a whimper instead of a bang. Nearly every concrete energy initiative withered in the face of politicians and interest groups who supported the status quo.

Why did momentum ebb? Partly it was the *fortuna* of history. In the fall of 1989 the border dividing the Western and Eastern halves of Germany crumbled, as peaceful revolutions rolled across Eastern Europe. In the Federal Republic, the fall of Communism sparked a campaign to reunify a divided nation and eclipsed nearly every other political issue.

Just as important, West Germany's ecological decade unfolded as market fundamentalism was spreading through the corridors of government. In the United States and Britain, neoliberalism was transforming from an intellectual movement into a "political order" under President Ronald Reagan and Prime Minister Margaret Thatcher. Internationally, new institutional arrangements emerged to protect markets from democratic intervention on a global scale. In Brussels, the European Commission began embracing similar views. Neoliberalism shaped policy across the Atlantic and reinforced the old paradigm by promising to deliver cheap energy. West Germany was not immune to this surging faith in markets, which constrained the type of state guidance of technology, prices, or investment that might unleash an energy transition.[9] At the same time, the incredible expansion of non-OPEC oil fields during the 1980s reshaped the hydrocarbon world and eroded the urgency that had put green energy on the agenda in the first place.

By 1990, when East and West Germans were imagining a new future together, the ecological hopes of the preceding years foundered. But beneath the surface, slow-moving developments were clearing the path for an energy transition to come. In this sense, the narrative of 1980s Europe that hinges on the sudden catastrophe of Chernobyl and the swift, unanticipated collapse of Communism overlooks more gradual changes unfolding during this critical decade. For West Germans, a new world of energy was becoming imaginable during these years. Just as important, boosted by the sole energy policy accomplishment of the decade, the cost of solar and wind began to fall within reach of the economically possible.

Recommunalizing the Grid

West Germany's ecological decade began with the federal election of 1983. Having come to power through a motion of no confidence the previous year, in his first major decision the new chancellor, Helmut Kohl, an ambitious CDU politician from the Rhineland, called early elections to reinforce his administration. Running on a platform of jobs and growth, Kohl's CDU and its liberal partner, the FDP, strengthened their ruling coalition. But the green lining was the surprising success of the Green Party, which rode momentum from peace marches and fears of Forest Death to enter national Parliament, the first new party to do so since 1953. Suddenly environmentalists had a powerful platform to shine a spotlight on their nation's problems.[10]

Their opening salvo came later that year, when the Green Party in Baden-Württemberg tried to overhaul the law that had governed the nation's electricity network—the single largest final user of energy—for a half century. Huge, carbon-intensive, heavily polluting corporations dominated the nation's supply of power. Yet this was not an iron law of nature but an artifact of history. In 1935 the Third Reich, with support from utility companies themselves, had passed an "Energy Economic Law" that granted monopoly zones to utilities in which they became the sole distributor of electricity, and that helped these massive companies expand vertically into the generation of power and even the extraction of coal. In return for monopoly rights, power companies paid lucrative concession fees to local municipalities. Postwar regulations subsequently guaranteed a steady profit. In exchange for avoiding blackouts and guaranteeing as much electricity as consumers could ever want, utilities became some of the largest companies in the country, as well as the biggest source of carbon emissions.[11]

The energy battles of the 1980s and 1990s would revolve around this electricity system. The Greens wanted it gone because it embodied the undemocratic concentration of corporate and political power that the party so viscerally opposed. Electricity was one of the most concentrated sectors in the nation: by the 1980s eight utilities produced over 90 percent of the country's power, with a single one—Rheinisch-Westfälisches Elektrizitätswerk (RWE) in the Ruhr—generating 40 percent. Utilities, moreover, were tightly interlinked with powerful companies in other energy spheres. RWE owned stakes in Ruhrgas AG, which controlled nearly two-thirds of Germany's natural gas imports. RWE's partners in Ruhrgas were none other than the global oil majors like Shell, Esso, and BP. To the Greens, utilities looked like the spearhead of a corporate constellation rooted in fossil fuels and nuclear energy. They kept the "municipal energy sector in total dependence," the party claimed, through their concession fees and their close relationships with mayors, city council leaders, national politicians, and above all the Economics Ministry, which had a revolving door of personnel with these power companies.[12] By the mid-1980s this patronage network took on a particularly odious sheen against the backdrop of the Flick Affair, West Germany's greatest corporate scandal, where allegations of bribery reached the highest levels, giving credence to complaints about the corrupt nexus of politics and business.[13]

In place of large electricity monopolies, the Greens wanted community ownership, or "recommunalizing" the grid. They believed that turning municipalities into the "central political location" for energy planning would democratize the nation *and* save the environment. In Parliament they doubled down on arguments about the legacy of the Nazis that had strengthened their movement in the 1970s. They cast the 1935 energy law as the "nucleus" of the Third Reich's "economy of destruction," which remained "hostile to democracy" today with its "huge public grey zones and politically intransparent fields of fog."[14] Reform should come through the municipalities, which guided land use patterns, authorized new infrastructure, and were themselves major consumers of energy. Municipalities, after all, could be run democratically. Electricity production and distribution were always planned, the Greens pointed out, whether by corporations or the state. "The question was rather by whom, according to which criteria, and how democratically will any planning be done." Why not put the electricity sector into the hands of local citizens?[15]

The Greens believed recommunalization would not only advance democracy, but also reduce the nation's ecological footprint. Here they drew on luminaries of the soft path movement like Amory Lovins, as well as West German experts, above all Peter Hennicke. Hennicke worked with the Eco-Institute in Freiburg advising municipalities as well as the Green Party on energy. In the 1980s he brought the American concept of Least Cost Planning (LCP) to West Germany. Developed by Lovins and applied first in California and the Pacific Northwest, LCP urged utilities to manage not only the supply but also the *demand* for electricity, since it often cost less to apply new techniques that could provide the same service—like refrigeration—with less electricity than it did to generate more electricity. In Lovins's and Hennicke's famous aphorism, policy should focus on "negawatts instead of megawatts." And less electricity meant burning fewer fossil fuels or running fewer reactors.[16]

The challenge was rewarding investment in efficiency instead of investment in new power plants. Hennicke thought this required total systemic reform, for under the current framework utilities passed the costs of new power plants on to consumers, regardless of how wasteful the electricity supply chain was. And it was very wasteful. From the point of generation to the point of use, conventional electricity networks lost 70 percent of the power they initially produced. Between 1973 and 1982 overall capacity in this sector was 30 to 40 percent higher than the maximum demand for electricity during the year, when best practice said it should be just 10 percent. And waste was expensive: in 1985 energy cost the average consumer 6–10 percent of their annual budget.[17] Through their political connections, moreover, the utilities suppressed alternative channels for consumers to secure heat or power. District heating that pumped warmth from power plants into factories or homes had gained momentum in the 1970s under the Social Democrats, but subsequently stagnated in the 1980s.[18] Utilities similarly blocked heat-power co-generation plants, as well as attempts by independent hydro-plants or wind turbines to feed electricity into the grid.[19]

New practices from California suggested that local authorities could change this framework. Municipalities could do regional planning and investment with efficiency in mind, and reduce energy demand through programs that improved the efficiency of heating buildings or running appliances. Such a program, the Greens hoped, would unleash an efficiency transition that would reshape society.[20]

In Baden-Württemberg the first attempt to apply these ideas failed. The other parties saw the Greens as a pariah, and the proposal read more like a "theoretical term paper" than a piece of legislation.[21] The Greens launched a second salvo the following year in the left-leaning state of Hesse. In 1984 the local Green Party, in a minor political revolution, joined the Social Democrats in an informal alliance to govern this central German state. Concrete policies proved elusive after the SPD rejected calls by the Greens to shut down the local nuclear reactor.[22] Instead, the two parties presented a law to transform the grid. Lobbying from the utilities torpedoed this too.[23] But two months before Chernobyl the Greens struck again, this time in the federal Parliament, where they proposed a national law for recommunalizing, decentralizing, and democratizing the grid. This too failed. The CDU argued that devolving power to municipalities would lead to higher prices and inequalities between regions; that it would jeopardize the stable supply of electricity by creating a patchwork of tiny grids; and that it would generate a flood of "prohibitions, rules, surveillance, and decrees" that bordered on dirigisme.[24]

The Greens expected such setbacks. They hoped less to pass a law than to change the debate around energy. And here they were succeeding, for while the CDU rejected recommunalization, the nation's other mass party embraced it. Under new, younger leadership following their fall from power, some in the SPD were trying to continue the party's transformation into a leader of ecology. In 1984 the party tasked Volker Hauff, former minister of technology, to lead a new environmental commission. Hauff's advisors included the renegade nuclear engineer Klaus Traube, a leading critic of large-scale technology, and Hans C. Binswanger, a pioneer of Ecological Modernization. Hauff drew on Green ideas, like strengthening municipalities.[25] But he wanted to reform more than just the grid, and by 1984 he began calling for the ecological modernization of society at large. Inspired by the ideas of young experts like Binswanger, he crafted plans for a broad-based investment drive into energy efficiency across all branches of the economy.[26]

A Rising Tempo: The Chernobyl Debate

Chernobyl accelerated these efforts to rethink the nation's energy system. As the cloud of radioactivity slowly dispersed during the summer of 1986, the

Federal Republic descended again into a vitriolic debate about nuclear power. The CDU-FDP government clung to their talking point that West Germany had the highest safety standards in the world. As the CDU chair put it, his party was "the security party of Germany," not only for geopolitics but also for energy. If the country closed its reactors, "then we would shut down the 19 safest ones while 336 others around us remain in operation." Chancellor Kohl called for a comprehensive international framework to enhance reactor safety everywhere along German lines.[27]

But for the populace these reassurances did not ring true, and by the summer, polls showed a vast majority of West Germans—86 percent—wanted to phase out atomic power. For the Greens, the reactor meltdown seemed like a political blessing by confirming risks their movement had been publicizing for years. But in practice Chernobyl deepened rifts in the party. In Hesse, Joschka Fischer emerged as the party's charismatic face. Once a feature of the radical student movement in Frankfurt, in the 1980s Fischer became a leader of the Greens' moderate wing and the party's first state officer, taking his oath as Hesse's minister of energy and the environment in jeans and running shoes. This Green Machiavelli epitomized the call for a long march through the institutions. By 1985 he believed an alliance with the SPD was his party's best hope of enacting change, and his own best hope of becoming a national figure. Fischer called for a gradual exit from nuclear power. But Chernobyl made gradualism look like a moral compromise for a party that built its reputation on condemning reactors. The opposing, fundamentalist wing of the Green party insisted that nuclear issues brooked no compromise. All "nuclear installations on earth," as one of Fischer's party opponents put it, "are declarations of war against us." In 1987 Hesse's Red-Green coalition collapsed over atomic tensions.[28]

While the Greens were busy fighting internally, in 1986 the more striking political move came from the SPD. Two years prior the party had declared nuclear power should only remain for a "transitional period," but after Chernobyl they began crafting plans to shut down all reactors.[29] Younger Social Democrats now positioned their party as an ecological leader. Hauff led the way for personal as well as political reasons, seeing a chance to boost his reputation by tying his career to energy. As one of the youngest party members to have held a ministerial post, Hauff was a poster boy of the new SPD. His earlier work on nuclear matters meant he understood the industry's problems and had credibility in atomic circles, but it also left him open to accusations of opportunism. After Chernobyl he spearheaded a party committee on nuclear exit that included many advocates of Ecological Modernization. That August, Hauff presented the committee's report to Parliament and published his own book that appropriated the Green concept of an "Energy Transition" for Social Democracy.[30]

Citing not only safety risks but the rising costs of nuclear power, Hauff called for a total exit within ten years. But his vision was more audacious than ending reactors. Drawing on studies he had overseen as technology minister, he suggested the nation could decouple growth from energy use. He made energy savings the SPD's main goal, the "most important energy source."[31] Hauff wanted more authority for municipalities, which he hoped could expand the efficiency programs the SPD had started in 1977. He proposed a billion DM energy investment program alongside a new plan for renewables, calling on utilities to open their monopoly to independently operated sources of electricity. "Instead of spending billions more on nuclear power," Hauff wrote in Die Zeit, "if we invest massively in solar power along the model of microelectronics then we will certainly make great strides. Here lays a great opportunity for our country in terms of industrial policy.... We stand not in the middle of the atomic age, but at the beginning of the solar age."[32]

Hauff's program was never just about ecology; building on ideas the SDP, Ecological Modernizers, and the Eco-Institute had crafted before 1982, he made it as much about exports and employment. By the mid-1980s these topics animated public debate more than ever as Western Europe suffered through accelerating deindustrialization. The iconic mass-production sectors that had made the continent wealthy—steel, textiles, shipbuilding—were shedding jobs at an alarming rate under the weight of high wages, high taxes, and competition from East Asia. In 1987 Europe's largest steelworks, in the Ruhr, announced its closure, threatening unemployment for six thousand workers. This structural crisis sapped the base of the SPD as abandoned factories increasingly dotted the country's industrial heartlands. At the same time, Western Europe was falling behind in the dynamic new field of information technology, as the United States and Japan stood poised to conquer the global digital market. But what if environmental, renewable, and energy savings technologies could be West Germany's new export niche, creating skilled jobs to replace those lost in older industries? This stood at the core of Hauff's vision. For new studies suggested that investment in solar, wind, or energy efficiency would raise employment in everything from construction to handicrafts and engineering, and position the Federal Republic as an export leader in these fields.[33]

Hauff's optimistic agenda dovetailed with an outpouring of nuclear phase-out studies that added concrete numbers to the question of whether the nation should shut down its reactors. Industrial groups claimed a full nuclear exit would destroy 200 billion DM worth of capital and drive up electricity costs. The CDU used such reports to put the cost of giving up the atom at a trillion DM.[34] But studies from the Eco-Institute and other new experts contradicted these incredible estimates, claiming the short-term economic costs would be in the single digit billions, and that over the long term a path of energy savings would provide the same level of welfare at the very same cost as the nation's current nuclear and fossil fuel trajectory.[35]

The most striking aspect of this debate was how little it featured questions of climate. The Greens played up the anti-democratic nature of nuclear power. The SPD homed in on employment and exports. Where environmental considerations arose, the parties focused on older concerns like smog and sulfur dioxide, the scourge of the nation's forests.[36] More generally, the debate centered on older fears of scarce resources spawned by the Club of Rome in 1972. When the government countered Hauff's proposal with an Energy Report, it opened the case for nuclear power by pointing to the long-term scarcity of fossil fuels. For the CDU, the main goal of energy policy remained what it had been since the 1960s: reducing dependence on the Middle East. Nuclear energy mattered less for being climate-friendly than for reducing West Germany's reliance on scarce, foreign oil for its export economy.[37]

The Politicization of Climate Change

But in the mid-1980s concern about too little fossil fuels began to transform into fears that the world was burning too much of them. The linchpin for this shift was global warming, which became a political issue in West Germany at the very moment Chernobyl was rocking the nation, unleashing a conjuncture of anxieties that focused on energy but that defied a clear resolution. For how could West Germany, or the world for that matter, stop using the atom *and* fossil fuels at the same time?

Knowledge about climate change was nothing new. As energy use exploded in the 1950s and 1960s, renowned scientists began to worry that burning fossil fuels might heat the atmosphere. But despite models that forecast warming, between 1940 and 1970 the Northern Hemisphere had actually cooled. Only in 1981, when Jim Hansen of NASA published a widely read report that gained coverage in the *New York Times*, did climate change enter the global spotlight. Other reports followed, and by 1985 leading climate scientists for the first time began calling on governments to prevent global warming.[38]

These apprehensions reached West Germany's public in early 1986, on the eve of Chernobyl, when the German Physics Society (*Deutsche Physikalische Gesellschaft—DPG*) published an account of the approaching "climate catastrophe." This eminent institution argued that humanity's current trajectory would lead carbon dioxide levels to rise from 340 parts per million to over 540 in the next half-century, increasing temperatures by 2 to 4 degrees Celsius. The DPG paired these figures with tangible predictions about how global warming would reshape the world: the Mediterranean and the Western United States would suffer desertification, while the ice caps would begin to melt. The DFG argued that humanity must "drastically" reduce fossil fuel use and begin to do so "immediately." If humanity delayed by even a decade or two, "it would in all likelihood already be too late."[39] If the DFG's terse prose was not enough, West Germany's leading magazine ran two major articles in 1986 dramatizing these findings for a lay audience. With a cover displaying the massive Cologne cathedral under water, *Der Spiegel* depicted a future Germany whose wine was too sweet to drink because its vineyards were too hot, whose alpine regions were covered by cypresses instead of pines, and whose farms were devasted by rising temperatures (see Figure 8.1).[40]

In just a few years, this emerging scientific consensus combined with sensational reports to turn global warming into the most important environmental issue in West Germany. Chancellor Kohl declared it so in his national address in 1987. That summer the United Nation's Brundtland Report on the environment underscored how the current pace of energy use threatened the planetary ecosystem.[41] In the fall, under pressure from the Greens and from their own MPs, the CDU government authorized a formal Parliamentary Inquiry into climate change. Members from all parties participated, soliciting work from Ecological Modernizers like Hennicke and Klaus Michael Meyer-Abich, the Social Democrat who had popularized energy savings in the 1970s. After working with fifty-one research institutes and publishing 3,300 pages of material, the Climate Inquiry concluded that global warming was real and dangerous, but also manageable. In 1990 the commission recommended that West Germany—the fourth largest carbon emitter in the world—begin working immediately to reduce carbon emissions 30 percent by 2005. Success would require nothing less than a "fundamental modernization" of the economy with "fast and far-reaching action in all areas," above all in the sphere of efficiency.[42]

This combination of Chernobyl and the "climate catastrophe" rippled across the nation, unleashing grassroots movements to reform local energy networks. Most renowned were the Schönau Power Rebels, a small community in Baden-Württemburg. In 1987 they began a campaign to conserve energy and take back control over their grid. "Make your relationship with your power meter a loving one," was their slogan, "visit it every day." On the French border, Saarbrücken already had a modest campaign to improve the heating efficiency of buildings. But after the mid-1980s the city developed the most cutting-edge municipal energy organization in the nation. By 1990 it produced most of its electricity with heat-power co-generation plants, offered energy advising to its citizens, performed thermographical analyses of buildings, and gave financial aid for locals to connect to district heating. The result was an astonishing three-quarter drop in the city's carbon emissions.[43]

Figure 8.1. Der Spiegel cover image: "Die Klima-Katastrophe [The Climate Catastrophe]" (1986). *Source: Der Spiegel,* "Die Klima-Katastrophe," 33/1986.

After 1986 the Greens and the SPD pointed to these local successes as they doubled down on efforts to score an energy victory at the federal level. Between 1986 and 1990 they introduced a slew of green energy legislative proposals. By now the two parties shared many goals and even the names of their respective agendas, which involved variations of ecological "reconstruction," "renewal," or "modernization."[44] But differences remained. The Greens believed West Germany could only solve its energy woes through structural reform on the local level. As late as 1990 they were still framing attacks on the 1935 energy law less as a solution to global warming than as a remedy for the legacy of the Third Reich. The SPD, by contrast, embraced global warming as the raison d'être for an energy transition, but they focused less on structural reform.[45] In 1988 Hauff presented yet another energy program that reflected the growing influence of Ecological Modernization, and the idea that the price of fossil fuels failed to account for their environmental damage. For the SPD, pricing energy correctly became the key to transforming the energy system. "The altered price level—where environmental products will be cheaper, ecologically harmful ones more expensive—will lead to structural change guided by the market economy. In the future this path will have priority whenever possible."[46]

1989: A Green CDU and a Carbon Tax?

Despite the nature of partisan politics, reform calls from the SPD and the Greens had a chance for success. For not only was the public transfixed by environmental crises, but the CDU itself was developing an ecological wing. Like the SPD, the anti-nuclear movement had forced the CDU to re-evaluate its stance on ecology. In 1978 it began to accept the importance of energy savings and qualitative instead of quantitative growth.[47] But it was the rise of sustainability as a global discourse that gave the CDU, a Christian Party, a unique angle into energy. In 1983 the United Nations commissioned the former prime minister of Norway to explore the relationship between development, resources, and the environment. Four years later Gro Harlem Brundtland published *Our Common Future*, which popularized sustainability with its famous call for development that "meets the needs of the present without compromising the ability of future generations to meet their own needs."[48] The CDU pounced on this idea of intergenerational responsibility, seeing the slogan as an opportunity to parlay its Christian ethos and differentiate itself from the Social Democrats. After 1987 the CDU foregrounded sustainability in its energy policy, underscoring how "Nature is not a quarry for our affluence, but rather a gift from God that we are obligated to pass down to future generations."[49] As a part of nature, humans must take responsibility for heavenly "creation." By the end of the decade, party leaders were speaking of an "ecological intergenerational contract."[50]

After assuming power, moreover, the CDU registered some ecological achievements. In 1984 it reduced sulfur emissions by mandating catalytic scrubbers for coal plants—the regulatory solution to Forest Death. After Chernobyl, Bonn gave environmental affairs its own ministry. The new ministry's second leader, Klaus Töpfer, vigorously promoted the party's sustainability ethos by, among other things, swimming across the Rhine, formerly one of the most polluted rivers in Europe but one that was being rehabilitated.[51]

By the mid-1980s, moreover, the CDU was suffering internal divisions that presented a potential for cross-party agreement. Most importantly, Helmut Kohl's influence faltered under the Flick corruption scandal when a Green MP accused the chancellor of lying about campaign donations. Kohl narrowly escaped the charges, but his government's poor handling of Chernobyl led to a disappointing result in the 1987 election. An opposition wing formed around the prime minister of the ecologically inclined Baden-Württemberg, and a group of CDU MPs signaled a willingness to think differently about energy.[52]

In 1988 this ecological rift came to the fore at the CDU congress in Wiesbaden, where members called on Bonn to transform the energy system. Pointing to global warming and the stunning inequality of energy consumption around the world, the CDU's ecological faction demanded "drastic measures to reduce energy consumption among industrial states." They hearkened back to the CDU's first reign, when it jump-started nuclear energy through a gigantic state program, and wanted a repeat performance with renewables: "massive and continuous support for research, development, and market launch," of solar, wind, and hydropower. And they drew on their own connections with Ecological Modernizers to integrate the new energy paradigm into the CDU program, arguing that "environmental costs must be reflected in the [market] price of an energy." Taking a cue directly from Lutz Wicke, CDU member and a leading environmental economist, these MPs demanded an "ecological 'Marshall Plan'" to save the world from climate change, funded by a tax on carbon.[53]

So in 1989 West Germany descended into a debate about its energy system. The SPD sparked matters with its ecological program. Since 1950 the rise in wage and income taxes had far outpaced the rise in taxation on energies.[54] The state, put differently, was contributing to environmental problems by making energy relatively cheaper than other inputs to production. The tax load on wages and salaries, meanwhile, was harming the nation's business sector as it struggled to compete with multinational companies, rendering West Germany uncompetitive. All of this had to change. The SPD's tool would be a tax on energy consumption, which would generate "incentives for environmental protection, energy savings, and rational energy use."[55] Their goal: to turn the price of gasoline into what the price of bread had been in the eighteenth century, "raising awareness for the invisible yet imminent global climate catastrophe." The rewards would not only be averting global warming; but also using the proceeds to reduce taxes on labor, thereby hopefully generating more jobs.[56]

The politics of global warming meant the CDU could not ignore these proposals. But they rejected a tax on all energy as a "poor simplification" of a complex problem. Töpfer deflected by suggesting differentiated taxes, above all on carbon, as a way to tackle climate change. In September 1990 his ministry proposed a carbon tax. He also began working to turn the Climate Inquiry's recommendations into a legal target to reduce greenhouse gas emissions. On the eve of the 1990 World Economic Summit in Houston, the heart of America's oil industry, Töpfer and the Economics Ministry announced a road map to dramatically reduce their nation's carbon footprint and position West Germany as a climate pioneer. The nation, it seemed, stood on the cusp of tackling humanity's greatest danger.[57]

Reunification and the Dashed Hopes of an Energy Transition

In 1990 West Germany did not become the energy leader as so many had hoped. Momentum ebbed and the nation's ecological decade ended with a disappointment, stymied by political, ideological, and economic developments. The carbon tax was driven into the ground by opponents. The nation's industrial associations protested, claiming it would damage their global competitiveness if West Germany went it alone. The conservative wing of the CDU agreed. In principle, the party's economic council thought prices should include the external costs of pollution, illustrating how far the ideas of Ecological Modernization had come. But there were limits. "If one includes all external costs in the price, as is often demanded," it concluded, "the price for some energies would rise to an unacceptable level on account of their carbon emissions."[58]

Energy savings also went into a tailspin. Under the SPD, West Germany had made strides in efficiency during the 1970s. But facing pressure from industrial organizations, the CDU let investment credits for building efficiency, long-distance heating, and heat-power co-generation lapse. Investment into energy savings plummeted. By the late 1980s the energy efficiency of the nation's building stock—the single largest consumer of final energy—began to deteriorate, while energy use among consumers and households rose. The Economics Ministry, meanwhile, watered down Töpfer's carbon road map, changing what had been a hard carbon reduction target into an "orientation."[59]

Why did West Germany's ecological decade end anticlimactically? At one level the answer is obvious: the drive to reunify a divided nation swept aside all other political goals. At the very moment the climate and energy debate was reaching a first crescendo, thousands of East Germans began fleeing across newly relaxed borders into Hungary and Czechoslovakia, and thousands more began demonstrating in Leipzig. On November 9, 1989, the spokesman for

Communist East Berlin announced that citizens were now free to cross into the Western half of the city. Crowds massed at border checkpoints demanding the gates open, and quickly began streaming westward, celebrating on top of the Berlin Wall. The German Democratic Republic (GDR), already teetering, began to crumble. With the Velvet Revolutions in Eastern Europe, the credibility of economic planning evaporated. But in Germany, 1989 had an emotional layer that extended beyond political economy to questions of identity. Were Germans really "one people," as crowds in the East began to chant that November? In the 1990 electoral campaign Chancellor Kohl answered that question with a resounding yes. That answer turned the prospect of a devastating election for the CDU into one of the party's greatest triumphs.[60]

Reunification supplanted nearly every other political current in 1990, sucking the oxygen out of ambitions for a green transition. When energy did capture the headlines, it was not a carbon tax, but the question of how to provide warmth and electricity to the former East. The GDR boasted the world's highest consumption of energy per person, fueled by dirty lignite coal and excessive subsidies that kept the price of electricity below market levels. Without this state support, in 1990 the eastern grid began to collapse, leading politicians to warn that "the winter of 1990–1991 could be quite cold in the new states." The cost of rebuilding this electricity network was enormous, something only the largest utilities could afford. And they would never spend the vast sums required unless they could extend their monopoly zones to the East. So in August 1990 Bonn brought the former East Germany into its existing electricity grid, extending the 1935 legal framework eastward and giving utilities a dumping ground for their excess power, along with millions of new clients.[61]

Even more disingenuously, the Economics Ministry outmaneuvered Töpfer to include in Germany's carbon reduction calculations the former GDR, the world's worst carbon-emitter per capita. Incorporating GDR emissions would allow Bonn to measure its future reductions from an artificially high starting point, making the newly reunified country look like a "model of restructuring." For the former East offered low-cost channels to lower emissions, above all by shutting down lignite coal powerplants. By 1994 reunified Germany's carbon emissions fell by over 10 percent, one of the best records in the world. But nearly all of this came from the East; in the West carbon emissions per person hardly budged.[62]

Reunification, meanwhile, threw the oppositional parties into disarray, further stifling prospects for a transition. The SPD was split over how quickly to bring East and West together, and the Party's chosen candidate for chancellor in 1990 remained ambivalent. Oskar Lafontaine, prime minister of the Saarland and an ally of the environmental movement, had helped center the SPD's program around energy and Ecological Modernization. But in doing so, he downplayed the momentous opportunity in 1990 to end the division of Germany, arguing for European not German unification. His strategy delivered the SPD's worst electoral performance since 1957.[63] But the Greens fared even worse. Given their antipathy to nationalism, most Greens argued that continued division was the price Germany must pay for the Holocaust as well as the precondition for European peace. The party's 1990 campaign focused on climate, but this was "met with blank stares" in the East, where the economy was cratering. The Greens suffered a horrendous electoral result and failed to enter Parliament that December. Many wondered if they would even continue to exist as a party.[64]

Free Market Revival

Reunification, however, was not the only obstacle to energy reform. Ideas mattered too. For by the 1980s a new faith in markets animated leaders in Britain and the United States, and began to spread around the world. Across much of the Atlantic, the crises of the 1970s had shattered trust in the ability of states to generate growth. After 1979 Margaret Thatcher and Ronald Reagan rode this sentiment into office, promising to restore prosperity with more markets. Both had imbibed the popular work of Friedrich Hayek, and both devoutly believed in the wonders of private entrepreneurship and free markets. In Britain, Thatcher sold off public housing, privatized state-owned enterprises, above all in the energy sector, crushed the coal unions, and cut taxes. Reagan, too, slashed taxes, froze civilian government hiring, appointed anti-state zealots to federal agencies, and fired 10,000 air traffic control workers to weaken union strength. This emerging neoliberal order brought the domestic momentum for a green transition to a standstill in these countries. Reagan, for instance, dismantled the solar subsidies of his predecessor and cut restrictions on hydrocarbon drilling.

Internationally, meanwhile, neoliberals began designing new institutional arrangements to protect markets and private investors from democratic pressures. Economists and lawyers working with the General Agreement on Tariffs and Trade (GATT) in Geneva, for instance, believed mass democracy, with its many interest groups and their claims on economic production, had caused the sclerosis of the 1970s. After 1980 these experts worked to insulate global trade and capital flows through international courts of arbitration, independent regulatory authorities, and bilateral investment treaties, which proliferated. These shifted key policy fields into the hands of "technocrat-guardians," who ensured that market prices and private capital guided the economy, not party politics or voters.[65] Even the European Community began drifting in this direction. After the stagflation of the 1970s, officials at the European Commission—the unelected administrative organ of the European Communities—doubled down on forging a single market. In 1985 Arthur Cockfield, appointed by Thatcher to be her man in the Commission, relaunched integration through the Single Europe Act, which aimed to promote free markets in goods, capital, peoples, and services. To do so, Cockfield and others in Brussels augmented the Commission's power to ensure members states followed the laws of competition that prioritized an even playing field for all.[66] As these ideas spread through organizations like GATT and the European Commission, they undermined the legitimacy of policies that might kick-start a green transition by favoring domestic renewables through price supports for national companies or state-led investment.

Continental Europeans looked on with a mixture of envy and trepidation. After 1981 both Britain and the United States enjoyed higher growth than West Germany. In Bonn, many in the CDU found features of Anglo-American policy useful for distinguishing themselves from Social Democrats. The conservative coalition, for instance, blamed high unemployment and rising public debt on statist policies from the 1970s. The crucial document that toppled the SPD government in 1982 was, after all, an emphatic memorandum by the FDP's Otto Graf Lambsdorff demanding that Bonn promote more markets. In the new coalition the FDP took command of the Economics Ministry, where Lambsdorff's ideas spread. Kohl, meanwhile, echoed aspects of Thatcher and Reagan's language in his first address to the nation, calling for "freedom, dynamism, and personal responsibility."[67]

Despite these neoliberal proclivities, however, most politicians in West Germany, even on the Right, thought Thatcher and Reagan placed too much faith in markets. The CDU instead portrayed itself as the party of Ordoliberalism and the social market economy, to create some distance from American Republicans and British Tories. Original Ordoliberals like Walter Eucken saw a limit to the market, whereas neoliberals like Hayek and Milton Friedman did not—limits defined by ethics and community, or at their worst, ethnicity and race. Wilhelm Röpke and Eucken, moreover, condemned economic concentration, while American neoliberals believed the size of an enterprise mattered little so long as it was efficient and profitable.[68]

Nevertheless, the space for a green transition narrowed in West Germany too, though not as much as in the United States. Informed by this more moderate faith in markets, the conservative wing of the CDU eschewed an activist state that would guide technological development. As its 1984 Party Principles explained, states should merely create a framework for prices and competition, and these would take the economy where it needed to go. Subsidies for specific technologies, by contrast, only led to "mismanagement and structural distortions, and hindered companies from developing their own solutions to a problem." The CDU concluded that "the state cannot and should not prescribe innovation."69 Even Töpfer shied away from state direction of technology. He believed environmental protection was a "civic duty." Change would best come from below, for ecologically-informed consumers could send signals that are "much more effective and faster than anything that we could achieve through laws, prohibitions, and orders."70

There were exceptions to this stance. And a handful of CDU leaders were more inclined to push new technology, above all technology minister Heinz Riesenhuber, a chemist by training who had worked in industry before politics. After Chernobyl, Riesenhuber internally called for more solar and wind power and greater attention to efficiency. But the Economics Ministry was responsible for market creation, and it blocked anything that smacked of subsidies, arguing that "[r]enewables' current lack of economic viability [can] not be made up for through permanent subsidies or taxes that compensate for the cost difference by burdening rival sources of energy."71

More generally, the CDU clung to the heritage of Ludwig Erhard, whose energy policy during the Economic Miracle emphasized the need for low costs to support the nation's export economy. As the party emphasized on numerous occasions, in no other industrial country "do so many jobs depend on the international competitiveness of the economy." Every third workplace in West Germany hinged on exports, compared to every tenth in the United States. West German industry must have "as inexpensive a supply of energy as possible" to sustain its "competitiveness on global markets, its growth, and its jobs," the CDU opined. And for this, markets "work perfectly, if they are not hampered by bureaucrats."72 Low energy prices would come from proven nuclear and fossil fuel technologies, not from the state pushing renewables. These the dominant wing of the CDU considered to be an expensive burden that would raise the cost of energy as a whole: "a vision for the coming century," not this one.73

In the 1980s, in sum, the CDU's Ordoliberalism and the old energy paradigm reinforced one another. What mattered were low energy prices in the present, for exports, provided by the market, not state-guided energies for the future. As a consequence of these views, federal investment into renewables and efficiency fell far short of what it had been under the SPD. Bonn did not organize a new energy research program until 1990.74 After Chernobyl, Riesenhuber gave minor subsidies to wind installations, culminating in the 100 MW Wind Demonstration Program. But its scope remained small and the program did nothing to help renewables break into the mass market that truly mattered, the grid.75

The Energy Countershock

Political events and ideological currents thus led West Germany to miss a window of opportunity for energy reform in 1990. But the window that had opened with the oil shock turned out to be a small one. For by the late 1980s the price of fossil fuels began to fall. Or rather, they plummeted like never before, in an oil "countershock" that rivaled 1973 in its impact by crushing the influence of OPEC. The countershock not only raised doubts about whether fossil fuels were as scarce as many claimed, it also created a new global geography of hydrocarbons and reinvigorated the belief that markets alone could deliver the cheap, stable energy prized by the old paradigm.

The oil shock had disrupted economies around the world, but it contained the seeds of its own undoing. For the rise in oil prices stimulated efficiency improvements everywhere. The Federal Republic and Japan led the way in decoupling energy from growth. But other countries followed to one degree or another, such that by the mid-1980s the non-Communist, non-OPEC world was producing more goods but consuming less oil than in 1973.[76]

The high price of oil also led consumers to switch to other energies, above all, natural gas. It also encouraged the oil majors to develop reserves outside of OPEC territory. Their initial prospecting in new regions, often in inaccessible locations above the Arctic Circle or underwater, seemed unpromising. But with their vast capital and their experience from the tidewaters of the United States, the oil majors mastered new technology like directional drilling and the cost of extraction came down. By the 1980s huge new fields came online in Alaska, Mexico, and Canada, but above all in the North Sea. This stormy body of water between Britain and Norway promised stunning amounts of crude, and lay directly next to Western Europe, far from the choke points that had throttled oil transportation in the 1950s and 1960s. Development there turned Norway and Britain into major players in the global petroleum market. These new discoveries dismantled the axiom of resource scarcity that had guided energy policy since 1973. Global oil reserves had plateaued in the 1970s, but by the 1980s they reached an all-time high.[77]

The aftershocks of 1973 thus created a new geography of petroleum. At its height in 1979, OPEC produced nearly two-thirds of the oil used by the non-Communist world. By 1985 this had fallen in half, the new fields undermining the cartel's ability to influence the global market, and putting downward pressure on the price of crude. In 1982 the spot price of oil began to fall, and that summer the first major cracks appeared in the long-term contracts offered by OPEC. To counter this, Saudi Arabia—the swing producer with the largest reserves in the world—cut output and OPEC organized a string of conferences to reduce supply. But global prices kept sliding until 1985, when Saudi Arabia abruptly changed tactics and began to claw back its lost market share by producing huge volumes of oil and letting prices do what they would. Crude prices crashed, falling 50 percent in just two months in the oil countershock (see Figure 8.2) Only at the end of 1986 did they bottom out.[78]

The oil countershock struck West Germany harder than America. Globally, crude was traded in dollars, and in the mid-1980s Federal Reserve policy let

Figure 8.2. Price of oil (West Texas Intermediate) in US dollars, adjusted for inflation, 1970–2020. *Source:* Data from Macrotrends. https://www.macrotrends.net/1369/crude-oil-price-history-chart.

the greenback plummet in value. Where a strong dollar had once kept West Germany's oil prices higher than in America, its weakness now led German oil prices to collapse nearly twice as rapidly as in the United States. West Germany, moreover, had the most liberal oil market in Europe. With prices uncontrolled by the state, it was an outlet for surplus petroleum products from France, Italy, the Netherlands, and Britain via the Rotterdam spot market.[79]

By the mid-1980s West Germany had the lowest petroleum prices in Western Europe, putting domestic refiners under immense pressure. Some retooled, others closed down. In 1981 Shell announced plans to close its Ingolstadt plant in Bavaria, which had brought oil to the south of the nation. BP too began shutting down refineries in the south and gave up on refining in Northern Germany altogether. By the end of the decade this conglomerate transformed into an oil trading company in West Germany, covering its needs entirely from Rotterdam. Other companies followed suit in selling or scrapping their refineries to become traders of oil like Gulf, Amoco, and Chevron. West Germany's national total refining capacity fell by half.[80]

The countershock fundamentally transformed the oil industry, which since the 1960s had extracted crude outside Europe, sold it through long-term contracts, and processed it in refineries across Europe. This was creative destruction at its best, but it created a new landscape that offered incredible benefits to certain groups, above all companies that pivoted to oil trading. For the exploitation of North Sea oil under Margaret Thatcher not only added to global supplies, it changed the way petroleum was priced and traded. The high volume, low tax regime, and plethora of companies in the North Sea created the conditions for oil futures market to emerge in the 1980s, at London's International Petroleum Exchange and at the New York Mercantile Exchange. These allowed oil trading companies to spread risk across time and space, and brought more consumers and non-OPEC producers together in short term markets and the spot market trade. This eroded the power of OPEC by circumventing the long-term contracts used by states like Saudi Arabia or Iraq. And it reaffirmed the neoliberal faith that markets could overcome energy challenges, as spot and futures markets came to dominate the global oil trade.[81]

This new geography and market structure of oil shattered many of the arguments for a Green energy transition. All of West Germany's parties had pointed to geopolitical insecurity whenever they called for energy reform. But after 1985 the prospect of another oil shock waned, while the notion that fossil fuels were fundamentally scarce seemed to be the quaint concern of a bygone decade. The share of West German oil coming from OPEC producers fell from 96 percent in 1973 to just 56 by the end of the 1980s, with a third of its petroleum coming from the North Sea. On a micro-level, the countershock made efforts to save energy much less rewarding. More than most other places in Europe, the Federal Republic was apparently enjoying a return to the paradise of cheap energy. By the end of the 1980s the nation was spending 20 billion DM a year less on energy than in the 1970s.[82] And this undermined the incentive to rationalize energy consumption. As the Greens themselves lamented, "Today it is not easy to encourage consumers to save energy and to make it clear to community leaders that the time for recommunalizing the energy system has arrived. Since we are apparently standing in a flood of energy."[83]

Renewables: A Distant Hope?

The countershock, lastly, wreaked havoc on renewables. Since 1973 interest in solar and wind had grown around the world. In the Federal Republic, the Technology Ministry began collaborating with large German companies to build photovoltaic (PV) cells, solar heating systems, and wind turbines. This continued in the early 1980s, with AEG establishing Europe's largest PV plant on an island in the North Sea, and West Germany's Aerospace Center collaborating with the IEA on a solar installation in Spain. German engineers mastered the basics of PV and were optimistic about the future. But their hopes resembled the large-scale technological visions spawned by the oil shock: building huge solar farms in the Sahara Desert, converting this energy into hydrogen to be pumped to Europe.[84] With wind power, too, research focused on gigantic endeavors that would support the nation's centralized grid. The epitome was GROWIAN—literally an acronym for Large Wind-Energy Installation. Working with West Germany's aerospace and shipping industry, GROWIAN engineers built the world's largest turbine, with a hub standing 100 meters high—a size that would not become commercially viable for another thirty years. But the entire project was online for just 420 hours.[85]

For some groups, failure was the point. As an RWE executive remarked internally, "We need GROWIAN to prove that wind power won't work." When comparing the cost of electricity from large-scale renewable installations against fossil fuel electricity, wind and solar were clearly *not* working. As energy prices reached rock bottom levels after 1986, solar and wind had little chance to succeed without state support. And this was not forthcoming once the Technology Ministry lost its struggles with the Economics Ministry.[86]

Nevertheless, renewable technology was making strides along an entirely different axis: among independent, small-scale engineers. Since the late 1970s energy innovation had flourished in Baden-Württemberg—heart of the antinuclear movement and center of a thriving craft economy. The protests in Wyhl encouraged a new cadre of innovator-activists to see energy not as an economic but a political project, where costs mattered less than the conviction that another type of energy was possible. "You have a solar system above all," noted *Energiewende* magazine, "to show that contrary to all other claims there is another way." In the region near Freiburg, home to the Eco-Institute, there sprouted an eclectic milieu of tinkerers and craftsmen who began building solar collectors in their garages. In 1976 they started the Sasbach Sundays, the nation's first solar exhibition. By 1980 the region boasted one of the largest assemblies of solar hot water collectors in the world, along with solar vocational programs. It was in Freiburg in 1981 that the physicist Adolf Goetzberger, impressed by warnings from the Club of Rome, established the Fraunhofer Institute for Solar Energy to combine applied research with market rollout.[87]

Small-scale wind producers also made strides during the 1980s, driven by many of the same convictions. Wind production had a long history in Northern Europe, which had been dotted by thousands of farm-based windmills into the nineteenth century. In the 1980s some farmers operated makeshift wind generators, but they stewed at the utilities' monopoly on electricity distribution. These conglomerates demanded exorbitant fees to connect wind turbines to the grid, with a hook-up costing as much as a new Volkswagen. For the power these turbines fed into the grid, utilities would then pay but a fraction of the retail price for electricity. As a wind pioneer put it, "This is what we mean by an abuse of monopoly power." Despite these hurdles, small-scale turbines were the center of action, and they drew as much on farming as on aerospace technology, with some installations generating more power than GROWIAN. These producers, moreover, made inroads into the largest world's largest wind market in Tehachapi, California, where new laws kick-started a "wind-boom" in 1980. Before it went bust, California's roaring wind sector became an experimental field for companies from Denmark and West Germany. And by the middle of the decade these producers formed their own interest groups and began pushing for small, decentralized wind parks back in the Federal Republic.[88]

The combination of top-down research and grassroots tinkering unleashed a positive learning curve for both wind and solar technology, with costs falling as the volume of PV cell and turbine production rose. Solar-generated electricity costs plummeted 90 percent during the 1980s, wind 75 percent.[89] But without state aid, renewables were chasing a moving target as fossil fuel and conventional electricity prices also plunged. This was so above all for solar. Compared to wind, silicon PV was a new technology and its initial costs were astronomical, literally, insofar as America's space program had driven solar development before the 1970s. Because it started from such high costs, PV-electricity remained exorbitantly expensive despite the gains of the 1980s. The cells were thick, the materials costly, and the entire process of building a solar system itself required too much energy. In 1983 a solar house near Munich was opened to great fanfare, covering the energy needs of a typical family entirely from the sun. But critics pointed out that the silicon roof alone cost 100,000 DM. The enormous capital required for solar deterred investors, and even with moderate tax write-offs very little capital flowed into solar in the 1980s. The mainstream media portrayed solar as an "activity for crafty hobbyists, wacked out hippies, and missionary do-gooders."[90]

Advocates like Goetzberger admitted they needed state help to make headway. But the CDU would only support the first phase of research, shutting off the spigot as soon as anything came close to being marketable. This locked solar producers into a vicious cycle in which high costs stopped consumers from purchasing solar panels en masse, which prevented engineers from learning by doing and reaping economies of scale, which in turn kept costs high. To break out of this cycle, argued solar activists, the state must cultivate the mass production of panels by creating market demand at a price that covered costs. They called on Bonn to guarantee a certain volume of solar cell sales, believing only this would unleash the "creative imagination of engineers."[91] Wind fared better, its cost declining to the cusp of being marketable under certain conditions and in certain regions. Where conventional electricity cost 18 cents/kWh, wind hovered between 15 and 45 by the end of the decade, and in the blustery North turbines could at times compete with conventional power. Despite gains, however, the falling price of conventional energy hindered a wider adoption of turbines.[92]

The Electricity Feed-In Law

Thus by the end of the 1980s renewable engineers were demanding a long-term, state-led investment strategy. As the Federal Association of Solar Energy put it, "one cannot just tolerate renewable energies. One has to really want them."[93] Even the Climate Inquiry argued that to avert global warming, Bonn must design a plan stretching over half a century to invest 4 percent of annual GDP into renewable energy, double what the nation currently spent on *all* energy infrastructure. Such a grand program did not materialize in 1990. But a minor one did. And it came from an unlikely quarter: CDU MPs, many of whom themselves had a material stake in hydropower.[94]

In 1979 and 1980 the Economics Ministry and utilities had forged an agreement whereby the latter were forced to accept into their grid power from independent generators. Modeled after America's Public Utility Regulatory Policies Act of 1978 (PURPA), the agreement focused on industrial co-generation plants that generated both heat and electricity. But its vague wording left open the door for renewable installations to also feed power into the grid. The critical question was the price at which utilities would remunerate industrial co-generators, wind farmers, or solar enthusiasts. "Why don't we just run the meter backwards," some asked, and have the utility pay the consumer price of electricity?[95] Regulators, however, interpreted the framework less generously than in the United States, and required utilities to pay merely the variable costs they avoided when buying power from an independent producer. In practice, these "avoided costs" amounted to the cost of fuel the utility saved when it imported power. But the framework did not cover any fixed costs of running conventional power plants or sustaining the transmission lines, which were a major share of a utility's total expense. As a result, the avoided costs paid to independent installations feeding electricity into the grid was pitifully low.[96]

Experts like Peter Hennicke pointed to California to argue that West Germany must be more generous to independent producers. Hennicke, the Eco-Institute, and the Greens, along with Hauff and the SPD, demanded that Bonn force utilities to pay independent producers more than the bare-bones avoided costs. Court cases failed to resolve the issue, and so embittered solar, wind, and hydro producers began lobbying for higher feed-in rates. In the meantime, municipalities began conducting their own experiments with renewables in Bremen, Hamburg, Düsseldorf, and Saarbrucken, the latter of which was paying 25 cents per kWh for electricity fed into the grid by the end of the 1980s.[97]

A different impulse for change came from a group of conservative MPs in southern Germany—land of mountains and fast-running rivers. Matthias

Engelsberger, a backbench MP in the CSU who would retire at the end of 1990, resented the existing utility framework. His family owned a small dam in Bavaria, and Engelsberger had struggled to get a better price from the utilities, but to no avail. With support from other CDU MPs, in 1988 Engelsberger called on Bonn for better feed-in rates, justifying his appeal by the "deterioration" in the cost of oil that had transformed the "overall world energy price level."[98]

The government showed little interest, Bavaria's leading politician calling it "foolish" to expect an energy solution from renewables.[99] But Engelsberger tried again in 1990, motivated not only by the countershock but by the "increasingly real climate catastrophe." And he drew directly on Ecological Modernization to make his case, pointing to the "external costs" of fossil fuels to demand an improvement in feed-in rates. The price should include not just the fuel costs of energy production, but also the costs of "forest damage, plant damage, damage to buildings, damage to human health, and what now appears especially urgent, climate-related damage through the emission of carbon dioxide." He estimated the social costs of fossil fuels to be 6 to 10 cents per kWh, which if incorporated into the market price of electricity would immediately make much wind and most hydro competitive. Engelsberger, in other words, broke with the mainstream CDU and demanded the state forge "market opportunities" for renewables where none existed before.[100]

Engelsberger and seventy other members of the CDU reached across party lines to gain support from a faction of the Greens and a handful of MPs from the SPD. The former feared they were losing their hold on renewables as a Green issue. Wolfgang Daniels, a Green MP and Bavarian scientist who knew Engelsberger personally, had already introduced several parliamentary energy initiatives, but he coordinated poorly with his own party, which wanted a broader law to advance efficiency, recommunalization, start-up financing, and an eco-tax.[101] The SPD leadership was even more critical of Engelsberger's law. At a time when Bonn was giving away billions to the utilities in East Germany, the CDU bill was "not a step in the right direction. This is not even a shuffle; it is a little toe wiggle, nothing more."[102] Nevertheless, in the SPD Engelsberger found support from Hermann Scheer, the party's leading advocate of renewables, who along with Volker Hauff had asked Parliament to launch a major new program into solar and hydrogen.[103]

Reunification almost pushed the feed-in bill off the agenda. CDU leaders only gave Engelsberger the chance as a favor for his years of loyalty to the party, and they refused to allow the Greens to add their name to the bill. Englesberger, moreover, had to contend with resistance from the Economics Ministry, which tried to torpedo his proposal as a market-distorting subsidy.[104] This led to an unorthodox process in which he and other CDU backbenchers wrote the legislation, not the government. Arguing that higher feed-in prices were not subsidies, but a way to manage the social costs of fossil fuels, they portrayed the bill as a small intervention that would have no tangible effect on the price of electricity.[105] Rallying support from the emerging wind lobby, Bavarian hydropower producers, the ecological wing of the CDU, along with factions of the opposing parties, the Electricity Feed-in Law passed Parliament on December 7, 1990. With it, West Germany's ecological decade came to a close, the hopes of the preceding years pinned on a minor law that increased compensation to wind and solar from 8.5 to 16.7 cents per kilowatt hour.[106]

Conclusion

In 1990 the Federal Republic did not emerge as the green energy champion like so many had hoped. For a decade defined by a nuclear meltdown and a dawning climate catastrophe, very little changed in the country's energy system on the surface. West Germany still satisfied 97 percent of its energetic appetite from hydrocarbons and nuclear power. Massive corporations still dominated the grid with an archaic law from the Third Reich. Decoupling seemed to have broken on the shoals of the oil countershock.[107] The postwar paradigm of extensive growth fueled by abundant energy seemed to withstand the earthquakes of the 1970s, reinforced by a revival of free market thinking, the countershock, and the rise of oil futures markets. The CDU's response to the Climate Inquiry captured the persistence of this older regime, for the party warned that the greater danger was not climate change but the "cost explosion" that would ensue if Bonn followed the Inquiry's recommendations. "We must regrettably conclude that renewable energy can help a little, but it will not be the workhorse that can free us from the worries of our future energy supply."[108]

By 1990 the CDU had governed for eight years and would do so for another eight. This sixteen-year reign stymied momentum for an energy transition, as shibboleths crafted during the Economic Miracle prevented the party's dominant wing from making a political case for energy savings or renewables. Helmut Kohl's CDU saw their country as an export nation that depended on inexpensive energy for its success. A green transition would have to be repackaged and sold as an engine of modernization and the seedbed for a skilled workforce that could provide an edge in the global marketplace. This the CDU failed to do. Instead, the party leaned on Ordoliberalism to portray federal aid for renewables and efficiency as a cost and a market distortion to be avoided.

Even the one successful piece of energy legislation seemed trivial. The Feed-in Law of 1990 covered a quarter of one percent of the nation's electricity. At first it cost utilities 50 million DM, a pittance for a sector that earned 90 billion DM a year. And the vast majority of this went toward hydropower. Critics panned it as a "bonbon" for the ecological movement at a time when utilities were reaping windfall profits in East Germany.[109]

Despite these pitfalls, the 1980s nurtured green energy seeds that had been planted in the 1970s. The Greens and the SPD fell closer in sync, finding common ground around ecological taxation, recommunalization, efficiency, and the need to frame energy policy as a political affair instead of an economic one. The new feed-in legislation, meanwhile, cracked the armor of the utilities just wide enough for wind and solar to gain a foothold in the grid.

Lastly, global warming became a political topic. Despite the focus on reunification, Germany's public came to appreciate that unseen pollution from fossil fuels not only caused regional problems, but global ones too. This had the potential to reshape the calculus of energy policy. For concern about climate change emerged at the very moment that many German politicians began adopting the ideas of Ecological Modernization. Before 1980, pricing the external costs of energy had been an obscure idea of environmental economists. But by 1990 these ideas took center stage in parliamentary debates about energy, even among the CDU. The coal turmoil of the 1950s and 1960s had led many to argue that the state must guide energy prices for social reasons; the oil crises of the 1970s did the same for geopolitical reasons. The climate catastrophe added a third layer to these calls to see the price of energy as a political one.

But going forward, Ecological Modernization and the new energy paradigm would face a novel challenge. For a new energy infrastructure was spreading across Europe—East and West—that promised to parry the criticisms of the old paradigm and give a new lease on life to fossil fuels. Perhaps the country did not need renewables, some argued, if it could rely on natural gas, a seemingly clean fuel, from a country that cultivated a special relationship with the Federal Republic: Russia.

9

The Energy Entanglement of Germany and Russia

Natural Gas, 1970–2000

All Socialist countries are waiting for natural gas. . . . All of Europe is waiting. The pipes for this will be bought in the Federal Republic. Once all of this is finished, then cooperation is guaranteed for 30, even 50 years.[1]

In September 2001, Vladimir Putin, Russia's new president, addressed the German Parliament in Berlin during his inaugural state visit to the Federal Republic. In the new glass dome on the banks of the river Spree, Putin spoke of a common security framework, of mechanisms for cooperation. But he also spoke of the economic ties that bound his nation to the Federal Republic. To emphasize these links, his next stop was Germany's industrial heartland of the Ruhr, where Burckhard Bergmann, chairman of Ruhrgas AG, presented the Russian president with a historic natural gas meter at the Düsseldorf trade fair (see Figure 9.1). It symbolized the greatest economic and physical link connecting the two countries and celebrated the nearly 400 billion cubic meters of gas that had flowed from Russia to Germany since the first pipeline between these two countries had opened in 1973.

Upon becoming president in 2000, Putin began to rebuild Russia's economy, which had cratered after the breakup of the Soviet Union. One pillar of this campaign was reasserting state control over Gazprom, the natural gas conglomerate that was Moscow's prime source of tax revenue, and solidifying sales to Germany, which bought more Russian energy than any other country on earth. A multi-tiered campaign ensued, spearheaded by Germans as much as by Russians. Putin toured Germany. Bergmann joined Gazprom's board of advisors, and Ruhrgas, Germany's largest gas corporation, began a publicity campaign to promote ties with Russia. In 2001 it gave 3.5 million dollars to rebuild the famed Amber

Figure 9.1. Putin (right) accepting a historic gas meter from Ruhrgas CEO Burckhard Bergmann (center) and North Rhine-Westphalian Minister President Wolfgang Clement (left) in 2001. *Source:* Reuters / Alamy Stock Photo. Image ID: 2D3M3HB.

Room of St. Petersburg, which had been gifted by King Frederick William I to Peter the Great, only to be looted by the Nazi Wehrmacht during World War II.[2]

Putin's Russia would become an international energy power; the country's growth, hard currency, state income, and geopolitical influence hinging on the export of hydrocarbons. Yet this was not a novel development: Russia began its evolution into a global energy force during the second half of the Cold War, when the Soviet Union surpassed the United States to become the world's largest producer of oil and natural gas.[3]

But the USSR and its Russian successor did not achieve this transformation alone. To tap its vast endowment of natural gas, Moscow needed capital and cutting-edge technology to extract and transport this fuel across incredible distances, as well as a market that could reliably buy the huge volumes it pumped out of the ground. It found all this in the Federal Republic of Germany. During the last three decades of the twentieth century, Russia's rise as an energy power was co-produced by Germany's energy transition to natural gas: the latter enabled the former and vice versa. The two countries built this symbiotic energy relationship on a material foundation of processing plants, pipelines, compressor stations, and storage systems that linked the gas fields of the Ural-Volga region and Western Siberia with Central Europe. Much like the oil supply chain that bound the Federal Republic to the Middle East after 1950, this physical network embodied an immense amount of capital and fixed the directional flow of energy from east to west. Thoroughly a product of the Cold War, it was profoundly shaped by, and in turn would itself shape, German domestic and foreign policy.[4]

Germany turned to natural gas because this hydrocarbon seemed to solve an array of challenges, many of which had nothing at all to do with energy. When it burst onto the European scene in the 1960s, gas was portrayed as a noble fuel. As the smallest hydrocarbon molecule—methane: one carbon and four hydrogen atoms—gas requires little refining to be useful. Unlike petroleum or coal, it leaves no tangible pollution when it combines with oxygen, just water and odorless, invisible carbon. But at first, natural gas was confined to the heating market and covered less than one percent of West Germany's energy needs. Beginning in the late 1960s, however, policymakers and boosters linked gas to key political issues, triggering the nation's fourth energy transition and transforming gas into Germany's second largest energy source by 2000. Social Democrats led the way, seeing this energy as a bridge across the Iron Curtain that would convince Moscow their nation was a reliable partner, and thus help the Federal Republic improve its relationship with its sibling state to the east. During the oil shocks of the 1970s, Social Democrats added security to their justification, portraying gas as a safe alternative to OPEC oil. After 1980 these same elites argued that the construction of gas infrastructure in Russia would sustain their country's ailing steel sector during a wrenching recession. Finally, in the 1990s new leaderships under the CDU re-envisioned natural gas as a prime sector to liberalize and expand in their quest to sustain the competitiveness of Germany's export economy. Gas became a neoliberal key that would provide cheap energy without, the party maintained, endangering the environment.

As Ecological Modernization and the new energy paradigm gained momentum after 1980, the natural gas transition helped the old energy paradigm retain its grip on the Federal Republic. Politicians from the conventional wing of the SPD and the CDU lauded gas as an abundant, inexpensive energy the nation must have to remain an export powerhouse. Conservative experts claimed that gas was clean, which let this energy sidestep the greatest criticism of the old paradigm, namely, that burning fossil fuels warmed the atmosphere. They used the old paradigm to justify Germany's deepening ties to the east: Germany needed energy to grow, Russia had energy to give. Gas, put differently, created a dual lock-in for Germany: to Russia and to the old energy paradigm. And this undermined Ecological Modernizers in their quest to build a new, green energy system. Why turn to renewables and efficiency when natural gas was abundant, inexpensive, supposedly clean, and already flowing without the need for a technological revolution?

Problems nevertheless emerged in Germany's natural gas transition. The infrastructure and capital requirements of pipelines meant gas became dominated by a small cluster of massive, insulated corporations that increasingly epitomized the technocratic, anti-democratic energy system which reformers despised. This new supply chain diminished Germany's dependence on the Middle East, but only by redirecting dependence to a new locus of fossil fuels, post-Soviet Russia. Meanwhile, new evidence came to light that natural gas might not be as beneficial for the climate as conventional wisdom suggested. In the new millennium, this criticism and acclaim would thrust gas, once a low-profile energy, into the center of the political battle for the energetic future of the Federal Republic.

Pipeline Conflicts and the Early Days of Natural Gas

The Federal Republic's energy relationship with the Soviet Union began not with natural gas, but with the fuel that powered Europe's transformation into a high-energy society. Since the 1880s Imperial Russia, then later the Soviet Union, had been one of the world's prime oil producers. But only after 1945 in the Cold War rivalry with the United States, the pioneer of hydrocarbon society, did the USSR begin putting oil at the heart of its economy. Between 1955 and 1965 it developed new fields in the Ural-Volga region that tripled its oil production. And to earn hard currency, in 1956 Moscow launched an oil offensive into Western Europe, with plans to build one of the longest pipelines in the world—"Friendship."[5]

The USSR, however, needed high-grade pipes that only Western Europe and the United States could produce at scale. Thus erupted one of the first energy conflicts of the Cold War. By 1959 Cold Warriors were warning of energy dependence if Europe became addicted to this "red oil flood" from an ideological opponent.[6] But while Britain embargoed Soviet oil and the European Community (EC) proposed a ceiling on its import, West Germany prioritized its own steel exporters. Over the next two years its companies would ship 700,000 tons of wide-diameter pipe eastward, transforming the Federal Republic into Moscow's largest commercial partner.[7] Mounting Cold War tensions, however, eventually altered the calculus in Bonn. First the German Democratic Republic (GDR) shocked the world by closing the border between East and West Berlin in 1961. A year later, just weeks after a major German-Soviet pipe deal, the world came palpably close to nuclear Armageddon during the Cuban Missile Crisis, leading

NATO to demand its members end strategic exports to the Soviet Union. Under pressure from the United States, Bonn began sanctions that December, formally denying export licenses to German companies exporting pipes to the Soviet Union.[8]

This first Cold War energy standoff created a pattern that would resurface over the coming decades, pitting American politicians against West German industry. In acrimonious confrontations, the CDU would side with the former, while the SDP would support the latter in the name of jobs. In 1962 the CDU only passed sanctions by walking out of Parliament to prevent quorum, allowing the embargo to take hold without a formal vote, tactics that Social Democrats castigated as a "blow to democracy." West German elites, moreover, drew lasting lessons from 1962: that sanctions hardly worked and that they came at the expense of their nation's economy. For despite the embargo, the "Friendship" pipeline opened with little delay, while German companies forfeited valuable contracts and German trade with the USSR plummeted by a quarter.[9]

Energy from the Eastern Bloc, put simply, was politicized from its inception. To the west, however, a different energy began to penetrate the Federal Republic at the very same moment, though with far less publicity. In the spring of 1959 a joint venture by the two largest global oil majors, Esso and Shell, found a natural gas field in the farmlands of the Netherlands. After two months the flow grew so large that flaring could be seen miles away in the town of Groningen. It soon became clear that Esso and Shell were sitting on massive natural gas reserves in the heart of Western Europe.[10]

Before Groningen, natural gas covered a tiny proportion of West Germany's energy supply, and was primarily a byproduct of oil. While the United States had started using this fuel in the late nineteenth century, in Western Europe only a handful of small gas fields were active, serving regional markets. Nevertheless, when Esso and Shell began working to export their gas into the Federal Republic, they encountered an existing ecosystem of pipelines crisscrossing Northwestern Europe that carried coal gas for heating and lighting. This network was particularly dense in the Ruhr, where it distributed gas generated by the coking plants that powered Germany's steel industry. In the 1920s Alfred Vögler, chairman of the largest steel conglomerate, had founded Ruhrgas AG in Essen to supply the gaseous byproduct of his business empire to local municipalities.

By the 1960s, however, Ruhrgas was struggling. Coal gas output was falling and its price declining as the majors dumped Middle East oil into West Germany. Within a few years, moreover, it was clear the Netherlands could never absorb the immense natural gas of Groningen. So after initially resisting Dutch gas, Ruhrgas realigned its business model and partnered with Esso and Shell to sell the energetic wealth of Groningen through its own pipelines. Esso and Shell bought the

German company Brigitta, through which they acquired holdings in Ruhrgas in return for the rights to use this company's distribution network. After mergers and acquisitions among Germany's other gas companies, by the mid-1960s the oil majors, chiefly Esso and Shell, gained access to West Germany's entire gas pipeline network. For the next generation natural gas would be run by a small, interlocked nexus of global energy giants and domestic conglomerates—a "gas club" of professionals who knew one another through intertwined careers. At the top of the hierarchy stood Ruhrgas, which would buy gas in twenty-year deals from Esso and Shell, move this energy through a network over which it exercised total control into regions in which it had monopoly power, and sell its gas to local companies—many of which were partially or fully owned by Ruhrgas itself—for distribution to the final consumer.[11]

In 1966 Groningen gas began flowing into Western Germany and a slow rolling energy transition commenced as Ruhrgas reoriented consumers to its new fuel. This transition evolved without much publicity: natural gas companies never developed a glamorous image like their oil counterparts, and few households even knew which distributor actually supplied their gas. Gas, moreover, was disconnected from the geopolitical upheavals that afflicted its hydrocarbon cousin, like the Suez Canal crises. After 1966 Ruhrgas and other gas companies aggressively marketed their product, first to the metallurgical and chemical industries, which required intense heat to smelt steel or produce plastics. They then turned to households, where a four-way competition between gas, fuel oil, hard coal, and electricity unfolded. Ruhrgas invested a billion DM to upgrade consumer appliances and hook-ups from coal gas to natural gas. It allied with other distributors to advertise natural gas through the mascot "Trixie Sunshine," portraying their product as the liberator of kitchens and housewives. In some towns natural gas heating grew tenfold in a decade. Gas stoves and gas room heating spread as this new fuel became a part of West Germany's high-energy society.[12]

But while natural gas was a clean, easy fuel for heating compared with many alternatives, its price remained high. Esso and Shell's business hinged on petroleum, including fuel oil, which households and industry also used for heat. After 1966 Esso and Shell worked with the Dutch government to elevate the export price of Groningen gas so it would not undermine their main energy and source of profits. Over the coming years fuel oil prices fell steadily while gas prices at the Dutch border hardly budged. German industry complained, arguing that this manipulation threatened their country's economy, where gas prices were 40 percent higher than in the Netherlands. State officials, meanwhile, began worrying the gas market was "caught up in the hands of a single monopolistic group."[13]

Ostpolitik: Accelerating the Gas Transition

So by the late 1960s interest in a different, less expensive source of gas beyond the control of the oil majors arose. That source lay on the other side of the Iron Curtain, across the Ural Mountains, in the heart of the Soviet Union, under the control of an entirely different political and economic system. Getting access to this gas would be an intensely political operation.

The operation began in 1966 when Willy Brandt, foreign minister in a new Grand Coalition with the CDU, crafted a novel approach to the Eastern Bloc. Having joined the SPD at the age of seventeen, Brandt was a committed, reform-minded Social Democrat who spent the war exiled in Sweden and Norway. Upon his return, he had worked in the former capital during the first Berlin crisis and became Berlin's mayor in 1957. His political career was made on the front lines of the Cold War: he was in Berlin when the wall rose, and he stood behind President Kennedy when the latter declared himself to be "a Berliner." Brandt feared the division of Germany was deepening, and that the CDU's hardline approach to the GDR—of refusing formal recognition—was counterproductive. Instead, Brandt wanted to normalize relations with the Eastern Bloc, and reassure East Berlin and Moscow that the Federal Republic had shed its revanchist ambitions. For in the 1960s Moscow still saw West Germany as a state run by a Nazified elite that wanted to regain the borders of 1937. With his new strategy of *Ostpolitik*, Brandt hoped to amend this view of his country, make the Iron Curtain more permeable, and ultimately increase contact between East and West Germans.

Central to Brandt's agenda was the belief that closer economic ties with the Soviet Union would advance political rapprochement by reassuring the powerholders in Moscow. Egon Bahr, Brandt's press officer in West Berlin and later secretary in the Foreign Office, had coined the expression "change through rapprochement (*Wandel durch Annäherung*)." After 1966 this became the foundation of Brandt's approach to the Soviet Union. As he bluntly put it, he wanted "more trade and less polemics." Gas would be a pillar of this trade agenda, and in the late 1960s Social Democrats linked this energy to their grand Ostpolitik strategy. In 1967 Bahr portrayed cooperation around gas as "proof of Germany's willingness to ease tensions," by appealing to Moscow's ambitions to export energy. Bahr also hoped a German-Soviet gas deal could generate "an immediate upswing in the steel and coal industry," and a boost to German-Soviet trade, which had stagnated since the 1962 embargo.[14]

On the other side of the Iron Curtain, Moscow elites were themselves becoming more interested in energy ties with Western Europe. Since its inception, the Soviet Union had been struggling to exploit the natural resources of its vast but inaccessible northern lands. In the early 1960s this campaign gained urgency, however, when Soviet geologists venturing up the Ob River and other Arctic-bound waterways made incredible discoveries. Western Siberia, they realized, contained the largest cluster of natural gas reserves in the world. The discovery of one new gas field followed another, climaxing in 1966 with a record-breaking find in Urengoy. The energetic magnitude of these discoveries rivaled those of the giant oil fields found in the Middle East at mid-century. Moscow created a Ministry for Gas and authorized export negotiations with Western European customers. For the new gas fields of Siberia presented a unique opportunity not only to support socialist industrialization with cheap energy, but to sell energy abroad in return for Western technology.[15]

The possibility of a grand East-West commercial exchange now materialized: Soviet gas for Western European technical equipment. A first tentative arrangement between Italy and the USSR failed. But in 1968 a second, with Austria, succeeded. West German industry and regional politicians in Bavaria saw the Austrian deal as a model, and worked to join in. Chief among these companies was Mannesmann AG, located on the outskirts of Düsseldorf in the Ruhr. Founded in 1886 with the world's first patent for seamless steel pipes, Mannesmann grew into one of Germany's leading steel manufacturers on the basis of its eponymous production process that built some of the world's most advanced pipes. Its pipes were stronger, and better able to withstand higher pressure and harsh conditions than those of its competitors. During the red oil wave of 1960, the company had recognized the business potential of the Soviet Union, only to lose $25 million when Bonn canceled its contracts. Once the possibility of a gas pipeline arose, Mannesmann lost no time wooing the Soviet trade delegation, who toured their Düsseldorf headquarters and nearby factories. The company teamed up with Thyssen AG, another of West Germany's great steel conglomerates, to win the bid to supply pipes for the Austrian-Soviet deal. And through Bahr, Mannesmann lobbied to add Bavaria to the deal in the hope of getting more Soviet contracts. The Economics Ministry torpedoed the idea, however, arguing that in his rush for a deal Bahr had ignored the possibility that Moscow might use a gas disruption for geopolitical ends.

Nevertheless, the Austrian contract minus Bavaria succeeded, and it wetted the appetite of Germany's steelmakers, for the Soviet Union represented a gigantic potential customer. At the heart of this deal and others to come was infrastructure—the exchange of technology and pipelines of incredible size, as well as the financial instruments to enable their construction, in return for the raw material that would flow through them.[16]

Two years later German-Russian gas cooperation would commence in earnest after a conjunction of developments created a more favorable environment.

In 1968 West Germany reorganized its ailing hard coal sector, once the nation's prime energy source, creating the space for new fuels to penetrate the heating market. In 1969, meanwhile, a new era of Cold War détente commenced when US president Richard Nixon began working with his counterpart in the USSR, Leonid Brezhnev, on reciprocal nuclear arms limitations. By then, the Soviet Union had violently crushed the Czechoslovakian reform movement to become more secure in its Eastern European empire, and more open to normalizing relations with the Federal Republic.[17]

Brandt hoped to capitalize on the softening Cold War, but strains with his CDU coalition partners during the 1969 election prevented a grand geopolitical bargain. Instead, Brandt turned to natural gas to move Ostpolitik forward. In January 1969 the Social Democratic state secretary of the Economics Ministry visited Moscow to feel out the Soviets, who responded at the Hannover Trade Fair that April by proposing a deal similar to the Austrian one, only larger. Discussion lasted through the fall, with Germany represented by private corporations in a complicated trilateral negotiation involving the sale of Soviet gas in return for pipelines and credit. As the largest German gas distributer, Ruhrgas muscled its way into the deal over the heads of a Bavarian gas group. Ruhrgas hoped to reduce its reliance on high-priced Dutch gas while the Economics Ministry, in turn, believed this conglomerate was the only company large enough to handle a potential Soviet disruption. Mannesmann and Thyssen, meanwhile, built on their experience in the Austrian deal to provide the pipelines.[18]

Because Brandt and Bahr saw natural gas as a spearhead of Ostpolitik, West Germany's government shaped matters from the beginning. That fall, state officials guided the negotiations, visited Moscow to keep dialogue flowing, and helped Ruhrgas and the Soviets settle on a gas contract price acceptable to both parties. Even more important, the state smoothed out the main hurdle in this first deal, credit. The Soviets were asking for an unprecedented loan in Western currency to purchase the huge volumes of steel they needed. Wilhelm Christians, chief officer of the Deutsche Bank, took the lead in organizing a consortium of West German banks to provide credit. As a board member of Mannesmann, Christians was connected to Ruhr industry and understood the importance of the deal. Keenly interested in raising his bank's foreign stature, he would become a key financial diplomat who brought West Germany closer to the USSR. Christians supported the pipeline deal, but the size of the loan was unprecedented—1.2 billion DM—and entailed major risk. Bonn consequently guaranteed half the credit, enabling the Deutsche Bank consortium to offer an absurdly low interest rate—3 percent below market.

West German allies roundly criticized this trilateral deal as "financial aid" for Communism, since Bonn's financial support was unprecedented in size and duration. The federal government nevertheless gave its blessing, and in November

1969 West German and Soviet negotiators signed the private contract, a prodigious agreement. For $400 million, Mannesmann and Thyssen would deliver 1.2 million tons of pipe to build a 1,500-mile line from Siberia to the West German–Czechoslovakian border. The deal would employ their pipe factories full time for two and half years. Ruhrgas, meanwhile, would buy up to 5 billion cubic meters (bcm) of natural gas a year once the line opened, enough to cover 3 percent of the country's primary energy needs.[19]

In February 1970, at the Kaiserhof Hotel in Essen, dozens of Soviet and West German leaders celebrated the signing of this historic contract in the presence of West Germany's minister for economics and finance. That July the first pipe destined for the Soviet Union left Mannesmann's factory on a flatbed truck, adorned with a symbolic wreath of fir branches and named Ludmilla. Because the agreement came to fruition after Brandt's landmark electoral victory in September 1969—the first time a Social Democrat became chancellor since the 1920s—it symbolized an Ostpolitik breakthrough. Going forward, West German and Soviet officials portrayed gas as the foundation for a new era of cooperation and a model that built confidence for the future. When Bahr began negotiating the political goals of Ostpolitik—recognizing borders and sovereignty—he and his Soviet counterparts opened by celebrating pipelines. Gas, in other words, paved the way for the famous accords that would normalize relations between West Germany and the Eastern Bloc: the Moscow and Warsaw Treaties of 1970, and the Basic Treaty with the GDR in 1972.[20]

The deal, however, and Ostpolitk in general were not without critics, and it began to change gas from a boring energy into a politicized one. Esso and Shell vigorously protested the contract and threatened to flood West Germany with oil to undercut Soviet hydrocarbons. Coal producers, too, lobbied against the agreement. American Cold Warriors, meanwhile, feared Germany was reasserting a more neutral and nationalistic stance in the heart of Europe. Most importantly, the CDU pushed back strongly. Out of power since 1969, the party lambasted Ostpolitik as a sacrifice of German interests, abstained from ratifying the Moscow and Warsaw Treaties, accused Bahr of treason, and came within a hair of achieving a vote-of-no confidence in Brandt's government. In the process, natural gas came under fire, above all the credit assistance, which the CDU castigated as a political price. More ominously, CDU energy experts warned that "when it comes to supplying life-critical products [like energy], you do not go to the hands of those who seek your life."[21]

Brandt nevertheless pushed on, since energy cooperation was finding support on both sides of the Iron Curtain. In 1971 Soviet premier Alexei Kosygin invited West German industrialists to tour Siberia, and is reputed to have toasted them with the remarks: "Gentlemen, you have seen the great possibilities. Please help yourselves." In 1972 the same consortium of Ruhrgas, Mannesmann, and

Deutsche Bank inked a second pipes-for-gas deal with the Soviets. Then in early 1973, Brezhnev himself visited Bonn to drum up support for a broader German-Russian campaign to develop Siberia's resource wealth. Born of a working-class family in Czarist Russia, Brezhnev rose through the party ranks after World War II to became general secretary of the Communist Party in 1964 after displacing Nikita Khrushchev, his former patron. Following the turbulence of his predecessor, Brezhnev followed a conservative agenda of political stabilization at home, alongside a pragmatic approach to foreign affairs. But as he slowly consolidated power, he worried about his country's economy. Despite reforms during and after Khrushchev's reign, Soviet growth rates and productivity were falling, and the technological gap with the United States was palpably widening in the latter's favor. Despite his efforts to forestall liberalization, under Brezhnev's watch the Soviet Union would become more entangled with the global economy than ever.[22]

In Bonn, Brezhnev boasted his country was becoming the world leader in hydrocarbons, surpassing even the United States. But by 1973 Moscow faced problems of technology and scale. Its older oil and gas fields in the Urals-Volga were declining, driving up the costs of energy even as the Soviet bloc needed ever more of this precious input. Siberia was to be the answer, but developing this distant region was challenging. The Soviets had to build entirely new cities from scratch to extract resources from the ground. Hot, humid summers brought forth vast swarms of insects that made for agonizing working conditions. In the winter, sub-zero temperatures and frozen tundra that thawed rapidly each spring presented hurdles for transporting gas over the staggering distances separating Siberia from the sites of consumption. The Soviets needed larger pipes with thicker walls than anything their industry could produce at scale. And they needed state-of-the-art compressor stations to condense gas at the appropriate pressures. Moscow soon found it had to sink ever more capital into this region for even a modest return.[23]

The Communist regime, put simply, was sitting on a treasure trove but it needed help tapping its riches. So the promise of Western technology motivated Moscow's pivot to the Federal Republic. A metallurgical engineer before World War II, Brezhnev held German industry in high regard, and believed the Federal Republic could provide aid now that it was governed by a sympathetic Social Democratic leader. In Bonn, Brezhnev laid out a vision of technology-for-energy cooperation that would last a generation. In return for its engineering prowess, West Germany would gain access to "natural resources that are nearly inexhaustible." Within a decade every fourth German household could be heated by Soviet gas.[24]

Brandt reciprocated, hoping to ensure that Moscow would look to West Germany instead of the United States for technology. To the German press,

Brandt hyped up a future of energy cooperation with the Soviet Union: he wanted more natural gas, more oil, and an "energy network" to link the two countries. This, his experts realized, would require long-term agreements and more state support to overcome the distances and the risks involved. The chancellor was happy to oblige. When the Federal Republic released its first formal Energy Program in September 1973, it projected a tripling of gas consumption over the coming decade: a rate of growth exceeded only by nuclear power.[25]

Oil Shocks and Security through Gas

On October 1, 1973, the director of Ruhrgas pressed the button in Waidhaus on the Czechoslovakian border, and the first Russian gas began flowing into West Germany. As *Die Zeit* declared: "the Russians are here." This particular gas came from Western Ukraine. But soon it would flow from Siberia.[26] For two weeks later the first oil shock erupted, accelerating Brandt and Brezhnev's vision of long-term energy cooperation.

With the oil crisis, the SPD added a new layer of justification for Soviet gas atop the old: security. The quadrupling of oil prices transformed West Germany's hyper-dependence on Middle Eastern petroleum into the prime geopolitical risk associated with energy. Brandt acknowledged this less than a month after the first oil price hike, at a press conference where he underscored that his nation must do everything it could to "reduce dependence of its energy supply on crude oil." Imported natural gas was high on his list of alternatives. As he later explained to Brezhnev, Soviet gas would "reduce our vulnerability to future blackmail." Overnight, the oil shock transformed West Germany's ideological opponent into a relatively reliable energy partner.[27]

Gas deals now came fast and furious. Just days after the OAPEC embargo, West Germany's economics minister spoke with Iranian leaders about buying their gas and transporting it through the Soviet Union to West Germany. Ruhrgas's executives had broached this idea earlier that spring, but now Moscow and Bonn got serious. In May 1974 the first trilateral meeting took place, with West Germany represented again by Ruhrgas. After a rocky start, the three countries concluded a huge deal the following spring, in which Iran would ship 13 bcm a year of gas to the Soviet Union, which would in turn send 11 bcm to Western Europe, all to start in 1981 following the completion of a new pipeline to be built by German companies.[28] The Economics Ministry, meanwhile, requested more funding for gas infrastructure. And in the fall of 1974 Bonn, again through Ruhrgas, struck a third bilateral gas-for-pipelines deal with the Soviet Union.[29]

Under Klaus Liesen, Ruhrgas now ramped up geopolitical arguments to market natural gas as a secure energy. Liesen, a lead negotiator in the Soviet deals, became head of Ruhrgas in 1976. One of the most influential men in Germany's economy, he epitomized the close ties between state and business that defined the energy industry, having worked at the Economics Ministry before coming to Ruhrgas, and then later joining the supervisory board of Mannesmann and VEBA, the nation's largest domestic energy conglomerate. Liesen argued that with enough investment, gas would advance "the stabilization and reliability of [West Germany's] energy supply" because its delivery was predictable. Not only were gas contracts concluded for timespans of twenty years or longer; the dense network that West Germany, France, Austria, Italy, and the Netherlands were constructing would "exert a powerful gravitational pull in the future. It permits the intake of natural gas from every direction and transfers this into one of the most interesting and efficient energy markets in the world." Exporters, in other words, had a huge incentive to sell to the world's most integrated gas market. Meanwhile, the pipelines stretching outward from Western Europe locked in the directional flow of gas, since the cost and specificity of this infrastructure meant the Soviet Union would be "unable to use the facilities" for "anything but deliveries to the other party of the contract."[30]

Natural gas, moreover, benefited from oil's surging price, which drove consumers in the heating market as well as industry to turn toward gas. Indeed, 1974 marked a watershed for this energy: that year, for the first time ever, West German gas imports exceeded domestic production.[31]

The Deal of the Century

From here the gap between domestic and imported gas would explode, for the biggest Soviet-German gas accord—the "deal of the century"—was still to come. Despite being an energy superpower, after the oil shock the Soviet Union faced its own quandaries. Soviet hydrocarbon reserves were growing, but demand for energy in the Soviet bloc was growing even faster. This was problematic, because the Soviet economy increasingly relied on the sale of energy abroad to pay for its imports. By 1978 energy exports accounted for 80 percent of Moscow's hard currency earnings. Brezhnev faced a dilemma: use Soviet gas to support development in Eastern Europe; or export it to the West in return for desperately needed technology and equipment.[32]

Siberia, however, could solve this dilemma, and in the late 1970s Brezhnev began a new campaign to extract gas from this inhospitable region, above all Urengoy, the world's largest gas field. Brezhnev looked first to American companies to build a pipeline and Liquified Natural Gas (LNG) Terminal in Murmansk. But cooling relations between the superpowers iced this Northstar project. So in 1978 the Kremlin flew Liesen to Moscow to discuss a new contract with Ruhrgas. The Soviets, however, wanted a guarantee that West Germany could actually buy the huge volumes needed to make this project viable. They envisioned a pipeline that would be exclusively dedicated to European exports, running over 4,000 kilometers across permafrost, mountains, swamps, and rivers. Liesen promised Ruhrgas could buy 20–25 bcm a year. Christians of the Deutsche Bank made his own visit to Moscow to give similar assurances. This set the stage for Brezhnev's return to Bonn in May to hash out an energy agreement with Helmut Schmidt, SPD chancellor since Brandt's resignation. Brezhnev wanted an agreement that would stretch into the new millennium.[33] Schmidt reciprocated. An avid proponent the old paradigm, Schmidt was at the same time recommitting to nuclear energy, against strong opposition from his own party. And like Brandt, he believed in Ostpolitik and thought the Russians "should be put into the position of losing their fear of Germany." Long-term agreements like what Brezhnev proposed would lead "the people in both countries to acquire a permanent interest in one another's economic welfare." The result that May was a twenty-five-year framework for economic cooperation between West Germany and the Soviet Union, based on the exchange of natural resources for technology.[34]

The new deal only took concrete form, however, after the 1979 revolution in Iran, which transformed one of the world's greatest hydrocarbon states into an Islamic Republic hostile to Western Europe. The revolution sparked a second oil crisis, and it capsized the tripartite gas deal between Iran, the Soviet Union, and the Federal Republic, throwing gas markets into turmoil.[35] West German leaders became even more worried about energy security, with Schmidt calling his nation's dependence on Middle Eastern oil "extremely dangerous"; that this was the crisis he "most feared." That May, a year after meeting with Brezhnev, Schmidt announced that following the Iranian Revolution, "all of us in all countries, not just in the United States, have to undertake all our efforts to open up other sources of energy than just oil."[36]

But in the early 1980s most sources of gas looked problematic. Analysts expected Dutch fields to be depleted within the next two decades. Algeria, another gas supplier, had unilaterally abrogated a major contract with several Western European countries. And as a member of OPEC it now demanded that gas prices be fully linked to oil, which torpedoed another major deal with Italy and France. Norway, meanwhile, the only other major source of gas for West Germany, suffered from the perception of reliability problems following a series of strikes by its gas workers.[37]

Soviet gas, in other words, seemed the best option for advancing two of West Germany's most important foreign policy goals: softening Cold War tensions and securing a more reliable source of hydrocarbons. So in 1980 the deal of the century advanced. Deutsche Bank leaders visited Moscow early that year to prepare the ground. Chancellor Schmidt followed in June. The potential contract was so immense that rival companies made bids, including Krupp and the hydrocarbon giant BP. They lost to the usual suspects of Ruhrgas, Deutsche Bank, Mannesmann, and now AEG Kanis, subsidiary of one of West Germany's largest engineering companies. The final matrix of deals—finance, pipes, and gas—was again superlative. Deutsche Bank spearheaded an international consortium that lent over $10 billion to the Soviets, ten times more than in the 1969 deal. Export contracts for the industrial consortium totaled 10 billion DM, with Mannesmann alone selling 3.5 billion DM in pipes.[38] In 1981 the parties signed a framework agreement during Brezhnev's final visit to the Federal Republic, with the pipeline to be named "Yamal" after new gas fields to the north of Urengoy. Kosygin lauded it as "the largest pipeline in the world." Schmidt saw it as the embodiment of Ostpolitik, a commercial contract that would outlast any political turbulence between the two geopolitical blocs.[39]

Criticisms surfaced immediately, above all the familiar reproach from American and European allies that the Deutsche Bank was again offering shockingly low interest rates to the Soviets—8 percent compared to a market rate of 12. Its consortium could afford this because German banks, residing in one of Europe's most liberal capital markets, were sitting atop a mountain of petrodollars following the oil shock. But also because the German state was again subsidizing much of the credit, making it nearly risk-free.[40]

Yet the geopolitical context of this mega deal differed from the first accord I 1969, for Cold War tensions were rising rather than falling. Already in the late 1970s regional conflicts and ideological disagreements over human rights began to strain East-West relations. In December 1979 these tensions burst into the open when Soviet forces landed in the capital of Afghanistan after a local coup threatened to form a regime hostile to Moscow. President Jimmy Carter called the Soviet invasion the "greatest threat to world peace since the Second World War." He expanded defense spending, enacted a grain embargo on the USSR, boycotted the Olympics, tightened restrictions on the export of technology, and began funding Afghan rebels. Western Europe's mild response to the invasion, however, infuriated Carter and his advisors, who worried that détente had made Europe too cozy with the Soviet Union. News that America's allies, led by the Federal Republic, were crafting an enormous energy deal with Moscow reopened rifts that had been simmering since the 1962 oil embargo.[41]

American hawks feared that Yamal, if completed, would deepen West German energy dependence and compromise Bonn's willingness to contain the Soviet Union. The electoral victory of Ronald Reagan—a devout anti-Communist who would famously call the Soviet Union an "Evil Empire"—ratcheted up tensions still higher. After 45 senators petitioned Reagan to stop Yamal, at the G7 summit in July 1981 Reagan asked West Germany to reject more gas from the Soviet Union, offering them instead American coal or support for nuclear reactors. As his assistant secretary of defense explained:

> There is the day-to-day influence that must flow, like the gas itself, through a pipeline to which there will be no practical alternative. Is there any doubt that our allies would listen more carefully to kings and rulers who supply them with energy than to those who do not?[42]

In late 1981 the pipeline debate became a diplomatic crisis when General Wojciech Jaruzelski declared martial law in Poland to suppress the nation's trade union movement. The Reagan administration responded by imposing sanctions on Poland and the Soviet Union. That December, as Ruhrgas, Mannesmann, Deutsche Bank, and AEG-Kanis were arranging their Yamal contracts, the United States suspended high-tech exports to the Soviet Union. In June 1982 the US added a second layer of sanctions, claiming "extraterritorial application of its export controls." This embargoed European firms that relied on technology licenses or components from American companies, hitting corporations that would build the compressor stations. West Germany's press exploded with indignation: the Americans, *Der Spiegel* complained, "were treating us as if we were not sovereign states."[43] Bonn and the rest of Western Europe told their companies to push on with the contracts. As Schmidt quipped in response to Reagan's sanctions, "the pipeline will be built."[44]

By 1982 transatlantic relations had been brought to a nadir by the question of energy security. According to American estimates, if Yamal went through the Soviet Union would provide 38 percent of West Germany's gas imports, even more for regions like Bavaria, and this was bound to influence Bonn's foreign policy.[45] West German officials disagreed. The issue of energy security had emerged for their nation earlier than the 1970s and extended far beyond the Soviet Union. In the 1960s German experts had developed a relativist approach to security: because the Federal Republic imported nearly all its hydrocarbons, it could only achieve security by balancing one energy or region against another, never letting a single source become dominant. In Chancellor Schmidt's words, "all energy entails a certain foreign policy risk." And he clearly believed OPEC was more volatile than the Soviet Union, which had, in his opinion, "always remained true to its contracts." For him, Gaddafi's Libya was a greater threat since it provided over 8 percent of West Germany's energy supply. Soviet gas, even if the American forecasts were accurate, would cover less than 6 percent. In any case, Schmidt's experts argued that the Federal Republic could handle a gas shutoff by ramping up imports from Norway or the Netherlands, slowing the industrial use of gas, investing in Germany's internal pipeline network, and tripling the reserve requirements for gas companies—all of which they incorporated into West Germany's third and final revision to its Energy Program.[46]

Gas, Exports, and Jobs

Yet security was not the only issue at stake. Beneath the surface another powerful motivation was moving West German leaders to finalize the "deal of the century": their country's hyper-dependence on exports. Like the Brazil nuclear deal five years earlier, in 1981 the quest for exports trumped transatlantic ties, for the Federal Republic was descending into a wrenching recession. Growth had slowed after the 1973 oil shock, but the country weathered the 1970s far better than the United States. This changed with the second oil shock. Unemployment rose, manufacturing profits slumped, investment shriveled, and private consumption fell. And, for the first time since 1950, the Federal Republic ran a major balance of payments deficit, spending more on imports than it earned from exports.[47]

The oil shocks were partly to blame. Since 1973 the cost of importing oil had risen from $3 billion to $20 billion.[48] Sky-high energy costs coincided with a global downturn that saw demand fade for West Germany's traditional exports, like steel. But in many ways, this was just the dramatic tail of a slow-moving restructuring that the country only began to feel in the 1980s. The prime casualties were many of the dirty, energy-intensive industries on which Germany had built its reputation, like steel and electrical engineering. During the Economic Miracle these had grown explosively in response to Europe's physical reconstruction. At its height, the steel industry employed 340,000 workers in the Federal Republic, and deep linkages with other sectors, like coal and shipbuilding, made steel the heart of an industrial cluster. But as reconstruction tapered off, the steel industry became more dependent on exports. And by the 1970s new competitors in Japan, South Korea, Taiwan, and Mexico entered the scene.

With the 1981 recession, deindustrialization, a scourge that first afflicted textiles and hard coal, began to ravage steel. That year 40 percent of steel capacity lay idle and the industry was shedding a thousand jobs a month. Long-term unemployment surfaced in Dortmund, Düsseldorf, and Duisburg. Steel workers—mostly older, skilled male workers—became bitter as they lost not only jobs but their identity. And the younger generation quickened their emigration from these industrial towns, leaving behind depressed communities. The federal government responded with a steel support program. But many believed the best salve was helping German steel do what it did best: exporting. As Schmidt argued to American negotiators, "the gas-pipe business is also about contracts for our industry, which cannot come from anywhere else."[49]

Yamal thus killed two birds with one stone—keeping energy prices low for growth while creating contracts for German industries teetering on the brink. For some companies the deal was existential. Mannesmann was hemorrhaging money. Since 1980 its export orders had evaporated and its profits collapsed. In 1981 the company lost $52 million. Yamal, management hoped, would be a game-changer, engaging 60 percent of its capacity and generating 2,500 jobs.[50] For AEG—which employed 100,000 workers in West Germany—the stakes were even higher. Its Kanis subsidiary had not turned a profit in six years and in 1981 lost 600 million DM. AEG as a whole, meanwhile, was on the cusp of bankruptcy, and in 1982 Bonn opened negotiations with a bank consortium to discuss a bailout. Kanis's bid to build the lion's share of compressor stations could determine the company's fate.[51]

By 1982, in other words, Yamal became more about German employment than Soviet gas. As the Economics Ministry put it, "we would have been able to survive very comfortably without the Soviet natural gas. . . . The pipeline contract was dictated by pure misery—jobs were the main consideration." The US embargo was a hammer blow. After it came, AEG's stock plummeted and management feared the end of Kanis, while Mannesmann began predicting another year in the red.[52]

Given the state of its economy, throughout the controversy Bonn prioritized its commercial interests over the Atlantic alliance. By aggressively supporting German companies and instructing them to defy American sanctions, by using targeted interventions to break logjams in the negotiations, by giving state support for the financing, Bonn did everything a liberal state could to facilitate the deal.[53] After months of negotiation, in November 1982 the United States signaled its willingness to end the embargo following a change in leadership at the State Department. In return, West Germany and its neighbors agreed to hold off on new Soviet contracts until the International Energy Agency (IEA) completed a study on East-West energy trade. Yet the agreement had little that was concrete, and Western Europeans declined to accept the IEA recommendations as binding. After the lifting of sanctions, West European firms signed another $1.5 billion worth of contracts with the Soviet Union. AEG-Kanis and Mannesmann expanded their existing deals and other West German companies joined the splurge.[54]

Reinforcing the Old Paradigm

West Germany thus rode into the 1980s on the crest of one of the largest energy-infrastructure projects of the twentieth century. Yamal was the fifth Soviet gas line built with German aid since 1969. When it ceremonially opened in late

1983, it could theoretically pump 40 bcm of gas a year. Together with other new lines, it altered the geography of Soviet energy by making Siberia the fossil fuel heart of the country. In 1983 the USSR surpassed the United States in the volume of gas produced, and Brezhnev's vision of transforming his country into an energy power took a leap forward.[55]

But the significance of the new infrastructure went beyond Soviet power to shape everything from European energy consumption to the potential for a green transition. At the beginning of the Yamal dispute, the US assistant secretary of commerce predicted that "developments in the next ten years will be crucial in determining the pattern of world energy markets for decades to come." What seemed like hyperbole was eerily accurate, for Yamal combined with other events to give the old paradigm a new lease on life. In the Federal Republic, on the plane of ideas, Ecological Modernization and decoupling were triumphing by the 1980s, driven forward by oil shocks, fears of resource scarcity, and global warming.[56]

But on the level of policy and infrastructure, Ecological Modernization was making little headway. The oil countershock of the 1980s was partly to blame. But just as important, even as criticism of the old paradigm gained momentum, the material infrastructure supporting this traditional set of energy ideas, above all gas infrastructure, was growing relentlessly around the world. Globally, gas reserves reached record heights as new fields came online not only in the Soviet Union, but also in Qatar, Norway, and Iran. In Europe the crown jewel was the gas network linking Central Europe to the energy reserves of the Ural Mountains and Siberia. These gas fields were so immense as to seem limitless, obviating earlier fears that the Federal Republic might run out of fossil fuels. In the North Sea, meanwhile, after 1986 new Norwegian fields came online. This flood of gas drove down prices. in the late 1980s the Soviet Union cut prices still further for distributors who could buy large volumes, to ensure that enough gas flowed through their new network to make it functional.[57]

In co-creating the gas link to the USSR, the SPD thus secured a new, non-OPEC source of cheap fossil fuels. While Ecological Modernizers and SPD reformers were calling on Germans to warm their homes with heating pumps and efficiency measures, consumers were instead turning to gas. By 1985, 7 million homes—a quarter of West German dwellings—were heated by natural gas. Under the SPD's watch, between the signing of the first Soviet contract and the opening of Yamal, gas consumption rose more than fivefold in the Federal Republic, far outpacing the growth of any other energy source. In 1969 natural gas covered 4 percent of the nation's energy needs; by 1990 more than 17 percent.[58]

Natural Gas, Neoliberalism, and Climate Change

Following Yamal, gas became less politicized. The CDU, governing Germany with the FDP after 1982, directed little attention to gas in its messaging about energy, particularly once Chernobyl drew the government's focus to nuclear power. But in the 1990s gas returned to the limelight when the CDU linked this energy to the rising creed of neoliberalism, arguing that market forces could advance gas to make German more globally competitive *and* fight climate change. Natural gas, put differently, became a neoliberal symbol of how markets could solve even the thorniest of problems.

Like Schmidt before him, CDU Chancellor Helmut Kohl was an advocate of the old paradigm, believing Germany's "economy depends on inexpensive energy if we want to survive in international competition." And Kohl's party saw economic policy as the quest to make the country fit for international business, something the chancellor emphasized in his first address to the nation and clung to for the coming decades.[59]

In the globalizing world of the 1990s, many in the CDU believed economic success required cheap energy. And affordable energy, the party argued, would come only with free markets composed of many buyers and sellers, who would compete to drive down prices. The gas market, however, was hardly a market at all. A handful of large conglomerates controlled domestic distribution in demarcated fiefdoms, barring entry to rivals. Local companies, many owned by the distributors themselves, held a virtual monopoly selling to final consumers. The situation resembled an "oligopoly confronting a monopsony." As a result, German gas prices were determined by opaque long-term contracts between large suppliers and even larger distributors. Ruhrgas stood atop the food chain: through a network of companies in which it held majority shares, this giant distributed over half of Germany's gas.[60]

Watching energy liberalization unfold in the United States and Great Britain, the CDU began advocating "as much markets as possible" for energy. Following reunification in 1989, Bonn pushed liberalization throughout the economy, calling for "deregulation, privatization, and market opening" in its first postreunification coalition agreement. It moved first to liberalize telecommunications and the postal service; gas and electricity were next on the docket.[61]

But affordability was not the only goal. The CDU framed energy policy as a dilemma between competing goals of low-priced power and environmental protection. Because the party had not fully adopted the tenets of Ecological Modernization, it still saw ecological preservation as being at odds with growth, a view expressed in its energy guidelines for the 1990s and its parliamentary discussions about liberalization. As the Economics Minister clarified, energy policy was about getting "an efficient and competitive energy supply for Germany. It is about cheap and environmentally friendly energy." In the wake of the oil countershock, energy security faded from view, particularly when it came to gas.[62]

Neoliberals and gas boosters argued that liberalizing gas could help Germany thread the needle of this dilemma, lowering energy prices while helping the country fight global warming. Since the 1970s and 1980s the government and Ruhrgas had portrayed gas as a clean energy. In contrast to coal, which was stoking fears of acid rain and Forest Death, gas generated neither sulfur nor nitrogen. In its energy programs the Federal government highlighted this feature, casting gas as "environmentally friendly." When pressed on the issue, the CDU's environmental minister admitted gas was not "absolutely" pollution-free. But he argued that it bested other fossil fuels in all categories of emissions when burned. And while scientists recognized that methane leakage from pipelines contributed to global warming, the best estimates put natural gas as just fifth among anthropogenic sources of methane, behind wet rice, livestock, burnt biomass, and landfills.[63] Those in the gas business went further, calling their product an "environmental winner," "almost perfect." Ruhrgas claimed the infrastructure it built would "carry huge quantities of energy from distant sources to consumers through invisible systems leaving the beauty of nature untouched"—at least in West Germany, the point of consumption.[64]

When Bonn proposed a natural gas tax in 1988 to reduce the budget deficit, Liesen led the industry in ramping up clean gas publicity. The gas lobby castigated the tax as a "stab in the back against environmental protection."[65] After 1990 researchers from Ruhrgas produced a flurry of studies suggesting gas was "the best substitute energy for protecting climate," the "greenest fossil fuel" because it generated roughly half the carbon as coal or oil when burned. They found support from neoliberals like Alfred Voss, director of the Institute for Energy Economics and Rational Energy Use in Stuttgart. Voss avidly supported nuclear power and the old paradigm, believing advanced economies needed immense energy use to thrive. He had circulated through the traditional energy think thanks, including the Nuclear Research Institute in Jülich, where he roundly criticized the soft energy path advanced by Ecological Modernizers. In the 1990s Voss earned a spot as expert advisor to Germany's Second Parliamentary Inquiry into Climate Change. He argued that alongside efficiency and nuclear power, which Germany should uphold, the country must use more natural gas as the least costly method of reducing greenhouse gases. By focusing on carbon over methane, this constellation of industrial and neoliberal experts concluded that gas could be "a bridge to a sustainable, environmentally and climatically benign future energy supply."[66]

In the 1990s the CDU deployed these arguments to promote gas for political as well as environmental motivations. For as climate change became a more pressing public concern, it gave momentum to Ecological Modernizers. These reformers wanted to tax energy heavily, and they aligned primarily with the political opposition of the SPD and the Greens. But with natural gas the CDU could retaliate and defend the old paradigm and its lobbies against criticism.[67]

The CDU's gas campaign hinged on the states of the former GDR, which after 1990 became a field of experimentation for neoliberal policies. There lignite, the dirtiest form of coal, powered the grid, warmed households, and spawned a sulfurous smog in cities like Berlin and Leipzig. The CDU wanted a total "restructuring of the energy supply" in this new territory by replacing its lignite plants with gas turbines.[68]

Gas's "special role" in the transformation of former East Germany and the fight against global warming heralded a dramatic expansion for this energy. Since 1973 the Federal Republic had discouraged the use of gas to generate electricity, to save this fuel for industry and households. The European Community had issued a similar directive in 1975, and as a result the power sector used precious little gas before 1990.[69] With the CDU's energy agenda for the East, this changed, because it coincided with the spread of a revolutionary new technology. In 1960 an Austrian utility company had deployed the first commercial combined cycle gas turbine (CCGT) to exploit the gas coming from Groningen. The new design produced electricity from a gas turbine, then captured the waste heat to power a steam boiler which generated still more electricity. Over the coming decades American companies honed this technology, and by the late 1980s falling gas prices combined with further advances to unleash a dash for gas, first in Britain, then the rest of Europe. The great advantage of CCGT was its small scale, which required far less capital and boasted much shorter build-times than reactors or the increasingly enormous coal plants.[70]

Though East Germany was no tabula rasa, reunification created fluid conditions for the Federal government to promote a new energy infrastructure. The former GDR, moreover, had a robust district heating network that could pump warmth across urban zones, warmth that before 1990 came from lignite. The region thus became ground-zero for a wave of investment in gas power plants, and gas use for district heating more than doubled in the seven years after reunification. Epitomizing the boom was the Nossener Brücke CCGT plant, built with aid from the European Investment Bank to provide power for much of Dresden while also heating 90,000 homes. Such projects helped gas break into the market for electricity and fueled gas consumption in the East, which in turn led gas demand nationwide to far surpass the growth in overall energy consumption.[71]

Post-Soviet Space and Gas Liberalization

But where to find more natural gas to fuel this boom, and how to ensure it remained affordable? This now became a pressing question. Bonn hoped liberalization would bring in new gas producers and distributors. But Ruhrgas vigorously protested liberalization, and after 1990 it moved into the new Eastern states by buying the dominant stake in the GDR's old gas distribution entity when it was privatized. Liberalization would be a hard nut to crack.[72]

The practical answer came from an unlikely place: the subsidiary of Germany's leading chemical company. In 1990 BASF, the largest chemical, plastic, and fertilizer conglomerate in the world, heir to the Nazi cartel IG Farben, and the single biggest consumer of natural gas in the Federal Republic, began worrying about the price of energy. Though it owned some gas reserves, BASF bought most of its supply from Ruhrgas, which was gouging the chemical conglomerate through its quasi-monopoly. BASF's management wanted to circumvent Ruhrgas by finding its own source of gas. After looking to Scandinavia, it turned its gaze eastward.

There BASF found a crumbling empire that owned 40 percent of global gas reserves. While the fall of the Berlin Wall symbolized for Germans the start of a new era, for Soviet citizens the collapse of their polity in 1991 marked an even greater caesura. When the presidents of Russia, Ukraine, and Belarus recognized each other's independence that December, the USSR ceased to exist. In its place rose a host of states—in Europe: Latvia, Lithuania, Estonia, Ukraine, Belarus, Moldova, and of course, Russia. For liberal elites across Western Europe and America, this heralded the birth of a "new world order," as American President George H. W. Bush put it, defined by democracy, peace, and capitalism. But the fracturing of what had once been the Soviet Union created political change that many in Moscow resented. The European Union began planning to expand eastward already in 1993. So too did NATO. The lands of the former Communist superpower, meanwhile, experienced a devastating economic collapse, as GDP declined by half and millions of people fell into poverty.[73]

Soviet disintegration profoundly shaped European energy affairs. Moscow privatized the Ministry of Gas, but through a combination of fortune and political acumen this institution avoided being torn asunder during the post-Communist market transition frenzy. In its wake emerged Gazprom, which would become the largest natural gas company in the world, with ownership of nearly all the gas fields in Russia. Gazprom, however, encountered its own problems. Soviet economic collapse led to an unprecedented decline in energy consumption, which dropped by a third, leaving Gazprom with a monumental gas surplus. The company responded by aggressively marketing its product across Europe, sparking a Russian gas bubble that flooded westward. Gazprom directors, moreover, feared they were losing out on the most lucrative part of the supply chain: distribution to final consumers. For the company's control of gas ended at the German border, where Ruhrgas took over and marked up this fuel for local consumption to earn enormous profits.[74]

In 1991 BASF struck a deal with Gazprom that would change Germany's energy landscape. Thirty years prior, BASF had bought Wintershall, a leading domestic gas producer. Wintershall now formed a joint venture with Gazprom, named Wingas, to deliver Russian gas directly to German consumers. In return for exclusive rights for Wingas to sell gas from Gazprom's Yamal fields, BASF agreed to build a chemical complex in Siberia. So began BASF's end-run around Ruhrgas, and for Gazprom, a model of joint ownership it would replicate across Europe. Yet Wingas did not own any pipelines in Germany, nor could it pump gas through the networks controlled by Ruhrgas. Wingas lobbied Bonn to require Ruhrgas to open its pipelines to third party access (TPA), launching a court case against its rival. "As long as there is no TPA," Wingas directors argued, "there will be no real competition." If Wingas wanted in the market, Ruhrgas responded, it had the "freedom to build [its own] pipelines."[75]

The legal challenge for TPA failed, so in 1992 Wingas began doing exactly what Ruhrgas suggested. Over the next four years it invested 4 billion DM to build pipelines and storage facilities across the Federal Republic: from the Czech border through Thuringia to Hesse; from the North Sea to Frankfurt and then Ludwigshafen, home of BASF. At a larger scale, Gazprom began building new pipelines from Siberia through Belarus and Poland, which would become one more physical link between the two countries. The new infrastructure unleashed a brutal price war in Germany. As Wintershall's chairman put it, this was "competition at its most primitive level." The industry would be "walking up to our knees in blood."[76]

Wingas's gamble suggested that liberalization might work, since prices declined wherever its new gas lines opened to rival old ones. The competition made gas far more widespread, particularly in eastern Germany where a gas boom accelerated. Yet this particular form of competition was immensely expensive, requiring newcomers to invest billions to build new lines from scratch. And it was restricted to only a handful of regions, as Wingas officials pointed out: "Now we have a two-class society: cities, consumers, and industries in a favorable position next to a Wingas pipeline; and those far away who have to pay extra." Far better, so CDU officials and their advisors realized, would be regulatory competition that forced distributors to open their network to third-party suppliers. ENRON Corporation, which was advancing free-wheeling energy markets in America, lobbied Bonn for TPA. So too did Germany's industrial gas consumers, since they thought price cuts would follow.[77]

In 1996 the Economics Ministry began crafting an energy liberalization law. By then a higher authority was moving in this direction as well. Brussels, de facto capital of Europe, was coming off the great success of passing the Treaty of Maastricht, which created a more tightly integrated European Union (EU). Nevertheless, many EU leaders were pessimistic about Europe's place in a globalizing world. EU growth rates had stalled, its investment rates had plunged, its unemployment rates remained stubbornly high, and its competitive position vis à vis American and Asian exporters was deteriorating. The Commission— executive organ of the EU—looked with envy across the Atlantic as America's economy unleashed a burst of digital innovation. The EU must reform or it would stagnate. Maastricht had advanced the transformation of Europe into a single market in which goods, services, and capital could cross borders. But the single market still did not extend to energy. Throughout Europe this sector remained dominated by monopolies operating in tightly protected national markets. So after 1993 the Commission turned its liberalizing gaze to energy. One model came from Britain, where Margaret Thatcher had broken apart the national power monopoly. The Commission hoped to replicate aspects of this groundbreaking policy on the continent, by bringing competition to sheltered markets, and above all, ensuring TPA so that new gas and power companies could access the networks controlled by traditional monopolies. In the spirit of the old paradigm, the Commission hoped these measures would bring down energy prices and make European companies globally competitive. As the Commission's 1995 Green Paper explained, "the search for competitiveness" lay at the heart of this agenda, which must "ensure that energy is made available in the most economic manner to end-users."[78]

After struggling against states that guarded their national energy champions, including Germany and France, in 1998 the Commission managed to push a gas liberalization directive through the European Parliament and the European Council. Ruhrgas protested that Brussels's framework would not work in Germany, which had no single state monopoly but an eco-system of small and large distributors. And it pointed to its foe, Wingas, to claim that competition was already unfolding without meddling from the EU. Nevertheless, in one of its last pieces of legislation before losing power, the CDU passed the Energy Industry Liberalization Act to comply with the EU. A follow-up act from a new government in 1999 gave newcomers access to the networks built and owned by the likes of Ruhrgas, at least in theory. It seemed that gas liberalization had arrived in Germany. Officials, big industry, and consumers groups expected gas prices to fall, gas consumption to surge, and the country to power ahead in the transition to a seemingly cleaner fossil fuel.[79]

Conclusion

By 2000 the Federal Republic was deeply entangled in one of the world's largest infrastructural networks, one it had co-produced with the Soviet Union and Russia over three decades. The architects included not only Soviet leaders like Leonid Brezhnev, the Ministry of Gas, and its successor Gazprom, but also Germany's SPD and CDU, neoliberals, and some of the nation's largest companies like Ruhrgas, Deutsche Bank, Mannesmann, Thyssen, AEG, and BASF. This physical supply chain that pumped gas from the Ural Mountains and Western Siberia to Central Europe locked Germany into a distinctive energy trajectory in two ways. It bound the Federal Republic to Russia even as it bound the country to the old paradigm by channeling fossil fuels into the nation's homes, factories, and now even power plants.

Germany came to depend on this incredible infrastructure less because gas was an inherently noble or inexpensive energy, and more because politicians and boosters linked gas to causes that resonated across the political spectrum, like Ostpolitik, employment, security, and liberalization. The birth of this physical network could not have happened without Willy Brandt's quest to soften the Iron Curtain. Its subsequent development hinged on the perception of gas as a secure fuel in a world riddled by oil shocks, and on the promise of contracts and jobs for German companies that would construct the pipelines and compressor stations. Where Ecological Modernizers looked to renewables and energy efficiency technology to be the job engines of the future, defenders of the old paradigm could point to gas infrastructure, which had helped keep an ailing steel sector alive. In the 1990s natural gas again thrived. After breaking into the market for electricity with CCGT technology and getting linked to a neoliberal vision of markets, gas, it seemed, could overcome the dilemma of the old paradigm by providing abundant, cheap energy without accelerating environmental damage.

Yet this neoliberal vision never came to full fruition; reality never matched theory. After growing explosively in the 1990s, gas consumption stalled during the following decade and problems materialized. Ruhrgas fought liberalization tooth and nail. Instead of creating an independent regulator to enforce TPA, the state had to opt for a looser liberalization—negotiated TPA—in which incumbents could effectively deny access to newcomers. By the first decade of the twenty-first century, gas still hardly resembled a market with transparent prices and many buyers and sellers. It remained dominated by five corporations who jealously guarded their customers and kept German gas prices among the highest in Europe.[80]

After 2000 the industry came to epitomize the large-scale, opaque power nexus—the hard energy path—vilified by Greens and Ecological Modernizers. While the electricity sector would begin a painful transformation, gas moved more slowly and the system of the 1960s and 1970s remained largely intact: an industry run behind closed doors through long-term contracts by a small cadre of professionals who enjoyed close ties to state officials. Even the limited competition that arose after 1990 required newcomers to mimic the incumbents they aimed to displace, since only enormous, capital-rich firms could afford to build their own pipelines. At the apex of the industry, Ruhrgas saw its power and profits grow relentlessly during these decades. In the 1970s its network managed half of all German gas, two decades later three-quarters—no less than 13 percent of the nation's entire primary energy supply. During the decade of Social Democratic rule that linked the Federal Republic to the Soviet Union, Ruhrgas's total profits grew sixfold. By the early 1990s they nearly tripled again. The company, moreover, retained ties to the other great symbols of the old energy paradigm. Oil majors like BP, Esso, and Shell retained commanding shares in Ruhrgas before they were bought out of this gas giant in a multibillion-dollar deal by Germany's largest utility, E.ON, successor to VEBA and an anathema to the Greens. In the twenty-first century, gas remained a cash-cow firmly anchored to the old paradigm.[81]

Even the myth of gas as a clean fuel began to crack, weakening this energy's defenders in their struggle against proponents of the new paradigm. When Ruhrgas and government officials touted the benefits of gas, they ignored the zones of extraction and transit where the real human and environmental toll was unfolding. In the rush to start the flow of gas, the Soviet Union drafted prisoners and probationary workers to speed construction. As Moscow prioritized haste, the quality of the infrastructure suffered in the difficult terrain of the Siberian permafrost. Poor welding led pipes to burst; blockages led to delays in transport; compressor stations broke down. The result was a network riddled with accidents, seasonal floods collapsing parts of the pipeline in Czechoslovakia and gas explosions killing dozens in Siberia. In 1989 an electric spark from a passing train caused the largest gas accident in Soviet history along the new lines, killing hundreds. The constellation of roads, settlements, refineries, drilling stations, and flaring sites at the source, meanwhile, strained Siberia's environment by polluting local air and water, and destroying the migratory routes of birds and reindeer.[82]

Most egregiously, this vast network was hardly airtight, something scientists began to realize already in the 1990s. Germany's second inquiry into climate change estimated that no less than 8 percent of all gas moving from Siberia and the Urals westward—40 billion cubic meters a year, equal to roughly half the gas Germany burned annually—leaked into the air. At the point of extraction, moreover, Russia flared more natural gas than any other nation on earth—tens of billions of cubic meters a year—transforming this fuel into carbon before it ever warmed a home or powered a factory. Contemporary experts, moreover, reckoned that methane accounted for 13 percent of anthropogenic climate change, likely a major underestimate given what we know today. Contrary to the remonstrations of German politicians, their nation's appetite for Russian gas was fueling global warming. The Eurasian infrastructure built with German aid and in German interests represented a massive climate pollutant.[83]

Finally, the dilemma that adherents of the old paradigm thought they faced was transforming into a trilemma. Politicians like Brandt and Schmidt alongside Ruhrgas directors believed they had found in the Soviet Union a secure energy partner. As Wintershall's directors put it, the Soviet Union (and later its Russian successor state) was a "safe bet." Yet Russian imports outpaced predictions: by 2000 Germany bought 35 percent of its gas from Russia, above the threshold deemed acceptable by Chancellor Schmidt in the midst of the Yamal negotiations. Gas imports rose just as the former Soviet Empire was crumbling into a new geopolitical landscape. Pipelines built during the Cold War followed the same route, but the space they transited was itself changing. Some Soviet successor states joined the EU while others gravitated toward Moscow, sowing the seed of future tensions. This harbored the potential to revive energy security as a problem. For most Russian gas after 1991 passed through a newly independent Ukraine. And almost immediately debt disputes between Moscow and Kiev led to temporary shutoffs that briefly threw European energy markets into turmoil. These disruptions soon passed, but they foreshadowed turbulence to come.[84]

After 2000 Germany's energy entanglement with Russia would deepen further, when Gazprom began discussing a pipeline to link Russia directly to Germany through the Baltic Sea: Nord Stream. All of the players who had forged the Russian-German gas nexus since 1970 participated. Gazprom—heir to the Soviet Ministry of Gas—spearheaded the project. E.ON, heir to VEBA and owner of Ruhrgas, along with BASF joined as major shareholders. Mannesmann again supplied the pipes, while Deutsche Bank was a lead financier for the project. In 2005 Vladimir Putin, having consolidated power over Russian energy, visited Berlin to preside over the formal declaration of intent to start building this new pipeline for the twenty-first century.[85]

Also presiding was the SPD Chancellor Gerhard Schröder. In 2003 Schröder had helped Putin inaugurate the reopening of the Amber Room. After losing the 2005 election, Schröder would join Nord Stream's board of directors, epitomizing the persistence of the old paradigm and the symbiosis that conventional energy companies maintained with leading state officials. But paradoxically, five years earlier this very same SPD leader oversaw the beginning of an energy transition with the potential to break the old paradigm for good. For Schröder, despite his qualms about green power, led a novel coalition that would transform the landscape of German energy for the twenty-first century.

10

Unleashing Green Energy in an Era of Neoliberalism

The Energiewende

> Do we want those who operate hydropower, wind, solar or biomass installations to make a profit? I want to put it even more clearly: Do we want the producers of solar and wind electricity to earn the same profits as RWE, Preussen-Elektra, and Bayernwerk, who destroy the environment with coal and nuclear and who, as we all know, have money coming out of their ears? Yes, is this what we want?[1]

In June 1992, at the Earth Summit in Rio de Janeiro, the world's most influential leaders promised to combat global warming "for the benefit of present and future generations." It was a remarkable statement given that a decade earlier climate change hardly featured as a political concern anywhere. But since 1980 a scientific consensus crystallized that fossil fuel consumption was warming the world, and politicians took notice. In his campaign for the US presidency in 1988, George H. W. Bush vowed to deploy the "White House effect" to combat the greenhouse effect. Four years later Bush joined over 100 world leaders to sign the United Nations Framework Convention on Climate Change (UNFCCC) in Brazil. Reunified Germany was at the forefront of this international campaign to tackle global warming. The Federal Republic lobbied wealthy countries to adopt carbon commitments. Klaus Töpfer, German environmental minister, became the "savior of Rio" for brokering hard compromises.[2] Three years later Germany hosted the follow-up conference in Berlin, where Chancellor Helmut Kohl proclaimed that Rio "remains a mandate and an obligation.... If we do not pursue an active climate policy, the temperature on earth is likely to increase by a global average of 1.5 to 4.5 degrees Celsius by the end of next century." This frenetic burst of international conferencing culminated in 1997 with the Kyoto

Protocols, where the United States, Japan, and the European Union agreed to slash their carbon emissions.[3]

Hermann Scheer, however, was not impressed. A member of Parliament since 1980, by the 1990s Scheer was Germany's leading apostle of renewable energy. Born at the end of World War II, Scheer came of age during the 1968 youth movement, and represented the SPD from Baden-Württemberg, heart of the ecological movement. A pentathlete and former army lieutenant, he never shied from clashing with his party colleagues, and for this reason he was passed over for important posts. An avid reader of green luminaries like Robert Jungk and E. F. Schumacher, in the 1980s Scheer devoted himself to energy, authoring five books and hundreds of articles on renewables and crafting solar legislation for parliament. Even many in the Green Party admitted he was, "from head to foot and with every nerve a committed champion of renewables"; to his friends he was the "solar pope"; to his opponents the "Stalin of renewables" (see Figure 10.1).[4]

For Scheer, Rio and Kyoto were window-dressing that let world leaders claim action as the destruction of the atmosphere proceeded. "Conference follows conference. . . . Report follows report—and the result is usually a contract for another report." The outcome: "nothing more than the metaphoric drop in the

Figure 10.1. Hermann Scheer (left) receiving the Alternative Nobel Prize in 1999, next to German author Günter Grass. *Source:* DPA Picture Alliance Archive / Alamy Stock Photo. Image ID: D524FG.

bucket, a single drip on the rapidly heating-up globe." In his bestselling *Solar Manifesto* (1993), Scheer argued that international summits could never succeed without a fundamental paradigm shift. Bush and Kohl wanted merely to modify fossil fuel energy systems. Instead, the world needed an "energy battle" and a "political revolution" to forge a new economy powered by renewables. Every major economic transformation "produces both winners and losers," and for the coming solar transition "the losers are already defined": oil, coal, gas, and nuclear power.[5]

Scheer put his finger on the paradox defining the 1990s: "The terrifying gulf between our knowledge of these present threats [global warming] and taking the political action to fend them off." The world knew with clarity that fossil energy was destroying the climate, yet did little to halt the onrushing catastrophe. As even the chairman of the IPCC admitted, enforcing Kyoto was "politically unrealistic." The protocols were littered with loopholes, oil companies continued to expand, power plants burned more fossil fuels, and renewables struggled to break into the market. In the fifteen years after Rio, global carbon emissions rose nearly 40 percent.[6]

Why, given knowledge of climate change, did the call for action fail in the 1990s? Reasons abound, from climate denialism to a deep belief in technological solutions.[7] At the center of this puzzle lies the striking grip of neoliberalism. After the fall of the Soviet Union this philosophy reached its zenith. Trust in the efficiency of free markets; skepticism in government's ability to steer the economy; confidence in profit-seeking corporations; the desire to safeguard markets from democratic politics—these hallmarks of neoliberalism became entrenched in institutions and states around the world, from the World Trade Organization to the European Commission. In America, Bill Clinton—a "different kind of Democrat"—scaled back public spending, deregulated finance, and pushed the unhindered flow of capital. His signature climate policy, taxing the heat content of fuels, died in the Senate. In Britain, Tony Blair cut corporate tax rates and public spending, and continued to privatize state assets. In Brussels, the European Union liberalized the electricity and natural gas sectors. Such market fundamentalism dampened hope for a green transition by disavowing the very policies that could orchestrate a shift off fossil fuels. The rise of oil and nuclear power had hinged on decisive state action. Why would replacing fossil fuels be any different?[8]

In the newly reunified Germany, the prospects for a green transition looked particularly bleak. For Germans the 1990s were "overloaded with events," from foreign wars to domestic turbulence. Most importantly, the grand project of reunification was failing; the "blooming landscapes" promised in the East by Chancellor Kohl were becoming sites of deindustrialization. As its economy stagnated, global warming waned in urgency.[9]

Nevertheless, in this era of neoliberalism, when Germany was weighed down by problems, Berlin—almost uniquely in the world—bucked the global trend of market fundamentalism to spark a green energy transition. How did this happen? Concern about global warming was not enough. Nor was this transformation driven by outsiders alone: a bottom-up transition "from the people, not the government." Such arguments reproduce a triumphalist narrative that misreads the nature of energy transitions by overlooking the critical role of state action, political insiders, policy linkages, and coalitions.[10]

Germany's green energy breakthrough was the product of a heterogenous alliance led by Scheer in the SPD and Michaela Hustedt and Hans-Josef Fell of the Green Party. It included not only ecologists and solar enthusiasts, but also farmers, craftsmen, metalworkers' unions, and even industrial associations. In 1998 these groups found common cause under a novel Red-Green coalition that ousted the CDU from power, and replaced the old energy paradigm with a new one that would use energy prices as a political lever. Drawing on the ideas of Ecological Modernization, this alliance reformed Germany's energy system not only for the sake of the climate, but because they believed reform would promote exports, modernize the economy, generate jobs, foster a more participatory democracy, and even render the geopolitics of hydrocarbons obsolete. This green transition, put simply, was a vessel into which the Eco-Alliance poured their hopes for a better society. And like the nuclear transition of the 1960s, it was guided by the strong hand of the state.

The Spirit of Rio in an Overburdened Decade

In 1992 the United Nations Conference on Environment and Development in Rio de Janeiro, known as the Earth Summit, marked a highwater mark for the global environmental movement. That summer 15,000 representatives from over 150 countries promised to balance the needs of present and future generations through the idea of sustainable development, and they elevated global warming to a first-order international concern by signing the UNFCCC. In Germany politicians and elites were euphoric. On the Left, Ernst Ulrich von Weizsäcker—nephew to the president and director of the new Wuppertal Institute for Climate, Environment, and Energy—spoke of a peace dividend following the end of the Cold War that could be redeployed to save the environment. Rio, he argued, illustrated "the explosive opportunities" for nations that led the fight against global warming.[11] From the Right, the CDU's new environmental minister, a little-known quantum chemist from the former East, broadcast the ideals of Rio. In 1995 Angela Merkel opened the Berlin Climate Conference by reminding Europeans and Americans that "we, the industrialised countries, must be the first to prove that we are bearing our responsibility in protecting the global climate. . . . After all, we are talking about preserving our *single* world."[12]

Despite attention to climate change across the political spectrum, the spirit of Rio remained stillborn as the newly reunified Germany struggled through a welter of challenges. Problems began even as the IPCC was drafting its first report in 1990, when 100,000 Iraqi troops invaded Kuwait. The ensuing crisis took four million barrels of oil off the market and threatened the global economy. When the United States retaliated against Iraq, Germany refused to join the coalition and its citizens vehemently protested the Gulf War. The uproar damaged transatlantic relations and unleashed a passionate debate about defining Germany's place the world. Less than six months later a second international crisis erupted in Yugoslavia, which descended into separatism and ethnic cleansing. Here too, Germany played an outsized role with its unilateral decision to recognize the new republics of Slovenia and Croatia. A third crisis followed when an influx of asylum seekers from Eastern Europe sparked a backlash of xenophobia throughout the Federal Republic, raising the specter of right-wing nationalism. In the West, conservatives stoked fears of a social service breakdown. In the East came violence, climaxing two months after the Earth Summit, when several thousand German youth stormed a hostel for immigrants in Rostock.[13]

Just as climate change was climbing the ranks of issues, in other words, other problems began demanding immediate attention. From a peak in the 1980s, German worries about climate change declined. New climate models showed a milder pace of global warming than earlier. Press outlets began to chastise environmentalists for fanning fears of apocalypse. The first climate change deniers began to emerge, alongside a new genre of eco-skeptical books. The percentage of Germans who saw the greenhouse effect as a major issue fell during the 1990s, and was lower than in either France or Italy.[14]

But more than foreign wars, immigration, or climate skepticism, it was the unexpectedly hard work of reunification that sapped energy from the spirit of Rio. In 1990 Chancellor Kohl had suggested that integrating East and West would be easy. His predictions failed miserably, and economic life in Eastern Germany ground to a halt when the former GDR joined the Federal Republic and found itself exposed to Western competition at the stroke of a pen. Bankruptcies spread. The effort to privatize state-owned companies lost billions of DM instead of turning a profit. The costs of rebuilding and supporting the unemployed led to soaring deficits: over 1 trillion DM of state funds flowed eastward during the 1990s. Bonn financed this unprecedented debt by hiking social insurance fees and pension contributions, raising the costs of German labor to one of the highest levels in Europe.[15]

In 1993 the cost of reunification drove Germany into recession. The price of electricity—twice as high as in the United States or Britain—only made matters worse. In an era when multinational companies were relocating production around the world, high wage and energy costs made Germany unattractive as a site for global capital. German manufacturing profits plummeted. Exports—the pride of the nation—stagnated as companies like Volkswagen outsourced to Eastern Europe. The Federal Republic ran a current account deficit for the first time since the 1950s. Joblessness rose. Even after emerging from recession the economy struggled: by 1998 the unemployment rate was nearly 10 percent and over four million people were out of work. Incomes stagnated and inequality surged as traditional manufacturing jobs disappeared. The country became the "sick man" of Europe. In 1997 Roman Herzog, the Christian Democratic president of Germany, captured this malaise in a broadcast to the nation, reflecting how "people in Germany sense that the growth to which they had become accustomed is now a distant memory.... For the first time ever, people who have never been threatened by unemployment are fearful about the future."[16]

Pushing the Old Paradigm

In the face of these challenges, attention to energy and climate waned. Kohl's government paid homage to the well-worn goal of making the economy more energy efficient, and the conservative coalition even launched a second parliamentary inquiry into global warming. But rhetoric did little to change the economy's energy profile. As Merkel's undersecretary at the Environmental Ministry admitted, the government developed no real detailed plan to meet its carbon targets. Through the 1990s the former Western states did not decrease carbon emissions one iota, and in fact generated more carbon than ever: all the gains in mitigating the nation's carbon footprint came instead from industrial collapse in the East.[17]

More generally, the CDU still saw environmental protection as a cost rather than an opportunity. Their prime concern was strengthening Germany as a site of capital investment and trade. The world, the party reflected in its 1993 program, was in a state of economic upheaval, defined by globalizing markets and surging international competition. Party leaders feared Germany—an "export-oriented industrial nation"—was missing the boat, reunification having distracted Bonn from grappling with globalization. In America and Japan a new digital revolution was unfolding. In Eastern Europe low-wage economies were becoming competitive destinations for investment. In East Asia rival export powerhouses were emerging.[18]

To meet this challenge the CDU doubled down on the old energy paradigm in the hope that abundant, low-priced energy would support the nation's struggling economy. It resisted calls from Ecological Modernizers to tax energy to slow pollution.[19] Merkel, having once suggested such a tax during local elections, withdrew her support under pressure from Kohl. She then helped craft the party's standard line of resistance against the European drive for a carbon tax by refusing to accept any "location disadvantages for the German economy."[20]

The CDU-FDP coalition instead tried to *lower* energy costs by expanding the two prime sources of conventional electricity—nuclear and lignite coal. They saw the former as a powerful, low-carbon energy in the fight against climate change. The CDU hoped to solve the Achilles heel of this fuel, nuclear disposal, by pouring money into storage programs and technology that could reprocess spent uranium. Here Merkel was cavalier about contamination, responding to critics that every form of energy poses an environmental risk, just as "in every kitchen, when baking a cake a little baking powder can go missing." She loosened the safety standards to entice companies into the atomic sector. And in 1995 she authorized the first Castor nuclear disposal shipments to go to Gorleben for interim storage.[21] This reinvigorated anti-nuclear protests, particularly after a scandal broke that the Castor containers returning from La Hague, France—which reprocessed nuclear waste from German reactors—were leaking radiation. But this did little to stem the generation of electricity from nuclear power plants, which rose during the 1990s.[22]

The CDU likewise pushed lignite as a "particularly competitive" energy, and portrayed the combination of reactors and coal as a match made in heaven. Both were domestic and free from foreign entanglements, and both were proven technologies that delivered reliable electricity.[23] While inefficient mines in the former East closed down, under the CDU's watch lignite mining expanded in the Rhineland to meet the needs of the nation's industrial hub next door in the Ruhr. In 1987 RWE, the nation's largest utility, inaugurated plans for a huge lignite surface mine on the outskirts of Cologne. The CDU endorsed this megaproject in the belief that open pit mining was less expensive than other extraction techniques. After hearings where the CDU and the traditional wing of the SPD supported the project, in 1995 RWE gained formal approval to begin construction on Garzweiler II, where it would deploy surface excavators of astonishing size to extract lignite, dispossessing thousands of people from their homes in the process (see Figure 10.2).[24]

Most importantly, the CDU tried to lower energy prices by imposing competition onto the legal monopolies that had governed the electricity sector since 1935. By the 1990s over 800 regional power companies dotted the country, but nine massive ones dominated the national market. Chief among them were RWE in the Ruhr and Preussen-Elektra, owned by VEBA, the sprawling, partly

Figure 10.2. Lignite coal bucket excavator near Garzweiler II. *Source:* Klaus Oskar Bromberg / Alamy Stock Photo. Image ID: B4NNN9.

state-owned conglomerate that Social Democrats had tried to forge into a national oil champion in the 1970s. These behemoths were vertically integrated, extracting coal from the ground, generating electricity, and distributing it to consumers (see Figure 10.3). Together they controlled nearly all of the grid, produced 80 percent of the country's electricity, and enjoyed handsome profits, since they could add the cost of new assets into the price of the electricity they sold (see Figure 10.3). More power, put simply, meant more profits. Competition in their territories was nonexistent. And this made the price of electricity high by international standards, because these companies had built far more capacity than they needed. By the 1990s Germany had an estimated 33,500 MW of unused generation, more megawatts than its entire fleet of reactors could produce.[25]

Since 1945 both the Left and the Right had accepted this system because it provided all the electricity that industry, workers, and consumers could want. But in a new age of global competition, the CDU fretted over the costs. Ideologically, moreover, the existing electricity framework hardly fit the emerging neoliberal philosophy that valorized competition and enterprise. Britain had liberalized electricity in 1989 by splitting up its national monopoly. The European Commission soon followed with a campaign to break up electricity monopolies around the continent to make Europe more competitive.[26] After Brussels passed a directive in 1996 requiring grid liberalization, the CDU moved to smash open utilities' fiefdoms through third party access (TPA), which required grid-owners

Unleashing Green Energy

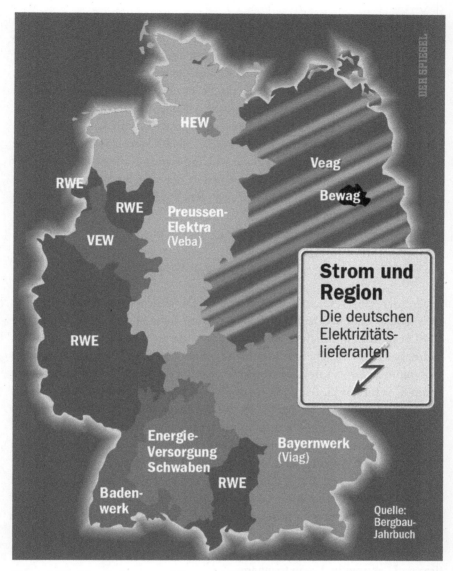

Figure 10.3. Monopoly zones of Germany's largest utilities. *Source: Der Spiegel*, "Strom und Region: Die deutschen Elektrizitätslieferanten," image on page 77 of the "Der Staat der 'Stromer,'" 46/1995.

like the Big Nine to grant access to any power generator who wanted in. The Economics Ministry believed competition would force utilities to deploy new technology, generate power more efficiently, and ultimately lower costs. As Merkel put it, "an ecologically oriented energy policy is compatible with deregulation and liberalization of the energy market."[27]

On April 29, 1998, Germany abolished the energy law that had been crafted by the Third Reich and that had stood at the core of the Federal Republic's economy. This made Germany one of the most liberal electricity markets in the world—at least in theory. Utilities could now penetrate each other's markets. And the excess capacity built up since the 1970s now came into play. So too did the vast overcapacity of electricity across the Rhine in France, a legacy of that country's aggressive push into nuclear energy after the oil shock. With too much power chasing too little demand, the Big Nine unleashed ruthless competition in a bid for customers, cutting costs along with thousands of workers. The director of the nation's second largest utility deemed it "the most brutal and chaotic development in the marketplace" he had ever seen. But Bonn achieved its goal, and power prices plummeted.[28]

Green Energy at Risk

Green energy, however, had no place in the CDU's agenda—not by a long shot, even though the CDU had launched Germany's foray into renewables with the 1990 Feed-in Law, which required utilities to connect small-scale wind, solar, hydro, and biomass generators to the grid. Lambasted by the SPD and Greens as a minor policy, to nearly everyone's surprise the Feed-in Law unleashed a boom in wind power. By requiring utilities to pay wind generators 90 percent of the consumer price of electricity, the law made it just barely profitable to build turbines in the windy north. From a macro perspective the gains seemed small: wind generated less than 2 percent of the nation's electricity. But on a micro level the law was remarkably successful. The industry grew 30–40 percent a year, and by the end of the decade Germany boasted ten times as much wind capacity as in 1990, with over 6,000 wind installations and the participation of 100,000 Germans, ranging from craftsmen to farmers. The Federal Republic became the global wind leader, home to some of the world's largest turbine manufacturers. A set of interest groups, a wind lobby, and more broadly, an entire wind community began to emerge.[29]

After 1990 solar grew too, though more slowly. The cost of producing electricity from photovoltaic (PV) cells was gradually falling. But in contrast to wind, the price of solar power remained an order of magnitude higher, and the rates set in 1990 never covered the costs of investing in this technology. Even with the Feed-in Law, a solar market only existed because of support from local governments. This kept alive the grassroots movement of tinkerers that had sparked interest in solar in the 1970s. Chief among them was Wolf von Fabeck, a nature-lover and officer who left the military after Chernobyl to turn to environmental protection. In the 1980s von Fabeck experimented with solar electricity, connecting a PV panel to his food processor, pooling money with his church parishioners to buy a dozen modules, and touring farmers' markets to illustrate how solar could run household appliances. In Aachen he founded the Solar Energy Development Association, which grew into a publicity mouthpiece for renewables. Hermann Scheer, meanwhile, after failing to pass solar legislation in the 1980s, founded the European Association for Renewable Energy—EUROSOLAR—to demystify the economics of green energy. By the 1990s solar momentum diversified and deepened, with pop stars and government officials touting the benefits of this technology.[30]

Instead of cultivating renewables, however, the CDU cultivated instability, for the party's own Feed-in Law came under assault by utilities and Bonn did little to defend it. The Big Nine feared the inroads made by independent generators into a system they once controlled from top to bottom. They excoriated the 1990 law as a dirigiste subsidy with no place in the social market economy. And they crafted what would become a common argument against green energy: that there were more efficient ways to reduce carbon emissions.[31] In 1995 a single utility stopped paying the mandated price to a hydro plant to provoke a legal challenge. As the case wound through the courts, the second largest utility, Preussen-Elektra on the windy North Sea coast, raised another suit, arguing that the burden of the law was spread unevenly across the country. Bonn did nothing to help renewables, instead taking up a proposal to cap the amount of feed-in power utilities had to absorb and fixing the ceiling at 5 percent in 1998. The wind lobby protested: "this can only be understood as a sales guarantee for 95 percent electricity from coal and nuclear energy."[32]

More generally, CDU leadership was hostile to renewables because they cost too much. In 1997 the Economics Ministry, supported by Merkel, tried to lower the rate paid to renewables. They were only stopped after metalworkers, farmers, and environmental groups demonstrated, and a small group of conservative MPs threatened to break with the CDU.[33] At a higher level, no less than the European Commission questioned the legality of feed-ins. Brussels, increasingly a bastion of neoliberal thought, argued that "energy prices ... should not be used as parafiscal instruments to support specific forms of energy." Feed-in tariffs in particular were "likely to result in distortions of trade and competition" that would undermine the even playing field the Commission was trying so hard to create. After 1996 the Competition Commissioner pressured Bonn to amend its framework, and at the end of the decade it opened a formal investigation.[34]

By 1998 pressure from utilities, the federal government, and Brussels placed Germany's green power at existential risk. With no fuel inputs, nearly the entire price of electricity from a wind turbine or a solar cell came from the cost of the capital. Securing investment, put simply, was everything, and investors seek stability. After 1995, however, legal turmoil wrecked any semblance of predictability for renewables. This was precisely the aim of the utilities, many thought: to heighten uncertainty, and make it hard for proponents of solar and wind to secure credit for their projects. By the mid-1990s banks were asking 30 percent down for loans dedicated to renewables.[35] For wind, the 5 percent cap sparked panic as investors rushed to build the last of the turbines before hitting the threshold. The period 1995–1996 was the peak of construction of new turbines. After that, wind companies stopped investing and many went bankrupt. Solar, too, slumped, the nation's largest module manufacturer closing its German plant and relocating to the United States.[36] Energy liberalization only added to the woes. The price collapse wreaked havoc since the 1990 Feed-in Law linked the rate received by wind and solar to the consumer price of electricity. As this plummeted, so did their remuneration. The press heralded the end of the wind boom.[37]

While Bonn and the European Union were suffocating the renewable transition, local actors kept momentum alive, and in fact municipal initiatives and nongovernmental organizations designed some of the most innovative policies. Saarbrucken began paying the highest rates in Germany for feed-in solar power, earning accolades at Rio. To the south, the Schönau power rebels of the Black Forest continued their campaign to buy their grid back from the regional utility, earning national acclaim with a plebiscite.[38] Most importantly, in the old Roman city of Aachen a model for stimulating solar gained prominence. Frustrated by the stubbornly high costs of panels—in the 1980s twelve panels cost as much as a small Volkswagen—von Fabeck argued that the local utility must pay the full cost of producing electricity from solar to anyone feeding power into the grid. A price that covered costs was, after all, what utilities were legally entitled to—why not extend this to renewables? In 1991 von Fabeck and his allies called on the local government to mandate 2 DM per kilowatt hour for wind and solar, twelve times the consumer rate. Under pressure from RWE the local utility refused, as did the federal Economics Ministry. But von Fabeck persisted, finding support from the city council and a lone rebel in the regional government. After stalling, in 1995 the opponents finally conceded, and a new model for funding solar power was born.[39]

Political Upheaval in 1998

In the 1998 elections, hardly more could have been at stake in the field of energy. Locally, renewables were crawling forward. Yet nationally, liberalization and legal challenges threatened solar and wind. Climate awareness was stagnating. Natural gas was advancing. The old energy paradigm of cheap power retained its hold in the corridors of power. And the country was mired in unemployment.

After sixteen years in office the CDU seemed out of ideas, the SPD blocking nearly every initiative with its slim majority in the upper house of Parliament. In 1997 "reform gridlock" became the word of the year.[40]

The Greens saw the election as a fantastic opportunity. They were now a completely different party. In the 1980s doctrinal purity had trumped electoral gains, with the party's strongest wing—the Fundis—aspiring to solve ecological problems through an "an anti-capitalist economy" that embraced the socialization of enterprises. But with their crushing electoral defeat in 1990, the Greens descended into an internal bloodletting. Prominent Leftists abandoned the party, while an infusion of dissidents from the GDR, who were skeptical of planning, energized moderates. The party also aged and lost some of its youth appeal.[41] Under Joschka Fischer's leadership, the Realos—reformist opponents of the Fundis—took command. Fischer had been a street-fighting dissident in the late 1960s, and aggressively critical of growth and technocracy in the 1970s. But his experience in Hesse's Red-Green coalition had brought him close to Social Democrats personally and had tempered his radicalism. After years of grinding away at politics, late in the 1990s he took up running in an effort to reinvent himself. As the only Green able to attract large audiences, Fischer became the party's parliamentary leader, dubbed the "superstar" by the press. By 1998 he believed the only way to advance Green ideals was to forge a Red-Green alliance at the federal level, and this meant opening the party to moderates through a message of incremental reform and markets.[42]

But the turn toward markets also derived from core ideas that had long animated the Greens. In condemning the centralized nature of Germany's energy system in the 1970s, the anti-nuclear movement had mirrored an older Ordoliberal critique about economic concentration. The call by Wilhelm Röpke, a founder of Ordoliberalism, for "decentralization in the widest and most comprehensive sense of the world" resonated with many ecologists. So too did the main challenge of political economy for Ordoliberals, which, to use the words of another Ordo forefather, was combating the "subservience of the individual to a massive state machine."[43]

Decentralization and individualism, in other words, had always been central tenets of the Green movement, and it took but a minor shift to argue that the party could achieve these goals through markets. The Eco-libertarian wing of the Greens first made this claim in the 1980s. They wanted to "return a maximum of responsibility to people themselves" by working through the market, which they described in strikingly Ordoliberal language as a self-regulating ecological system.[44] Thomas Schmid was their champion, a friend of Fischer's from Frankfurt who, after flirting with the Red Army Faction, became a pragmatist and editor of the pro-market newspaper *Die Welt*. In purging the Fundis, the Realos aligned with Eco-libertarians and adopted their optimism about capitalism, which after the fall of Communism was enjoying a moment as, seemingly, the only way to organize society. In 1989 Fischer laid out this vision in *The Transformation of Industrial Society*, where he admitted that "capitalism has won, socialism has lost."[45]

After reunification, the Greens' policy evolution accelerated, ironically, because of the 1990 Feed-in Law. After initially excoriating this legislation, the surprising surge of wind power led the party to adopt feed-ins as a model. Here Michaele Hustedt epitomized the party's trajectory toward a pragmatic, market-friendly orientation. Born in Hamburg, Hustedt studied chemistry and biology and taught secondary school. In the late 1970s she had participated in the peace and ecological movements, and even joined the Marxist Student Association Spartacus to spend several months in East Germany. But by the 1980s environmental issues brought her into local politics. After joining the Greens in 1990, Hustedt became speaker for ecological and energy affairs, and entered federal Parliament in 1994. In Bonn she tried to buy wind power for her home from her constituents. She fought the local utility for months, only to find the grid fees cost four times more than the power she was purchasing. There had to be a better way to advance renewables. So Hustedt turned to the feed-in as a policy that not only seemed to work, but one that would also appeal to the middle class. Since 1990, she enthused, this framework had created an entirely "new, small and medium-sized economic branch" that benefited normal Germans. Most importantly, it "mobilized private and middle-class capital," something Hustedt and other Realos came to appreciate would be essential for any energy transition.[46] Gradual or incremental reform was the best way forward, and Hustedt urged the party to accept compromise and seize whatever chance was available for change. For her the decisive question was:

> How do you turn opponents into supporters or at least into neutrals, and where are the social groups that are not yet supporters, but that can be won over? . . . We must get out of the niche and use our strengths to form innovative alliances. . . . This is work, it takes time and effort and requires a high capability of dialogue. But there is no other way.[47]

Harnessing small-scale capital, unleashing innovation, embracing profit, forging alliances: by 1998 these were watchwords of the reformed Greens, who had fully adopted the new energy paradigm. Under the leadership of Fischer and Hustedt, they looked to the SPD as their best hope for an ally. Since the 1980s the SPD's reform wing had embraced Ecological Modernization and had much in common with the Greens. The parties had collaborated in state governments, while the SPD's leaders, Gerhard Schröder and Oskar Lafontaine, came from the 1968 generation, had marched in the peace protests, and knew many Greens personally.[48]

But there were hurdles to any Red-Green federal alliance, including disagreement with the SPD's traditional wing over coal. Since falling from power, moreover, Social Democrats were themselves divided over whether to defend the working-class or adapt to the new realities of globalization. Lafontaine, a provocative leader dubbed "Red Oskar" by the press, hewed to the first path and saw Social Democracy as a program for the working classes. Nevertheless, he was a powerful advocate of Ecological Modernization and an ally of the Greens.[49] Schröder, by contrast, embraced the second path. An ambitious ladder climber, he had led the SPD's youth association in the 1970s, wore a sweater to Parliament instead of a suit, and had led a Red-Green state coalition in the 1990s. But since becoming minister president of Lower Saxony, Schröder shifted to the right, joining the supervisory board of Volkswagen. Once in office, he began to adopt the "neoliberal lite" ideas that pervaded Left-Center parties everywhere in the 1990s.[50] As the SPD's candidate for chancellor in 1998, Schröder promised to tackle the nation's malaise through more markets. In a widely read pamphlet coauthored with Britain's Tony Blair, he argued that Social Democracy must embrace "creativity and innovation." Government could never substitute for enterprise; rather, it must create "conditions in which existing businesses can prosper" and cultivate "entrepreneurial independence and initiative."[51]

The SPD's rift—epitomized in their slogan "Innovation and Justice"—threatened their campaign. Energy, however, offered a path to overcome these cleavages and connect with the Greens. Both parties cared about climate change and believed an energy transition would combat global warming. But they also saw energy reform as a multifaceted tool to modernize the economy and address unemployment, the single most important issue in 1998. Framing the debate in terms familiar to the public, reform Social Democrats and Greens argued that Germany's future lay not in mature industries but in new sectors like green technology.[52] According to a variety of studies, no other market was growing as fast as the one for environmental protection, energy efficiency, and renewables. Drawing on arguments made by the SPD a decade prior, Hustedt argued that Germany, with its "top ranking in exports," stood poised to take the global lead in renewables. It was already an "[e]xport world champion in environmental technology," a field that was growing 6 percent a year.[53] The right policies could turn this, alongside renewables and efficiency, into a jobs engine. "'Made in Germany,' in the future should have a double meaning: technologically first rate and at the same time efficient.... Environmental protection means jobs."[54] From the SPD, Scheer agreed, arguing that countries who led the climate fight would reap enormous economic benefits. Germany must "strive to bolster national competitiveness by being the first to make the next breakthrough."[55] A transition would come not from international agreements, Scheer believed, but from countries pushing the technological envelope and forcing others to adapt. The main question for Scheer was which nation would take the lead. Whoever first created a mass market for renewables would capture, "next to telecommunications, the largest technological market of the twenty-first century."[56]

Together, the reform wing of the SPD and the Greens formed an "Eco-Alliance" that shared common goals and personal connections. The alliance placed Ecological Modernization at the heart of their domestic agenda. They slammed the CDU for raising the cost of wages to pay for reunification, arguing that this burdened labor at the expense of the environment. They would make labor cheap and energy expensive by overhauling Germany's fiscal system with an eco-tax. They would promote renewables and efficiency as forces for exports abroad and jobs at home. Ideas about energy, put simply, were glue for parties that held very different views on everything from foreign policy to family values.[57]

Despite missteps in the Green campaign, in 1998 fatigue with conservative rule led to the CDU's electoral downfall. That fall, Social Democrats and Greens joined together in coalition at the helm of a new government. The media declared the old Bonn Republic history and a new Berlin one dawning, bringing a younger generation to power that had come of age during the anti-nuclear debates of the 1970s. In matters of energy and climate, this marked nothing less than a miniature political revolution.[58]

Energy Battle Royale

The Red-Green coalition rode into Berlin steeled for battle. Scheer captured this combative sentiment in his popular books and speeches. Since the Industrial Revolution, energy had become the "ghost in the machine" that drove modern economies, he argued. Countries in the Global North depended on an intricate supply infrastructure that moved fossil fuels over vast distances and that could only be managed by the largest transnational corporations. Governments supported these megafirms, creating what Scheer called a "political-power industry complex." Of the fifty largest European corporations, over forty were related to fossil fuels. Any green transition must confront this bloc head on. Germany thus stood, Scheer emphasized, "at the beginning of a fundamental energy conflict."[59] In his mind, there could be no transition "without creative destruction," since "the goal must be to dismantle the traditional energy system completely." Failure meant apocalyptic climate change. Success would realize the "most comprehensive economic restructuring since the beginning of the Industrial Revolution." And it would only come by making priority for renewable power "a political norm."[60]

The Greens agreed that transition meant conflict, above all with the utilities that controlled the grid. These mega-firms had stood at the center of the Green critique of politics since the 1970s, because they operated the nation's reactors and because they exercised great influence over local and national politics. RWE, the largest, was known for its dirty tactics of bribing officials and threatening to stop investment if state authorities resisted its projects.[61] Contemporaries called the Big Nine utilities the "princes of power," and even federal officials admitted they "were enormously harmful to the economy."[62] But the Greens fixated on them as the root of all evil, the spearhead of a nuclear and fossil fuel constellation that was ruining the environment and German democracy; the epitome of the old paradigm that must be abolished. Companies like RWE and VEBA reaped enormous profits from expanding into the former East where they hiked the price of power. They made still more from their reactors, since regulators let them build the future costs of atomic waste storage—costs which were utterly incalculable—into what they charged their customers. Through such gimmicks VEBA recovered nearly the full cost of its Brokdorf reactor that had kick-started the anti-nuclear protests of 1976. More generally, the Greens feared these utilities were taking over Germany's entire infrastructure. After VEBA was fully privatized in 1987, it restructured and expanded into new spheres outside energy. By the late 1990s VEBA was involved in oil, gas, chemicals, cable, software, telecommunications, logistics, and nuclear waste disposal. Cash-rich power companies like VEBA used the lucrative sale of electricity to keep growing and swallowing other enterprises. Its managers and those of RWE, meanwhile, advised powerful German politicians on energy. If left unchallenged, Hustedt feared, this energy-power constellation "would result in a Germany, indeed in a Europe, as a single large corporation in which democracy and preventative politics—such as climate change and environmental protection—falls [sic] by the wayside." The Greens wanted to tear them apart. This required a political battle royale.[63]

To sell this conflict to the public, the Eco-Alliance deployed the language of Ordoliberalism so familiar to Germans, who prided themselves on a robust *Mittelstand* and an identity built on the Economic Miracle. While sharing similarities with the market fundamentalism prevailing in the United States and Britain, Ordoliberalism condemned concentrations of economic power, valorized small-scale producers and communities, and centered the consumer-citizen in a way neoliberalism did not. The Eco-Alliance played on all of these distinguishing attributes. The state must impose competition on oversized utilities like RWE or VEBA by ensuring access to the grid for any citizen who wanted to produce power. This would unleash a solar economy that would, in turn, foster "stable regional business structures, cultures, and democratic institutions" and advance local "self-sufficiency and independence." In this rendering, all Scheer wanted was "competitive and consumer equality" for green power. How could Ordoliberal Germany say no?[64] The Greens, too, wove these tenets into their call for reform, demanding competition in the power market to stimulate the "innovative capacity of small and middle-class companies." They vowed to "mobilize more markets for the energy transition," by unbundling electricity generation from transmission.[65] All of this would create a "decentralized, citizen-oriented solar economy," rooted in normal Germans like those in Saarbrucken, Aachen, or Schonau. Cheering on this Ordoliberal framing were a range of organizations from Green Budget Germany, with its Adam Smith Prize, to many of the nation's leading newspapers.[66]

By 1998 the need for a green transition seemed more urgent than ever. Not only was Germany missing its carbon targets, but the CDU's energy liberalization in 1998 actually strengthened utilities instead of weakening them. Though requiring these conglomerates to grant third party access (TPA), utilities retained formal ownership of the grid, which meant they could favor electricity from their own power plants, as Hustedt had experienced firsthand in her struggle to buy local wind power.[67] Without an independent regulator to separate the grid owner from the power generator, Hustedt and Fischer argued, the CDU's liberalization would create an "Energy Wild West" where big firms dumped electricity to drive out their rivals.[68] They were right. After 1998 power prices plummeted and a wave of mergers swept the sector, turning the Big Nine into four mega-utilities. VEBA—the nation's second largest power company—merged with Bavaria's main utility to form E.ON, which became the largest publicly listed company in all of Europe. RWE retaliated by buying Germany's sixth largest electricity company. These new corporate formations became some of the richest companies in Germany. Together they supplied 70 percent of the nation's electricity and controlled 90 percent of the grid.[69]

Not only was concentration on the rise, after 1998 energy security reared its thorny head again. Since the Suez crisis in 1956, German leaders had worried their nation was acutely dependent on foreign energy. Rock bottom oil prices in the 1980s had quelled these anxieties, but now they returned. The first Gulf War set a bad precedent. Then in the late 1990s turbulence returned to oil markets. East Asia's swift development sucked petroleum into the region, leading prices to surge. In 1997 the Asian financial crisis tanked the oil market, but a year and half later it rebounded as China prepared to join the WTO and continued its incredible growth, pulling crude prices to heights not seen since the 1970s. Talk of global resource exhaustion, dormant for almost two decades, resurfaced. In 1998 C. J. Campbell, a geologist who managed the largest independent oil database in the world, published an influential essay suggesting that within a decade the demand for oil would far outstrip conventional supply. Raising public awareness still higher was the publication in Germany that same year of *Blood for Oil*, which depicted this energy as one defined by war and exploitation.[70]

The Eco-Alliance used these fears to portray security as a threat and an opportunity. High oil prices were the new normal, the Greens claimed, which meant market volatility and new global conflicts. Events seemed to bear this out. Between late 1998 and the summer of 2000 crude oil prices in Germany quadrupled, truck drivers protested at the pump, and the nation's oil-industry association fretted. To exploit this situation, Hustedt and her colleague in Parliament, Hans Josef Fell, rehabilitated the powerful trope of an "oil crisis." Like Hustedt, Fell had been a teacher before entering local politics in Northern Bavaria. The oil turbulence of the 1970s had impressed on him just how dependent Germany was on foreign energy. In the 1980s he built an eco-house for his family made of wood, replete with a grass roof, a co-generator powered by rapeseed oil, and later solar panels. He lived the Green hope of local energy self-sufficiency, in other words, and in the early 1990s he had helped his town beat Aachen in becoming the first to establish cost-covering feed-in rates for solar. Fell, like Scheer, wanted a world that ran entirely on renewables, arguing for this not only on ecological but also on security grounds. After 1998, as a Green MP in Parliament, Fell suggested Germany was entering a new energy crisis that offered a "once in a century chance" to free itself from foreign fuels. The country could solve its dependence, he argued, by replacing hydrocarbon imports with domestic engineering, innovation, technology, and labor.[71] From the SPD Scheer joined the chorus, pointing to conflicts in the oil-rich Caspian involving Azerbaijan, Armenia, Chechnya, and Russia as evidence of a coming battle for fossil fuels.[72]

The time thus seemed ripe to assault the utilities. But immediately upon coming to power, the Red-Green coalition was derailed by an actual military conflict in Europe, as Kosovo struggled to separate itself from what remained of the former Yugoslavia. America pressured Germany to join its air campaign and Fischer, now foreign minister, consented even though his party embraced pacifism. In 1999 Germany's first military engagement since World War II commenced. At the very same time, the two wings of the SPD descended into a confrontation over taxes, and before the first bombs fell on Belgrade, Finance Minister Lafontaine left the government and quit the party leadership.[73]

Four Successes

The Red-Green coalition, in other words, seemed mired in chaos as it embarked on its first and most ambitious energy initiative: phasing out Germany's atomic reactors. Nineteen of these installations produced a third of the country's power.

Closing them had been the lodestar for the Greens, because reactors epitomized the centralized technology they despised. As Jürgen Trittin, Green environmental minister after 1998, put it, "For us, the fight against nuclear power plants was the reason to the found the party." And while many Social Democrats questioned the need to close a sector that employed thousands, the reform wing had been pushing nuclear exit since the 1980s.[74]

But problems emerged from the start. Shortly after the election Chancellor Schröder met with Trittin and Economics Minister Werner Müller to conceptualize a nuclear exit. A more ill-fitting duo was hard to imagine. Trittin was a sarcastic firebrand, the rebel to Fischer's statesman. He hailed from the defunct Fundi wing of the Greens, and to many Social Democrats he was like a red flag to a bull. Müller, by contrast, was a businessman through and through. He came from the very utility sector the Greens vilified, having worked not just for RWE but also for VEBA/E.ON. Above all, he wanted a slow atomic exit. The chief negotiators may have agreed on ending nuclear power, in other words, but they disagreed vehemently on the details and the pace. Trittin muddied the waters by demanding that reprocessing spent fissile material be outlawed. This dismayed France and Britain, who feared losing billions of DM for handling Germany's nuclear detritus in La Hague and Sellafield. More importantly, banning reprocessing would leave industry with only a socially charged method of managing the disposal of atomic fuel, namely, storage at Gorleben, a center of protest. But this was Trittin's point, to force the utilities' hands. Schröder responded by excluding Trittin from the initial talks. A year in, it seemed the prestige project of the Red-Green coalition might fail.[75]

Yet the talks continued, Trittin and Schröder building on plans that aimed to avoid paying an indemnity to shut off reactors as the Federal Republic had done when it started closing its hard coal mines in the 1960s. Over time, Trittin accepted that his party must compromise, and the Greens lengthened their timeline to twenty-five years, then even longer. The Eco-Alliance, meanwhile, pressured utilities by raising their financial liability in the event of an accident, and by threatening the tax-free status of their capital reserves. After twenty months of negotiation the government reached a compromise in June 2000. The utilities agreed to shut down their reactors in thirty-two years. They could, though, close some reactors earlier in order to operate others for longer. Berlin inscribed this deal into a new atomic law that took hold in 2002, and by the end of the year the oldest reactor in the nation, in the north of Baden-Württemburg, would shut down for good.[76]

The compromise caused an outcry among the Green Party base, which feared a reversal if the CDU regained power. Many left the party for good. Atomic advocates, by contrast, lamented the demise of technical knowhow that had taken decades to master and predicted skyrocketing energy prices. The deal was bound to disappoint because it entailed such heavy compromises from all parties. But polls indicated a majority of Germans supported a gradual phaseout. And Realos portrayed this as a remarkable achievement, for, as Hustedt pointed out, "ending an entire branch of technology is a complicated and ambitious undertaking in a complex society" like Germany's. Now might be the Greens' only chance to make progress on this sacred goal, for who could say the party would return to power in the future.[77]

Ending an energy, however, is not the same as starting a new one, and the Eco-Alliance saw the reduction of atomic power as part of a holistic transition. Atomic exit, they hoped, would provide an "incentive for rational and economical use of energy as well as renewable power and thus for the necessary reorientation of energy policy" as a whole.[78] To achieve this broader transition, Berlin mobilized the entrepreneurial state that had kick-started atomic energy in the first place. Since the 1960s the Federal Republic had spent 14 billion DM of public funds on nuclear technology. With atomic power on the way out, the state now redirected public funds and attention to green energy.[79]

But the Eco-Alliance realized a transition would also need private capital—lots of it. The key was using the state to guide capital by eliminating risk and increasing returns for renewables and energy efficiency. The CDU and the SPD had done this for nuclear energy by assuming the astronomical cost of insuring this technology against the risk of a meltdown. The Eco-Alliance now wanted to do the same for green energy, by signaling a "clear, unequivocal, and decisive yes" for investors to sink money into solar and wind.[80] Building a green energy system, they noted,

> requires high investments, which the budgets of the federal government, the states, and the municipalities can only raise to a limited extent. But utilities and the mass of middle-class investors can achieve this quite well. We want to mobilize this capital for the energy transition through more markets.[81]

With the burning need for private capital in mind, the coalition started its second initiative in late 1998, calling for solar panels on 100,000 roofs. Scheer had authored a similar legislative proposal in 1994, only to be shot down by the CDU. But his motivation for the program was shared by renewable enthusiasts, like von Fabeck, who with Scheer believed a solar breakthrough no longer required deep research, but a mass market and mass production to radically reduce costs. At this point solar companies still used manual work throughout the production process, and there was the potential for economies of scale along the entire supply chain. As Scheer argued to Parliament, progress would only come with a "transition to mass production, in order to break through the vicious cycle: no broad market, because it is too expensive—too expensive, because there is no broad market."[82] Since 1994, Japan had started a similar program while President Clinton had proposed a project to install solar systems on one million buildings in America. Unless Germany acted now, Scheer worried it would fall behind in the solar race. His new program would spend 1 billion DM of public funds to subsidize the installation of 300 MW of solar over six years. He hoped less to reduce carbon than to create a mass market that would lower the price of producing PV panels.[83]

The 100,000 Roofs program started in January 1999, giving public grants that covered 12.5 percent of the cost of any solar installations. But far more important, it guaranteed low-interest solar loans through the nation's state investment bank, the Kreditanstalt für Wiederaufbau (KfW). Since capital was by far the largest cost-component in solar panels, this was a gift to small investors who could now go to any bank, fill out a credit application, and receive a loan to install panels. Scheer initially secured a 4.5 percent fixed rate for ten years, the lowest cost of capital on the market. But during the rollout, he and Hustedt impressed on the Finance Ministry the need to go further, and the KfW responded by offering interest-free loans. With inflation near 2 percent, the state, in other words, was literally paying investors to sink money into solar power.[84]

This was the first step in Berlin's campaign to retool the KfW to advance green development, the coalition's third initiative. Unlike nuclear exit, green finance evolved under the radar since the KfW was far from the public eye. Yet by 2000 this was a powerful institution. Founded after World War II to distribute Marshall Plan aid, by 2000 the KfW had become one of the largest development banks in the world, with over 500 billion dollars in assets, more than the World Bank. With its impeccable credit rating and state support, the bank could lend money far below market rates. Since 1990 it had focused on rebuilding the former East, but now the Red-Green coalition redirected this financial bazooka to the energy system. After using it to support their solar initiative, in October 2000 they unveiled a new climate program to upgrade energy efficiency in buildings, echoing the SPD's pioneering efficiency agenda of the 1970s.[85] Modernizing the nation's 20 million buildings was the single most important step the country could take to fight global warming, since providing heat and power to the built environment generated 40 percent of the nation's carbon. The coalition authorized the KfW to lend out huge sums of low-interest loans for efficiency renovations in buildings—over 50 billion DM through the next decade.[86]

While groundbreaking, these initiatives could not spark a transition on their own. The nuclear phaseout would take decades. Upgrading the built environment likewise required time and huge sums of capital. And despite its generosity, the 100,000 Roofs program flopped in its first year. Even with subsidies and cheap credit, experts worried solar might still be too expensive to mass produce.[87]

The Eco-Alliance realized they needed more. Specifically, Scheer, Hustedt, and Fell wanted to revamp the feed-in framework before it succumbed to liberalization or court challenges. They believed the 1990 law had been "decisive" for wind's breakthrough. Giving higher rates to more expensive forms of green power was now the key to everything—"high enough that operating renewable installations is profitable." But they also realized they must decouple these rates from the volatile consumer price of energy. As Scheer put it, "the primacy of renewables" must be "anchored in law."[88] Fell had achieved this in Bavaria, von Fabeck in Aachen. The Eco-Alliance wanted to replicate this on the federal level.[89]

In the summer of 1999 the Eco-Alliance separated this feed-in revision from a broader legislative package to overhaul the entire grid—a Green dream since the 1980s—in the hope of achieving immediate results. They found support in a diverse coalition that, at first glance, had little in common. It included the usual suspects: Ecological Modernizers from institutes in Freiburg, Wuppertal, and Berlin; the Federal Wind Association; and EUROSOLAR.[90] But now new actors joined the fray. Since 1990, farmers had installed many of the nation's turbines and they worried that liberalization would turn these into stranded assets. The German Farmers Association thus began calling for higher, more predictable feed-in rates. Wind and biogas could turn farmers into the "oil sheiks of tomorrow," the Greens noted, and bring forth a new political base for the party.[91] Large companies that had entered the solar industry, meanwhile, were now demanding better rates to safeguard their investments. With a new factory in Gelsenkirchen, Shell Oil believed solar had "great potential for exports and for generating jobs in Germany." But the nation's ability to produce solar modules actually outstripped what could be sold on the market. The company, put simply, might run out of customers. Shell and other big firms considering solar manufacturing consequently aligned with Scheer and called on Berlin to forge a larger domestic market by radically updating the Feed-in Law.[92]

Most important were two organizations that before the 1990s had shown little interest in a green transition. IG Metall, representing metalworkers, was the country's most influential union and was deeply interested in sales abroad. Renewables, moreover, were labor-intensive, and by the union's count, 30,000 jobs now relied on these technologies. And while many industries opposed a new Feed-in Law, the country's single largest industrial organization broke ranks and supported the Eco-Alliance. Like IG Metall, the Equipment and Machinery Producers Association (*Verband Deutscher Machinen- und Anlagenbau e.V.—VDMA*) had a vested interest in keeping the flow of turbines going, since this engaged so many of its members. Numbering 3,000 firms with a million employees, the VDMA was also focused on exports. And this is precisely how the Eco-Alliance courted both the VDMA and IG Metall, through the promise of sales abroad. Pointing to studies that predicted the greatest future growth would come in the energy sectors of developing nations, they argued that now was "the once-in-a-lifetime chance to become the global market leader" in solar and wind technology. In 1998 both organizations came out forcefully in favor of green energy.[93]

Still, some Social Democrats hesitated to embrace the reform. Opponents argued that higher-priced green power would burden consumers and do little to reduce carbon. Müller's Economics Ministry, meanwhile, refused to write the legislation. But lobbying from this unorthodox coalition helped push the feed-in update through Parliament. Hustedt conceptualized the law and publicized its export potential. Fell penned the actual legislation. Scheer, with his experience in Parliament, ushered it through difficult negotiations. At a dicey meeting with the SPD's coal wing, he even allowed mine gas onto the list of energies receiving preferential rates, bringing onboard the party's coal wing.[94]

On April 1, 2000, the Renewable Energy Act went into force. The law gave unprecedented financial support to renewable power. It required utilities to purchase all electricity produced by wind turbines, solar cells, geothermal or biogas facilities, waste, sewer, and mine gas installations, and small hydro-plants that sought entry into the grid. It fixed the price per kilowatt hour for renewable power, making the rate entirely independent from the consumer price of power. Wind received an average of 16.6 cents per kWh, higher than what it was getting in the late 1990s; biogas got 20 cents at a time when Germany's consumer price of power was in the teens. The law, in other words, paid above-market prices, gave investors unbelievable stability, and eliminated nearly all the risk of putting capital into renewables. But the most astonishing was what it offered to solar: no less than 99 cents per kWh at first—7 to 8 times the consumer rate for electricity. Alongside the subsidies and free loans under the 100,000 Roofs, it was now impossible not to make money investing in solar.[95]

Ecological Modernization and the New Energy Paradigm

While the Eco-Alliance spoke the language of Ordoliberalism, its package of energy policies was thoroughly the product of a different theoretical tradition. That even the once far-left Greens now put profits, markets, and private capital into their agenda illustrates not only their transformation as a party, but also how Ordoliberalism as a philosophy became hollowed out by the 1990s; more a language used to make political claims than a distinct theory of economic life. Fell,

Scheer, and Hustedt professed the need for competition, and after 2000 the big utilities had to contend with a new eco-system of small-scale power producers. But for those investing in renewables there was hardly anything competitive or free about the new environment: one was practically guaranteed to make money over a twenty-year time-horizon. The market, furthermore, did not determine the price of electricity, experts did when they inscribed the rates into law. For solar, the price of money itself was controlled by the KfW. The state's new framework, in other words, did not create an even playing field for energies, but one tilted dramatically in favor of renewables. This policy mixture functioned less as a market device than a device for raising capital; the entrepreneurial state assuming the risk of plunging headlong into a new energy system.[96]

Ecological Modernizers like Hans-Christoph Binswanger, Udo Ernst Simonis, and Lutz Wicke had crafted these ideas in the 1980s. With its policy package the Red-Green coalition adopted these ideas wholesale. As Fell put it during the battle over the Renewable Energies Act, "The market can do a lot, it can break up rigidities, ensure more economic efficiency, and give citizens new opportunities to influence the energy industry. But we cannot foster renewable energy through the market alone."[97] Fell and his allies set a politically determined price for energy. They aggressively steered capital into green energy and efficiency through state institutions and federal law. And they did so in the name of forcefully guiding technology. "Innovation needs a direction," argued Hustedt, in this case an ecological one.[98] Hence her party's quest for an energy price that rewarded the good and punished the bad, and one that would provide an "impulse for accelerated structural change, for a wave of innovations as well as for new branches of production and sustainable jobs."[99]

The entire justification for these price policies hinged on the theory of externalities, the bedrock of Ecological Modernization. To be sure, some in the Eco-Alliance shunned this concept, including von Fabeck, who thought it the "height of cynicism to translate flood-deaths in Bangladesh... or thyroid disease victims in Ukraine into pennies per kilowatt hour."[100] But the Eco-Alliance used externalities in its proposals, including the text of the Renewable Energy Act itself, which argued that "the external costs of producing electricity from conventional energies is not reflected in the price, but is born by the general public and future generations."[101]

Despite Scheer's nod toward Ordoliberalism, most self-identified proponents of this philosophy rejected his agenda. By the late 1990s, many Ordoliberals were adopting a more aggressive attitude about markets, akin to Neoliberals, and in the sphere of energy Ordoliberalism and neoliberalism converged in condemning political pricing. They wanted this sphere removed from the influence of political parties, and controlled instead by legally encased markets. Walter Hamm, editor of the journal *ORDO* and member of the Kronberger Kreis, a think tank devoted to free market ideas, counted the feed-in framework among Germany's "List of Sins." Chief among his concerns were the high energy prices that he predicted would result from this legislation, a burden for consumers and exporters. Having experts set the price of something as important as energy went against everything Ordoliberalism stood for. Hamm saw feed-ins as a distortion to the organic signals of the market that would lead only to windfall profits for energies favored by the Eco-Alliance. If fighting climate change was the goal, he and others argued, pumping money into photovoltaics was one of the least cost-effective strategies. Ordoliberals instead joined neoliberals in calling for emissions trading and auctions as the best tools for climate policy.[102]

Hamm and his Ordoliberal allies reserved their greatest ire for the Eco-Alliance's ecological tax reform. More than any other policy, this stood at the heart of Ecological Modernization and the new energy paradigm. Greens and reform Social Democrats believed an eco-tax was the most elegant climate policy since it would encourage companies and people to use energy more efficiently, while the state could deploy the proceeds to lower the tax burden on labor. With an eco-tax they aimed to reverse decades of rising wage costs and mounting energy use by making labor less expensive and energy more so.[103] And they would favor renewables over hydrocarbons and atomic power. Such price guidance, they hoped, would nudge companies to employ more workers, emit less carbon, and ultimately, put this "classical industrial country on a foundation of ecological innovation."[104]

The Greens had aggressively pushed this tax during the 1998 election. But they went too far in their campaign, proposing a gasoline levy that would triple the price of this precious fuel in ten years. Conservative media called it the "Green Nightmare" for assaulting the car, a foundation of German consumer society. The vitriol shocked the Eco-Alliance, and underscored how a green transition would challenge routine behaviors and entrenched interests. Many Social Democrats pulled back from taxing energy, including Chancellor Schröder.[105]

Nevertheless, after the election the Red-Green alliance put an eco-tax front and center in their coalition agreement, sparking a fierce confrontation. Since the 1990s momentum for such a tax had been growing, supported by a welter of studies, including a statement signed by 2,000 of America's leading economists. In Germany, advocates ranged from progressive institutes to the nation's largest environmental group, the Federal Association of Young Entrepreneurs, and preeminent Social Democrats like Lafontaine. They predicted an energy tax would generate hundreds of thousands of new green jobs while reducing the nation's carbon footprint. As a director of Berlin's Institute for Economic Research reflected, "everyone knows that the government will not meet its climate targets with the existing means. And everyone knows that Bonn is looking for new sources of income." An Eco-tax was the clear answer.[106]

Arrayed against this coalition was another one of equal power, which worked to crack open the divisions in the Red-Green government. The CDU and the FDP spearheaded this opposition, amplifying arguments previously made by Merkel that Germany could never accept an energy tax without similar policies in other wealthy countries, since the burden on its exporters would be too great. They found intellectual support among Ordoliberals and neoliberals, who argued that an energy tax epitomized all that was wrong with Ecological Modernization. Utilities, car manufacturers, and oil companies all joined in condemning this fiscal tool. When the Eco-Alliance rolled out their formal proposal the tabloid press jumped, claiming the government would create a two-class society by discriminating against low-income commuters who could no longer afford gasoline.[107]

Within the Red-Green coalition, debate hinged on gasoline and the question of exempting key industries. Schröder, with ties to Volkswagen, drew the lesson from the election campaign that he must not touch the holy car industry. He fought every penny levied on road fuels. Trittin, meanwhile, resisted all exemptions from the tax, while Economics Minister Müller pushed to exclude energy-intensive companies. The infighting became so tense that the SPD called for Trittin to resign. In 1999 the coalition finally reached an agreement. But the tax remained low and excluded twenty-seven sectors. That summer the Greens made a second push to raise the tax higher. Yet at this very moment global crude prices were being pulled to dramatic heights by China's growth, and in Germany gasoline prices broke the symbolic 2 DM/liter threshold. The Eco-Alliance responded with paeans about fighting global warming and spurring innovation. But these arguments found little resonance in the context of sky-high oil prices. It proved hard to explain to the public, moreover, why the tax proceeds went to something entirely unrelated to climate, like labor.[108] Payroll taxes did fall. But even sympathetic experts noted how arguments like, "look, driving may cost you more but your pension contribution fell by a fantastic 1 percent last year," found little purchase.[109]

In 2000 Berlin passed a second law that gradually raised taxes on electricity and road fuels. But the rates still disappointed the Eco-Alliance. Because so many industries remained exempt, the burden fell far more on households than on companies. By 2001 Schröder called off further energy reform, and the Greens steeled themselves for another federal election. The grand hope of overhauling the fiscal system ended with a whimper.[110]

Conclusion

In 2002 Schröder, after successfully defending the Red-Green coalition in the first election of the new millenium, pivoted to labor market reform, sparking a new round of political battles. Climate change lost urgency, and it seemed the Eco-Alliance had missed its chance to build a new economy. The end of the oil age, as Hans-Josef Fell predicted, had not yet come. The nation continued to burn fossil fuels. In 2005 construction commenced on the natural gas pipeline Nord Stream. The following year Garzweiler II formally opened, one of the largest open-pit lignite mines in Europe.[111]

Nevertheless, within a few years it would become clear that the Eco-Alliance had achieved something remarkable. After the European Court of Justice ruled in favor of feed-in tariffs in 2001, the risk of investing in solar and wind evaporated entirely. For the first time in history, it became profitable to sink capital into solar power on a mass scale. The 100,000 Roofs and the Renewable Energy Act complemented each other in a way that surpassed the wildest hopes of the Eco-Alliance. Applications for solar loans soared and capital flooded into PV. The nation would begin its fifth energy transition since World War II, as a boom in renewable power began to transform the economy.

In an age of neoliberalism, Germany launched this *Energiewende* through state-guided market creation, pure and simple. Hermann Scheer, Wolf von Fabeck, Michaela Hustedt, and Hans-Josef Fell countered the global call for unfettered markets and instead steered them with a strong hand. Their entrepreneurial state set a political price for energy, assumed the risk of building a green power network, gave powerful incentives for investing in renewables and efficiency, and through the KfW provided a third of the funding for Germany's green transition.[112]

And while Germany was laying the foundation for a national energy system that might someday run on the wind and the sun, international climate cooperation faltered. In 2001 President George W. Bush, son of the president who had sign the world's first climate agreement, pulled the United States out of the Kyoto protocols. In Europe, the EU would unveil its mechanism for implementing the carbon targets of Kyoto only in 2005. But Europe's Emissions Trading System— the largest cap and trade market for carbon in the world—would prove wildly ineffective in dampening the continent's appetite for fossil fuels.

Scheer, in retrospect, was vindicated in putting his faith in a national breakthrough. His confrontational approach captured the stakes of a green transition. He and the Eco-Alliance accepted that the campaign to forge a new energy system would create winners as well as losers, and they mobilized a diverse coalition to take on conventional energy producers. This monumental package of energy legislation was not only, or even primarily, about fighting global warming. It succeeded because politicians and experts linked energy to issues that were dear to the German public, from exports and jobs to security and democracy. But the potential gains would be enormous and would stretch across the world.

Coda

German Energy in the Twenty-First Century

In the new millennium the paradoxes of German energy intensified. A green transition unfolded, yet it generated contradictions even as conventional ideas about the need for fossil fuels retained a grip on German society. The country became the center of a revolution in renewable power. But it also burned more lignite coal than ever, stalling the decarbonization of its economy. Berlin accelerated its nuclear power phaseout. Yet the Federal Republic stumbled deeper into dependence on natural gas from a belligerent geopolitical neighbor.

By the 2010s it was as though Germany had two different energy systems: a new one based on the flow of the sun and wind; and an older one based on stocks of fossil fuels that had accumulated over eons. Each was justified by a distinct paradigm. The new paradigm held that growth could continue without ever more energy, that energy prices must be high, that renewables and efficiency technology could cover society's needs; the old: that growth required cheap, abundant energy and stability that solar and wind could not provide. In striking ways, the success of the new strengthened what remained of the old, because of the daunting challenge of handling the intermittency of the sun and the wind. As in the twentieth century, after 2000 Germany was profoundly shaped by the tense interplay between these two systems and two ways of thinking about energy, in everything from its foreign policy and domestic economy, to the nature of its democracy. This dichotomy was reflected in the passionate language with which Germans spoke of the *Energiewende*. Some lauded it as "one of the greatest social experiments" in German history; the Federal Republic's "gift to the world." Others castigated these policies as a "crime against future generations"; a "lunatic gamble" that would deindustrialize the nation.[1]

This *Energiewende* unfolded in the shadow of a long history of social confrontation over energy. The coal crises of the 1950s and 1960s had politicized energy for West Germans, and ever since, experts, politicians, and the public came to accept that energy markets were not purely commercial, but a profoundly contested political arena. Germans built on their turbulent history to forge a corporatist energy policy that included insider stakeholders as well as grassroots outsiders. Corporatist negotiations had guided the Federal Republic through four major energy transformations since 1945, and after 2000 the state deployed these practices to tackle the looming political battle over green energy. For that is what unfolded, as the Social Democrat Hermann Scheer so aptly captured: a stark confrontation between two different ways of thinking about energy.

A combination of market forces, the state and the parties that ruled it, and social groups had fashioned the Federal Republic's energy trajectory since 1945, and they continued to do so in the new millennium. The specific coalitions driving new energy systems changed over time. The Christian Democrat Ludwig Erhard, exporters, and oil majors began by pushing oil in the 1950s and 1960s, triggering one of the most rapid energy transitions of all time. After this postwar oil wave, the 1970s and early 1980s stand out as the most transformative moment in German energy history. During these years a powerful anti-nuclear movement emerged with its own experts, which demanded the ecological renovation of economy and society. To counter the rise of this new green political force, Social Democratic reformers popularized the idea that efficiency and renewables could be the next great engines of modernization and exports for German capitalism. After the 1973 oil crisis they passed pioneering energy efficiency legislation. This, together with West Germany's liberal regime that translated the oil shock into consumer prices, led the country to decouple energy use from growth earlier than other large industrial countries, laying the foundation for an *Energiewende* to come. At the same time, paradoxically, Social Democrats led West Germany in co-producing a vast natural gas infrastructure that linked the Federal Republic with the Soviet Union, joining the fate of these two countries and extending the old energy paradigm.

Every transition in the Federal Republic, moreover, was connected to a compelling vision of the future that bound together coalitions, like the dream of a car-powered, consumerist society propelled by oil, or the aspiration for a decentralized, more democratic society that sparked energy reform after 1973. Transitions, moreover, were advanced by savvy actors who could successfully exploit crises and play on the fear that their hydrocarbon-poor nation would stumble deeper into energy dependency on regions of the world over which Bonn or Berlin had little influence.

But the most important factor determining the success of those demanding, or fighting against, transitions was their ability to link energy to other goals. Developing new technology, promoting exports, improving democracy, enhancing geopolitical security, fighting climate change: these became rallying points that experts, politicians, unions, corporations, and grassroots actors connected to energy.

After 1998 a novel Red-Green coalition drew on all of these linkages to pass revolutionary legislation, justifying its policies through Ecological Modernization, a philosophy which aimed to aggressively guide markets through state action. Their policies set the Federal Republic apart from other large industrial countries, by kick-starting a green transition and reducing energy use at a time when market fundamentalism still governed much of the world.

But while German reformers succeeded in building up renewables, they failed to dismantle the fossil fuel foundation of the economy. Here other countries and institutions took the lead. At the European level, the EU Commission constructed an extensive carbon trading system with the hope of making it costly to burn excessive amounts of fossil fuels. Across the Rhine, the French state placed its finger on the scale to keep electricity prices among the lowest in the OECD. Between 2000 and 2010 its total energy consumption did not fall like Germany's, while its solar and wind lagged far behind. Nevertheless, France became an alternative energy model for the EU because it boasted one of the greatest nuclear industries in the world, never having turned against this energy as Germans had in the 1970s. A handful of massive French corporations supported a full nuclear cycle, mining uranium in Niger and Namibia and reprocessing spent fuel from across Europe in La Hague. With its faith in the atom, France achieved one of the lowest carbon emissions per capita in Europe.

The most defining developments for Britain and the United States, meanwhile, was the dash to gas. This began in Britain, after its companies exploited their country's North Sea gas fields, after Margaret Thatcher aggressively shrunk the coal sector and smashed its unions, and after liberalization let utilities find the cheapest source of power. The result was an acceleration of natural gas consumption, driving coal out of the market for electricity. Then in 2008 London passed a far-reaching climate act that required carbon capture and storage for new coal plants. Together with its relative deindustrialization, the decline of coal made Britain another global leader in decarbonization. In the United States, policy did little to alter the nation's energy structure. Most major legislative initiatives failed in a gridlocked Congress, including President Barack Obama's push for a national carbon trading system. Electricity market liberalization, meanwhile, gave America some of the lowest power prices in the world, and by 2010 American households were consuming more than twice as much electricity as in the 1980s. The defining energy development of the new millennium came not from Washington but from the hydrocarbon industry, which developed hydraulic fracking techniques that, spurred by a burst of drilling permits from President George W. Bush, unleashed a shale gas revolution. After 2009 gas flooded the market, driving down coal use as well as the country's carbon footprint.[2]

Germany, in other words, is not the only model for transforming an energy system. But its history does offer a way to understand how states, citizens, organizations, experts, and companies can drive forward an energy transition by building coalitions and linking energy to other pressing issues. Since 2000, some of linkages that spurred the *Energiewende* have waned while others have waxed, and the political coalitions around energy have shifted. But by 2022 a new conjuncture of crises, anxieties, and hopes emerged that created the potential, perhaps, for the new energy paradigm to triumph over the old.

Green Energy Triumphant?

After 2000 the new energy paradigm gained momentum despite the fall from power of the Green Party. On November 22, 2005, Germany's Red-Green coalition came to an end when President Horst Köhler swore in the nation's first woman chancellor. Physicist by training, Easterner, and former environmental minister, Angela Merkel of the CDU would govern Germany through a turbulent era that witnessed the near collapse of the euro, the influx of a million refugees, the rise of a radical right-wing party, and the first major hot conflict in Europe since the 1990s. Before any of these challenges erupted, though, Merkel presided over an energy transition with immense global import. Upon taking office, she earned a reputation as the "climate chancellor" during her time as EU Council president, setting some of the world's most aggressive carbon targets. And during her first six years in office Germany unleashed a technological revolution in solar power.[3]

Credit for these achievements, however, belonged less with Merkel than with the Red-Green coalition she ousted from office in 2005. Having passed painful labor market reforms in 2004, Social Democrats and Greens lost seats in 2005, and after long negotiation the SPD joined a Grand Coalition with the CDU. But Sigmar Gabriel, protégé of the previous chancellor, Gerhard Schröder (SPD), retained the Environmental Ministry, and with this his party's signature energy legislation. Merkel, who had initially opposed the Renewable Energy Act (*Erneuerbare Energien Gesetz—EEG*) as too costly, changed her tune and began promoting green energy as a force for modernization.[4]

After 2005, in other words, Germany's groundbreaking energy framework remained in place, its effects snowballing and reshaping the country's energy system through a boom in green investment. The high price guaranteed by the EEG to anyone producing solar, wind, or biogas; massive investment support from the state's Kreditanstalt für Wiederaufbau (KfW); subsidies from the 100,000 Rooves Program: these initiatives gave incredible profits to anyone who

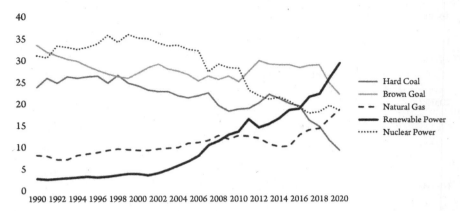

Figure C.1. How Germany makes its electricity: percentage of total power production by energy type. *Source:* Data from AEG Energiebilanzen e.V. Einsatz von Energieträgern zur Stromerzeugung.

sank capital into renewables. "Profit margins were huge," the head of a German solar company reflected, "customers snatched everything out of our hands that only barely looked like a photovoltaic wafer." Returns on wind investment could top 20 percent, those on solar reached 40–50 percent. Capital poured into renewables and investors began to ramp up the scale of production—just what the architects of the *Energiewende* had hoped for. Scheer and Michaele Hustedt (Greens) had believed mass production of wind and solar components would revolutionize the cost of these technologies. They were right. By the first decade of the twenty-first century Germany's *Mittelstand* companies began producing the world's first standardized, turnkey solar modules in a frantic effort to keep pace with domestic demand.[5]

This led the Federal Republic through its fifth energy transition since 1945 (see Figure C.1) After 2000 wind power expanded 14 percent a year, and by the end of the decade Germany boasted a third of global wind capacity. The success of solar was even more shocking. Sky-high profits drew huge amounts of capital into this sector. Solar capacity grew 55 percent a year, and by 2010 Germany was home to nearly half of all photovoltaics (PV) being installed globally each year, was generating 45 percent of the earth's solar power, and was the largest module producer in the world (see Figure C.2). An integrated cluster of component manufacturers, researchers, and producers emerged as Germany became the center of a European supply chain for renewable technology. This drove the cost of PV modules down at an unprecedented rate. Between 2006 and 2012 the cost of a solar system fell by two-thirds. As Hal Harvey, one of America's leading green experts put it, "Germans were not really buying power—they were buying a price decline."[6]

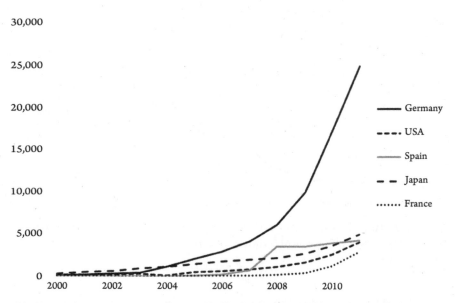

Figure C.2. Cumulative installed solar PV capacity in select IEA countries, in MW, 2000–2011. *Source:* Data from IEA, *Trends in Photovoltaic Applications*. Report IEA-PVPS T1-21 (2012), Table 2, p. 5.

The benefits of this revolution spread beyond the Federal Republic. Germany installed so many panels that its manufacturers could not keep pace, sparking a global silicon shortage. The nation began sucking in PV components from around the world, creating a "gold rush" for exporters and stimulating solar production in North America and Asia. Germany became the demand engine that drove the great cost digression which made solar power competitive.[7]

Transformation of the country's energy system encompassed nuclear power as well. For the Red-Green coalition, phasing out reactors had been intimately linked with broader reform, because Scheer, Hustedt, Hans-Josef Fell, and Jürgen Trittin believed nuclear exit would create the space for a new, locally oriented energy system to thrive. Many in the CDU, by contrast, wanted to keep nuclear power alive, arguing as they had since Chernobyl that this technology was good for security, good for the climate, and good for energy prices. In the Grand Coalition Merkel did not alter the nuclear phaseout law of 2002. But after forming a new conservative coalition in the 2009 elections with the FPD, the CDU recast nuclear technology as a bridge fuel and reversed the Red-Green effort to end nuclear power, allowing existing reactors to stay online.[8]

Half a year later, however, events on the other side of the world altered Germany's energy trajectory as they so often had in the past—in 1950, 1956, 1967, 1973, 1979, and 1986. On March 11, 2011, an earthquake off the coast of Japan triggered a tsunami that flooded the Fukushima Daiichi nuclear plant, causing the worst meltdown since Chernobyl. The catastrophe sparked a global reckoning with nuclear energy, but one that was particularly sensationalized in Germany, where it was filtered through a culture that had been passionately debating the atom since 1975. Germany's press portrayed the meltdown as a global apocalypse, hyperbolically declaring Japan "done for" and proclaimed the "End of the Atomic Age."[9]

Within days Merkel executed an entirely unplanned U-turn. The context was not only Fukushima, but the CDU's loss to the Greens in state elections in Baden-Württemberg. With little consultation, Merkel announced on national television that Fukushima had created "a new situation." "The only honest response is to accelerate our path towards the age of renewable energies."[10] She declared a moratorium on atomic power, temporarily closed the oldest reactors, and convened an Ethics Commission for a Safe Energy Supply to examine nuclear energy. Following the corporatist energy tradition first established by the Parliamentary Inquiry into coal in 1960, the Ethics Commission analyzed energy from a transdisciplinary perspective and included social scientists, natural scientists, philosophers, corporate directors, and church leaders. After deliberating for nearly two months, it recommended ending nuclear power in a decade and radically expanding renewables and efficiency. "Much like the investment in a child's future," one member reflected, "investment in a renewable energy based electricity system would lead to a better future for next generations."[11]

Following a stirring speech from Merkel, Parliament that June revised the Atomic Law with immense cross-party support and public approval. Reactors temporarily offline now closed for good, with the remaining ones slated to shut down by December 2022. The "ideological Thirty Years' War" that had started in Wyhl finally seemed to end. *Energiewende* advocates believed this would make Germany the model for a new sort of energy system by accelerating renewables and efficiency. As Peter Hennicke, a pioneer of Ecological Modernization, hopefully put it, Germany's response to Fukushima would "trigger a global domino effect" to a greener future.[12]

With Fukushima, the new paradigm crafted after the oil shock by anti-nuclear protestors, Green Party MPs, reform Social Democrats, and new experts seemed triumphant. The year 2011 marked a pinnacle of this vision, which held that growth could be delinked from the consumption not just of fossil fuels, but of energy in general; that renewables and efficiency would be a force for modernization; that the *Energiewende* would not only save the climate, but forge a more democratic society.

Indeed, since reunification Germany boasted one of the best records of growing its economy while limiting energy consumption, despite an industrial sector that was more than twice as large as that of France or Britain. Between 1990 and 2011 the energy intensity of Germany's economy fell by a third, reviving the momentum of the 1970s when Germany departed from America, Britain, and France in decoupling growth from energy use. In purchasing power parity, German energy intensity was now lower than the world average, lower than the OECD average, and lower even than the European average.[13] Renewables and efficiency, in turn, were proving to be agents of modernization. Once a critic of the EEG, Merkel now spoke of a Third Industrial Revolution.[14] Germany led the world in exporting wind turbines, solar technology, and efficiency equipment. At home, green power became an employment engine through a new ecosystem of companies that produced, installed, or maintained everything from silicon panels to insulation. By 2010 an estimated two million Germans owed their jobs to environmental protection, over half of which stemmed from the *Energiewende*.[15]

The *Energiewende*, moreover, seemed to be advancing the soft, decentralized democracy green activists had been pining for since the 1970s. Millions of Germans gained a stake in this new system. Beyond those whose jobs hinged on renewables, even more became "prosumers": consumers who produced power from panels on their roofs or turbines on their farms. Local energy cooperatives proliferated, and hundreds of communities bought their grid back from utilities. By 2011 the four big utilities owned less than 6 percent of the nation's renewable capacity; private citizens and farmers together held over half.[16]

Conventional utilities, meanwhile, found themselves on the defensive. Since the 1970s green activists had seen these corporations as the heart of a fossil fuel–nuclear nexus that was destroying the environment. But the dramatic expansion of renewables sent wholesale electricity prices—the price at which utilities bought and sold power before taxes or surcharges—plummeting. The share prices of E.ON, one of the richest companies in Germany, collapsed by three-quarters. RWE, after E.ON the nation's second largest utility, posted its first loss in sixty years.[17]

A False Dawn?

Hermann Scheer passed away six months before Fukushima, but until his death he believed the "systemic requirements" of nuclear power and fossil fuels—capital-intensive, centralized technological systems—were fundamentally "incompatible with those of renewables"; that it was an illusion to believe a transition could proceed by accommodating the interests of the old paradigm. By 2011 the profit collapse, job losses, and economic woes among conventional energy companies suggested the energy conflict that Scheer and his allies predicted was tipping in a green direction.[18]

But even as the new paradigm gained momentum, cracks appeared, along with new paradoxes stemming from its very success. Abroad, new rivals arose who could manufacture solar panels even cheaper than Germans could. After the 2007 financial crisis, China threw stimulus at its solar sector, which, combined with inexpensive labor and lax environmental regulations, let Chinese companies leapfrog Germany. Between 2008 and 2012 its solar manufacturing capacity grew tenfold and China became the new center of this technological supply chain. Europe retaliated with anti-dumping laws, but not before Chinese modules decimated Germany's solar sector, with manufacturers laying off tens of thousands. By 2013 one of the most powerful arguments linking the *Energiewende* to other goals evaporated, namely, the claim that green power was an engine of exports.[19]

The onslaught of renewables, meanwhile, sparked power surges along Germany's windy North Sea coast that destabilized Germany's energy market as well as neighboring ones. On blustery days wind farms produced so much power they turned the price of electricity negative. Germany exported these surges to France, the Netherlands, Poland, and Austria, to the point where these countries began rethinking their own grid. Berlin hoped to master the spatial challenge of power surges with high voltage national transmission conduits. But these stalled as locals protested the construction of "monster pylons" in their backyard.[20] Above all, the monetary costs of the *Energiewende* mounted. The architects of the *Energiewende* believed reform required a high energy price to encourage efficiency and steer innovation in an ecological direction. High energy prices were, put simply, *the* guiding principle of the new paradigm. By 2010 the prime lever forcing up consumer prices was the surcharge on electricity levied by the EEG. While wholesale electricity prices plummeted, the final consumer price rose because of this surcharge, which grew to fund the huge volume of solar, wind, and biogas coming online. Between 2010 and 2014 this fee more than doubled, becoming larger than the price per kWh of electricity itself.[21]

Surging consumer energy prices strengthened the voices of the old paradigm. Crafted by Ludwig Erhard and the Federal Republic's first generation of energy experts, this paradigm held that the cost of energy must be as low as possible for an export nation like Germany to thrive. Merkel herself clung to this view even after her volte-face with Fukushima.[22] Others went further. Hans Werner Sinn, a neoliberal and national figure after the Eurocrisis, predicted high energy prices would drive industry out of the Federal Republic. Social Democrat Wolfgang Clement, economics minister from 2002 to 2005 who later joined RWE's board of advisors, condemned policies he had once overseen. "We are in an industrial battle in the middle of a period of globalization and high energy prices mean we have a real problem."[23]

Arguments about deindustrialization were specious, since the EEG gave exemptions to its surcharge for energy-intensive companies. The cheap wholesale prices generated by the *Energiewende* gave large German manufacturers roughly the same cost of electricity as those in Texas.[24] Nevertheless, the old paradigm proved resilient, and during the 2013 election the debate over energy exploded when Environmental Minister Peter Altmaier (CDU) claimed that the *Energiewende* would cost a trillion euros. The press ran with these unsubstantiated remarks, *Der Spiegel* speaking of energy turning into a "luxury item." Merkel, too, wavered, admitting the quest to ground Germany's energy system on renewables was a "Herculean task."[25]

The *Energiewende*, lastly, hardly seemed to be helping Germany achieve its carbon targets, a priority that was gaining urgency. In 2007 the IPCC and former US vice president Al Gore won the Nobel Peace Prize for spreading climate awareness. In 2009 the Copenhagen Climate Summit—successor to Rio and Kyoto—briefly gave hope that the global community might agree on binding carbon targets that could slow the warming of the world.[26] But Germany's climate progress was stalled. By 2014 its energy sector was emitting roughly as much carbon as in 1999, while other countries like France and Britain decarbonized.[27]

German emissions remained high because the CDU refused to tackle the fossil fuel industry directly. Even after Fukushima, Merkel shied away from this and instead charted the middling course that Scheer vilified. "We have to make sure the basic providers of energy," she opined, referring to conventional utilities, "do not collapse under the burdens caused by the *Energiewende*." For Scheer, collapse was precisely the point.[28]

This refusal to embrace the logic of conflict appeared most clearly in the revival of lignite, which drove German emissions. While the Federal Republic had phased out hard coal mining since the 1960s, the excavation of surface lignite continued, and in many ways became symbiotic with Germany's nascent green energy system. By 2014 renewable capacity roughly equaled the electricity capacity of lignite, natural gas, and nuclear power combined, and during the best moments the nation hardly needed any non-renewable sources. But on cloudy, windless days the country relied heavily on traditional energies. Here lignite stood out as the leading option. Nuclear power was scheduled for phaseout, and reactors could not quickly ramp up to cover periods when solar and wind flagged. Natural gas, meanwhile, was dispatchable but expensive. Lignite mining, by contrast, was cheap, dispatchable, and an important source of employment in the Rhineland and in the east, which gave its advocates political clout.[29]

In 2010 Germany still burned as much lignite as in 2000. Over the next five years this figure rose, making Germany the lignite king of the world. Old paradigm stalwarts argued that nuclear exit was to blame: as reactors closed, coal plants had to fill the void. Defenders of the new paradigm countered by blaming the European Union. To achieve the climate targets of Kyoto, in 2005 Brussels had organized the largest carbon trading system in the world: the Emissions Trading System (ETS). Brussels hoped this cap-and-trade scheme would put a price on carbon, but the devil lay in the details. To pass this controversial legislation, which Berlin initially opposed, the Commission let member states allocate as many carbon certificates as they wanted, and to hand them out gratis. This created a massive surplus of pollution rights that pushed the price of carbon down to a pitifully low level. German utilities received certificates for free, with RWE earning an estimated 5 billion euros in windfall profits. After 2006 RWE and E.ON used this money to build new lignite power plants. Just as Merkel began phasing out reactors, in other words, a fleet of modern coal facilities commissioned well before Fukushima were coming online. With this unintended push from the ETS, RWE became Europe's main engine of global warming, owning four of Europe's top ten polluting installations and emitting more carbon than any other European institution.[30]

Rolling Back Ecological Modernization

Factors both inherent *and* external to the *Energiewende* were thus fueling Germany's appetite for lignite, pitting advocates of the old and new paradigms against one another. But with the 2013 elections the old paradigm regained the upper hand as the SPD, once a force for reform, rejected Scheer's ideas and joined the CDU in a second Grand Coalition. Like left-wing parties across Europe, the SPD had seen its membership wither since 2000. In the 1970s and 1980s a vibrant reform-wing had emerged under Erhard Eppler, Volker Hauff, and later Scheer. But now conventional Social Democrats came to dominate the party and worked to safeguard the lignite industry, a traditional voting bloc. They were led by Sigmar Gabriel who, now as economics minister, protected lignite with the argument that Germany "can't simultaneously get out of nuclear and coal."[31]

These developments coincided with a triple challenge that sapped strength from Europe's broader fight against global warming. The Eurozone crisis, roiling the continent since 2010, exacerbated tensions within the EU. In the midst of this fiasco came Russia's invasion of Crimea. Then in 2015 over a million refugees flowed into Europe, stimulating a populist backlash. Commentators spoke of the unraveling of the EU, and as in the early 1990s, the warming of the world was trumped by more immediate troubles.

To navigate the Eurozone disaster, the new Grand Coalition turned toward neoliberalism and the old energy paradigm. Berlin not only lectured its neighbors about the need for austerity, but practiced austerity itself. By limiting public investment and borrowing, the Grand Coalition undermined a key policy instrument that Ecological Modernizers hoped would accelerate a green transition. Berlin, moreover, aggressively supported its exporters by trying to lower the costs of production, including energy costs.[32] With SPD now on board, in 2014 Merkel reformed the EEG by limiting the growth of renewables and drastically reducing the rates paid to solar and wind.

Most importantly, the Grand Coalition replaced feed-in-rates—the core of the *Energiewende*—with auctions, in which producers competed to offer the lowest bid, with the winner receiving a one-off subsidy. This devasted the hopes of local power cooperatives and municipalities—because auctions required major sunk costs up front, they favored big companies that could take on risk.[33] In a stroke, this policy shift undermined the *Energiewende* as a force for democracy. Citizen-owned renewable projects lost out to large corporations. Big wind farms and solar parks with complex financing models, capital in the hundreds of millions of euros, and a support team of analysts and lawyers became the norm. Germany now had an "*Energiewende* without citizens," and a second argument linking energy with other goals—green power as force for democracy—vanished.[34]

A Dangerous World

But as the *Energiewende* lost momentum, one of the longest-standing issues linked to energy resurfaced in dramatic fashion: security. After demonstrations in the capital of Ukraine toppled the nation's pro-Russian president in 2014, Moscow retaliated by sending soldiers into Crimea and fomenting insurgency in Eastern Ukraine. So began the Russo-Ukrainian war, which sparked an energy as well as a geopolitical crisis. Europe imported huge volumes of hydrocarbons from Russia, with the Federal Republic relying on this country for a quarter of its natural gas. As Germany closed its reactors and tried to dampen its appetite for lignite, Berlin saw natural gas as a bridge to meet its climate targets, since gas released just half as much carbon as coal when burned. But after 2014 natural gas became a geopolitical hostage, hampering Berlin's ability to face a revanchist Vladimir Putin. Could Germany afford to impose the sanctions demanded by its allies if Russia might shut off the tap?

The Federal Republic's gas relationship with Russia dated to 1969, but the two countries built most of the sprawling network that bound them together after the oil shock through the 1980s. Ironically, security worked then in the USSR's favor, since German leaders saw this country as a reliable alternative to OPEC hydrocarbons. But intimations that this physical infrastructure might pose a geopolitical problem surfaced in 2006 and 2009, when disputes between Ukraine and Russia over gas destined for Western Europe led Russia to briefly shut off the tap. Greens like Jürgen Trittin saw this as the "prelude to the great energy conflicts that will define the coming decades."[35] But his fears were drowned out by other voices pointing out how the Soviet Union and its Russian successor had reliably supplied gas during even the tensest Cold War and post–Cold War moments. As the CEO of E.ON, a company immersed in the world of Russian hydrocarbons, claimed, "there hasn't been one single day . . . the country has used gas as a strategic weapon."[36]

Merkel and her SPD allies hoped to extend Germany's playbook from the 1970s, by softening Russian revanchism with natural gas. And both the CDU and the SPD publicized the Nord Stream gas line through the Baltic Sea as a commercial project that would bring East and West together—a matter of private commerce, not geopolitics.[37]

But critics abounded. In Germany, the world of gas increasingly epitomized the old paradigm and the anti-democratic energy system that reformers had fought so hard to dismantle. While the nuclear industry declined, gas remained an opaque, centralized technological system run by an interlocking network of corporate behemoths that Green reformers vilified: BASF, Germany's largest chemical conglomerate; Ruhrgas, a huge enterprise with ties to international oil; and E.ON, the nation's second largest utility that purchased Ruhrgas in 2003. These companies had long-standing ties with each other and with state leaders, symbolized by former chancellor Schröder's chairmanship of Nord Stream I and his cozy relationship with Putin.[38]

In concrete terms, however, the Grand Coalition did little to change Germany's security calculus. Neither the EU nor Germany included energy in their sanctions on Russia. Europe tried to minimize its dependence with LNG terminals, storage, and an enhanced European gas grid. Germany, however, participated with half a heart. Its Grand Coalition hardly discussed the security implications of energy or the potential exposure to blackmail by Russia in the event of a hot conflict. Instead, after a year Berlin doubled down on Russian gas by approving a second pipeline through the Baltic: Nord Stream 2. The United States, many Eastern European countries, and now even Brussels condemned the project. But Berlin pushed ahead, with Gabriel visiting Putin's personal dacha to reassure the Russian leader on Germany's need for gas.[39]

Commercial energy interests, put simply, were trumping Germany's transatlantic and European relations. In this Berlin was following a pattern that dated to the 1970s, when the Federal Republic pursued bilateral contracts with OPEC instead of a collective European agenda, when it prioritized a lucrative reactor deal with Brazil over American fears of nuclear proliferation, and when it built the Yamal gas pipeline against the wishes of Washington.

But security could cut in other directions. Since 1973 reformers had used the fear of hydrocarbon dependence to advance the new paradigm, highlighting how green energy was free from foreign entanglements. SPD reformers had made this case when pushing efficiency in the 1970s. The Red-Green coalition returned to this logic in 1998 to launch the *Energiewende*. In 2014 the Greens reanimated these ideas, arguing that Crimea gave new urgency to reform, that the nation needed the *Energiewende* for security. As Trittin and a young Green leader, Annalena Baerbock, argued before Parliament, "energy saving, more energy efficiency, and the forceful switch to renewable energies are the best way to free ourselves from energy resource dependency."[40]

This security debate overlapped with a new wave of environmental mobilization that rivaled the protests of the 1970s. Grassroots organization has been a part of every German energy transition since 1945. Now as climate change entered a new phase of urgency, mobilization again surged, though with coal as the target instead of reactors. In 2012 "forest defenders" occupied woods next to the nation's largest open pit lignite mine, sparking a repressive crackdown from the owner, RWE. Then in the run-up to the 2015 Paris Climate Conference—a landmark agreement that aimed to limit global warming—a new organization formed to intensify civic mobilization. *Ende Gelände*—"here and no further"—occupied the Garzweiler II lignite mine with over 1,500 protesters. They too met with violent reprisals from RWE and local police. But the organization expanded across the Federal Republic and Europe, demanding an end to coal. At the UN Climate Conference in 2017 in Bonn, thousands of Germans of all ages rallied to condemn Chancellor Merkel and her support for coal (see Figure C.3). By 2019 this anti-coal movement found support in institutions that ranged from Greenpeace to Fridays for Future, a new youth-led global climate movement.[41]

By then global warming topped immigration as the greatest concern among Germany's public. This, combined with grassroots pressure, led the government to launch a coal exit commission. Following Germany's tradition of energy corporatism, the Coal Commission brought together political representatives, scientists, union leaders, environmental organizations, and energy company delegates to find a way to achieve the carbon targets of Paris. Yet the commission suffered from internal rifts as energy companies protested a fast exit and demanded major compensation. The rise of a right-wing, climate-denying party based in Germany's, the *Alternativ für Deutschland*, only heightened fears that lignite phaseout might trigger political radicalization in Germany's eastern coal regions. After delay, the Commission issued its report, charting a coal exit that would close a third of capacity by 2022, another third by 2030, with final closure in 2038. Critics complained the pace was too slow. The report, moreover, epitomized the tension between old and new paradigms. In a statement that could have come from Ludwig Erhard himself, it argued that an "export nation" like Germany needed "internationally competitive electricity prices" to remain a "site of global business and economics." But in the next breath it drew on the heritage of reformers like Eppler, Scheer, and Hustedt to argue that efficiency,

Figure C.3. Climate March in Bonn, November 2017. *Source:* Author's Photo.

renewables, and climate technology would be a "driver of new business fields," and unleash "modernization of all sectors of the German economy."[42]

From a political standpoint, however, the Coal Commission was an achievement. For the third time since 1960, the Federal Republic orchestrated the beginning of an *end* of an energy. True, lignite mines would only close in 2038. But Germany accomplished what few coal nations besides Britain had ever attempted: planning a formal path to end a major fossil fuel. It did so by accepting this transition would create winners and losers, and paying the latter for their losses. Just as the Federal government had bought out hard coal mines in the 1960s, and cushioned the closure of reactors for utilities in 2000, so too would it pay vast public funds to close the country's lignite mines and revitalize mining regions.[43]

More Transition to Come?

Coal exit, however, exacerbated the greatest challenge besetting the *Energiewende*: intermittency. With nuclear power slated to end and lignite to steadily ramp down, how would Germany run its grid on cloudy windless days? For old regime stalwarts, the answer was natural gas. Russia's annexation of Crimea, however, damaged the case for this fuel. It received a second blow in 2021, when a dire IPCC report illustrated how colossal amounts of methane escaped into the atmosphere before natural gas ever reached the powerplant. Natural gas, the IPCC estimated, was responsible for 30 percent of global warming and could no longer be seen as a climate bridge.[44]

A third blow came six months later, when Russia invaded Ukraine a second time in eight years, precipitating the greatest upheaval in Europe since the collapse of Communism. On the one hand, Russia's violent assault tore open rifts that had defined German energy policy for decades. Germany quickly became the "fulcrum" for Putin's pressure on Europe. Predictions of recession and social turmoil from sky-high energy prices if Moscow shut off the tap made Berlin reluctant to end its energy ties with Russia overnight. Old paradigm stalwarts in heavy industry and the traditional unions staunchly opposed boycotting Russian gas. The CEO of BASF—a huge consumer of energy and an architect of Germany's energy relationship with Moscow—argued there was no alternative to Russian gas; that 40,000 of his firm's jobs depended on the flow of this precious fuel. Regional leaders in Saxony and Bavaria even suggested freezing the war in Ukraine and thinking of the needs of German citizens.

Critics responded by condemning Berlin for succumbing to its dependence on Russian gas, fearing the Federal Republic would prioritize its energy and economy over Ukraine's continued existence. Many argued that Merkel's Grand Coalition had erred massively by never developing a clear policy to handle energy security. Some stridently called for reviving nuclear power to reduce Germany's reliance on Russian gas, or at least not shuttering the nation's three remaining reactors as planned. Polls, in fact, showed a surge in support for atomic energy, as the visceral debate that seemed closed with Fukushima reopened. Yet here, too, the government—led since late 2021 by a novel coalition of Social Democrats, Greens, and Liberals—delayed. The Greens in particular argued that the atom was not the panacea for their country's looming energy crisis, since the country lacked stockpiled uranium and increasingly technicians who could operate reactors. But even utility companies admitted the debate over nuclear phaseout, the object of political confrontation for half a century, was too emotional to revisit.[45]

On the other hand, however, the catastrophe in Ukraine created a conjuncture with the potential to accelerate the green transition Germany had launched in 1998, by wedding fears of security with mounting anxiety that the world was warming catastrophically. For in the midst of Putin's invasion the IPCC issued its starkest warning yet: that global warming was accelerating faster than anticipated; that the world lay on the cusp of irreversible tipping points. Heatwaves, unprecedented flooding, and forest fires ripped through Europe that summer, stunning Germans and confirming the stark words of the IPCC. A fraught new vision emerged: without progress on the *Energiewende*, Germans would face a volatile Russia as well as a volatile climate.

The year 2022 thus produced a powerful linkage between security and climate with the potential to reinvigorate the *Energiewende*. Because a system based on renewables and efficiency could, in theory, fight both dependence and global warming, these linkages appealed to many groups in Germany—political insiders and outsiders; ecologists, experts, defense hawks, and the general public, which rallied to support Ukraine. The country entered a "watershed" moment, as Germany's new Social Democratic chancellor Olaf Scholz put it, that rivaled 1945 or 1989 in significance; a crisis with the potential to break old modes of thought about fossil fuels and fundamentally change the country's energy foundation, if the political will could be found.[46]

National energy systems are vast, complex physical infrastructures, intricately connected with other regions of the world. Historically, they have changed slowly. But change has come, and it has often begun in dramatic fashion, spurred by crisis moments. As the history of Germany illustrates, energy transitions require more than just better technology and falling prices for the new fuel. They are profoundly political affairs defined by conflict, and historically shaped as much by concerns over security, social stability, growth, or power as by fears of climate change. German energy history is the story of a continual clash between two competing paradigms, the relative influence of which has ebbed and flowed as energy issues interacted with everything from geopolitics to unemployment, economic development, and even historical memory.

Because energy is so important to everyday life, so interconnected with all spheres of society, changing an energy system requires refashioning daily habits and altering the relative power of social groups. Germany's history shows how transitions produce losers—economically, socially, or politically—and therein lay the heart of the challenge. The country's leaders navigated the politics of transitions through corporatist negotiation, at times even buying off representatives of the losing energy at great public expense. In the Federal Republic change came when political outsiders and insiders worked together to link energy to other issues dear to the public, build diverse and wide-ranging coalitions, deploy convincing narratives of the future backed by expertise, and exploit crises. All of these strategies will be needed if a new, sustainable energy system is ever to triumph over the old.

ACKNOWLEDGMENTS

In the winter of 2008–2009 I was living in Berlin, finishing my PhD, when suddenly the news in Europe turned toward energy. Russia temporarily shut off the supply of natural gas through Ukraine, accusing the latter of illegally siphoning off some of this precious fuel that was bound for Western Europe. After a brief spell, when fears of energy shortages and a cold winter dominated headlines across Central and Eastern Europe, the crisis passed and things returned to normal; meaning questions of hydrocarbons largely disappeared from the front page. But this affair sparked my interest in European energy, and planted the seed of what would become *Energy and Power*. For in reading about and discussing energy policy with friends and colleagues, it soon became clear that "energy" was a word that came loaded with history in Germany. Germans hold strong opinions on this subject, historically disagreeing intensely over whether to retain or end nuclear power, how to promote renewable energy, how quickly to phase out coal, and how to fight global warming. My conversations and research convinced me that the Federal Republic's experience with energy not only was especially turbulent. But it offers lessons that can be useful for other countries as the world finally—hopefully—turns its attention in earnest to the fight against climate change by orchestrating one of the most challenging energy transitions in human history, off fossil fuels. The raucous debate over energy and climate in the United States in particular—a country richly endowed with all types of fossil fuels—could benefit from deeper knowledge about how other parts of the world experienced energy transitions, how they navigated energy crises, and why their leaders and citizens might think differently about energy and climate change.

My archival work for this book began in 2015, and the final product owes much to the many people who helped me along the way. I would like to thank my colleagues at the Center for European and Mediterranean Studies (CEMS) and the History Department at New York University, who provided a supportive and lively intellectual community in which to test out many of these ideas, and who read many portions of the book: Larry Wolff, Mary Nolan, Hadas Aron, Christian Martin, Tamsin Shaw, Guy Ortolano, Stefanos Geroulanos, Steven Hahn, Jasmine Samara, John Shovlin, Ed Berenson, Kim Phillips-Fein, Peter Baldwin, Andrew Sartori, Karl Appuhn, Christiane Lemke, Thomas Zittel, Julie Livingston, Yanni Kotsonis, and others. Special thanks goes to Andrew Needham, with whom I co-taught graduate and undergraduate courses on energy history, where I learned an immense amount from him as well as our wonderful students. This book would have been impossible without the truly incredible administrative and research assistance that I benefited from at CEMS and History: Mikhala Stein Kotlyar, Anastasia Skoybedo, Melissa Graves, Guerline Semexant, Karin Burrell, Alexander Maro, Christopher VanDemark, Margaret Miller, Kyle Shybunko, Sven Van Mourik, and Tim O'Donnell. Beyond NYU, others who contributed to the intellectual environment out of which this project emerged—through reading drafts, participating in conferences and workshops, or otherwise—include Dolores Augustine, Duccio Basosi, Joseph Bohling, Eva Oberloskamp, Henning Türk, Giuliano Garavini, Stephen Milder, Rüdiger Graf, Troy Vettese, Frank Laird, Carol Hager, Adam Tooze, Christopher Shaw, Venus Bivar, Carl Wennerlind, Fredrik Albritton Jonsson, Stephen Macekura, Christopher Jones, Elizabeth Chatterjee, Joel Isaac, John Connelly, Astrid Eckert, Malgorzata Mazurek, and Michael Farrenkopf. I would also like to thank the *Journal of Modern History* for permitting an adaptation of a longer article I first published with them, in March 2023, to appear as part of Chapter 7. The financial support for the archival research, the writing, and even for taking classes on energy and environmental economics was provided by the Institute for New Economic Thinking, the Carnegie Corporation of New York, and the Andrew Mellon New Directions Fellowship. In the end, I dedicate this to my wife and two children, who have been with me through all the ups and down that accompany research, and who are an inspiration in so many ways.

—Stephen G. Gross,
New York University

ARCHIVES AND ABBREVIATIONS

AAPBD   Akten zur Auswärtigen Politik der Bundesrepublik Deutschland. Series (Berlin: De Gruyter Open)
ACDP   Archiv für Christlich Demokratische Politik. Konrad-Adenauer-Stiftung, Bonn
AEI   Archive of European Integration (online), University of Pittsburgh
AGG   Archiv Grüner Gedächtnis. Heinrich-Böll Stiftung, Berlin
ASD   Archiv der Sozialen Demokratie. Friedrich-Ebert-Stiftung, Bonn
BAK   Bundesarchiv Koblenz
BBA   Bergbau-Archiv Bochum
BSB   Deutscher Bundestag Stenographische Berichte
DIW   Deutsches Institute für Wirtschaftsforschung, Berlin
DS   Druckschriften (referring to BSB above)
DTM   Deutsches Technisches Museum, Berlin
EWI   Energiewirtschaftliches Institut, Cologne
HAEU   Historical Archives of the European Union, Florence
KaW   Die Kabinettsprotokolle der Bundesregierung: Kabinettsausschuss für Wirtschaft. Series
KpBr   Die Kabinettsprotokolle der Bundesregierung. Series
MTA   Margaret Thatcher Archive (online)
PAAA   Politisches Archiv des Auswärtigen Amt, Berlin
SKE   Steinkohleeinheit/-en—Tons of Hard Coal Equivalent
TNA   The National Archives, London (online)
WBA   Willy Brandt Archiv, at the Archiv der Sozialen Deokratie. Friedrich-Ebert-Stiftung, Bonn

NOTES

Introduction

1. Hermann Scheer, *Solar Economy: Renewable Energy for a Sustainable Global Future*, translated by Andrew Ketley (London: Earthscan, 1999/2002), 3.
2. Mark Braly, "Renewables 2004 Conference Opens in Germany," *Renewable Energy World*, June 2, 2004; Edgar Wolfrum, *Rot-Grün an der Macht: Deutschland 1998–2005* (Munich: C.H. Beck, 2013), 235.
3. AGG, Kristin Heyne, 106, 1/2, "Das Ende des Erdölzeitalters ist in Sicht," September 6, 2000.
4. Solar King quotation from Kate Connolly, "Hermann Scheer Obituary," *The Guardian*, October 18, 2010, https://www.theguardian.com/world/2010/oct/18/hermann-scheer-obituary; second quotation from Scheer, *Solar Economy*, xiv–xv.
5. AGG, Kristin Heyne, 104, 2/2, Hans Josef Fell quoted in "Solarwirtschaft und Wissenschaft unterstützen rot-grüne Energiepolitik," October 24, 2000; AGG, Hans Josef Fell, 02, 2006, BSB, "Für eine sichere Energieversorgung im 21. Jahrhundert," DS 16/579, February 7, 2006.
6. Staffan Jacobsson and Volkmar Lauber, "Politics and Policy of Energy System Transformation—Explaining the German Diffusion of Renewable Energy Technology," *Energy Policy* 34/3 (2006), 256–276; Volkmar Lauber and Staffan Jacobsson, "Lessons from Germany's Energiewende," in Jan Fagerberg et al. (eds.), *The Triple Challenge for Europe* (New York: Oxford University Press, 2015), 173–203; Travis Bradford, *Solar Revolution: The Economic Transformation of the Global Energy Industry* (Cambridge, MA: MIT Press, 2006), 102–104, 135–137; Craig Morris and Arne Jungjohann, *Energy Democracy: Germany's Energiewende to Renewables* (Cham: Palgrave Macmillan, 2016).
7. First quotation from Noam Chomsky and Robert Pollin, *Climate Crisis and the Global Green New Deal* (New York: Verso Press, 2020), 109; second quotation from IEA, *Market Report 2013*, 149; last quotation from "Solar Power: Gradually, then all at once . . ." *The Economist. Technology Quarterly*, January 9, 2021, 5–6.
8. Quotation from Elizabeth Grenier, "Could Germany Be Charged with Ecocide?" *Deutsche Welle*, November 16, 2020, https://www.dw.com/en/could-germany-be-charged-with-ecocide/a-55585186; Kajsa Lindqvist, "Europe's Most Polluting Power Plants," *Acid News* No. 2 (June 2012), 14; "Mapped: How Germany Generates Its Electricity," *Carbon Brief*, September 20, 2016, https://www.carbonbrief.org/how-germany-generates-its-electricity/; Deutscher Bundestag, "Klimaschutzbericht 2019," DS 19/22180, August 19, 2020.
9. Quotation from "Tod im Treibhaus," *Der Spiegel* Nr. 4 (1986); 15–16; Nicola Jones, "How the World Passed a Carbon Threshold and Why it Matters," *Yale Environment 360*, January 26, 2017, https://e360.yale.edu/features/how-the-world-passed-a-carbon-threshold-400-ppm-and-why-it-matters; "Climate Milestone: Earth's CO2 Level Passes 400 ppm," *National Geographic Resource Library*, March 29, 2019, https://edhub-ui.nationalgeographic.org/resource/climate-milestone-earths-co2-level-passes-400-ppm/.

10. Ulrich Beck, *Risk Society: Towards a New Modernity*, translated by Mark Ritter (London: Sage Publications, 1986/1992).
11. Quotation from Andreas Malm, *Fossil Capital: The Rise of Steam-Power and the Roots of Global Warming* (London: Verso, 2016), 358; IEA, "World Energy Investment 2019" (May 2019); Daniel Yergin, *The Quest: Energy Security, and the Remaking of the Modern World* (New York: Penguin Books, 2012), 401; Hannah Ritchie and Max Roser, "Energy," *Our World in Data* (July 2018), https://ourworldindata.org/energy.
12. Quoted in Andrew Nikiforuk, *The Energy of Slaves: Oil and the New Servitude* (Berkeley, CA: Greystone Books, 2012), 65; figures from Ritchie and Roser, "Energy"; Vaclav Smil, *Energy and Civilization: A History* (Cambridge, MA: MIT Press, 2018), 295–363.
13. Travis Bradford, *The Energy System: Technology, Economics, Markets, and Policy* (Cambridge, MA: MIT Press, 2018), 436–437; David Wallace-Wells, *The Uninhabitable Earth: Life after Warming* (New York: Tim Duggan books, 2019), 1–41; Kelly Levine, "Carbon Dioxide Emissions from Fossil Fuels and Cement Reach Highest Point in Human History," *World Resources Institute*, November 22, 2013.
14. Andreas Malm, "Who Lit This Fire? Approaching the History of the Fossil Economy," *Critical Historical Studies* 3/2 (2016), 215–248.
15. First quotation from Eric Hobsbawm, *The Age of Extremes: A History of the World* (New York: Vintage Books, 1994), 5; second quotation from Mark Mazower, *Dark Continent: Europe's Twentieth Century* (New York: Vintage Books, 1998), xi, xiv–xv.
16. Ian Kershaw, *The Global Age: Europe 1950–2017* (New York: Viking Press, 2018); Tony Judt, *Postwar: A History of Europe since 1945* (New York: Penguin Press, 2005); Konrad Jarausch, *Out of Ashes: A New History of Europe in the Twentieth Century* (Princeton, NJ: Princeton University Press, 2016).
17. Quotations from Hobsbawm, *Age of Extremes*, 262 and 474; other influential works that pay little attention to energy include Barry Eichengreen, *The European Economy since 1945: Coordinated Capitalism and Beyond* (Princeton, NJ: Princeton University Press, 2007); Jeffry A. Frieden, *Global Capitalism: Its Fall and Rise in the Twentieth Century* (New York: W. W. Norton, 2006); Robert Brenner, *The Economics of Global Turbulence: The Advanced Capitalist Economies from Long Boom to Long Downturn, 1945–2005* (New York: Verso, 2006).
18. Konrad H. Jarausch and Michael Geyer, *Shattered Past: Reconstructing German Histories* (Princeton, NJ: Princeton University Press, 2003).
19. Quotation from Heinrich August Winkler, *Germany: The Long Road West*, Volume 2: *1933–1990*, translated by Alexander J. Sager (New York: Oxford University Press, 2000/2007), 1.
20. Timothy Garton Ash, *In Europe's Name: Germany and the Divided Continent* (New York: Vintage Press, 1994); Konrad Jarausch, *After Hitler: Recivilizing Germans, 1945–1995* (New York: Oxford University Press, 2006); Mary Fulbrook, *A History of Germany, 1918–2014: The Divided Nation* (Malden, MA: Wiley Blackwell, 2015); Helmut Walser Smith, *Germany: A Nation in Its Time: Before, during, and after Nationalism, 1500–2000* (New York: W. W. Norton, 2020).
21. Anselm Doering-Manteuffel and Lutz Raphael, *Nach dem Boom: Perspektiven auf die Zeitgeschichte seit 1970* (Göttingen: Vandehoeck & Ruprecht, 2008); Niall Ferguson et al., *The Shock of the Global: The 1970s in Perspective* (Cambridge, MA: Harvard University Press, 2010); Frank Bösch, *Zeitenwende 1979: Als die Welt von Heute Begann* (Munich: C. H. Beck, 2019).
22. Examples include Rüdiger Graf, *Oil and Sovereignty: Petro-Knowledge and Energy Policy in the United States and Western Europe in the 1970s*, translated by Alex Skinner (New York: Berghahn Books, 2018); Henning Türk, *Treibstoff der Systeme: Kohle, Erdöl und Atomkraft im geteilten Deutschland* (Berlin: Be.bra, 2021); Sophie Gerber, *Küche, Kühlschrank, Kilowatt. Zur Geschichte des privaten Energiekonsums Deutschlands, 1945–1990* (Bielefeld: Transcript Verlag, 2014).
23. Helmut Walser Smith, *The Continuities of German History: Nation, Religion, and Race across the Long Nineteenth Century* (New York: Cambridge University Press), 16.
24. BSB, "Dritter Bericht der Enquete-Kommission: Vorsorge zum Schutz der Erdatmosphäre...," May 24, 1990, DS 11/8030.

25. Peter J. G. Pearson, "Past, Present and Prospective Energy Transitions: An Invitation to Historians," *Journal of Energy History/Revue d'Histoire de l'Énergie* 1(2018), 2–21; Christophe Bonneuil and Jean-Baptiste Fressoz, *The Shock of the Anthropocene* translated by David Fernbach (New York: Verso Press, 2013); Kate Brown, *Plutopia: Nuclear Families, Atomic Cities, and the Great Soviet and American Plutonium Disasters* (New York: Oxford University Press, 2013); Christopher Jones, *Routes of Power: Energy and Modern America* (Cambridge, MA: Harvard University Press, 2014); Gabrielle Hecht, *Being Nuclear: Africans and the Global Uranium Trade* (Cambridge, MA: MIT Press, 2012); Andrew Needham, *Power Lines: Phoenix and the Making of the Modern Southwest* (Princeton, NJ: Princeton University Press, 2014).
26. Quotations from Bruce Usher, *Renewable Energy: A Primer for the Twenty-First Century* (New York: Columbia University Press, 2019), 3, 4, 19; E. A. Wrigley, *Energy and the English Industrial Revolution* (New York: Cambridge University Press, 2010); Robert Allen, "Backward into the Future: The Shift to Coal and Implications for the Next Energy Transition," *Energy Policy* 50/C (2012), 17–23; Roger Fouquet, *Heat, Power and Light: Revolutions in Energy Services* (Cheltenham, UK: Edward Elgar, 2008).
27. Quotations from Malm, *Fossil Capital*, 91, 124; Timothy Mitchell, *Carbon Democracy: Political Power in the Age of Oil* (London: Verso, 2013).
28. Edmund Russell, James Allison, Thomas Finger, John K. Brown, Brian Balogh, and W. Bernard Carlson, "The Nature of Power: Synthesizing the History of Technology and Environmental History," *Society for the History of Technology* 52 (2011), 246–259.
29. Paul Sabin, *Crude Politics: The California Oil Market, 1900–1940* (Berkeley: University of California Press, 2005).
30. Quotations from Arnulf Grubler, "Energy Transitions Research: Insights and Cautionary Tales," *Energy Policy* 50/C (2012), 8–16; Vaclav Smil, *Energy Transitions: Global and National Perspectives* (Santa Barbara, CA: Praeger, 2017), 12.
31. Quinn Slobodian, *Globalists: The End of Empire and the Birth of Neoliberalism* (Cambridge, MA: Harvard University Press, 2018); Alisdair Roberts, *The Logic of Discipline: Global Capitalism and the Architecture of Government* (New York: Oxford University Press, 2010); David Harvey, *A Brief History of Neoliberalism* (New York: Oxford University Press, 2005); Naomi Klein, *This Changes Everything: Capitalism vs. the Climate* (New York: Simon & Schuster, 2014); Frank Bösch et al., *Grenzen des Neoliberalismus: Der Wandel des Liberalismus im späten 20. Jahrhundert* (Stuttgart: Franz Steiner Verlag, 2018).
32. Gary Gerstle, *The Rise and Fall of the Neoliberal Order: America and the World in the Free Market Era* (New York: Oxford University Press, 2022); David Nye, *Consuming Power: A Social History of American Energies* (Cambridge, MA: MIT Press, 1998); Meg Jacobs, *Panic at the Pump: The Energy Crisis and the Transformation of American Politics in the 1970s* (New York: Farrar, Straus and Giroux, 2016); Simon Pirani, *Burning Up: A Global History of Fossil Fuel Consumption* (London: Pluto Press, 2018); Dieter Helm, *Energy, the State, and the Market: British Energy Policy since 1979* (Oxford: Oxford University Press, 2010); Michael Bess, *The Light Green Society: Ecology and Technological Modernity in France, 1960–2000* (Chicago: University of Chicago Press, 2003); Gabrielle Hecht, *The Radiance of France: Nuclear Power and National Identity after World War II* (Cambridge, MA: MIT Press, 2009).
33. Quotation from Paul Nolte, "A Different Sort of Neoliberalism? Making Sense of German History since the 1970s," *Bulletin of the German Historical Institute* 64 (2019), 9–27, here 20; David Blackbourn, "The Culture and Politics of Energy in Germany: A Historical Perspective," *Rachel Carson Center: Perspectives* 4 (2013), 1–31; Frank Uekötter, *The Greenest Nation? A New History of German Environmentalism* (Cambridge, MA: MIT Press, 2014).
34. Quotation from Morris and Jungjohann, *Energy Democracy*, 351; Carol Hager, "The Grassroots Origins of the German Energy Transition," in Carol Hager and C. H. Stefes (eds.), *Germany's Energy Transition: A Comparative Perspective* (New York: Palgrave Macmillan, 2016), 1–26; Josephine Moore and Thane Gustafson, "Where to Now? Germany Rethinks Its Energy Transition," *German Politics & Society* 36/3 (2018), 1–22.
35. Quotation from Slobodian, *Globalists*, 266.
36. Mariana Mazzucato, *The Entrepreneurial State: Debunking Public vs. Private Sector Myths* (New York: Public Affairs, 2015).

Chapter 1

1. Friedrich Spiegelberg, *Energiemarkt im Wandel: Zehn Jahre Kohlenkrise an der* Ruhr (Baden-Baden: Nomos, 1970), 13; Christoph Nonn, "Kohlenkrise," in Michael Farrenkopf and Stefan Przigoda (eds.), *Glück auf! Ruhrgebiet: Der Steinkohlenbergbau nach 1945* (Bochum: Deutsches Bergbau-Museum, 2010), 235–328; Franz-Joseph Brüggemeier, *Grubengold: Das Zeitalter der Kohle von 1750 bis heute* (Munich: C. H. Beck, 2018).
2. Anne Brownley Raines, "Wandel durch (Industrie) Kultur: Conservation and Renewal in the Ruhrgebiet," *Planning Perspectives* 26/2 (2011), 183–207.
3. Peter R. Odell, *Oil and World Power* (London: Penguin Books, 1986), 120.
4. J. R. McNeill and Peter Engelke, *The Great Acceleration: An Environmental History of the Anthropocene since 1945* (Cambridge, MA: Harvard University Press, 2016).
5. Anthony Nichols, *Freedom with Responsibility: The Social Market Economy in Germany 1918–1963* (Oxford: Oxford University Press, 2004); Alfred C. Mierzejewski, *Ludwig Erhard: A Biography* (Chapel Hill: University of North Carolina Press, 2004); Mark E. Spicka, *Selling the Economic Miracle: Economic Reconstruction and Politics in West Germany, 1949–1957* (New York: Berghahn Books, 2007).
6. First quotation from Judt, *Postwar*, 13; second quotation cited in Konrad Jarausch, *Broken Lives: How Ordinary Germans Experienced the 20th Century* (Princeton, NJ: Princeton University Press, 2018), 237; Ulrich Herbert, *A History of Twentieth-Century Germany*, translated by Ben Fowkes (New York: Oxford University Press, 2019), 443–450.
7. BBA, 20, 416, Amerikanische Bergleute über den Ruhrbergbau: Robinson-Bericht/Robinson Report (1953).
8. Martin F. Parnell, *The German Tradition of Organized Capitalism: Self-Government in the Coal Industry* (New York: Oxford University Press, 2004), 80; Konrad Adenauer, *Memoirs 1945–1953* (Chicago: Regnery, 1966), 63; Judt, *Postwar*, 13–40; Jarausch, *Out of Ashes*, 404.
9. BBA, 12, 785, Andre Philip, "Die europaeische Gemeinschaft für Kohle und Stahl" (1952); BBA, 26, 67, Helmuth Burckhardt, "Langfristige Kohlen- und Energiepolitik," *Arbeit und Sozialpolitik* 10/10 (1956), 310–313; Robinson Report (1953); Brüggemeier, *Grubengold*, 338–41.
10. Herbert, *Germany*, 492–494; Jarausch, *Out of Ashes*, 417.
11. Nye, *Consuming Power*, 190–207; BBA, 160, 842, Burckhardt, "Eindrücke von einer Reise durch die USA," May, 1955.
12. Adam Tooze, *Wages of Destruction: The Making and Breaking of the Nazi Economy* (New York: Penguin Books, 2006), 454; Dietrich Eichholtz, *War for Oil: The Nazi Quest for an Oil Empire*, translated by John Broadwin (Washington, DC: Potomac Books, 2012).
13. Wolfgang Sachs, *For Love of the Automobile: Looking Back into the History of Our Desires*, translated by Don Reneau (Berkeley: University of California Press, 1992), 68; Michael Farrenkopf, "Zur Geschichte der Verwendung von Steinkohle," in Farrenkopf and Przigoda, *Glück auf*, 159–165; Ivan Berend, *An Economic History of Twentieth Century Europe* (New York: Cambridge University Press, 2006), 251–254; Hans Ulrich Wehler, *Deutsche Gesellschaftsgeschichte. 1949–1990* (Munich: C. H. Beck, 2008), 78–83.
14. Figures from https://sites.fas.harvard.edu/~histecon/energyhistory/; U.S. Energy Information Administration, "March 2019 Monthly Energy Review" (March 26, 2019).
15. Alan Milward, *The Reconstruction of Western Europe, 1945–1951* (Berkeley: University of California Press, 1986), 350–359.
16. Mierzejewski, *Erhard*, 64–68, quotation from 33; Hartmut Berghoff, Jürgen Kocka, and Dieter Ziegler (eds.), *Business in the Age of Extremes* (New York: Cambridge University Press, 2013), 139–152.
17. Ludwig Erhard, *Deutsche Wirtschaftspolitik: der Weg der sozialen Marktwirtschaft* (Düsseldorf: Econ Verlag, 1962), 102–103; Nichols, *Freedom*, 244, 275–279; Milward, *Reconstructing*, 299–335.
18. Erhard, *Wirtschaftspolitik*, 132.
19. Ludwig Erhard, *Prosperity through Competition* (New York: Fredereck A. Praeger, 1958), 37; Werner Abelshauser, *Der Ruhrkohlenbergbau seit 1945: Wiederaufbau, Krise, Anpassung*

(Munich: Beck, 1984), 70–71; Mark Roseman, *Recasting the Ruhr, 1945–1958: Manpower, Economic Recovery, and Labour Relations* (New York: Berg, 1992), 136–140.
20. James C. Van Hook, *Rebuilding Germany: The Creation of the Social Market Economy, 1945–1957* (Cambridge: Cambridge University Press, 2004), 214. Roseman, *Recasting*, 125–136; Carolyn Eisenberg, *Drawing the Line: The American Decision to Divide Germany, 1944–1949* (New York: Cambridge University Press, 1996).
21. BBA, 26, 67, H. Hembeck, Essen, "Ist die Kohle zu teuer?," *Arbeit und Sozialpolitik* 10/10 (1956), 317–319. Spiegelberg, *Energiemarkt*, Ch. 2; Parnell, *Coal Industry*, 83; Karl Hardach, *The Political Economy of Germany in the Twentieth Century* (Berkeley: University of California Press, 1980), 165–168; Hook, *Rebuilding*, 196–199; Wendy Carlin, "West German Growth and Institutions, 1945–1990," in Nicholas Crafts and Gianni Toniolo (eds.), *Economic Growth in Europe since 1945* (New York: Cambridge University Press, 1996), 455–497.
22. BBA, 26, 62, Franz Grosse, "Die Neuordnung an der Ruhr: Wirtschafts- und politische Ausblicke," lecture in Dusseldorf, October 9, 1952; Werner Abelshauser, "Korea, die Ruhr, und Erhards Marktwirtschaft: Die Energiekrise von 1950/51," *Rheinische Vierteljahrsblätter* 45 (1981), 287–316; Parnell, *Coal Industry*, 23, 82; Werner Abelshauser, *Deutsche Wirtschaftsgeschichte seit 1945* (Munich: C. H. Beck, 2004), 170; Christoph Nonn, *Die Ruhrbergbaukrise: Entindustrialisierung und Politik 1958–1969* (Göttingen: Vandenoeck & Ruprecht, 2001), 63–67.
23. First quotation from Erhard, *Wirtschaftspolitik*, 103; Volker Berghahn, *The Americanisation of West German Industry, 1945–1973* (New York: Cambridge University Press, 1986), 163.
24. Eisenberg, *Drawing the Line*, 139–149, 374–379; Berghahn, *Americanisation*, 95–104.
25. Nichols, *Freedom*; 277–82; Hook, *Rebuilding*, 190–212; Parnell, *Coal Industry*, 85; Roseman, *Recasting*, 125–128, 131–135.
26. Werner Abelshauser, "Dokumentation: Ein Briefwechsel zwischen John McCloy und Konrad Adenauer während der Korea-Krise," in *Vierteljahresheft für Zeitgeschichte* 30/4 (1982), 715–769.
27. Hans-Peter Schwarz, *Konrad Adenauer: A German Politician and Statesman in a Period of War, Revolution, and Reconstruction* (Providence, RI: Berghahn Books, 1995), 596–600; quotation from Hook, *Rebuilding*, 223.
28. Abelshauser, *Wirtschaftsgeschichte*, 169–175.
29. Roseman, *Recasting*, 140–141, 151; Abelshauser, *Ruhrkohlenbergbau*, 79; Falk Illing, *Energiepolitik in Deutschland: die energiepolitischen Massnahmen der Bundesregierung 1949–2013* (Baden-Baden: Nomos, 2012), 66; *KpBr*, 118 Kabinettssitzung (1956), 162.
30. Quotation from Abelshauser, *Wirtschaftsgeschichte*, 202–203; Erhard, *Deutsche Wirtschaftspolitik*, 121.
31. BBA, 26, 62, Theobald Keyser, "Die Bildung von 'Einheitsgesellschaften' im Ruhrbergbau nach dem Gesetz Nr. 27" (1952); ibid., Franz Grosse, "Die Neuordnung an der Ruhr: Wirtschafts- und politische Ausblicke," October 9, 1952; ibid., Theobald Keyser, "Vorschlag zur Bildung eines Bundskohlenverbands" (1952); Martin Martiny and Hans-Jürgen Schneider (eds.), *Deutsche Energiepolitik seit 1945: Vorrang für die Kohle* (Cologne: Bund Verlag, 1981), doc. 22.
32. Brüggemeier, *Grubengold*, 342–347; Ivan Berend, *The History of European Integration. A New Perspective* (New York: Routledge, 2016), 63–77; Ernst B. Haas, *The Uniting of Europe: Political, Social, and Economic Forces, 1950–1957* (South Bend, IN: Notre Dame Press, 1958/2004), 197; Brian Shaev, "Coal and Common Market: Forecasting Crisis in the Early European Parliament," in Lars Bluma et al. (eds.), *Boom-Crisis-Heritage: King Coal and the Energy Revolutions after 1945* (Oldenbourg: De Gruyter, 2022), 71–80; Jean Monnet, *Memoirs*, translated by Richard Mayne (New York: Doubleday, 1978), 293.
33. Fritz Burgbacher and Theodor Wessels, *Die Energiewirtschaft im Gemiensamen Markt* (Baden-Baden: August Lutzeyer, 1963), 224–244; N. J. D. Lucas, *Energy and the European Communities* (London: Europa Publications, 1977), 5–10; Roseman, *Recasting*, 141; BBA, 12, 785, Andre Philip, "Die europäische Gemeinschaft für Kohle und Stahl" (1952); Nonn, *Ruhrbergbaukrise*, 36–38.
34. Quotation BBA, 26, 67, H. Hembeck, "Ist die Kohle zu teuer?" in *Arbeit und Sozialpolitik* 10/10 (1956), 317–319; BBA, 32, 3686, Aktennotiz über die Kohle/Heizoel-Besprechung on November 10, 1955; Roseman, *Recasting*, 151–154.

35. Erhard, *Prosperity*, 48; Abelshauser, *Ruhrkohlenbergbau*, 73–75; Rainer Karlsch and Raymond Stokes,"*Faktor Öl*": *die Mineralölwirtschaft in Deutschland 1859–1974* (Munich: Beck, 2003), 280–281; BAK, 136, 7701, Deutscher Bauernverband to Adenauer, March 29, 1951, and Schaeffer to Brentano, February 16, 1951; Manfred Horn, *Die Energiepolitik der Bundesregierung von 1958 bis 1972* (Berlin: Duncker und Humboldt, 1977), 240–263.
36. BBA, 26, 67, in Theobald Keyser, "Für eine produktionsorientierte Kohlenpolitik," July 14, 1956, and Theobald Keyser, press conference on July 13, 1956; *Unternehmensverband Ruhrburgbau in Essen. Jahresbericht, 1955–57*; BBA, 32, 3686, Aktennotiz über die Kohle/Heizoel-Besprechung am 1. Juli 1955 in Hamburg; BBA, 26, 67, Helmuth Burckhardt, "Langfristige Kohlen- und Energiepolitik," *Arbeit und Sozialpolitik* 10/10 (1956), 310–313.
37. Nonn, "Kohlenkrise."
38. BBA, 26, 67, Helmuth Burckhardt, "Langfristige Kohlen- und Energiepolitik," *Arbeit und Sozialpolitik* 10/10 (1956), 310–313; Abelshauser, *Wirtschaftsgeschichte*, 68–83; Nonn, *Ruhrbergbaukrise*, 28–31; *Jahresbericht des Unternehmensverband Ruhrburgbau in Essen*, 1955–1957.
39. Odell, *World Power*, 121.
40. Benn Steil, *The Marshall Plan: Dawn of the Cold War* (New York: Simon and Schuster, 2018), 12; Robert Keohane, *After Hegemony: Cooperation and Discord in the World Political Economy* (Princeton, NJ: Princeton University Press, 2005), 135–141, 151–179; Tyler Priest, "The Dilemmas of Oil Empire," *Journal of American History* 99/1 (2012), 236–251.
41. David S. Painter, "Oil and the Marshall Plan," *Business History Review* 58/3 (1984), 359–383; Karlsch and Stokes, *Faktor Öl*, 258–263; Raymond Stokes, "Oil as a Primary Source of Energy," in Carol Fink et al. (eds.), *1956: European and Global Perspectives* (Leipzig: Leipziger Universitäts Verlag GmbH, 2006), 245–264.
42. Stokes, *Faktor Öl*, 273–274, 282–284; Heiner Radzio, *Unternehmen Energie: Aus der Geschichte der Veba* (Düsseldorf: Econ Verlag, 1979), 223; Burbacher and Wessels, *Energiewirtschaft*, 79–109.
43. Heinz W. Vollrath, *Die Mineralölwirtschaft in der Bundesrepublik: Ihr Aufbau und ihre Entwicklung seit 1945* (Munich: Uni-Druck, 1959); Karl-Heinrich Von Thümen, *Die deutsche Mineralölwirtschaft* (Hamburg: Industrieverlag von Hernhaussen, 1956); "Ölstadt Hamburg," *Die Zeit*, July 21, 1949.
44. Stokes, *Faktor Öl*, 273–274, 282–284; Vollrath, *Mineralölwirtschaft*.
45. BAK, B102, 5448, Boecker to Hessische Ministerium für Arbeit, Wirtschaft und Verkehr, November 5, 1953; Karlsch and Stokes, *Faktor Öl*, 69–79; Horn, *Energiepolitik*, 242–244; Willem Molle and Egbert Wever, *Oil Refineries and Petrochemical Industries in Europe: Buoyant Past, Uncertain Future* (Brookfield, VT: Gower, 1984).
46. OEEC, *Europe's Growing Needs of Energy: How Can They Be Met?* (Paris: OEEC, 1956); W. G. Jensen, *Energy in Europe, 1945–1980* (London: Foulis, 1967), 34.
47. Quotation from BBA, 32, 3686, Report to Fachverband Kohlechemie E.V. by Heinz Nedelmann, concerning heating oil, November 7, 1955; Heinrich Gutermuth, "Probleme der Bergbauwirtschaft in gewerkschaftliche Sicht, May 18, 1956, Doc. 25, in Martiny and Schneider, *Energiepolitik*; BAK, B102, 5429 2/2, Boecker to Hübner, Tätigkeitsbericht der Bundesregierung 1956, October 31, 1956; BSB, 2. Deutscher Bundestag—174 Session, Coal Debate on November 29, 1956; BSB, Energiebilanz des Bundesgebiet, June 24, 1957, DS 3665.
48. Fritz Baade, *Weltenergiewirtschaft: Atomenergie—Sofortprogramm oder Zukunftsplanung?* (Hamburg: Rowohlt, 1958), 70.
49. *Jahresbericht des Unternehmensverband Ruhrburgbau in Essen*, 1958–1960; Abelshauser, *Wirtschaftsgesichte*, 200–201; *KaW*, 33. Sitzung on August 3, 1955, and 24. Sitzung on February 14, 1955; *KaW*, 45. Sitzung on March 9, 1956; BSB, 2. Deutscher Bundestag—174 Session, Coal Debate on November 29, 1956.
50. Bundesregierung, *KaW*, 45. Sitzung on March 9, 1956; BAK, B136, 2506, "Zur Frage der künftigen Entwicklung des Energiebedarfs des Bundesgebietes und seiner Deckung," January 4, 1956; Nonn, *Ruhrbergbaukrise*, 34–51.
51. BSB, "Energiebilanz des Bundesgebiet."

52. Bundesregierung, *KaW*, 45. Sitzung on March 9, 1956; on the SPD see BSB, 2. Deutscher Bundestag—174, Coal Debates on November 29, 1956.
53. BBA, 32, 4146, Auszug aus dem Entwurf des Protokolls der 45. Tagung des Ministerrats in Luxemburg, October 8, 1957.
54. Horn, *Energiepolitik*, 201; Eichengreen, *European Economy*, 93.
55. Abelshauser, *Wirtschaftsgeschichte*, 202; *KaW*, 45. Sitzung on March 9, 1956; Parnell, *Coal Industry*, 89; BSB, 2. Deutscher Bundestag—174 Session, Coal Debates on November 29, 1956; Nonn, *Ruhrbergbaukrise*, 45.
56. Astrid Kander, Paolo Malanima, and Paul Warde, *Power to the People: Energy in Europe over the Last Five Centuries* (Princeton: Princeton University Press, 2013), 260; Theodor Wessels, "Struktur und Entwicklungstendenzen der deutschen Energiewirtschaft in der Sicht der Enquete-Ergebnisse," in Wessels (ed.), *Energie-Enquete*, 12–25; Bardou et al., *Automobile Revolution*, 172; Farrenkopf, *Gluck Auf*, 540; Chick, *Electricity*, 10.
57. BBA, 32, 3686, Aktennotiz über die Kohle/Heizöl-Besprechung in Essen July 19, 1957; BBA, 26, 67, Helmuth Burckhardt, "Langfristige Kohlen- und Energiepolitik"; "Helmuth Burckhardt," *Der Spiegel*, March 11, 1959.
58. BBA, 160, 651, Burckhardt, "Die Steinkohle im Wettbewerb," lecture in Basel, January 16, 1958.
59. BBA, 32, 4146, Auszug aus dem Entwurf des Protokolls der 45. Tagung des Ministerrats in Luxemburg, October 8, 1957.
60. B102, 5444, Aussenstelle Hamburg to Bonn, March 8, 1957; BBA, 26, 67, "Suezkrise und Kohlenproblem in den Diskussionen der OEEC," December 12, 1956; Ethan B. Kapstein, *The Insecure Alliance: Energy Crises and Western Politics since 1944* (New York: Oxford University Press, 1990), 96–125.
61. BBA, 160, 651, Helmuth Burckhardt, "Die Steinkohle im Wettbewerb," lecture in Basel, January 16, 1958; BBA, 26, 67, Theobald Keyser, press conference of the UvRb on July 13, 1956.
62. Quotation from BBA, 26, 62, Helmuth Burckhardt, "Langfristige Kohlen- und Energiepolitik."
63. Quotation from BBA, 26, 67, Theobald Keyser, "Vor der Kohlendebatte im Bundestag," in *FAZ*, November 13, 1956.
64. BBA, 160, 651, Helmuth Burckhardt, "Die Steinkohle im Wettbewerb," lecture in Basel, January 16, 1958; BBA, 32, 3686, Aktennotiz über die Kohle/Heizöl-Besprechung in Hamburg," July 11, 1955.
65. Erhard quoted in Abelshauser, *Wirtschaftsgeschichte*, 202–203; *KpBr*, 196. Sitzung, October 9, 1957; BBA, 32, 4146, Keyser and Söhngen to Erhard, September 16, 1957; Mierzejewski, *Erhard*, 161–162.
66. Quotations from BBA, 32, 4146, Auszug aus dem Entwurf des Protokolls der 45. Tagung des Ministerrats in Luxemburg, October 8, 1957.
67. BAK, B102, 5444/1, Erhard to Heinrich Fassbender, February 8, 1957; BAK, B102, Report from Neef to BWM, November 22, 1956; BAK, B102, 5429, Bundeswirtschaftsministerium, "Tätigkeitsbericht der Bundesregierung 1957," November 7, 1957.
68. BAK, B136, 7702, Bockelmann of BP Benzin und Petroleum AG; van Drimmelen, Deutsche Shell AG, and Geyer of ESSO AG to Erhard, November 17, 1958.
69. Quotations from BBA, 32, 3686, Aktennotiz über die Kohle/Heizöl Besprechung in Hamburg, June 3, 1954; S. Jonathan Wiesen, "Miracles for Sale: Consumer Displays and Advertising in Postwar West Germany," in David Crew (ed.), *Consuming Germany in the Cold War* (New York: Berg, 2003), 151–178; "Ökonomische Vernunft: Besser Leben–Rationalisierungsausstellung in Düsseldorf," *Die Zeit*, July 9, 1953.
70. BBA, 32, 3686, Arbeitsgemeinschaft Erdöl-Gewinnung und -Verarbeitung, "Feste und flüssige Brennstoffe in der Energieversorgung Westdeutschlands," November 10, 1953.
71. BAK, B102, 5444/2, Heinrich Köhn, "Blick in die Zukunft: Die Aufgaben der westdeutschen Mineralölindustrie bis 1965," and "Der westdeutsche Energiebedarf bis 1965 und seine Deckung," March 20, 1956.
72. Quotations from Graf, *Öl und Souveraenitaet*, 35–43; Ferdinand Mayer, *Erdöl-Weltatlas* (Braunschweig: Georg Westermann Verlag, 1966); M. A. Stahmer, "Neue Märkte—neue

Technik," *Die Zeit*, July 20, 1950; "Heizöl quilt nach oben," *Die Zeit*, March 29, 1956; "Modernes Erdöl-Labor," *Die Zeit*, October 10, 1957; "Siegeszug der Erdölmänner," *Der Spiegel*, July 30, 1952.
73. United Nations Economic Commission for Europe, *The Price of Oil in Western Europe* (Geneva: United Nations, 1955); Francisco Parra, *Oil Politics: A Modern History of Petroleum* (London: I. B. Tauris, 2004), 64–69.
74. BAK, B102, 73362, "Ist Unser Benzing zu teuer," *FAZ*, August 27, 1955.
75. Quotation from BAK, B102, 73362, BP Benzin- und Petroleum-Gesellschaft to Dr. Neef, August 19, 1955; ESSO to Economics Minister, August 17, 1955, and Bundesamt für Gewerbliche Wirtschaft, Hamburg, to Boecker, July 12, 1955, in ibid.
76. BAK, B102, 73362, ESSO to Economics Minister, August 17, 1955, and "Ist Unser Benzin zu teuer," *FAZ*, August 27, 1955.
77. BAK, B102, 73362, "Ist Unser Benzin zu teuer," *FAZ*, August 27, 1955.
78. Parra, *Oil Politics*, 36–50.
79. Shell, *Weltenergieverbrauchs*.
80. M. A. Adelman, "My Education in Mineral (Especially Oil) Economics," *Annual Review of Energy and the Environment* 22 (1997), 13–46; M. A. Adelman, *The Genie out of the Bottle: World Oil since 1970* (Cambridge, MA: MIT Press, 1995), 47–68; M. A. Adelman, *The World Petroleum Market* (Baltimore, MD: Johns Hopkins University Press, 1972); Mitchell, *Carbon Democracy*, 166–170.
81. Parra, *Oil Politics*, 63.
82. Adelman, *Petroleum Market*, 177; Yergin, *Prize*, 391–412.
83. BAK, B102, 73362, Mineralöl Zentralverband, "Mineraloel Nachrichten," August 4, 1955.
84. *KaW*, 45. Sitzung, March 9, 1956; BBA, 26, 67, Europäischer Wirtschaftsrat, "Energiekommission: Berichtsentwurf. Vertraulich," March 9, 1956.
85. BBA, 32, 3686, "Aktennotiz über die Kohle/Heizöl-Besprechung in Hamburg," January 23, 1958, "Notizen für das Kohle-/Heizölgespräch auf dem Petersberg," May 30, 1958, and Heinz Nedelmann to H. W. von Dewall, of Bergwerksgesellschaft Hibernia AG, May 23, 1958.
86. Quotation from "Kohlenkrise: Gute alte Zeit," *Der Spiegel*, December 16, 1959. Parnell, *Coal Industry*, 96; BAK, B136, 2508, Burckhardt to Adenauer, October 30, 1959; Nonn, "Kohlenkrise," Nonn, *Ruhrbergbaukrise*, 102; "Kohlenkrise: Gute alte Zeit," *Der Spiegel*, December 16, 1959; Abelshauser, *Ruhrkohlenbergbau*, 87–105.
87. BBA, 32, 3686, "Aktennotiz über die Kohle/Heizöl-Besprechung in Hamburg," January 23, 1958, and "Notizen für das Kohle-/Heizölgespräch auf dem Petersberg," May 30, 1958; *Jahresbericht des Unternehmensverband Ruhrburgbau in Essen*, 1958–1960; Hardach, *Political Economy*, 179–183; Parnell, *Coal Industry*, 96; Illing, *Energiepolitik*, 70–71.
88. BAK, B136, 2508, Burckhardt to Adenauer, October 30, 1959; Heinrich Gutermuth, "Bergbauwirtschaft heute und morgen," June 8/13, 1958, doc. 29 in Martiny and Schneider, *Energiepolitik*.
89. Theodor Wessels, "Wandlungen in der Energiewirtschaft," in *Kyklos* (1962); Adelman, *Petroleum Market*, 357; Bundesregierung, *KaW*, 12. Sitzung, November 18, 1958; *Jahresbericht des Unternehmensverband Ruhrburgbau in Essen*, 1958–1960; Spiegelberg, *Energiemarkt*, CH 1; BAK, B136, 2508, Burckhardt to Adenauer, October 30, 1959; BBA, 32, 3686, Nedelmann to Dewall, "Niederschrift über die Kohle/Heizoel-Besprechung," July 10, 1958.
90. First quotation cited in Nonn, *Ruhrbergbaukrise*, 106. Second quotation from Bundesregierung, *KpBr*, 23. Sitzung, April 30, 1958.
91. Martiny and Schneider, *Energiepolitik*, Introduction; Abelshauser, *Wirtschaftsgeschichte*, 202.
92. BBA, 32, 4146, "Niederschrift über eine Besprechung Staatssekretär Westrick," May 5, 1958.
93. Quotation, Erhard paraphrased by reporter, Bundesregierung, *KpBr*, 23. Sitzung, April 30, 1958. Kabinetsprotokoll; Abelshauser, *Wirtschaftsgeschichte*, 202–204.
94. John A. Hassan and Alan Duncan, "Integrating Energy: The Problems of Developing an Energy Policy in the European Communities, 1945–1980," *Journal of European Economic History* 23/1 (1994), 159–176; Shaev, "Common Market"; Lucas, *European Communities*, 31–33; Brüggemeier, *Grubengold*, 347–355, 387–390.
95. *KpBr*, 23. Sitzung, April 30, 1958; BSB, 3. Wahlperiode, 59 Session, Coal Debates, January 29, 1959.

96. Quotation from BSB, 3. Wahlperiode, 59 Session, Coal Debates, January 29, 1959; BBA, 32, 3686, Nedelmann to Dewall, "Niederschrift über die Kohle/Heizöl-Besprechung," June 10, 1958; Bundesregierung, *KaW*, 11. Sitzung, October 23, 1958.
97. *KaW*, Kabinetssprotokoll, 18. Sitzung, September 9, 1959.
98. Gutermuth, "Bergbauwirtschaftspolitik heute und morgen," speech in Munich, June 8–13, 1958, doc. 29 in Martiny and Schneider, *Energiepolitik*.
99. BAK, B136, 7702, Helmuth Burckhardt, "Die Lage des Steinkohlenbergbaus Ende 1958," November 10, 1958; Hans-Christoph Seidel and Klaus Tenfelde, "Arbeitsbeziehungen und Sozialpolitik im Ruhrbergbau nach 1945," and Stefan Goch, "Der Weg zur Einheitsgesellschaft Ruhrkohle AG," in *Gluck Auf*, 284–316; Nonn, *Ruhrbergbaukrise*, 47–58, 81–83.
100. Bundesregierung, *KpBr*, 23. Sitzung, April 30, 1958.
101. BBA, 55, 969/1, Aktennotiz über die Besprechung mit Bundeskanzler Adenauer, December 19, 1958.
102. Bundesregierung, *KpBr*, 76. Sitzung, August 28, 1959.
103. *KpBr*, 45. Sitzung, December 3, 1958, and *KpBr*, 76. Sitzung, August 28, 1959.
104. *KaW*, 12. Sitzung, November 18, 1958; *KaW*, 10. Sitzung, September 11, 1958; Parnell, *Coal Industry*, 98–101.
105. Quotation Heinrich Gutermuth, "Wirtschaftliches Stalingrad an der Ruhr," speech from January 25, 1959, doc. 35, in Martiny and Schneider *Energiepolitik*; Nonn, *Ruhrbergbaukrise*, 102; Abelshauser, *Ruhrbergbau*, 87–91.
106. BSB, 3. wahlperiode, 59 Session, Coal Debates, January 29, 1959.
107. BBA, 55, 969/1 unsigned memo from November 29; BAK, B136, 7702, Helmuth Burckhardt, "Die Lage des Steinkohlenbergbaus Ende 1958," November 10, 1958.
108. Mierzejewski, *Erhard*, 34–35; Heinrich Gutermuth, "Wirtschaftliches Stalingrad an der Ruhr," speech on January 25, 1959, doc. 35, in Martiny and Schneider *Energiepolitik*; BBA, 55, 969/1 Nieland to Burckhardt, Dewall, Dütting, Kemper, Keyser, Betrifft: Vereinbarung Kohle-Oel, December 19, 1958; BBA, 32, 3868, "Niederschrift über die Kohle/Heizöl-Besprechung in Hamburg," November 12, 1958; Parnell, *Coal Industry*, 99.
109. BBA, 32, 3686, "Ergebnisprotokoll über das Kohle/Heizöl-Gespräch auf dem Petersberg," July 7, 1959; Bundesregierung, KaW, 17. Sitzung, August 13, 1959; Abelshauser, *Ruhrkohlenbergbau*, 95–96.
110. Quotations from Erhard, "Die Sorgen des Steinkohlenbergbaus, in *Deutsche Wirtschaftspolitik*, 441; Parnell, *Coal Industry*, 101; Abelshauser, *Ruhrkohlenbergbau*, 112.
111. Quotation from Westrick, *KaW*, 17. Sitzung, August 13, 1959.
112. *KpBr*, 76. Sitzung, August 28, 1959; *KaW*, 18. Sitzung, September 9, 1959; Parnell, *Coal Industry*, 101.
113. Bundesregierung, *KpBr*, 76. Sitzung, August 28, 1959.
114. Quotation from ESSO AG, *Die Problematik der Heizölsteuer* (Hamburg: Esso, 1959); BAK, B136, 2506, Herbert Fischer-Menshausen of ESSO to Bundeskanzleramt, October, 26, 1960; BAK, B136, 7702, Bockelmann of BP Benzin und Petroleum AG; van Drimmelen of Deutsche Shell AG, and Geyer of ESSO AG to Erhard and Adenauer, November 17, 1958.
115. Quotation from Heinrich Gutermuth, "Bergbauwirtschaftspolitik heute und morgen," speech in Munich, June 8–13, 1958, doc. 29, in Martiny and Schneider, *Energiepolitik*, 129; Niederschrift über die Sitzung im Bundeskanzleramt," from August 6, 1958, doc. 32, in Martiny and Schneider, *Energiepolitik*; H. Burckhardt, "Die Ergebnisse der Energie-Enquete aus der Sicht des deutschen Steinkohlenbergbaus," in Wessels, *Energie-Enquete*, 101–118.
116. BAK, B136, 7666, Vialon to Staatssekretär, February 13, 1959, Burgbacher to Dr. Hellwig, March 19, 1959.
117. Quotations from Bundesregierung, *KpBr*, 76. Sitzung, August 28, 1959; Bundesregierung, *KaW*, 12. Sitzung, November 18, 1958.
118. "Niederschrift über die Sitzung im Bundeskanzleramt," from August 6, 1958, doc. 32, in Martiny and Schneider, *Energiepolitik*; BAK, B136, 2508, Burckhardt to Adenauer, October 20, 1959.
119. Burckhardt, "Ergebnisse der Enerige-Enquete," 116.

Chapter 2

1. Theodor Wessels, "Present State of Methods of Forecasting Demand for Energy," in World Power Conference, *15th Sectional Meeting of the World Power Conference* (Tokyo: Japan Power Association, 1966), 539–554, 539.
2. Michel Foucault, *The Birth of Biopolitics: Lectures at the Collège de France 1978–1979* (New York: Palgrave MacMillan, 2008), translated by Graham Burchell, 85–86; Sally Razeen, *Classical Liberalism and International Economic Order* (New York: Routledge, 2002); Ralph Ptak, "Neoliberalism in Germany: Revisiting the Ordoliberal Foundations of the Social Market Economy," in Philip Mirowski and Dieter Plehwe (eds.), *The Road from Mont Pelerin: The Making of the Neoliberal Thought Collective* (Cambridge, MA: Harvard University Press, 2015), 98–138.
3. Jenny Andersson, "The Great Future Debate and the Struggle for the World," *American Historical Review* 117/5 (2012), 1411–1430, here 1414; Elke Seefried, "Steering the Future: The Emergence of 'Western' Futures Research and Its Production of Expertise, 1950s to early 1970s," *European Journal of Futures Research* 2/29 (2014), 1–12, online.
4. Gabriele Metzler, *Konzeptionen politischen Handelns von Adenauer bis Brandt. Politische Planung in der pluralistischen Gesellschaft* (Paderborn: Schöningh 2005); Tim Schanetzky, *Die grosse Ernüchterung. Wirtschaftspolitik, Expertise und Gesellschaft in der Bundesrepublik 1966 bis 1982* (Berlin: Akademie Verlag, 2007); Alexander Nützenadel, *Stunde der Ökonomen: Wissenschaft, Politik und Expertenkultur in der Bundesrepublik 1949–1974* (Göttingen: Vandenhoeck & Ruprecht, 2005); Elke Seefried, *Zukünfte: Aufstieg und Krise der Zukunftsforschung 1945–1980* (Oldenbourg: De Gruyter, 2015).
5. Matthias Schmelzer, *The Hegemony of Growth: The OECD and the Making of the Economic Growth Paradigm* (New York: Cambridge University Press, 2016); Reinhard Steurer, *Der Wachstumsdiskurs in Wissenschaft und Politik: Von der Wachstumseuphorie über 'Grenzen des Wachstums' zur Nachhaltigkeit* (Berlin: VWF, 2002).
6. Mitchell, *Carbon Democracy*, first quotation from 124, second from 132.
7. H. W. Arndt, *The Rise and Fall of Economic Growth: A Study in Contemporary Thought* (Melbourne: Longman Cheshire, 1978); Robert Collins, *More: The Politics of Economic Growth in Postwar America* (Oxford: Oxford University Press, 2000); Stephen Macekura, *The Mismeasure of Progress: Economic Growth and Its Critics* (Chicago: University of Chicago Press, 2020); Kander et al., *Power to the People*; Philip Mirowski, "Review: Energy and Energetics in Economic Theory," *Journal of Economic Issues* 22/3 (1988), 811–830; David Pearce, "An Intellectual History of Environmental Economics," *Annual Review of Energy and Environment* 2000/27 (2002), 57–81.
8. Werner Bonefeld, "Freedom and the Strong State: on German Ordoliberalism," *New Political Economy* 17/5 (2012), 633–656; Daniel Stedman Jones, *Masters of the Universe: Hayek, Friedman, and the Birth of Neoliberal Politics* (Princeton, NJ: Princeton University Press, 2014).
9. Eucken quoted in Helge Peukert, "Walter Eucken and the Historical School," in Peter Koslowski (ed.), *The Theory of Capitalism in the German Economic Tradition: Historicism, Ordo-Liberalism, Critical Theory, Solidarism* (Heidelberg: Springer Verlag, 2000), 93–145; Manfred Streit and Michael Wohlgemuth, "The Market Economy and the State. Hayekian and Ordoliberal Conceptions," in ibid., 224–271; Keith Tribe, *Strategies of Economic Order: German Economic Discourse, 1750–1950* (New York: Cambridge University Press, 1995).
10. Franz Böhm, "The Non-State 'Natural' Laws Inherent in a Competitive Economy," in Stützel et al., *Standard Texts*, 107–115.
11. Razeen, *Classical Liberalism*, 105–131; Ptak, "Neoliberalism in Germany"; Volker Berghahn and Brigitte Young, "Reflections on Werner Bonefeld's 'Freedom and the Strong State: On German Ordoliberalism' and the Continuing Importance of the Ideas of Ordoliberalism to Understand Germany's (Contested) Role in Resolving the Eurozone Crisis," *New Political Economy* 18/5 (2013), 768–778.
12. Wilhelm Röpke, *Economics of the Free Society*, translated by Patrick M. Boarman (Chicago: Henry Regnery, 1937/1963); Alfred Muller Armack, "The Social Aspect of the Social Market Economy (1947)," in Wolfgang Stützel et al. (eds.), *Standard Texts of the Social Market Economy: Two Centuries of Discussion* (New York: Gustav Fischer, 1982), 9–23.

13. Walter Eucken, "A Policy for Establishing a System of Free Enterprise (1952)," in Stützel et al., *Standard Texts*, 115–133, here 116; Streit and Wohlgemuth, "Market and State."
14. Quote from Wilhelm Röpke, *The Social Crisis of Our Time* (Chicago: University of Chicago Press, 1942/1950), 21; Müller-Armack, "Social Aspect."
15. Quotation from Röpke, *Free Society*, 257.
16. Quotation from Wilhelm Röpke, "Is the German Economic Policy the Right One? (1950)," in Stützel et al., *Standard Texts*, 37–48, here 46; Wilhelm Röpke, *Civitas Humana: A Humane Order of Society*, translated by Cyril Fox (London: Hodge, 1948); Foucault, *Biopolitics*, 147.
17. Quotation from Angus Burgin, *The Great Persuasion: Reinventing Free Markets since the Depression* (Cambridge, MA: Harvard University Press, 2012), 114; For Röpke, a healthy liberal order required a "necessarily 'hierarchical' composition, where each individual has the good fortune of knowing his position." Röpke, *Social Crisis*, 10; Wilhelm Röpke, "The End of an Era (1933)?" in *Against the Tide*, translated by Elizabeth Henderson (Chicago: Henry Regnery, 1969); Quinn Slobodian, "The World Economy and the Color Line: Wilhelm Röpke, Apartheid, and the White Atlantic," *Bulletin of the German Historical Institute* 54/10 (2014), 61–90.
18. Röpke, *Free Society*, 239.
19. Müller-Armack, "Social Aspect," 11.
20. Quotation from Eucken, "System of Free Enterprise (1952)," in Stützel et al., *Standard Texts*, 116.
21. Italics added. Röpke, *Free Society*, 34.
22. Ziegler, "Regulated Market Economy"; Tooze, *Wages of Destruction*.
23. Mierzejewski, *Erhard*, 62; Röpke, "'Repressed Inflation': The Ailment of the Modern Economy," in Röpke, *Against the Tide* (1947), 181; Ptak, "Neoliberalism in Germany."
24. Tribe, *Economic Order*, 235–237.
25. Ludwig Erhard, *Prosperity through Competition*, translated by Edith Temple Roberts and John B. Wood (New York: Praeger, 1958); Mark Spicka, *Selling the Economic Miracle: Economic Reconstruction and Politics in West Germany, 1949–1957* (New York: Berghahn Books, 2007); Slobodian, "Color Line"; Walter Eucken, *Die Grundlagen der Nationalökonomie* (Jena: G. Fischer, 1940).
26. Nützenadel, *Stunde der Ökonomen*; Metzler, *Konzeptionen politischen Handelns*; Foucault, *Biopolitics*.
27. Röpke, *Free Society*, 128 and 148.
28. Quotation from Helm, *British Energy*, 18; N. J. D. Lucas, *Energy in France: Planning, Politics, and Policy* (London: Europa Publications, 1979), 6; Martin Chick, *Electricity and Energy Policy in Britain, France and the United States since 1945* (Cheltenham: Northampton, MA: Edward Elgar, 2007), introduction.
29. Alfred Müller-Armack, *Auf dem Weg nach Europa* (Stuttgart: C. E. Poeschel, 1971), 63–65; Wilhelm Röpke, Europäische Investitionsplanung: Das Beispiel der Montanunion," *ORDO: Jahrbuch für die Ordnung von Wirtschaft und Gesellschaft* 7 (1955), 71–102.
30. Wilhelm Throm, "Die Kohle und der Wettbewerb," *ORDO: Jahrbuch für die Ordnung von Wirtschaft und Gesellschaft* 11 (1959), 241–255.
31. Uekotter, *Greenest Nation*, 59–80; David Blackbourn, *The Conquest of Nature: Water, Landscape, and the Making of Modern Germany* (New York: W. W. Norton, 2006), 322–335.
32. Quotation from Andersson, "Great Future Debate," 1414; Seefried, "'Western' Futures Research"; Philip Mirowski, *Machine Dreams: Economics Becomes a Cyborg Science* (New York: Cambridge University Press, 2002).
33. Fritz Baade, *Weltenergiewirtschaft: Atomenergie—Sofortprogramm oder Zukunftsplanung* (Hamburg: Rowohlt, 1958), 9 and 24.
34. Louis Armand, "Some Aspects of the European Energy Problem," Report Prepared for the OEEC, June 1955, 1–17, here 2; Jensen, *Energy in Europe*, 32–34; Lucas, *European Communities*, 10–19.
35. Hartley Report—OEEC, *Europe's Growing Needs of Energy: How Can They Be Met?* (Paris: OEEC, 1956); George Gonzalez, *Energy and the Politics of the North Atlantic* (Albany: State University of New York Press, 2013).

36. *Primary energy* is the amount of energy contained in raw materials, such as anthracite or crude oil, before they enter the economic system and are converted into usable energy; *final energy* is the amount of energy consumed by end users after it has been converted into a usable form, like electricity or gasoline.
37. Robinson Report. OEEC, *Towards a New Energy Pattern in Europe* (Paris: OEEC, 1960), 33.
38. Classic examples being the use of coking coal for steel production, or refined petroleum in automobiles; OEEC, *Europe's Growing Needs of Energy*; quotation from OEEC, *New Energy Pattern*, 34.
39. Italics added. OEEC, *New Energy Pattern*, 34–35.
40. Hartley's experts, for instance, initially found energy to be more elastic than they were comfortable with, and so they winnowed down their elasticity estimates to more cautious levels. Harold Lubell, "Survey of Energy and Oil Demand Projections for Western Europe," *US Air Force Project Rand Research Memorandum RM-2385* (May 21, 1959); Jensen, *Energy in Europe*, 32–39.
41. Müller-Armack quoted in Christian Watrin, "Alfred Müller-Armack—Economic Policy Maker and Sociologist of Religion," in Peter Koslowski (ed.), *The Theory of Capitalism in the German Economic Tradition: Historicism, Ordo-Liberalism, Critical Theory, Solidarism* (Heidelberg: Springer Verlag, 2000), 192–220, here 197; Nützenadel, *Stunde der Ökonomen*; Philipp Lepenies, *The Power of a Single Number: A Political History of GDP*, translated by Jeremy Gaines (New York: Columbia University Press, 2016).
42. Adolf Arndt in Bundestag Debate, Deutsche Bundestag, *Stenographische Berichte*, 30. Sitzung, March 16, 1966, 1342.
43. BAK, B136, 2508, "Europas Energie-Bedarf; Sein Anwachsen—Seine Deckung," June 26, 1956; KaW, 58. Sitzung on September 10, 1956, page 273; BSB, "Energiebilanz des Bundesgebiet"; BBA, 26, 67, Report from Wirtschaftsvereinigung Bergbau, November, 1956, "Zukünftige Entwicklung des Rohenergiebedarfs unterteilt nach Energietraegern"; BAK, B136, 2506, Circular from Steinhaus from March 15, 1956.
44. BAK, B136, 7666, Antrag der Fraktion der SPD, 3. Wahlperiode, "Entwurf eines Gesetzes über eine Untersuchung der Erzeugungs- und Absatzbedingungen der deutschen Kohlewirtschaft," from November 26, 1957.
45. BAK, B136, 7666, Kurzprotokoll der 3. Sitzung des Wirtschaftsausschusses, February 12, 1958.
46. The DIW and RWI were also involved. Rüdiger Graf, *Öl und Souveränität: Petroknowledge und Energiepolitik in den USA und Westeuropa in den 1970er Jahren* (Berlin: De Gruyter, 2014), 73–74; BAK, B136, 7666, Report on first meeting in Cologne of under-committee for the Enquete, on February 12, 1958; Report from Dr. Burgbacher on the under-commission of the Enquete, from March 19, 1959; Theodor Wessels (ed.), *Die Energie-Enquete: Ergebnisse und wirtschaftspolitische Konsequenzen. Vorträge und Diskussionsbeitraege der 12. Arbeitstagung am 14. und 15. Juni 1962 in der Universitaet Koeln* (Munich: Oldenbourg, 1962), preface.
47. Fritz Burgbacher, "Eine Würdigung zur Vollendung des 65. Lebensjahres," in Fritz Burgbacher and Theodor Wessels (eds.), "Ordnungsprobleme und Enwticklungstendenzen in der deutschen Energiewirtschaft. Theodor Wessels zur Vollendung seines 65. Lebensjahres gewidmet" (Essen: Vulkan-Verlag, 1967), 2–5.
48. BAK, B136, 7666, executive summary of Enquete Report from March 5, 1962.
49. BAK, B136, 7666, executive summary of Enquete Report from March 5, 1962; H. Burckhardt, "Die Ergebnisse der Energie-Enquete aus der Sicht des deutschen Steinkohlenbergbaus," and Dr. H. Theel, "Die Ergebnisse der Energie-Enquete aus der Sicht der deutschen Mineralölwirtschaft," in Wessels (ed.), *Die Energie-Enquete*, 101–119 and 119–137; in Wessels (ed.), *Die Energie-Enquete*; Unternehmensverband Ruhrburgbau, *Jahresbericht 1961–63* (Essen: Ruhrbergbau, 1963); Discussion by Schmucker in Parliamentary Coal Debates, BSB, 5. Wahlperiode, 30. Session, March 16, 1966.
50. BAK, B136, 7666, executive summary of Enquete Report from March 5, 1962, and Dr. Prass to Chancellor summarizing the Enquete, January 20, 1962.
51. Stefan Cihan Aykut, "Energy Futures from the Social Market Economy to the *Energiewende*: The Politicization of West German Energy Debates, 1950–1990," in Andersson, *Struggle for the Long-Term*, 63–91; Unternehmensverband Ruhrburgbau, *Jahresbericht 1961–1963*; BAK,

B136, 7666, President of Deutscher Sparkassen und Giroverband, to Adenauer, May 9, 1962; Siegfried Balke, "Die Entwicklungschancen der Atomenergie und die Ergebnisse der Energie-Enquete," in Wessels (ed.), *Die Energie-Enquete*, 26–38; Burckhardt, "deutschen Steinkohlenbergbau," in Wessels (ed.), *Die Energie-Enquete*; BAK, B136, 7666, executive summary of Enquete Report from March 5, 1962.

52. Joachim Radkau, *Aufstieg und Krise der deutschen Atomwirtsschaft, 1945–1975* (Hamburg: Rohwohlt, 1983), 123–125.
53. Theel, "deutschen Mineralölwirtschaft," 130–131; Dr. Blömer of EWI, in "Diskussion zu den Referaten von Generaldirektor, Dr.-Ing. Burckhardt und Direktor Dr. Theel," in Wessels (ed.), *Die Energie-Enquete*, 137–165.
54. J. R. McNeill, *Something New under the Sun: An Environmental History of the Twentieth Century World* (New York: W. W. Norton, 2001), 236; Matthias Schmelzer, "The Growth Paradigm: History, Hegemony, and the Contested Making of Economic Growthmanship," *Journal of Ecological Economics* 118/C (2015), 262–271.
55. Schmelzer, *Growth Paradigm*, 87 and 167; Daniel Speich Chassé, "The Use of Global Abstractions: National Income Accounting in the period of Imperial Decline," *Journal of Global History* 6/1, 7–28; Lepenies, *Single Number*; Diane Coyle, *GDP: A Brief but Affectionate History* (Princeton, NJ: Princeton University Press, 2014); Lepenies, *Single Number*; Laura Belmont, "Selling Capitalism: Modernization and US Overseas Propaganda," in *Staging Growth: Modernization, Development, and the Global Cold War* (Amherst: University of Massachusetts Press, 2003), 107–128.
56. Peter Wiles quoted in Macekura, *Mismeasure*, 69; Arndt, *Economic Growth*, 62–64; Schmelzer *Growth Paradigm*, 167.
57. Quotation from Mitchell, *Carbon Democracy*, 139, more generally, 136–143; Paul Kennedy, *The Rise and Fall of the Great Powers: Economic Change and Military Conflict 1500–2000* (New York: Random House, 1987); Stanley Jevons, *The Coal Question; an Inquiry Concerning the Progress of the Nation, and the Probable Exhaustion of our Coal-Mines* (London: Macmillan, 1865/1906); Macekura, *Mismeasure*, 14–16, 35.
58. Quotation from Lewis Mumford, *Technics and Civilization* (Chicago: University of Chicago Press, 1934/2000), 375; Mirkowski, "Energy and Energetics"; Nikifouruk, *Energy of Slaves*, 68, 135–140; Vaclav Smil, *Energy in World History* (Boulder, CO: Westview Press, 1994), 203.
59. Agnar Sandmo, "The Early History of Environmental Economics," *Review of Environmental Economics and Policy* 9/1 (2015), 43–63; Pearce, "Environmental Economics"; Clive L. Spash, "The Development of Environmental Thinking in Economics," *Environmental Values* 8 (1999), 413–435; Kenneth Boulding, "The Economics of the Coming Spaceship Earth," H. Jarrett (ed.), *Environmental Quality in a Growing Economy* (Baltimore, MD: Johns Hopkins Press, 1966), 3–14; Robert Constanza, "The Early History of Ecological Economics and the International Society for Ecological Economics," in *International Society for Ecological Economics Internet Encyclopaedia of Ecological Economics* (April 2003).
60. Galbraith cited in Arndt, *Economic Growth*, 125; Spash, "Environmental Thinking."
61. Harrod cited in Theodore W. Schultz, "Connections between Natural Resources and Economic Growth," in Joseph J. Spengler (ed.), *Natural Resources and Economic Growth* (Washington, DC: Resources for the Future, 1961), 1–9, here 3–4.
62. Schultz, "Natural Resources," in Spengler (ed.), *Natural Resources*; Paley Report—The President's Materials Policy Commission, *A Report to the President by the President's Materials Policy Commission* (1952).
63. First quote from Joseph Spengler, "Summary, Synthesis, Interpretation," in Spengler (ed.), *Natural Resources*, 275–304, here 277; second quote from Schultz, "Natural Resources," 5, both in Spengler (ed.), *Natural Resources*.
64. Harold J. Barnett and Chandler Morse, *Scarcity and Growth: The Economics of Natural Resource Availability* (Baltimore, MD: Johns Hopkins Press, 1963), quotations from 10, 236, 240; Pearce, "Environmental Economics"; Arndt, *Economic Growth*, 119–125.
65. Quotation from Macekura, *Mismeasure*, 69; P. S. Dasgupta and G. M. Heal, *Economic Theory and Exhaustible Resources* (Cambridge: Cambridge University Press, 1979), 1.
66. Quotation from Röpke, *Social Crisis*, 67; Lepenies, *Single Number*.

67. Sally Razeen, "Ordoliberalism and the Social Market: Classical Political Economy from Germany," *New Political Economy* 1/2 (1996), 234; Nützenädel, *Stunde der Ökonomen* 18–19, 90–114; Steurer, *Wachstumsdiskurs*, 368; Arndt, *Economic Growth*, 62–64; Illing, *Energiepolitik in Deutschland*, 36–37.
68. ACDP, 08–003, Entwurf für eine Grosse Anfrage der CDU/CSU-Fraktion betreffend Strom- und Gasversorgung, October 23, 1968.
69. Vaclav Smil, *Energy Transitions: Global and National Perspectives* (Santa Barbara, CA: Praeger, 2017).
70. BAK, B136, 7666, executive summary of Enquete Report from March 5, 1962; BBA, 26, 80, Esso AG, "Energiepolitik in Deutschland: Eine Bestandsaufnahme" (1968); Graf, *Öl und Souveränität*, 60–62; Hans Diefenbacher and Jeffrey Johnson, "Energy Forecasting in West Germany: Confrontation and Convergence," in Thomas Baumgartner and Atle Midttun (eds.), *The Politics of Energy Forecasting: A Comparative Study of Energy Forecasting in Western Europe and North America* (Oxford: Clarendon, 1987), 61–84; Farrenkopf and Przigoda (eds.), *Glück auf!*, Table 9, p. 548.
71. Theodor Wessels, *Die volkswirtschaftliche Bedeutung der Energiekosten* (Munich: Oldenbourg, 1966), 57.
72. Alexander Rüstow, "Kritik des technischen Fortschritts," *ORDO: Jahrbuch für die Ordnung von Wirtschaft und Gesellschaft* 4 (1951), 373–407; Wessels, *Energiekosten*, part II.
73. Figures from Wessels, *Energiekosten*, 19; Theodor Wessels, "Wandlungen in der Energiewirtschaft," in *Kyklos* 15/1 (1962), 317–323; Theodor Wessels, "Die Investitionen in der Energiewirtschaft im Rahmen der volkswirtschaftlichen Entwicklung," in Theodor Wessels (ed.), *Investitions- und Finanzierungsprobleme in der Energiewirtschaft* (Munich: Oldenbourg, 1968), 12–31.
74. Siegfried Balke, *Investitionspolitik im Grossraum Europa am Beispiel der Energiewirtschaft* (Berlin: Duncker & Humblot, 1963), 7.
75. Wessels, *Energiekosten*, 57.
76. Baade, *Weltenergiewirtschaft*, 72.
77. Theodor Wessels, "Die Marktstellung des Mineralöls in der deutschen Energiewirtschat," in Wilfred Schreiber et al. (eds.), *Festschrift für Wilfried Schreiber: Der Mensch in sozio-ökonomischen Prozess* (Berlin: Duncker & Humblot, 1969), 199–214; BBA, Nachlass Keyser, 26, 79, Gerhard Bischoff, "Weltenergieüberfluss—wie lange?" (1966).
78. Wessels, "Investitionen," in Wessels (ed.), *Investitions- und Finanzierungsprobleme*; Wessels, *Energiekosten*, 56–62.
79. Hans K. Schneider, "Zur Konzeptionen einer Energiewirtschaftspolitik," in Fritz Burgbacher (ed.), *Ordnungsprobleme und Entwicklungstendenzen in der deutschen Energiewirtschaft: Festschrift für Theodor Wessels* (Essen: Vulkan-Verlag, 1967), 19–45, here 25.
80. Manfred Liebrucks, "Corrigendum zum Energiegutachten von 1961," *Vierteljahreshefte zur Wirtschaftsforschung* 2 (1966), 179–199; BAK, B136, 7666, *Corrigendum zum Energiegutachten*, May 1966; Wessels, "Methods of Forecasting" in *World Power Conference*; DIWA, Report to BWM by Manfred Liebrücks, H. W. Schmidt, and D. Schmitt, *Sicherung der Energieversorgung der Bundesrepublik Deutschland* (1972).
81. Quotation from Hans-Joachim Burchard, *Methoden und Grenzen der Energieprognosen* (Hamburg: BP Benzin und Petroleum AG, 1968), 37; Graf, *Öl und Souveränität*, 73–80; Diefenbacher and Johnson, "Energy Forecasting."
82. BSB, Wahlperiode, 30. Session on May 16, 1962; Diefenbacher and Johnson, "Energy Forecasting"; FES, Bechert, 107–251, Eduard Grueber, "Steigender Energiebedarf kein Naturgesetz," September 20, 1973.
83. Heinz W. Vollrath, *Die Mineralölwirtschaft in der Bundesrepublik: Ihr Aufbau und ihre Entwicklung seit 1945* (Munich: Uni-Druck, 1959), 145–158; Wessels, "Die Marktstellung des Mineralöls; Daniel Yergin, *The Prize: The Epic Quest for Oil, Money, and Power* (New York: Free Press, 2008), 174, 483; Lucas, *Energy in France*.
84. Wessels, *Energiekosten*, 57–59; Wessels in "Internationale Zusammenhänge in der Energiewirtschaft: Bericht ueber die 11. Arbeitstagung des Energiewirtschftlichen Instituts an der Universitaet zu Koeln, am 6. und 7. April 1960," *Erdöl und Kohle. Erdgas, Petrochemie* 13 (August 1960), 606.

85. Wessels, "Investitionen," in Wessels (ed.), *Investitions- und Finanzierungsprobleme*, 23.
86. Wessels, *Energiekosten*, 58–60.
87. Italics added. ACDP, 07–001, 9037, Vermerk für Herrn Russe Betr. Sitzung der Kommission "Kohle und Stahl" on March 17, 1967.
88. ACDP, 07–001, 9037, Vermerk für Herrn Russe Betr. Sitzung der Kommission "Kohle und Stahl" on March 17, 1967; Theodor Wessels, "The Reorganization of the Hard-Coal Mining Industry in the Federal Republic of Germany," *German Economic Review* 8/1 (1970), 81–88.
89. Richard H. K. Vietor, *Energy Policy in America since 1945: A Study of Business Government Relations* (New York: Cambridge University Press, 1984); Pratt et al., *Voice of the Marketplace*.
90. Cited in William J. Barber, "The Eisenhower Energy Policy: Reluctant Intervention," in Goodwin, *Energy Policy*, 205–286, here 224; Martin Melosi, *Coping with Abundance: Energy and Environment in Industrial America* (Philadelphia: Temple University Press, 1985).
91. William J. Barber, "Studied Inaction in the Kennedy Years," and James L. Cochrane, "Energy Policy in the Johnson Administration: Logical Order versus Economic Pluralism," in Goodwin, *Energy Policy*, 287–336 and 337–395; Megan Black, *The Global Interior: Mineral Frontiers and American Power* (Cambridge, MA: Harvard University Press, 2018), 126–128.
92. Tyler Priest, "Hubbert's Peak. The Great Debate over the End of Oil," *Historical Studies in the Natural Sciences* 44/1 (2014), 37–79, here 59.
93. Allais cited in Chick, *Energy Policy*, 17, see also 12–16; Lucas, *Energy in France*, 41–45.
94. Quotation from Harold Lubell, "Security of Supply and Energy Policy in Western Europe," *World Politics* 13/3 (1961), 400–422, here 414; Lucas, *Energy in France*, introduction.
95. Farrenkopf and Przigoda, *Glück auf*, Table 9, p. 548; Odell, *World Power*, 120–121.
96. Quotation from BAK, B136, 7703, Economics Minister to the Chancellor, March 8, 1965; Farrenkopf and Przigoda, *Glück auf*, Tables 8 and 9, pp. 546–548; BSB, "Energiebilanz der Bundesgebietes."
97. Theodor Wessels, "Die Sicherheit der Nationalen Versorgung als Ziel der Wirtschaftspolitik," *Zeitschrift für die gesamte Staatswissenschaft / Journal of Institutional and Theoretical Economics* 120/4 (1964), 602–617, here 602.
98. First quotation "Helmuth Burckhardt," *Der Spiegel* 11 (1959), https://www.spiegel.de/politik/helmuth-burckhardt-a-2e268738-0002-0001-0000-000042624766; last quotation BBA, 26, 79, Anonymous, "Was Kostet eine Sichere Energieversorgung?" *WID: Energiewirtschaft*, November 9, 1967; Burckhardt, "Die Ergebnisse der Energie-Enquete aus der Sicht des deutschen Steinkohlenbergbaus," in Wessels (ed.), *Die Energie-Enquete*, 101–119.
99. Wessels, "Sicherheit," in *Zeitschrift für die gesamte Staatswissenschaft*, 602–603; Theodor Wessels, "Die Marktstellung des Mineralöls," in Schreiber et al. (ed.), *Festschrift*, 199–214.
100. Italics added, Wessels, "Struktur und Entwicklungstendenzen," in Wessels (ed.), *Die Energie-Enquete*, 23.
101. Wessels, "Sicherheit"; Heinz Jürgen Schürmann, "Veba–Geslenberg–Deminex: Unternehmerische Konsquenzen aus der Energiekrise," *Zeitschrift für öffentliche und gemeinwirtschaftliche Unternehmen* 4/1 (1981), 19–50.
102. Wessels, "Sicherheit," 617.
103. Balke, *Grossraum Europa*, 7; Eckert Teichert, *Autarkie und Grossraumwirtschaft in Deutschland, 1930–1939: Aussenwirtschaftspolitischen Konzeptionen Zwischen Wirtschaftskrise und Zweiten Weltkrieg* (Munich: Oldenbourg, 1988).
104. Wessels, "Sicherheit"; Fritz Burgbacher and Theodor Wessels, *Die Energiewirtschaft im Gemeinsamen Markt* (Baden-Baden: August Lutzeyer, 1963), 243–244;
105. Albrecht Mulfinger, *Auf dem Weg zur gemeinsamen Mineralölpolitik* (Berlin: Duncker & Humblot, 1972), 187.
106. Italics added, quotation from Schneider, "Konzeptionen einer Energiewirtschaftspolitik," 32; Urs Dolinski and Hans-Joachim Ziesing, *Die regionalen Entwicklungstendenzen des Energieverbrauchs in Baden-Württemberg und seinen Regierungsbezierken bis 1980* (Berlin: DIW, 1970); Manfred Liebrucks und Hildebrand Kummer, *Grundlagen einer regionalwirtschaftlich orientierten Energiepolitik im norddeutschen Raum* (Berlin: Duncker & Humblot, 1972).
107. Quotation from Schneider, "Konzeptionen einer Energiewirtschaftspolitik," 22.
108. Schneider, "Konzeptionen einer Energiewirtschaftspolitik."

109. Italics added. Urs Dolinski and Hans-Joachim Ziesing, *Die regionalen Entwicklungstendenznen des Energieverbrauchs in Bayern und seinen Regierungsbezirken bis 1985* (Berlin: DIW, 1975), introduction.
110. DIWA, Report to BWM by Manfred Liebrücks, H. W. Schmidt, and D. Schmitt, *Sicherung der Energieversorgung der Bundesrepublik Deutschland* (1972).

Chapter 3

1. Siegfried Balke, West Germany's Second Minister of Nuclear Energy, *Grossraum Europa*, 14.
2. Odd Arne Westad, *The Global Cold War: Third World Interventions and the Making of Our Times* (New York: Cambridge University Press, 2007), 123–126; Ethan B. Kapstein, *The Insecure Alliance: Energy Crises and Western Politics since 1944* (New York: Oxford University Press, 1990), 118; OEEC, *Europe's Need for Oil: Implications and Lessons of Suez Crisis* (Paris: OEEC, 1958).
3. BAK, B102, 5444, Circular from Dr. Raabe from August 11, 1956.
4. Quotation from Malm, *Fossil Capital*, 358; Smil, *Energy Transitions*, 200–201.
5. Kander et al., *Power to the People*, 8, also 28–29; Brian Black, *Crude Reality: Petroleum in World History* (New York: Rowman & Littlefield, 2012), 155–183; Nye, *Consuming Power*.
6. Quotation from Christopher Jones, *Routes of Power: Energy and Modern America* (Cambridge, MA: Harvard University Press, 2014); Russel et al., "The Nature of Power."
7. Quotation from A. H. Stahmer, *Erdölvorkommen der Welt* (Heidelberg: Hütig und Dreyer, 1952), 7; Giuliano Garavini, *The Rise & Fall of OPEC in the Twentieth Century* (New York: Oxford University Press, 2019).
8. Nathan Rosenberg, *Exploring the Black Box: Technology, Economics, and History* (New York: Cambridge University Press, 1994), 161–164; Malm, *Fossil Capital*, 358–360.
9. Odell, *World Power*, 21, 87–88, 200–202; Daniel Yergin, *The Prize: The Epic Quest for Oil, Money, and Power* (New York: Free Press, 2008), 373–413; Melosi, *Abundance*, 245; Mulfinger, *Mineralölpolitik*, 250–278; Morris A. Adelman, *The World Petroleum Market* (Baltimore, MD: Johns Hopkins University Press, 1972), 305, tables II-E-6 and II-E-7; E. Kratzmüller, "Die Investitionsproblematik der Mineralölwirtschaft unter Berücksichtigung der internationalen Verflechtungen," in Theodor Wessels (ed.), *Investitions- und Finanzierungsprobleme in der Energiewirtschaft* (Munich: Oldenbourg, 1968), 115–133.
10. Adelman, *World Petroleum*, 322–324, table III-B-1; Odell, *World Power*, 11, 94; Yergin, *Prize*, 373; Pirani, *Burning Up*, 79; Garavini, *OPEC*, 88.
11. OEEC, *Suez Crisis*; Zachary Cuyler, "Tapline, Welfare Capitalism, and Mass Mobilization in Lebanon, 1950–1964," in Atabaki Touraj et al., *Working for Oil: Comparative Social Histories of Labor in the Global Oil Industry* (New York: Palgrave Macmillan, 2018), 337–368.
12. Adelman, *World Petroleum*, 105; OEEC, *Oil, Recent Developments in OEEC Area* (Paris: OEEC, 1960), 63; Michael Miller, *Europe and the Maritime World: A Twentieth Century History* (Cambridge: Cambridge University Press, 2012), 308–315.
13. Morris A. Adelman, *Genie out of the Bottle: World Oil since 1970* (Cambridge, MA: MIT Press, 2008); Anna Rubino, *Queen of the Oil Club: The Intrepid Wanda Jablonski and the Power of Information* (Boston: Beacon Press, 2008), 156–163, 189–195; William J. Barber, "The Eisenhower Energy Policy: Reluctant Intervention," in Goodwin and Barber (eds.), *Energy Policy*, 205–286; Yergin, *Prize*, 196–201, 481–482, 517–520; Francisco Parra, *Oil Politics: A Modern History of Petroleum* (London: I. B. Tauris, 2004), 68 and 87.
14. Kander et al., *Power*, 292–294; Black, *Crude Reality*, 60–95, 160–161; Christopher W. Wells, "The Rise of the Model T: Culture, Road Conditions, and Innovation at the Dawn of the American Motor Age," *Technology and Culture* 48/3 (2007), 497–523.
15. Glenn Yago, *The Decline of Transit: Urban Transportation in German and U.S. Cities, 1900–1970* (New York: Cambridge University Press, 1984), 33–34; Thomas Zeller, "Building and Rebuilding the Landscape of the Autobahn, 1930–1970," in Christoph Mauch and Thomas Zeller (eds.), *The World Beyond the Windshield: Roads and Landscapes in the United States and Europe* (Athens: Ohio University Press, 2008), 125–142; Bernard Rieger, *The People's Car: A Global History of the Volkswagen Beetle* (Cambridge, MA: Harvard University Press, 2013).
16. Karlsch and Stokes, *Faktor Öl*, 69–79; Horn, *Energiepolitik*, 242–244.

17. Quotations from Barbara Schmucki, *Der Traum vom Verkehrsfluss: Städtische Verkehrsplanung seit 1945 im deutsch-deutschen Vergleich* (Frankfurt: Campus Verlag, 2001), 118–120, 123–124, 259–267; *Der Spiegel*, "Leber-Plan: Ärmel hoch," February 10, 1967, https://www.spiegel.de/politik/aermel-hoch-a-bcfd9a8c-0002-0001-0000-000046289916; Dietmar Klenke, *"Freier Stau für freie Bürger": Die Geschichte der bundesdeutschen Verkehrspolitik, 1949–1994* (Darmstadt: Wissenschaftliche Buchgesellschaft, 1995), 64; Hans Bernhard Reichow, *Die autogerechte Stadt: Ein Weg aus dem Verkehrs-Chaos* (Ravensburg: Otto Maier Verlag, 1959); Thomas Zeller, *Driving Germany: The Landscape of the German Autobahn, 1930–1970*, translated by Thomas Dunlap (New York: Berghahn Books, 2007), 183–185.
18. Quotations from Sachs, *Automobile*, 83 and 71, also 74.
19. Quotation in Sachs, *Automobile*, 71, 79–81.
20. Heinz Ortlieb quoted in Victoria De Grazia, *Irresistible Empire: America's Advance through 20th-Century Europe* (Cambridge, MA: Harvard University Press, 2005), 359; Wehler, *Deutsche Gesellschaftsgeschichte*, 76–80; Berend, *Economic History*, 255–253; Christian Pfister, "The '1950s Syndrome' and the Transition from a Slow-Going to a Rapid Loss of Global Sustainability," in Frank Uekötter (ed), *The Turning Points of Environmental History* (Pittsburgh: University of Pittsburgh Press, 2010), 90–118; Herbert, *Germany*, 657–662.
21. Odell, *World Power*, 120; Klaus-Dieter Fischer, "Struktur und Entwicklungstendenzen der Energiewirtschaft in der Bundesrepublik Deutschland," in Fritz Burgbacher and Theodor Wessels (eds.), *Ordnungsprobleme und Entwicklungstendenzen in der deutschen Energiewirtschaft* (Essen: Vulkan-Verlag, 1967), 61–107; Sachs, *Automobile*, 71; Zeller, *Driving*, 184; Molle and Wever, *Petrochemical Industries*, 84; James Bamberg, *British Petroleum and Global Oil, 1950–1975* (Cambridge: Cambridge University Press, 2000), 243–245.
22. Quotation Kratzmüller, "Investitionsproblematik," 116–117; BAK, B102, 126250, H. Streicher (BP) to Schmidt in BWM, "Volkswirtschaftliche Bedeutung der Mineralölraffinieren."
23. Molle and Wever, *Petrochemical Industries*, 41–43, 164; Odell, *World Power*, 115; Mulfinger, *Mineralölpolitik*, 67–68; BAK, B102, 126250, E. Bockelmann and H. Streicher, "Ursachen und Wirkungen der veränderten Standortstruktur . . ."; BAK, B102, 126250, H. Streicher (BP) to Schmidt, "Volkswirtschaftliche Bedeutung . . . ."
24. BAK, B102, 126250, E. Bockelmann and H. Streicher (BP), "Ursachen und Wirkungen der veränderten Standortstruktur. . . ."
25. "West Germany Report," *Oil and Gas Journal*, December 31, 1956, 132–133; *KaW*, 8th session of the committee on June 24, 1959, 97–100; Michael Hascher, *Politikberatung durch Experten: Das Beispiel der deutsche Verkehrspolitik im 19. Und 20. Jahrhundert* (Frankfurt: Campus Verlag, 2006), 249–253.
26. BAK, B102, 126250, H. Streicher (British Petroleum) to Schmidt, "Volkswirtschaftliche Bedeutung der Mineralölraffinieren"; OEEC, *Recent Developments*; figures from Fischer, "Energiewirtschaft"; quotation from Molle and Wever, *Petrochemical Industries*, 49.
27. Quotation from BAK, B102, 126250, H. Streicher (BP) to Schmidt, "Volkswirtschaftliche Bedeutung der Mineralölraffinieren."
28. Quotation from BAK, B102, 126250, speech by Mr. W. R. Stott, Executive VP of Standard Oil Company (NJ) at Ingolstadt, February 22, 1964; BAK, B102, 126250, H. Streicher (BP) to Schmidt, "Volkswirtschaftliche Bedeutung der Mineralölraffinieren."
29. BAK, B102, 126250, E. Bockelmann and H. Streicher (BP), "Ursachen und Wirkungen der veränderten Standortstruktur . . ."; Kratzmüller, "Investitionsproblematik"; BAK, B136, 7702, Walter Bauer (Deutsche Shell), "Stellung des Mineralöls in der deutschen Energiewirtschaft," October 9, 1961; BBA, 26, 80, Report from ESSO, "Energiepolitik in Deutschland: Eine Bestandsaufnahme."
30. BAK, B102, 126250, E. Bockelmann and H. Streicher (BP), "Ursachen und Wirkungen der veränderten Standortstruktur . . ."; Fischer, "Energiewirtschaft," 103–106; Vincent Lagendijk, *Electrifying Europe: The Power of Europe in the Construction of Electricity Networks* (Amsterdam: Askant Academic, 2008), 168–169; Urs Dolinski and Hans-Joachim Ziesing, *Die regionalen Entwicklungstendenzen des Energieverbrauchs in Bayern und seinen Regierungsbezirken bis 1985. Teil III* (Berlin: DIW, 1975), 217–218; Abelshauser, *Wirtschaftsgeschichte*, 316–318; "Die Entwicklungstendenzen des Energieverbrauchs . . . ," *DIW Wochenbericht*, May 26, 1971, 157–160.

31. Quotation in BAK, B102, 126250, Westrick to Guillaumat, April 29, 1963; BAK, B102, 126250, E. Bockelmann and H. Streicher (BP), "Ursachen und Wirkungen der veränderten Standortstruktur...."
32. Robert Vitalis, *America's Kingdom: Mythmaking on the Saudi Oil Frontier* (Stanford, CA: Stanford University Press, 2007).
33. BAK, B102, 5444/2, Boecker to Reichardt, October 3, 1956; KaW, 59th meeting, on November 10, 1956, 287; BAK, B102, 5444/2, circular from July 31, 1956, and Bundesamt für gewerbliche Wirtschaft, Hamburg to BWM, August 8, 1956.
34. BAK, B102, 5444/2, "Mineralölversorgung der Bundesregierung unter Berücksichtigung der Suezkrise," September 26, 1956; *KaW*, 59th meeting, November 10, 1956, 287; *KpB*, special meeting, November 8, 1956, 696; BAK, B102, 5444, Biersack to Erhard and Berger to Erhard, both from December 3, 1956; Kapstein, *Insecure Alliance*, 118–120.
35. Quotation from BAK, B102, 5446, Neef to Erhard, November 29, 1956; BAK, B102, 73459, Kling to Prass, January 26, 1957, Erhard Statement from November 22, 1956, Müller-Armack draft statement to OEEC, November 22, 1956, and Krautwig to Erhard, November 23, 1956; HAEU Online Archives, OEEC-162, Executive Committee Minutes 1956; Rolf Kowitz, *Alfred Müller-Armack: Wirtschaftspolitik als Berufung* (Cologne: Deutscher Instituts-Verlag, 1998), 225, 289.
36. BAK, B102, 73459, Kling to Prass, January 26, 1957; BAK, B102, 5444/1, Erhard to Fassbender, February 8, 1957; Paul Frankel, "Oil Supplies during the Suez Crisis: On Meeting a Political Emergency," *Journal of Industrial Economics* 6/2 (1958), 85–100; Stokes, "Oil," 252; Joseph Pratt, *Voices of Marketplace: A History of the National Petroleum Council* (College Station: Texas A&M Press, 2002), 44–48; Yergin, *Prize*, 474–476; OEEC, *Suez Crisis*, 31–37; Vessela Chakarova, *Oil Supply Crisis: Cooperation and Discord in the West* (Lanham, MD: Lexington Books, 2013), 47–49.
37. KAW, 60th meeting, November 19, 1956, 300–302.
38. BAK, B102, 5446, Report from Kling, November 30, 1956; BAK, B102, 5446, Krautwig to Erhard, November 23, 1956; BAK, B102, 73459, Kling in BWM to Prass, January 26, 1957.
39. BAK, B102, 5444/1, Erhard to Fassbender, February 8, 1957.
40. BAK, B102, 5429/2, Kling to Dr. Huebner, November 7, 1957; BSB, Schmücker's (CDU) report to Parliament, 4. Wahlperiode, 148 Session, December 2, 1964; Graf, *Souveranität*, 53–55; Henning Türk, "The Oil Crisis of 1973 as a Challenge to Multilateral Energy Cooperation among Western Industrialized Countries," *Historical Social Research* 39/4 (2014), 209–230.
41. Cantoni, "NATO and the EEC"; Karlsch and Stokes, *Faktor Öl*, 306–310; Mulfinger, *Mineralölpolitik*, 192–193.
42. BAK, NL 1229, Schiller, Arendt to Schiller, "Energiepolitik im Rahmen einer freiheitlich-sozialen Wirtschaftsordnung," November 16, 1964.
43. Memorandum der IGBE zur Lage im Westdeutschen Steinkohlenbergbau, April 7, 1962, Doc. 42 in Martiny and Schneider, *Energiepolitik*.
44. Quotations from "Die Erdölzufuhr muss gesichert werden ...," May 1965, doc. 49 in Martiny and Schneider, *Energiepolitik*; Wessels, "Sicherheit."
45. Quotation from BAK, NL 1229, Schiller, 176, Report from Energy Policy Unit from November 12, 1964; BAK, B136, 7703, Economics Ministry to Chancellor's Office, March 3, 1965; Karlsch and Stokes, *Faktor Öl*, 318–322; Horn, *Energiepolitik*, 39, 44–49; Theodor Wessels, "Marktstellung des Mineralöls in der deutschen Energiewirtschaft," in Wilfred Schreiber et al. (eds.), *Festschrift für Wilfried Schreiber: Der Mensch in sozio-ökonomischen Prozess* (Berlin: Duncker & Humblot, 1969), 199–214.
46. Quotation from BAK, NL 1229, Schiller, 176, "Zur Wirtschaftspolitik im Bereich der Energiewirtschaft," April 23, 1965; see also in ibid., Report from Energy Policy Unit from November 12, 1964.
47. BAK, NL 1229, Schiller, 176, Report from Energy Policy Unit from November 12, 1964.
48. Molle and Wever, *Petrochemical Industries*, 35; Fischer, "Energiewirtschaft," 76, 79, 89; Farrenkopf and Przigoda, *Glück auf!*, 542.
49. BAK, B136, 7702, Walter Bauer (Deutsche Shell), "Stellung des Mineralöls in der deutschen Energiewirtschaft," October 9, 1961.

50. Fischer, "Energiewirtschaft," 106; BBA, 26, 79, "Was Kostet eine Sichere Energieversorgung?" in WID: Energiewirtschaft, 6/44; Stefan Goch, "Der Weg zur Einheitsgesellschaft Ruhrkohle AG," in Farrenkopf and Przigoda, *Glück auf!*, 284–316.
51. BSB, 4. Wahlperiode 148th session, December 2, 1964; Adelman, *World Petroleum*, 399–403; BAK, B102, 126250, Unternehmensverband Ruhrbergbau to Dr. Gerhard Woratz, in BMW, November 14, 1966; BAK, B136, 2508, Prass to Chancellor, June 26, 1961; BAK, NL 1229, Schiller, 176, Unternehmensverband Ruhrbergbau, "Heizölpreis in den EWG-Ländern," September 9, 1965.
52. BAK, NL 1229, Schiller, 176, 1964 Jahresbericht: Rationalisierungsverband des Steinkohlenbergbaus" and "Zur Wirtschaftspolitik im Bereich der Energiewirtschaft," April 23, 1965; D. Müller, "Die Absatzlage des Steinkohlenbergbaus und ihre Rückwirkungen...," in Wessels (ed.), *Investitions- und Finanzierungsprobleme*, 101–114; Wessels, "Investitionen," 12–31; Spiegelberg, *Energiepolitik*, 95; Unternehmensverband Ruhrburgbau, *Jahresbericht*, 1961–63, 58.
53. Unternehmensverband Ruhrburgbau, *Jahresbericht*, 1961–63, 42; Horn, *Energiepolitik*, 299; Müller, "Steinkohlenbergbaus," 107.
54. Abelshauser, *Ruhrkohlenbergbau*, 108–109, 124; Horn, *Energiepolitik*, 79.
55. Lucas, *European Communities*, 34–39; Hassan and Duncan, "Integrating Energy."
56. Heinz Kegel, "Die Künftige Bedeutung der Kohle...," July 6–7, 1961, doc. 40 in Martiny and Schneider, *Energiepolitik*; Unternehmensverband Ruhrburgbau, *Jahresbericht, 1958–1960*; Wessels, "Enquete-Ergebnisse."
57. Goch, "Ruhrkohle"; Abelshauser, *Ruhrkohlenbergbau*, 106–108; Nonn, *Ruhrbergbaukrise*, 192–208; Illing, *Energiepolitik*, 73–75; BSB, 4. Wahlperiode, 30th Session, May 16, 1962—Coal debates.
58. First quote from Goch, "Ruhrkohle," 289; second from Abelshauser, *Ruhrkohlenbergbau*, 106–107; Illing, *Energiepolitik*, 64–65; Farrenkopf and Przigoda, *Glück auf!*, 534.
59. Christoph Nonn, "Kohlenkrise–Das erste Jahrzehnt 1958–1968," in Farrenkopf and Przigoda, *Glück auf!*, 235–283, here 262; Abelshauser, *Ruhrkohlenbergbau*, 106; BBA, 11, DEA "Geschäftsbericht 1967."
60. BSB, 4. Wahlperiode, 148th Session, December 2, 1964; BAK, B136, 7702, Report to Schmücker, December 1, 1964, and Schmücker to Chancellor's Office, December 3, 1964; BAK, B136, 7704, Economics Ministry to Westrick in Chancellor's Office, March 4, 1966, and Krink, report to Cabinet, February 7, 1967; *KpB*, 67th Meeting, February 22, 1967; Farrenkopf and Przigoda, *Glück auf!*, 542; BBA, 26, 80, Report from ESSO, "Energiepolitik in Deutschland: Eine Bestandsaufnahme."
61. Karl Hardach, *The Political Economy of Germany in the Twentieth Century* (Berkeley: University of California Press, 1980), 199–200; Carlin, "West German Growth; Winkler, *Long Road*, 213.
62. Quotation from BBA, 26, 79, Keyser, "Kohlenbergbau- Ruhrgebiet/Sorge und Hoffnung," March, 1967; BAK, NL 1229, Schiller, 129, DEA "Erklärung des Vorstandes," July 1, 1965; Nonn, *Ruhrbergbaukrise*, 265; Farrenkopf and Przigoda, *Glück auf!*, 534; Goch, "Ruhrkohle"; Parnell, *Coal Industry*, 106–110.
63. Nonn, *Ruhrbergbaukrise*, 261.
64. Rudolf Wawersik, *Ausbeute eines Bergmannslebens* (Essen: Glückauf, 1981), 163.
65. First quotation from BBA, 26, 79, Keyser, "Kohlenbergbau- Ruhrgebiet/Sorge und Hoffnung," March 1967; second quotation from Stellungnahme der IGBE, 1967, doc. 58 in Martiny and Schneider, *Energiepolitik*; Überlegungen der IGBE zur energiepolitischen Situation, October 31, 1966, doc. 54 in ibid.
66. Nonn, *Ruhrbergbaukrise*, 271; Goch, "Ruhrkohle."
67. Schmidt's report in Farrenkopf and Przigoda, *Glück auf!*, 266; Parnell, *Coal Industry*, 111; BAK, B136, 7666, Entwurf eines Gesetzes zur Sicherung des Steinkohlenabsatzes in der Elektrizitaetswirtschaft, April 6, 1966; Unternehmensverband Ruhrburgbau, *Jahresbericht*, 1964–1966.
68. BAK, NL 1229, Schiller, 176, DEA "Erklärung des Vorstandes," July 1, 1965.
69. Winkler, *Long Road*, 217–220.
70. BAK, NL 1229, Schiller, 176, DEA "Erklärung des Vorstandes," July 1, 1965.

71. Quotation from BAK, B136, 7704, Goerg von Opel to Erhard, August 26, 1966; Ernst Rueckwarth to Erhard, August 25, 1966, in ibid.; BAK, NL 1229, Schiller, 176, DEA "Erklärung des Vorstandes," July 1, 1965; BAK, B136, 7704, Schmücker to Chancellor's Office, September 27, 1966.
72. BAK, NL 1229, Schiller, 176, DEA "Erklärung des Vorstandes," July 1, 1965.
73. Horn, *Energiepolitik*, 54–55; Mulfinger, *Mineralölpolitik*, 83–86; BAK, NL 1229, Schiller, 176, DEA "Erklärung des Vorstandes," July 1, 1965.
74. BAK, B136, 7704, Schmücker to Chancellor's Office, September 27, 1966; Radzio, *Veba*, 273; Karlsch and Stokes, *Faktor Öl*, quotation from 355.
75. KpB, 26th Session, May 11, 1966, 213.
76. Quotation in Bamberg, *British Petroleum*, 245; Karlsch and Stokes, *Faktor Öl*, 353–358; BAK, B136, 7704, Schmücker to Chancellor's Office, September 27, 1966; KpB, 26th Session, May 11, 1966, 213.
77. M. S. Daoudi and M. S. Dajani, "The 1967 Oil Embargo Revisited," *Journal of Palestine Studies* 13/2 (1984), 65–90, here 69.
78. Kapstein, *Insecure Alliance*, 142–144.
79. Italics added, quotation from BAK, B102, 313603, Woratz to Neef, June 27, 1967; Kling to Neef, July 5, 1967, and Neef to Economics Minister, in ibid.; Kapstein, *Insecure Alliance*; 147–148; BAK, B102, 313603, Woratz to Economics Ministry, July 31, 1967.
80. BAK, B102, 313603, BWM and Foreign Office to Vogel, June 26, 1967, Neef to Economics Minister, June 13, 1967, Woratz to Neef, July 13; Woratz to Neef, July 11, 1967, and Woratz to Neef, July 31, 1967; BBA, 26, 79, Erich Schieweck, "Die energiepolitischen Folgen der Nahost-Krise"; Kapstein, *Insecure Alliance*, 148–149; Daoudi and Dajani, "1967."
81. BAK, B136, 7703, Economics Minister to Chancellor's Office, March 8, 1965; OEEC, *Suez Crisis*; Graf, *Petroknowledge*, 74–76; Cantoni, "NATO and the EEC"; Beltran, "France," 43–52.
82. Quotation from Commission, "First Guidelines"; Hassan and Duncan, "Integrating Energy"; Lucas, *European Communities*, 46–51.
83. BAK, B136, 2506, Report on German-French Oil Discussions, November 17, 1959, Burckhardt to Kattenstroth, November 26, 1959; Vialon to Chancellor, November 30, 1959; Cantoni, "NATO and the EEC."
84. Chick, *Energy Policy*, 18–21; Lucas, *France*, 8–20, 41–48; Lubell, "Security of Supply"; Wessels, "Security."
85. BAK, B136, 7704, Knauss, "Neuordnung der deutschen Mineraloelindustrie," November 28, 1968.
86. First quote from Heinz Kegel, "Forderung nach einer Neuordnung der Bergbau- und Energiewirtschaft . . . ," September 23–24, 1965, doc. 50, in Martiny and Schneider, *Energiepolitik*; second quote from "Überlegungen der IGBE . . ." October 31, 1966, in ibid.; BAK, NL 1229, Schiller, 176, Arendt to Schiller, June 30, 1965.
87. Quotation from "Stellungnahme der IGBE . . . ," Fall 1967, doc. 58 in Martiny and Schneider, *Energiepolitik*; BAK, NL 1229, Schiller, 176, Arendt to Schiller, November 16, 1964; "Der Erdölzufuhr muss gesichert werden . . . ," May 1965, doc. 49 in Martiny and Schneider, *Energiepolitik*; Abelshauser, *Ruhrkohlenbergbau*, 124–125.
88. Quotations from BAK, NL 1229, Schiller, 176, DEA "Erklärung des Vorstandes," July 1, 1965; Horn, *Energiepolitik*, 226–228.
89. BAK, NL 1229, Schiller, 176, Report from Energy Policy Unit, November 12, 1964.
90. BAK, 102, 126250. Schmidt to Woratz, May 18, 1966.
91. Quotation from BAK, NL 1229, Schiller, 176, excerpt from Schiller's speech in Munich, July 2, 1965; Nützenadel, *Stunde der Ökonomen*, 131–132.
92. Quotation from BAK, NL 1229, Schiller, 176, "Brief an die Kumpels: Vorstand der SPD nimmt Stellung," November 4, 1964; Abelshauser, *Ruhrkohlenbergbau*, 128–132; ACDP, 07-001, 9037, Russe to Bilke, Vorbereitung Aktionsprogramm—Kommission 2 Kohle/Stahl, October 19, 1967; also in ibid.: report to Dr. Heck from May 29, 1967, Aktionsprogramm Kohle 2. Entwurf, April 7, 1967, and Alfred Müller-Armack, "Ein neuer Vorschlag zur Überwindung der Kohlekrise," May 5, 1967.
93. BSB, 5. Wahlperiode, 30th Session, March 16, 1966, Coal Debates.

94. Christophe Nonn, "Das Godesberger Programm und die Krise des Ruhrbergbaus. Zum Wandel der deutschen Sozialdemokratie von Ollenhauer zu Brandt," *Vierteljahrshefte für Zeitgeschichte* 50/1 (2002), 71–97; Abelshauser, *Ruhrkohlenbergbau*, 124–130.
95. BBA, 55, 983, Hermann J. Abs and others to Schiller, May 11, 1968; Abelshauser, *Ruhrkohlenbergbau*, 124–129; Goch, "Ruhrkohle"; Horn, *Energiepolitik*, 267.
96. BAK, NL 1229, Schiller, 176, excerpt from Schiller speech, July 2, 1965; *KpB*, Special Session on November 7, 1967, 521; BBA, 55, 983, Ruhrkohle AG Vorstand to Kommission EG, August 22, 1969, 3rd Attachment; *KpB*, Special Session on November 7, 1967, 521.
97. Quotations from BSB, 5. Wahlperiode, 131st Session, November 8, 1967; BBA, 26, 79, Report on the Steinkohlentag of 1967, November 6, 1967, in *Zechen-Kurier*; Abelshauser, *Ruhrkohlenbergbau*, 133; Nonn, "Kohlenkrise."
98. Quotation from BBA, 26, 80, Kurt H. Biedenkopf, "Energiepolitik nach dem Kohleanpassungsgesetzt"; Goch, "Ruhrkohle," 298–299.
99. Abelshauser, *Ruhrkohlenbergbau*, 145–161; Parnell, *Coal Industry*, 180; Illing, *Energiepolitik*, 114–117; BBA, 55, 983, Ruhrkohle AG Vorstand to Kommission, August 22, 1969, 3rd Attachment.
100. Quotations from BBA, 32, 332, VEBA Geschäftsbericht über das Geschäftsjahr vom 1. Oktober 1967 bis 30. Sept. 1968; BBA, 26, 80, "Ausführungen von Herrn Oberbergrat Keyser zur Hauptversammlung der Gelsenkirchener Bergwerks-Aktiengesellchaft"; BBA, 11, DEA, "Geschäftsbericht 1968," DEA, "Geschäftsbericht 1969"; Horn, *Energiepolitik*, 102.
101. Abelshauser, *Ruhrkohlenbergbau*, 145–149; Illing, *Energiepolitik*, 116; Horn, *Energiepolitik*, 298–299.
102. BSB, 5. Wahlperiode, 131st Session, Coal Debate, November 8, 1967; Karlsch and Stokes, *Faktor Öl*, 372;
103. BAK, B136, 7666, Neemann to Kiesinger, "Energiepolitischen Leitsätze des Deutscher Gewerkschaftsbund," February 15, 1968; Winkler, *Long Road*, 249–251.
104. Radzio, VEBA, 223–243; Ingrid Neumann, "Deutsches Zebra in der Wüste," *Die Zeit*, April 28, 1967.
105. Horn, *Energiepolitik*, 44; Radzio, *VEBA*, 243–244, 277–281; *KpB*, 82nd Session, June 6, 1967, 311; BAK, B136, 7704, Knauss, "Neuordnung der deutschen Mineralölindustrie," November 28, 1968, and Report to Bundeskanzler from Prass, September 27, 1968, Report on GBAG, October 2, 1968.
106. Karlsch and Stokes, *Faktor Öl*, 359–364.
107. Quotation from Walter Cipa, director of GBAG, cited in "Viel Geld fuer Wenig Oel," *Die Zeit*, February 25, 1972, https://www.zeit.de/1972/08/viel-geld-fuer-wenig-oel; Karlsch and Stokes, *Faktor Öl*, 374; Heinz Jürgen Schürmann, "Veba —Gelsenberg — Deminex: Unternehmerische Konsequenzen aus der Energiekrise," *Zeitschrift für öffentliche und gemeinwirtschaftliche Unternehmen* 4/1 (1981), 19–50; BAK, B102, 281076, "Richtlinien über die Gewährung von Darlehen...," January 12, 1970.
108. Quotation from "Viel Geld fuer Wenig Öl"; Graf, *Petroknowledge*, 60–61, 76–77; Raymond Vernon, "An Interpretation," Joel Darmstadter and Hans H. Landsberg, "The Economic Background," and Edith Penrose, "The Development of the Crisis," in Raymond Vernon (ed.), *The Oil Crisis* (New York: W. W. Norton, 1976), 1–15, 15–38, and 39–58; Frank Bösch and Rüdiger Graf, "Reacting to Anticipations: Energy Crises and Energy Policy in the 1970s: An Introduction," and Nuno Luis Madureira, "Waiting for the Energy Crisis: Europe and the United States on the Eve of the First Oil Shock," in *Historical Social Research* 39/4 (2014), 7–21, 70–93.
109. BAK, B102, 206541, Schmidt on the "Bundesrohölreserve...," September 15, 1972.
110. BAK, B102, 281077, "Ziele der Mineralölpolitik und ihre Verwirklichung," January 20, 1972; Graf, *Petroknowledge*, 74.
111. Quotation from BAK, B102, 281077, Lantzke to Rohwedder, February 16, 1972; also ibid., Lantzke to Rohwedder, "Abu Dhabi-Projekt," February 10, 1972.
112. BAK, B102, 206541, Schmidt on the "Bundesrohölreserve...," September 15, 1972; BAK, B102, 206540, Schmidt to Lantzke and Kling, January 13, 1971, Schmidt to Lantzke and Kling, January 18, 1971, Deminex meeting in Düsseldorf, January 18, 1971, Deminex,

"Verhandlungskonzept für den Einfkauf von Bevorratungsrohoel im Iran, March 24, 1971; BAK, B102, 281077, Lantzke to Rowedder, February 16, 1972; Bamberg, *British Petroleum*, 277–279; Karlsch and Stokes, *Faktor Öl*, 369–370.
113. Penrose, "Development of Crisis"; Petrini, "Squeezed Sisters."
114. Quotations from BSB "Energieprogramm 1973," DS 7/1057; and BAK, B136, 7668, Draft Energy Program, August 22, 1973.
115. BAK, B102, 313597, Untitled memo on Supply Crisis from August 20, 1971, Petersen of the Mineralölwirtschaftsverband to Kling, November 2, 1971, and report from Plessner, December 7, 1971.
116. Quotation from Theodor Wessels, "The Reorganization of the Hard-Coal Mining Industry in the Federal Republic of Germany," *The German Economic Review* 8 (1970), 81–88, here 82.
117. Martin Meyer-Renschhausen, *Das Energieprogramm der Bundesregierung: Ursachen und Probleme staatlicher Planung im Energiesektor der BRD* (Frankfurt/Main: Campus Verlag, 1981), 4–6.

Chapter 4

1. Heinz Krekeler, West Germany's delegate to Euratom, speech at John Hopkins University, March 16, 1959. AEI.
2. Leo Brandt (SPD), "Atomenergie als wirtscahftliche Kraftquelle auch fuer Deutschland?" September 17, 1955, BAK, B136, 6099.
3. DTM, FA AEG-Telefunken I, JB 0859, quote from Protokoll über die Besprechung des zukünftigen Aufsichtsrates der Kraftwerk Union...," February 21, 1969; ibid., "Besprechung der zukünftigen Aufsichtsräte und Vorstände der Kraftwerk Union...," December 23, 1968; DTM, FA AEG-Telefunken, GS 5145, "Ein Lehrstück industrieller Konzentration," June 11, 1979; "Konzentration in der Atomindustrie," *Atomwirtschaft*, June 1969, 285.
4. DTM, FA AEG-Telefunken, GS 5415, KWU Press Conference, October 15, 1969.
5. Quotation from DTM, FA AEG-Telefunken, GS 5415, "An vierter Stelle der Weltrangliste," *Wese Kurier*, October 17, 1969; ibid., Gemeinsame Presseerklärung von AEG-Telefunken & Siemens, March 27, 1969, KWU Report: "Die Kraftwerk Union an der Schwelle eines Neuen Jahrzehnts," March 15, 1970, and "Kraftwerk Union fast ein Wunderkind," *Mannheimer Morgen*, October 17, 1969.
6. Vernon Ruttan, *Is War Necessary for Economic Growth? Military Procurement and Technological Development* (New York: Oxford University Press, 2006); Fred Block and Matthew R. Keller, *State of Innovation: The US Government's Role in Technological Development* (London: Paradigm, 2011).
7. David Blackbourn, *The Long Nineteenth Century: A History of Germany, 1780–1918* (New York: Oxford University Press, 1998); Jeffrey Fear, "German Capitalism," in Thomas K. McCraw (ed.), *Creating Modern Capitalism: How Entrepreneurs, Companies and Countries Triumphed in Three Industrial Revolutions* (Cambridge, MA: Harvard University Press, 1997), 133–182.
8. Mazzucato, *Entrepreneurial State*.
9. Quotation from BAK, B102, 5429, 1/2 Helle to BWM, "Tätigkeitsbericht der Bundesregierung 1957," November 21, 1957; Joachim Radkau, *Aufstieg und Krise der deutschen Atomwirtschaft, 1945–1975* (Hamburg: Rohwohlt, 1983); Hecht, *Radiance of France*; R. G. Hewlett and J. M. Holl, *Atoms for Peace and War: 1953–1961* (Berkeley: University of California Press, 1989).
10. Ernst Bloch, *The Principle of Hope*, translated by Neville Plaice, Stephen Plaice, and Paul Knight (Cambridge, MA: MIT Press, 1953/1986). Quotations from the introduction, 660, and 664; Leszek Kolakowski, *Main Currents of Marxism*, translated by P. S. Falla (New York: W. W. Norton, 2005), 1124–1147.
11. First quotation from Dwight D. Eisenhower to UN General Assembly: https://www.iaea.org/about/history/atoms-for-peace-speech; second from Louis Armand, "Some Aspects of the European Energy Problem," https://www.cvce.eu/content/publication/1999/1/1/6761172f-1f18-45b0-a247-e50faedb0e5d/publishable_en.pdf; Morris and Jungjohahn, *Energy Democracy*, 304; Ulrich Herbert, *Geschichte Deutschlands im 20. Jahrhundert* (Munich: C. H. Beck, 2014), 801.

12. Dolores Augustine, *Taking on Technocracy: Nuclear Power in Germany, 1945 to the Present* (New York: Berghahn Press, 2018), 21; Wolfgang Müller, *Geschichte der Kernenergie in der Bundesrepublik Deutschland* (Stuttgart: Schäffer Verlag, 2001).
13. BAK B136, 6099, Heisenberg to Adenauer, April 29, 1953, and Schmid to Globke, November 4, 1952; Peter Fischer, *Atomenergie und staatliches Interesse: Die Anfänge der Atompolitik in der BRD, 1949–1955* (Baden-Baden: Nomos, 1994), 31–35, 55–58; Cathryn Carson, *Heisenberg in the Atomic Age: Science and the Public Sphere* (New York: Cambridge University Press, 2010), 220–228; Cathryn Carson, "Nuclear Energy Developments in Postwar West Germany: Struggles over Cooperation in the Federal Republic's First Reactor Station," *History and Technology* 18/3 (2002), 233–270.
14. BAK, B136, 6099, Report to Chancellor, January 6, 1955; Carson, *Heisenberg*, 232–233.
15. Quotation from Ernst Telschow, BAK, B136, 6099, "Was tut uns Not?" November 15, 1955; ibid., Report from Grau to Chancellor, March 4, 1955; ibid., Physikalisches Studiengesellschaft to Adenauer, September 16, 1955; ibid., "Gegenstand der Besprechung und kurze Stellungnahme zu den Besprechungsthemen," March 23, 1955; BAK, B135, 6105, Hess to Adenauer, October 3, 1955.
16. On Strauss, Harry Trimborn, "Volatile Bavarian Premier," *Los Angeles Times*, October 4, 1988, https://www.latimes.com/archives/la-xpm-1988-10-04-mn-3518-story.html; last quotation from BAK, B136, 6105, Report to press from Minister for Atomic Power, January 26, 1956; ibid., Strauss to Chancellor's Office, December 16, 1955; ibid., Strauss to Globke, January 18, 1956; BAK, B136, 6099, excerpt from minutes to 99th Cabinet Session, October 6, 1955; Illing, *Energiepolitik*, 78–80.
17. Quotation from Konrad Adenauer, *Erinnerungen, 1955–1959* (Stuttgart: Deutsche Verlags-Anstalt, 1967), 296.
18. Thomas-Durell Young, "Force, Statecraft, and German Unity: The Struggle to Adapt Institutions and Practices" (Working Paper, US Army War College, 1996); Hans-Peter Schwarz, *Konrad Adenauer: A German Politician and Statesman in a Time of War, Revolution and Reconstruction*, translated by Louise Willmot (Providence, RI: Berghahn Books, 1997), 263–277.
19. Schwarz, *Adenauer*, 263–264.
20. Quotations from Adenauer, *Erinnerungen*, 296 and 299; Radkau, *Krise*, 94–95; Schwarz, *Adenauer*, 219; Hermann Laupsien, "Public Relations in der Atomwirtschaft," *Atomwirtschaft* 1/12 (December 1956), 404; Augustine, *Technocracy*, 20–25.
21. First quotations from ACDP, Balke, 01–175, 006/2 Letter to Balke from Hahn, Weizsäcker, Heisenberg, and others, November 19, 1956; Georg-August-Universität Göttingen, "Text des Göttinger Manifests": http://www.uni-goettingen.de/de/54320.html; last quotation from the communiqué, cited in Schwartz, *Adenauer*, 269–270; Carson, *Heisenberg*, 329; Radkau, *Krise*, 98–99.
22. Lucas, *European Communities*, 13–14; Francois Duchene, *Jean Monnet: The First Statesman of Interdependence* (New York: W. W. Norton, 1994), 264–268; AEI, Louis Armand, Franz Etzel, Francesco Giordani, "A Target for Euratom" (1957); "Möglichkeiten Europäischen Zusammenarbeit," in *Atomwirtschaft*, 1956).
23. Quotations from ACDP, Balke, 01–175, 012/2, "Aide-Memoire from Armand, Etzel, and Giordani . . . ," February 4, 1957; Louis Armand, Franz Etzel, Francesco Giordani, "Target for Euratom," (1957); AEI, Euratom Commission, "Agreement for Cooperation between the European Atomic Energy Community and the United States of America," November 8, 1958.
24. John Krige, "The Peaceful Atom as Political Weapon: Euratom and American Foreign Policy in the Late 1950s," *Historical Studies in the Natural Sciences* 38/1 (2008), 5–44; Berend, *European Integration*, 43–48, 71–82.
25. H. Fischerhof, "Euratom: Der Vertrag—und Fragen seiner Verwirklichung," *Atomwirtschaft* 5 (1957), 147–148; "Der Vertrag zur Gründung der europäischen Atomgemeinschaft," *Atomwirtschaft* 5 (1957), 152–155; Andrew Moravscik, *The Choice for Europe: Social Purpose and State Power from Messina to Maastricht* (Ithaca, NY: Cornell University Press, 1998), 95–99; Duchene, *Monnet*, 290–299; Radkau, *Krise*, 170–178.
26. Christian Deubner, "The Expansion of West German Capital and the Founding of Euratom," *International Organization* 33/2 (1979), 203–228; Christian Deubner, "The Final Crisis of

Euratom?" *Current Research on Peace and Violence* 2/2 (1979), 53–65; Christian Deubner, *Die Atompolitik der westdeutschen Industrie und die Gründung von Euratom* (Frankfurt a.M.: Campus-Verlag, 1977), 11–12, 44–61; Hecht, *Radiance*, 60–78; Radkau, *Krise*, 49, 100–105; Duchene, *Monnet*, 298–299; BBA, 32, 3460, Menne cited in Tagungsbericht über 3. Internationales Atom-Seminar: "Atomkernenergie als Wirtschaftsfaktor," May 19–23, 1958.

27. Quotation from AEI, Speech by Euratom Commissioner Sassen, March 16, 1961; Deubner, "Crisis of Euratom"; Duchene, *Monnet*, 298–300.
28. Siegried Balke, "Atomtechnik und Atompolitik: Zum Regierungsentwurf des Atomgesetztes," *Atomwirtschaft* 2 (1957), 35–38.
29. Cited in Krige, "Euratom"; ACDP, Balke, 01–175, 012/2, "Aide-Memoire from Armand, Etzel, and Giordani ...," February 4, 1957.
30. "Der Vertrag zur Gründung der europäischen Atomgemeinschaft," *Atomwirtschaft* 5 (1957), 152–155; ACDP, Balke, 01–175, 008/1, Vertrages zwischen Euratom und den Vereinigten Staaten von Amerika, May 10, 1958.
31. BSB, "Energiebilanz des Bundesgebiet."
32. First quotation from Hans-Josef Strauss, "Aufbau und Aufgaben des Bundesministeriums fuer Atomfragen," *Atomwirtschaft* 1 (1956), 2–5; second from ACDP, Balke, 01–175, 007/1, Report on Formative Meeting of the Atomic Commission, January 26, 1956; Horn, *Energiepolitik*, 184–187.
33. BAK, B102, 5429, 1/2 Helle to BWM, "Tätigkeitsbericht der Bundesregierung 1957," November 21, 1957; BAK, B136, 6105, Resolution of the Deutschen Atomforums e.V.zur Förderung des Reaktorbaus," April 1961; Radkau, *Krise*, 165–167.
34. Quotations from BAK, B102, 41043, Günther Grüneberg, "Atomfinanzierung auf der langen Bank," in *Der Volkswirt*, February 7, 1059; ibid., "Ziel: Export von Atomkraftwerken," *Frankfurter Rundschau*; BBA, 75, 47, "Zwei Jahre Reaktorbau in Karlsruhe," in *FAZ*, April 11, 1959.
35. Quotation from BAK, B136, 6099, Kriele to Vialon, September 8, 1961; BAK, B136, 6099, Balke to Etzel, November 21, 1958.
36. Hans Fischerhof, "Atomwirtschaft und Gesetzgeber: ...," *Atomwirtschaft* 3 (1956), 93–98; Hans Fischerhof, *Rechstfragen der Energiewirtschaft* (Frankfurt a.M.: Verlag für Sozialwissenschaft, 1956); Edgar Salin, *Ökonomik der Atomkraft: Vor einer neuen Etappe der industriellen Revolution* (Köln: Sigillum, 1955); Rudolf Greifeld, "Kernreaktor Bau- und Betriebs-Gesellschaft Karlsruhe," *Atomwirtschaft* 9 (1956), 293–294.
37. Rudolf Hahn, "Atomenergie in Grossbritannien" *Atomwirtschaft* 2 (1956), 54–55; Michael T. Hatch, *Politics and Nuclear Power: Energy Policy in Western Europe* (Lexington: University of Kentucky Press, 1986); Chick, *Energy Policy in Britain*.
38. First quotation from "Godesberger Programm: Grundsatzprogramm der Sozialdemokratischen Partei Deutschlands," November 15, 1959; second from Ludwig Ratzel, "Atomenergie—kein Tummelplatz fuer Spekulationen ...," *SPD Pressedienst*, January 22, 1960; "Atomenergie—Privat oder Gemeinwirtschaftlich?," *SPD Pressedienst*, January 20, 1956.
39. Quotation from Franz Josef Strauss, "Aufbau und Aufgaben des Bundesministerium fuer Atomfragen," *Atomwirtschaft* 1 (1956), 2–5; Rudolf Greifeld, "Kernreaktor Bau- und Betriebs-Gesellschaft Karlsruhe," *Atomwirtschaft* 9 (1956), 293–294; BBA, 75, 47, "Zwei Jahre Reaktorbau in Karlsruhe," in *FAZ*, April 11, 1959; Radkau, *Krise*, 132–138.
40. Quotation from Balke, "Atomtechnik und Atompolitik: Zum Regierungsentwurf des Atomgesetztes" *Atomwirtschaft* 2 (1957), 2; "Siegfried Balke," *Atomwirtschaft* 12 (1956), 411; BAK, B102, 41043, Memorandum: Der Fachkommission V "Wirtschaftliche, Finanziele und Soziale Probleme" der Deutschen Atomkommission, December 1957.
41. Franz Josef Strauss, "Aufbau und Aufgaben des BM für Atomfragen," *Atomwirtschaft* 1 (1956), 2–5.
42. BAK, B136, 6099, Brandt, "Atomenergie als wirtschaftliche Kraftquelle auch für Deutschland?" September 17, 1955.
43. Quote from Rolf Raiser, "Die Versicherung der Atomrisiken: der heutige Stand der Probleme in Deutschland," *Atomwirtschaft* 6 (1956), 213–216; Illing, *Energiepolitik*, 79–83; Fischer,

Atompolitik, 179-182; BAK, B136, 2508, "Hintergründe des Scheiterns des Atomgesetzes," *Stuttgarter Ztg*, August 1, 1958.
44. ACDP, Balke, 175, 007/1, Cämmerer, "Bericht über die Tätigkeit der Fachkommission I," April 21, 1961; "Atomgesetz" *Bundesgesetzblatt* Nr. 56, December 31, 1959.
45. Morris and Jungjohann, *Energiewende*, 305; Horn, *Energiepolitik*, 184-187; Christoph Wehner, *Die Versicherung der Atomgefahr. Risikopolitik, Sicherheitsproduktion und Expertise in der Bundesrepublik Deutschland und den USA 1945-1986* (Göttingen: Wallstein, 2017).
46. Quotation from BAK, B102, 41043, Position Paper of Economics Ministry, April 1, 1958; quotation from BAK, B102, 5429 1/2, Helle to BWM, "Tätigkeitsbericht der Bundesregierung 1957," November 21, 1957; BAK, B102, 41043, Memorandum: Der Fachkommission V "Wirtschaftliche, finanziele und soziale Probleme" der Deutschen Atomkommission, December 1957; ACDP, Balke 01-175, 007/1, Deutschland im Wiederaufbau 1957 . . . ," November 1957.
47. Siegfried Balke, "Atomtechnik und Atompolitik: Zum Regierungsentwurf des Atomgesetzes" *Atomwirtschaft* 2 (1957), 35-38; BAK, B102, 41043, Memorandum: Der Fachkommission V "Wirtschaftliche, finanziele und soziale Probleme," December 1957, Vermerk Betr.: Atomprogramm der Deutschen Atomkommission., February 11, 1958, and Brandl to Schmidt-Amelung, December 10, 1959.
48. BAK, B136, 6099, Kattenstroth to Atomic Minister, April 28, 1958; BAK, B102, 41043, Economics Minister to Ministry for Atomic and Water Power, Stellungnahme zur Investitionspolitik auf dem Gebiet der Kernenergie, April 21, 1958, also in BAK, B136, 6099; BAK, B102, 41043, Westrick to Balke, October 7, 1960.
49. ACDP, 01-175, 007/1, "Deutschland im Wiederaufbau 1957," November 1957; BAK, B102, 41043, Vermerk Betr.: Atomprogramm der Deutschen Atomkommission, February 11, 1958; Radkau, *Krise*, 45-50.
50. Figures from ACDP, Balke 01-175, 007/1, Bericht über die Taetigkeit der Fachkommission V, by Dr. Menne, April 21, 1961; BAK, B102, 41043, Kattenstroth to Westrick, September 26, 1960.
51. BAK, B102, 41043, Referat from Kühne, March 8, 1960, and Vermerk: Siemens-Mehrzweckreaktor, October 1, 1960; BAK, B102, 41043, Kattenstroth to Minister for Atomic and Water Power, March 4, 1958, Kattenstroth to Westrick, September 26, 1960, Brandle to Schmidt-Amelung, December 10, 1959, and Westrick to Balke, October 7, 1960; Radkau, *Krise*, 149-153; Illing, *Energiepolitik*, 79-83.
52. BAK, B136, 6099, Meeting about Supplier Industries, Finke, August 5, 1958; figures from Wessels, "Investitionen"; Peter Becker, *Aufstieg und Krise der deutschen Stromkonzerne: Zugleich ein Beitrag zur Entwicklung des Energierechts* (Bochum: Ponte Press, 2011), 48-56; Lutz Metz, *RWE: Ein Riese mit Ausstrahlung* (Cologne: Kiepenheuer & Witsch, 1996).
53. Quotation from ACDP, Balke, 01-175, 06/2 Schöller to Riegel, February 17, 1958; ACDP, Balke, 012/2, RWE Directorate to Balke and to Winnacker, including essay "Aussichten für Wirtschaftliche Atomenergie," July 23, 1958; BAK, B136, 6099, Minutes on Meeting about Supplier Industries by Dr. Finke, August 5, 1958; figures from Baade, *Weltenergiewirtschaft*, 120-122; "Stand und Aussichten der weiteren Entwicklung auf dem Gebiet der Kernenergie, *DIW Wochenbericht* August 14, 1959; Morris and Jungjohann, *Energiewende*, 315-318.
54. ACDP, Balke, 01-175, 012/2 "Gemeinsames Kommunique des State Dept., des Präsidenten der AEC und des Euratom Ausschusses," February 8, 1957; Krige, "Euratom"; Deubner, "Founding of Euratom"; Radkau, *Krise*, 172-176.
55. ACDP, Balke, 01-175, 012/2, Balke to Grau, February 28, 1957; Krige, "Euratom"; Radkau, *Krise*, 172-176; BBA, 32, 3460, Menne's statement in "Atomkernenergie als Wirtschaftsfaktor," Wiesbaden, May 19-23, 1958; ACDP, Balke, 01-175, 012/2, Report from Schnurer, December 12, 1959; Deubner, "Founding of Euratom"; Radkau, *Krise*, 172-176.
56. ACDP, Balke, 01-175, 008/2, Balke: Aktenvermerk für Abteilungsleiter II, April 28, 1959.
57. Quotation from BBA, 75, 47, "Zwei Jahre Reaktorbau in Karlsruhe," in FAZ, April 11, 1959; BAK, B102, 41043, Kattenstroth to Westrick, September 26, 1960, Vermerk Betr.: Siemens-Mehrzweckreaktktor, October 1, 1960, and Westrick to Balke, October 7, 1960; Radkau, *Krise*, 47-50, and 191-194.

58. Heinrich Mandel, "Die Planung des RWE auf dem Atomsektor," *Atomwirtschaft* 10 (1956), 332–334; ACDP, Balke, 01–175, 007/1, Mandel, "Das RWE-Projekt zum Bau eines Versuchs-Atomkraftwerkes," December 6, 1957; Becker, *Stromkonzerne*, 204–207; Stephan Deutinger, "Eine 'Lebensfrage für die bayerische Industrie': Energiepolitik und regionale Energieversorgung 1945 bis 1980," in Thomas Schlemmer and Hans Woller (eds.), *Bayern im Bund Band 1: Die Erschliessung des Landes 1949 bis 1973* (Munich: Oldenbourg, 2001), 74–77.
59. ACDP, Balke, 007/1, Menne, Bericht über die Tätigkeit der Fachkommission V, April 21, 1961.
60. BAK, B136, 6105, *Atom-Informationen*, July 12, 1962; BAK, B102, 113812, article clipping from *Handelsblatt* January 30, 1962; ibid., "Soll der Bund Atomkraftwerke bauen?" *Die Welt*, January 30, 1962; BBA, 75, 47, "Die deutsche Atomenergietechnik holt auf" *FAZ*, August 25,
61. Quotation from ACDP, Balke 01–175, 007/1, Resolution des Deutsche Atomforums e.V. zur Förderung des Reaktorbaus, April 1961; ACDP, Balke, 01–175, 007/1, Kurzprotokoll 14. Sitzung deutsche Atomkommission, July 11, 1962; ACDP, Balke 01–175, 007/1, Resolution des Deutsche Atomforums e.V. zur Förderung des Reaktorbaus, April 1961.
62. BAK, B136, 6099, Balke to Etzel, November 21, 1958; ACDP, Balke, 01–175, 007/1, Menne, Bericht über die Tätigkeit der Fachkommission V, April 21, 1961; BAK, B136, 6105, "Atomforum fordert 500 Millionen" *Deutsche Zeitung—Wirtschaftszeitung*, April 22, 1961, and Bechtolsheim to Bundeskanzler, Deutsche Atomkommission, May 12, 1961.
63. ACDP, Balke 01–175, 007/1, Winnacker, Bericht über den Bau eines grossen Kernkraftwerk, July 11, 1962.
64. ACDP, Balke, 01–175, 007/1, Karl Kaissling, Bericht über den Bau eines grossen Kernkraftwerk, July 11, 1962; BAK, B136, 6105, Atom-Informationen, July 12, 1962.
65. BAK, B102, 113812, Mandel to Finke, January 3, 1963.
66. BAK, B102, 113812, Schmidt-Amelung, Vermerk, Betr.: Förderung des Baus von Atom-Grossrkaftwerken, August 20, 1962; ACDP, Balke, 01–175, 007/1, Menne, Bericht über den Bau eines grossen Kernkraftwerk, July 11, 1962; BAK, B136, 6105, Atom-Informationen, July 12, 1962; BBA, 75, 47, "Atomstrom jetzt konkurrenzfähig? ...," *FAZ*, October 15, 1961Heinrich Mandel, "Kernkraftwerk Gundremmingen—seine Stellung in der deutschen Atomwirtschaft" *Atomwirtschaft* 11 (1965), 564–565.
67. Peter Strunk, *Die AEG: Aufstieg und Niedergang einer Industrielegende* (Berlin: Nicolaische Verlagsbuchhandlung, 1999), 93–95; Morris and Jungjohann, *Energiewende*, 318–320.
68. BSB, 4. Wahlperiode, 118. Sitzung, March 4, 1964; BSB, 4. Wahlperiode, DS IV/1211, April 24, 1963; Christoph Vogel, *Deutschland im internationalen Technologiewettlauf: Bedeutung der Forschungs- und Technologiepolitik fuer die technologische Wettbewerbsfaehigkeit* (Berlin: Duncker & Humblot, 1999), 158–162; Carson, *Heisenberg*, 253–254; Illing, *Energiepolitik*, 86–88; Jürgen-Friedrich et al., "The German Energiewende—History and Status Quo," *Energy Policy* 92/3 (2015), 532–546.
69. Kallenbach, "Kernkraftwerk Obrigheim—die Konzeption des Bauherrn," *Atomwirtschaft* 6 (1965), 266–271; Radkau, *Krise*, 178–179.
70. Quotation from Mazzucato, *Entrepreneurial State*, 81, more generally see ch. 4; Ruttan, *War*; Fred Block, "Innovation and the Invisible Hand of Government," and Andrew Schrank, "Green Capitalists in a Purple State," in Block and Keller (eds.), *State of Innovation*, 1–27 and 96–108; John Perlin, *From Space to Earth: The Story of Solar Electricity* (Ann Arbor, MI: Aalec, 1999), 49–55.
71. BSB, 5. Wahlperiod, 64th Session, October 12, 1966, first quotation from Gerhard Stoltenberg, 3087; ibid., Dr. Schober of CDU, 3082; AEI, "The European Policy for Research and Technology," by Bersani, September 12, 1968; ACDP, Balke, 01–175, 012/1, OECD, "Science and Policy: the Implications of Science and Technology ...," March 4, 1963.
72. ACDP, Balke, 01–175, 006/2 Beantwortung der Grossen Anfrage der CDU/CSU und FDP betreffend Förderung der Forschung zur wirtschaftlichen Nutzung von Kernenergie ..., October 12, 1966.
73. ACDP, Balke, 07–001, 9037, Hermann Josef Russe to Karl-Heinz Bilke, October 19, 1967; BSB, 5. Wahlperiod, 64th Session, October 12, 1966, quotation from Gerhard Stoltenberg, 3095.

74. Quotations from ACDP, Balke, 01–175, 010/4, Deutscher Forschungsdienst: Sonderbericht Kernenergie, 10. Jahrgang. 24/1965.
75. Quotation from ACDP, 07–001, 9037, Hermann Josef Russe to Bilke, October 19, 1967; W. Koeck, "Die Technik von morgen muss heute entwickelt werden: ...," *Atomwirtschaft* 1 (1965), 21–23.
76. Joachim Raffert, "Rückenwind fuer Stoltenberg? ...," *SPD Pressedienst*, October 14, 1966.
77. Klaus-Dieter Fischer, "Struktur und Entwicklungstendenzen der Energiewirtschaft in der Bundesrepublik Deutschland," in Fritzburgbacher and Theodor Wessels (eds.), *Ordnungsprobleme und Entwicklungstendenzen in der deutschen Energiewirtschaft* (Essen: Vulkan-Verlag, 1967), 61–107; Manfred Liebrucks, "Corrigendum zum Energiegutachten von 1961," *Vierteljahreshefte zur Wirtschaftsforschung* 2 (1966), 179–199.
78. Liebrucks, "Corrigendum"; Karl Winnacker, "Der Markt der Atomenergie in Europa. Bundesrepublik Deutschland," *Atomwirtschaft* 9 (1965), 442–447; Krekeler of Euratom cited in BBA, 75, 57, "Durchbruch der Atom-Elektrizitaet," *FAZ* January 29, 1963; "Die voraussichtliche Entwicklung der westdeutschen Elektrizitätswirtschaft ...," *DIW Wochenbericht*, 35. Jahrgang, January 11, 1968; BSB, 5. Wahlperiode, 64th Siztung, October 12, 1966; ACDP, Balke, 01–175, 007/1, Report from Winnacker, April 5, 1967.
79. Quotation from "Ohne billige Energie keine Zukunft: ...," *SPD Pressedienst*, November 21, 1969; BSB, "Zukunftsaspekte der Kernenergie," July 14, 1972, DS IV/3661.
80. Quotation from "Regierungserklärung der Grossen Koalition, December 13, 1966; Brandt, "Regierungserklärung vor dem Deutschen Bundestag," October 28, 1969, Friedrich-Ebert Stiftung online; "Godesberger Programm."
81. BSB, 5. Wahlperiode, 106th Session, April 27, 1967, 4942; "Ohne billige Energie keine Zukunft: ...," *SPD Pressedienst*, November 21, 1969.
82. Becker, *Stromkonzerne*, 40–56, 264; Morris and Jungjohann, *Energiewende*, 59–60, 133–135; Horn, *Energiepolitik*, 82–85, 102; E. Kelstch, "Zukunftsaufgaben der Elektrizitaetswirtschaft in Deutschland," *Atomwirtschaft* 1 (1969), 18–20; Nikolaus Eckardt, Margitta Meinerzhagen, and Urlich Jochemson, *Die Stromdiktatur: von Hitler ermächtigt—bis heute ungebrochen* (Hamburg: Rasch und Röhring, 1985), 103.
83. Keltsch, "Zukunftsfragen"; ACDP, 07–001, 9037, "Starthilfe für deutsche Atomkraftwerke?" *Atomwirtschaft* 1 (1967); BAK, B102, 113812, "Kernenergie wird Bevölkerungsprobleme lösen," May 23, 1967.
84. Joachim Radkau and Lothar Hahn, *Aufstieg und Fall der deutschen Atomwirtschaft* (Munich: Oekom, 2013), 125–134.
85. BAK, B102, 113812, Schmidt-Amelung to Woratz, April 12, 1967; Joachim Raffert, "Rückenwind für Stoltenberg? Bundestagdebatte über Kernenergie und Weltraumforschung," *SPD Pressedienst* October 14, 1966; BSB, 5. Wahlperiode, 64th Session, October 12, 1966; KpB, 106. Sitzung, December 13, 1967; Horn, *Energiepolitik*, 298–299.
86. BAK, B102, 113812, Josef Brandl, "Mitwirkung der öffentlichen Hand bei der Entwicklung der deutschen Kernenergiewirtschaft," and "Kern-Kraftwirtschaft und Subvention," in *Zechenkurier, Unternehmensverband Ruhrbergbau*, December 22, 1965.
87. Radkau and Hahn, *Fall*, 135–137; Radkau, *Krise*, 214–215.
88. Radkau and Hahn, *Fall*, 134–140; Radkau, *Krise*, 207–208; Radzio, *Veba*, 273–286.
89. Hans Matthöfer, *Interviews und Gespräche zur Kernenergie: "Den Unsterblichen Tiger am Schwanz gepackt"* (Karlruhe: C. F. Müller, 1976), 51–59.
90. Radkau, *Krise*, 217–219; Berend, *Integration*, 106–111.
91. Quotation from Brandt, "Regierungserklärung (1969)"; DTM, FA AEG-Telefunken, JB 0859, Protokoll über die Besprechung des zukünftigen Aufsichtsrates, February 24, 1969; DTM, FA AEG-Telefunken, GS 5415, "Kraftwerk Union fast ein Wunderkind," *Mannheimer Morgen*, October 17, 1969.
92. DTM, FA AEG-Telefunken, JB 0859, Protokoll über die Besprechung des zukünftigen Aufsichtsrates, February 24, 1969; DTM, FA AEG-Telefunken, GS 5415, Press Conference in Mülheim on October 15, 1969, and "Kraftwerk Union fast ein Wunderkind," *Mannheimer Morgen*, October 17, 1969; DTM, FA AEG-Telefunken, JB 0859, Protokoll über die Besprechung des zukünftigen Aufsichtsrates, February 24, 1969.

93. Quotation from DTM, FA AEG-Telefunken, GS 5415, "Kraftwerk Union fast ein Wunderkind," *Mannheimer Morgen*, October 17, 1969; ibid., "KWU auf Platz 4 in der Welt" and "Gemeinsame Presseerklaerung von AEG-Telefunken und Siemens," March 27, 1969; DTM, FA AEG-Telefunken, JB 0859, Protokoll über die Besprechung des zukünftigen Aufsichtsrates, February 24, 1969.
94. DTM, FA AEG-Telefunken, GS 5415, "Kraftwerk Union fast ein Wunderkind," *Mannheimer Morgen*, October 17, 1969; BSB, 5. Wahlperiode, 106th Session, April 27, 1967; Radkau and Hahn, *Fall*, 180–184; *KpB*, 111. Sitzung, January 31, 1968.
95. Kruse and Müller, "Rohstoffe als Kostfaktor der Atomenergie," *Atomwirtschaft* 2 (1956), 64–65.
96. Quotation from AEI, Etienne Hirsch, president of Euratom, speech to European Parliament, June 29, 1961; Baade, *Weltenergiewirtschaft*, 120–122; BBA, 26, 67, Herbert Grund, "Vorkommen und Gewinnung von Uran- und Thoriumerzen . . . ," December 22, 1956; see also ACDP, Balke, 01–175, 006/1, "Zwischenbilanz der Uranprospektion in Westdeutschland," *Neue Zürcher Zeitung*, September 4, 1959.
97. Quotation from PAAA, B22, 1, Conversation with Balke, May 24, 1957; Siegfried Balke, "Atomtechnik und Atompolitik: Zum Regierungsentwiruf des Atomgesetztes," *Atomwirtschaft* 2 (1957), 35–38; ACDP, Balke, 01–175, 006/2, "Jahresbericht 1961 des Bundesministerium für Atomkernenergie."
98. BAK, Schiller, NL 1229, 176, FU Bischoff to Brandt, November 5, 1964; BBA, 26, 79, Bischoff, "Weltenergieüberfluss—wie lange?" November 25, 1966; BBA, 75, 47, Robert Gerwin, "In Karlsruhe brüte man schnell," August 30, 1963; Schober, BSB, 5. Wahlperiode, 64th Session, October 12, 1966, 3085; AEI, Eduard R. von Geldern,"A European Assessment of Nuclear and Conventional Fuel Costs," November 15, 1965; H. J. Brüchner, "Wirtschaftliche Aspekte einer europäischen Uran-Anreicherungsanlage," *Atomwirtschaft* 2 (1969), 72–75; Gabrielle Hecht, *Being Nuclear: Africans and the Global Uranium Trade* (Cambridge, MA: MIT Press, 2012), 56–60, 61–65.
99. *KpB*, 147. Sitzung, November 21, 1968; H. J. Brüchner, "Wirtschaftliche Aspekte einer europäischen Uran-Anreicherungsanlage," *Atomwirtschaft* 2 (1969), 72–75; Ole Pedersen, "Developments in the Uranium Enrichment Industries," *IAEA Bulletin* 19/1 (1977), 40–51.
100. Frank von Hippel, "Overview: The Rise and Fall of the Plutonium Breeder Reactors," and Thomas B. Cochran, Harold A. Feiveson, and Frank von Hippel, "Fast Reactor Development in the United States," in Thomas B. Cochran et al. (eds.), *Fast Breeder Reactor Programs: History and Status*, Research Report 8 for the International Panel on Fissile Materials, 1–16 and 73–88.
101. Radkau and Hahn, *Fall*, 45–50.
102. ACDP, Balke, 01–175, 008/2, "Stand der Entwicklung von schnellen und thermischen Brutreaktoren in der Welt," June 8, 1961; Gerwin, "In Karlsruhe brüte man schnell," August 30, 1963.
103. BBA, 75, 47, "Durchbruch der Atom-Elektrizität," in *FAZ*, January 29, 1963.
104. BSB, 5. Wahlperiode, 64th Session, October 12, 1966.
105. BAK, B102, 113812, Josef Brandl, "Mitwirkung der öffentlichen Hand bei der Entwicklung der deutschen Kernenergiewirtschaft"; Morris and Jungjohann, *Energiewende*, 309; W. Marth and M. Koehler, "The German Fast Breeder Program (A Historical Overview)," *Energy* 23/7–8 (1998), 593–608; Radkau, *Krise*, 64–71.
106. ACDP, Balke, 01–175, 008/2, "Stand der Entwicklung von schnellen und thermischen Brutreaktoren in der Welt," June 8; Radkau and Hahn, *Fall*, 148.
107. Fermi was not restarted for another five years because a later sodium explosion caused delays in the repair. Hippel, "Breeder Reactors"; Cochran, Feiveson, and Hippel, "Fast Reactor."
108. See the discussions in the Atomic Ministry on June 8, 1961, in ACDP, Balke, 01–175, 008/2.
109. Quoted in Herbert, *20 Jahrhundert*, 804; Radkau, *Fall*, 145.
110. Radkau, *Fall*, 200–201; Morris and Jungjohann, *Energiewende*, 317; BSB, "Die Energiepolitik der Bundesregierung," October 3, 1973, DS 7/1057.
111. Quotation from BSB, "Bundesbericht Forschung IV," March 13, 1972, DS VI/3251; figures from Horn, *Energiepolitik*, table 4 298–299.

112. ACDP, Balke, 01-175, 007/1, Formative Meeting of the Atomic Commission, January 26, 1956; Radkau, *Krise*, 144-146; Carson, *Heisenberg*, 168-169.
113. W. A. Menne, "Atomnutzung im Blickfeld der Wirtschaftspolitik," *Atomwirtschaft* 1 (1956), 6-8.
114. BAK, B136, 10901, "Genscher über mögliche Standorte von Kernkraftwerken," *Atom-Information* (May 1974).

Chapter 5

1. Heinrich Mandel, "Die Kernenergie im Spannungsfeld der Energiepolitik," *Atomwirtschaft* 1 (1974), 18-22.
2. ASD, WBA, A3 538, speech on November 27, 1973.
3. ASD, WBA, A3, 538, speech on November 23, 1973; Jens Hohensee, *Der erste Oelpreisschock 1973/74* (Stuttgart: Franz Steiner Verlag, 1996), 151-156.
4. Fiona Venn, *The Oil Crisis* (New York: Loman, 2002), 7-15; Garavini, *OPEC*, 216-253.
5. First two quotations, Hohensee, *Oelpreisschock 1973/74*, 111-112; third quotation from "Gehen in Europa die Lichter aus? *Die Zeit*, November 9, 1973; last quotation from Marion Gräfin Dönhoff, "Zuruck zur Bescheidenheit," *Die Zeit*, December 7, 1973.
6. Graf, *Petro-Knowledge*, 204; BSB, 7. Wahlperiode, 73th Session, January 17, 1974.
7. Quotation from Doering-Manteuffel and Raphael, *Nach dem Boom*, 15; Michel Crozier, Samuel P. Huntington, and Joji Watanuki (eds.), *The Crisis of Democracy: Report on the Governability of Democracies to the Trilateral Commission* (New York: New York University Press, 1975); Claus Offe, "Ungovernability: On the Renaissance of Conservative Theories of Crisis," in Jürgen Habermas (ed.), *Observation on "The Spiritual Situation of the Age"* (Cambridge, MA: MIT Press, 1984), 67-88; Hobsbawm, *Extremes*, 408-409; Schmelzer, *Hegemony of Growth*.
8. Quotation from Jacobs, *Panic*, 3; BAK, B102, 206541, Schmidt on Federal Oil Reserve, September 15, 1972; Graf, *Petro-Knowledge*, 57-58, 68-70; Frank Bösch and Rüdiger Graf, "Reacting to Anticipations: Energy Policy in the 1970s," and Nuno Luis Madureira, "Waiting for the Energy Crisis: Europe and the United States on the Eve of the First Oil Shock," *Historical Social Research* 39/4 Special Issue, 7-21 and 70-93.
9. BAK, B102, 313596, confidential memorandum in Economics Ministry, August 10, 1971; BAK, B102, 313597, unlabeled memorandum, August 20, 1971; BAK, B102, files 206540-02; Graf, *Petro-Knowledge*, 69; Wessels, "Mineralöls."
10. Quotation from BSB, "Die Energiepolitik des Bundesregierung," October 3, 1973, DS 7/1057; Graf, *Petro-Knowledge*, 72; BAK, B136, 7668, Lantzke to Schmidt and Friderichs, July 10, 1973.
11. Quotation from BSB, "Die Energiepolitik des Bundesregierung" (1973); Meyer-Renschhausen, *Energieprogramm*, 3, 16-26; DIW, *Sicherung der Energieversorgung für die Bundesregierung Deutschland* (Berlin: DIW, 1972).
12. BAK, B102, 313603, Schmidt to Sames, "Aktuelle Situation im Mineraloelbereich," October 24, 1973; BSB, "Erste Fortschreibung des Energieprogramms der Bundesregierung," October 30, 1974, DS 7/2713.
13. First quotations from Helmut Schmidt, "The Energy Crisis: A Challenge for the Western World . . . 13 March 1974," in Hartmut Soell (ed.), *Helmut Schmidt, Pioneer of International Economic and Financial Cooperation* (New York: Springer, 2014), 39-51; second quotation in Hohensee, *Oelpreisschock*, 130-131; Graf, *Petro-knowledge*, 204-208.
14. ASD, WBA, A3 538, television speech from November 23, 1973.
15. BSB, "Sondergutachten des Sachverständigenrates zu den geseamtwirtschaftlichen Auswirkungen der Ölkrise," December 19, 1973, DS 7/1456.
16. Hatch, *Nuclear Power*, 130.
17. BAK, B102, 313604, "Report: Aktuelle Situation im Mineralölbereich," November 3, 1973, "Kleinen Arbeitsgruppe" to Lantzke, November 3, 1973, and "Entwurf eines Gesetztes zur Sicherung der Energieversorgung bei Gefaehrdung oder Stoerung der Einfuhren von Mineralöl oder Erdgas," November 7, 1973; Graf, *Petro-Knowledge*, 210; Hohensee, *Oelpreisschock*, 119-120.

18. Jonathan Carr, *Helmut Schmidt: Helmsman of Germany* (New York: St. Martin's Press, 1985), 78–79), 18–19, 24–26, 61–62; Horst Mendershausen, *Coping with the Oil Crisis; French and German Experiences* (Baltimore, MD: Johns Hopkins University Press, 1976), 83, FN. 26; Helmut Schmidt, "Kernenergie unverzichtbar, aber politische nicht akzeptiert," *Atomwirtschaft* 6 (1979), 319–324.
19. Quotation from "Nische für Liberale," *Der Spiegel* 6 (1973), 65; BAK, B102, 313604, "Kleinen Arbeitsgruppe" (Mineralölindustrie, wissenschaftliche Institute) to Lantzke, November 3, 1973; Helene Miard-Delacroix, *Willy Brandt: The Life of a Statesman*, translated by Isabelle Chaize (London: I. B. Tauris, 2016), 143–149; Graf, *Petro-Knowledge*, 224.
20. Quotations from ASD, WBA, A3 539, Brandt's Speech to Parliament, November 29, 1973.
21. PAAA, Zwischenarchiv, B71, 105693, Berichts der Gruppe "Energie," April 13, 1973; Helmut Schmidt, "Washington Energy Conference, 11 February 1974," in Soell (ed.), *Helmut Schmidt*, 33–39; Mattias Schulz, "The Reluctant European: Helmut Schmidt, the European Community, and Transatlantic Relations," in Mattias Schulz (ed.), *The Strained Alliance: US-European Relations from Nixon to Carter* (New York: Cambridge University Press, 2010), 279–309; Hartmut Soell, "Helmut Schmidt: Zwischen reaktivem und konzeptionellem Handeln," in Konrad Jarausch (ed.), *Das Ende der Zuversicht? Die siebziger Jahre als Geschichte* (Göttingen: Vandehoeck & Ruprecht, 2008), 279–295.
22. PAAA, Zwischenarchiv, B 71, 113893, Rohwedder to Grabert, December 13, 1973; ibid., Energiepolitik im Rahmen der Copenhagener Gipfelkonferenz, December 13, 1973.
23. Henning Türk, "The Oil Crisis of 1973 as a Challenge to Multilateral Energy Cooperation among Western Industrialized Countries," *Historical Social Research* 39/4 (2014), 209–230; Venn, *Oil Crisis*, 105–110; Kapstein, *Insecure Alliance*, 177–199.
24. Romano Prodi and Alberto Clo, "Europe," in Raymond Vernon (ed.), *The Oil Crisis* (New York: W. W. Norton, 1976), 91–112; Marloes Beers, "The OECD Oil Committee and the International Search for Reinforced Energy-Consumer Cooperation, 1972–1973," in Elisabetti Bini, Giuliano Garavini, and Federico Romo (eds.), *Oil Shock: The 1973 Crisis and Its Economic Legacy* (London: I. B. Taurus, 2016), 142–171; Lucas, *European Communities*, 46–76; Kiran Klaus Patel, *Project Europe: A History*, translated by Meredith Dale (New York: Cambridge University Press, 2020).
25. Quotation from PAAA, Zwischenarchiv, B 71, 105693, Brussels to Bonn, May 24, 1973; ibid., EG Ratstagung über Energiefragen, on May 22 and 23, 1973; ibid., Friderichs to Brandt, December 11, 1973, Department 4 to State Secretary and Foreign Minister, November 2, 1973; and Friedrichs to Brandt, December 11, 1973.
26. PAAA, Zwischenarchiv, B71, 105693, Jelonek to Buro Staatssekretäre about oil crisis and the EC, November 6, 1973, and Friderichs to Brandt, December 11, 1973; Graf, *Petro-Knowledge*, 245–248; Venn, *Oil Crisis*, 99–115.
27. Quotation from Lucas, *European Communities*, 76; BAK, B136, 7668, Vorbereitung des Besuchs des algerischen Industrie- und Energieministers . . . , January 4, 1974; PAAA, Zwischenarchiv, B 71, 113893, Algiers to Bonn, February 15, 1974.
28. Garavini, *OPEC*, 107–109; Yergin, *Prize*, 511–512.
29. Bennigsen-Foerder paraphrased in "Öl: Amerikas Krise schlägt auf Europa durch," *Der Spiegel*, June 24, 1973; second quotation from BAK, B102, 281077, Lantzke to Rohwedder, February 22, 1972; BBA 32, 333 VEBA, Lecture from Kempfer on August 19, 1971; Monopolkommission, *Zusammenschlussvorhaben der Deutschen BP AG und der VEBA AG* (Baden-Baden: Nomos, 1979); Radzio, *VEBA*, 293–302; Hardach, *Germany*, 154–155.
30. ASD, WBA, A3, 540, Meeting between Brandt, Palme, and Kreisky, December 2, 1973; Radzio, *VEBA*, 281–284; Soell, "Helmut Schmidt."
31. BAK, B136, 7668 report on the financial effects of the "Energiekonzept," July 4, 1973; ibid., draft of energy program from August 22, 1973; ibid., draft of energy program, July 4, 1973; BSB, "Die Energiepolitik des Bundesregierung (1973)."
32. Schmidt, "Washington Energy Conference, 11 February 1974"; Hohensee, *Oelpreisschock*, 166; Black, *Crude Reality*, 73–77.
33. Quotation from "Potenter Partner," *Der Spiegel*, September 30, 1973; ACDP, 08-003, 021/1 Marx to Cartens, Windeien, and Luda, July 19, 1973; Radzio, *VEBA*, 281–290; BAK, B136, 7668, "Aufzeichnung zu den finanziellen Wirkungen des Energiekonzepts," July 4,

1973; Schürmann, "Veba–Geslenberg–Deminex"; Hans Otto Eglau, "Bonn muss tief in die Tasche greifen," *Die Zeit*, October 26, 1973; "Rudolf von Bennigsen, German Executive, 63," *New York Times*, November 10, 1989; "Kartellstreit: Das Unwirksame Veto," *Die Zeit*, January 18, 1974.
34. Quotation from BSB, "Erste Fortschreibung des Energieprogramms der Bundesregierung," (1974); BAK, B102, 281077, report on Deminex, October 5, 1973, Buff to Dr. Schill, January 23, 1974, and Herbert Lögters to Friderichs, January 10, 1974.
35. First quotation from BSB, "Erste Fortschreibung des Energieprogramms der Bundesregierung," (1974); second from BAK, B102, 281077, Lantzke and Engelmann to Friderichs, February 16, 1974; also ibid., Lantzke to Economics Minister, February 16, 1974, and report on measures to secure Germany's oil supply, March 26, 1974.
36. Mendershausen, *Oil Crisis*, 71–78; Radzio, *VEBA*, 304–308; Hans Otto Eglau, "Energie Versorgung: Der Schah lockt mit Öl und Gas," *Die Zeit*, October 19, 1973; BAK, B102, 206541, Lantzke to Rohwedder, March 7, 1973; BSB, "Erste Fortschreibung des Energieprogramms," (1974); Meyer-Renschausen, *Energieprogramm*, 29.
37. "News in Brief: West Germany," *Petroleum Economist*, December (1974), 472; Charles William Carter, "The Importance of Osthandel: West German-Soviet Trade and the End of the Cold War,1969–1991" (Ohio State University, dissertation, 2012), 69–70.
38. Schürmann, "Veba–Gelsenberg–Deminex," 44–46.
39. BAK, B102, 281077, Lantzke to Rohwedder, April 11, 1974, Friderichs to Schmidt, March 10, 1974, and report on measures to secure Germany's oil supply, March 26, 1974; BAK, B136, 7668, Draft Energy Program circulated by Friderichs, August 22, 1973; Mendershausen, *Oil Crisis*, 78; "Deminex exploriert weltweite," *Oel—Zeitschrift für Mineralölwirtschaft* (June 1974), 260–264; "VEBA-Öl: Mehr als ein neuer Name," *Oel—Zeitschrift für Mineraloelwirtschaft* (April 1979), 88–89; BSB, "Erste Fortschreibung des Energieprogramms der Bundesregierung" (1974).
40. Frank Bösch, "Energy Diplomacy: West Germany, the Soviet Union and the Oil Crises of the 1970s," *Historical Social Research* 39/4 (2014), 165–184.
41. Quotation from Wolfgang D. Müller, "Die Kernenergie hat sich durchgesetzt," *Atomwirtschaft* 5 (1974), 225; D. Smidt, "Engpass Genehmigungsverfahren?" *Atomwirtschaft* 3 (1973), 116; H. Borsch et al. (ed.) *Nutzen und Risiko der Kernenergie* (Jülich: KFA, 1975); Reinhard Kallenbach, "Die wirtschaftliche Nutzung der Kernenergie aus der Sicht der Energiewirtschaft," in Kurt Naumann (ed.), *Kraftwerk 2000: Ein energiepolitisches Forum* (Stuttgart: Seewald Verlag, 1975), 67–75.
42. Quotations from Hans Matthöfer, *Interviews und Gespräche zur Kernenergie: "Den unsterblichen Tiger am Schwanz gepackt"* (Karlsruhe: Juristischer Verlag, 1976), 98 and 109; Werner Abelshauser, *Nach dem Wirtschaftswunder: Der Gewerkschafter, Politiker und Unternehmer Hans Matthöfer* (Bonn: Dietz, 2009).
43. BAK, B136, 10868, Friderichs on current political questions, November 14, 1976; also BSB, "Erklärung der Bundesregierung zur Lage der Energieversorgung," 7. Wahlperiode 73th Session, January 17, 1974; ASD, WBA, A3, 539, BMFT, "Beitrag zur Erklärung des Bundeskanzlers zu Energiefragen," November 11, 1973; H. Michaelis, "Zur Wettbewerbslage der Kernenergie in der BRD," 11 (1973), 518–520.
44. Quotation from Matthöfer, "Zur Forschungs- und Energiepolitik der Bundesregierung," in Naumann (ed.), *Kraftwerk 2000*, 11–19; ASD, Matthöfer Nachlass, 276, speech at Georgetown University, October 29, 1976.
45. Quotation from BAK, B136, 10868, Parliamentary report from 7. Wahlperiode, Drucksache 7/3871, from July 16, 1975; ibid., *Nachrichten* from November 14, 1976; Heinrich Mandel, "Die Kernenergie im Spannungsfeld der Energiepolitik," *Atomwirtschaft* 1 (1974), 18–22; J. Rembser, "Das 4. Atomprogramm der Bundesrepublik Deutschland für die Jahre 1973–76," *Atomwirtschaft* 4 (1973), 162–167; Smidt, "Engpass."
46. BAK, B136, 10868, Parliamentary report from 7. Wahlperiode, Drucksache 7/3871, July 16, 1975; Matthöfer, *unsterblichen Tiger*, 60–95.
47. Quotations from BSB, "Erste Fortschreibung des Energieprogramms der Bundesregierung," (1974); BSB, "Die Energiepolitik des Bundesregierung (1973); Meyer-Renschausen, *Energieprogramm*, 39–42; Smidt, "Engpass"; DTM, FA AEG-Telefunken I.2.060 Ü, GS 5416,

"Kraftwerk Union 1973: Geschäftsbericht 1973," July 1974; ASD, WBA, A3, 540, Speech to BASF on December 5, 1973.
48. Quotations from BAK, B136, 12453, BMI notes on delays in nuclear power authorization, May 31, 1976; DTM, FA AEG-Telefunken I.2.060 Ü, GS 5416, "Kraftwerk Union 1973: Geschäftsbericht 1973," July 1974.
49. Norman Gall, "Atoms for Brazil, Dangers for All," *Foreign Policy* 23 (1976), 155–201; Erwin Häckel, "The Politics of Nuclear Exports in West Germany," in Robert Boardman, *Nuclear Exports and World Politics* (New York: St. Martin's Press, 1983), 62–78.
50. Helmut Schmidt, "Report, Labor Party 1974," in Soell (ed.), *Helmut Schmidt*, 55–61; Helmut Schmidt, "Interview by James Reston," and "Address to the Council of Foreign Relations, 6 December 1974," in ibid., 51–55, 64–65; William Glenn Gray, "Learning to 'Recycle': Petrodollars and the West, 1973–5," in Bini et al., *Oil Shock*, 172–197; Robert Brenner, *The Economics of Global Turbulence* (New York: Verso Books, 2006); Frieden, *Global Capitalism*, 363–372; Eichengreen, *European Economy*.
51. Quotation from Matthöfer, *Unsterblichen Tiger*, 105; BMFT, *Fourth Nuclear Program 1973 to 1976 of the Federal Republic of Germany* (Bonn: BMFT, 1974).
52. Quotation from DTM, KWU, 19, "Kernenergieexport- Atombombe für Entwicklungsländer?" July 1976; ACDP, 1–956, Riesenhüber, 063/1, Pressedienst from June 27, 1975; BBA, 122, 23, Schilling to Peters, April 1, 1976.
53. BAK, B136, 10950, speech on German-Brazilian treaty, June 27, 1975; ACDP, 01–560 Riesenhüber, 063/1, Vermerk: zur gegenwärtigen Problematik des deutschen nuklear Exports, January 31, 1977; Frieden, *Global Capitalism*, 413–444.
54. Quotation from Gall, "Atoms for Brazil"; DTM, KWU, 19, "Kernenergieexport—Atombombe für Entwicklungsländer?" July 1976; William Glenn Gray, "Commercial Liberties and Nuclear Anxieties: The US-German Feud over Brazil, 1975–1977," *The International History Review* 34/3 (2012), 449–474; BSB, "Erste Fortschreibung des Energieprogramms der Bundesregierung," (1974).
55. DTM, KWU, 19, "Kernenergieexport—Atombombe für Entwicklungsländer?" July 1976.
56. Quotation from BAK, B136, 10950, "Atom-Industry: Kuschen vor Carter," *Der Spiegel*, December 6, 1976; ibid., KWU to Genscher, December 20, 1976; Gall, "Atoms for Brazil"; Hatch, *Nuclear Power*, 83–84.
57. Quotation from ACDP, 01–560 Riesenhüber, 063/1, Huyn on German-Brazilian nuclear deal, March 1, 1976; BAK, B136, 10950, Heinz Schulte, "Die Zusammenarbeit im Kernenergiebereich . . . "; Matthöfer, *Unsterblichen Tiger*, 102–115; Gall, "Atoms for Brazil"; Gray, "Nuclear Anxieties"; Häckel, "Nuclear Exports."
58. BBA, 75, 47, Gerwin, "In Karlsruhe brüte man schnell," *Christ und Welt* 35/16 (1963); Robert Jungk, *The Nuclear State*, trans. Eric Mosbacher (London: John Caldor, 1979), 29–31; Bill Keepin and Brian Wynn, "The Role of Models—What Can We Expect from Science?" in Thomas Baumgartner and Atle Midttun (eds.), *The Politics of Energy Forecasting: A Comparative Study of Energy Forecasting in Western Europe and North America* (Oxford: Clarendon, 1987), 33–57; Wolf Häfele and Alan S. Manne, "Strategies for a Transition from Fossil to Nuclear Fuels," *Energy Policy* (March 1975), 3–23.
59. Quotation from Häfele and Manne, "Strategies"; Wolf Häfele and Wolfgang Sassin, *Zukünftige Energieversorgung: Optionen und Strategien* (Essen: Glückauf, 1976); Wolf Häfele, *The Fast Breeder as a Cornerstone for Future Large Supplies of Energy* (Laxenburg: IIASA, 1973); Club of Rome, *The Limits to Growth: A Report for the Club of Rome's Project on the Predicament of Mankind* (New York: Universe Books, 1972).
60. Quotation from Wolf Häfele, "Hypotheticality and the New Challenges: The Pathfinder Role of Nuclear Energy," *Minerva: A Review of Science, Learning and Policy* 12/3 (1974), 303–322.
61. Mandel, "Kernenergie"; Wolf Häfele, "Energy Choices That Europe Faces: A European View of Energy," *Science* 184/4134 (1974), 360–367.
62. Häfele, "European View"; Häfele and Manne, "Strategies"; BMFT, *Fourth Nuclear Program*; Matthöfer, *Unsterblichen Tiger*, 51–59.
63. BBA, 122, 23, Peters to Ziegler, March 9, 1976; BBA, 122, 4, Ruhrgas, Ruhrkohle, and STEAG to BMFT, January 21, 1974; BBA, 122, 5, Saarbergwerke AG to BMFT, February

15, 1974; BBA, 122, 5, Kernforschungslage Jülich to Bergbau-Forschungs GmbH, November 25, 1974.
64. BSB, "Die Energiepolitik des Bundesregierung (1973); BAK, B136, 7668, Junghans to Brandt, August 28, 1973; "Forderung nach einem langfristigen Energiekonzept der Bundesregierung," doc. 70, September 24-28, 1972, in Martiny and Schneider (eds.), Energiepolitik, 356-358.
65. Quotations from Adolf Schmidt, "Gründung und Entwicklung der Ruhrkohle . . . ," doc. 72, July 1, 1976, in Martiny and Schneider, Energiepolitik, 360-367; Farrenkopf and Przigoda, Glück auf, table 4, p. 540; Parnell, Coal Industry, 135; BSB, "Erste Fortschreibung des Energieprogramms der Bundesregierung" (1974).
66. Quotation from BBA, 160, 231, "Doppelte Sicherheit," Study commissioned by Horst Ludwig Riemer, November 12, 1976; BBA, 122, 4, Ruhrgas, Ruhrkohle, and STEAG to BMFT, January 21, 1974; ASD, WBA, A3, 539, Brandt to Parliament, 67. Sitzung, November 29, 1973; ASD, WBA, A3, 540, Brandt Speech to BASF, December 5, 1973; BMFT, Energieforschung 1974-1977 (Bonn: BMFT, 1974); BMFT, Technologies Program, 1977-1980 (Bonn: Federal Min. for Research and Technology, 1977).
67. BBA, 122, 5, Jülich to Bergbau-Forschungs GmbH, November 25, 1974; BBA, 122, 4, Ziegler to Bergwerksverband, December 3, 1974.
68. Matthöfer, Unsterblichen Tiger, 17-38; Hatch, Nuclear Power, 86.
69. Matthöfer, Unsterblichen Tiger, 98-115; BSB, "Erste Fortschreibung des Energieprogramms der Bundesregierung" (1974); ASD, Matthöfer Nachlass, 276, Speech at Georgetown University, October 6, 1975; BAK, B102, 313603, Lantzke to Economics Minister, October 23, 1973.
70. ASD, WBA, A3, 539, Brandt to Parliament, 67. Sitzung, November 29, 1973; BMFT, Fourth Nuclear Program; Rembser, "4. Atomprogramm"; BBA, 122, 4, Ruhrgas to BMFT, January 28, 1975; BBA, 122, 24, Report on Kohleveredlung, May 23, 1975; BMFT, Energieforschung 1974-1977; BAK, B102, 313608, BMFT program for Cabinet Meeting on January 9, 1974; Hohensee, Ölpreisschock, 215.
71. Venn, Oil Crisis, 103-104; Robert B. Stobaugh, "The Oil Companies in the Crisis," in Vernon, Oil Crisis, 179-202; Bamberg, British Petroleum, 220, 479-490.
72. First quotation from Anthony Sampson, The Seven Sisters: The Great Oil Companies and the World They Made (New York: Viking Press, 1975), 261; second quotation from Raymond Vernon, "Introduction" in Vernon, Oil Crisis, 7; Stobaugh, "Oil Companies."
73. Radzio, VEBA, 308-315.
74. BSB, "Dritte Fortschreibung des Energieprogramms der Bundesregierung," May 11, 1981, DS 9/983; "Deminex—Cooperation in Exploration and Production," OIL GAS—European Magazine 2 (1978), 42; Tyler Priest, "Shifting Sands: The 1973 Oil Shock and the Expansion of Non-OPEC Supply," in Bini et al. (eds.), Oil Shock, 117-141.
75. "Europe's Unprofitable Markets," Petroleum Economist (August 1977), 294-295; W. N. Scott, "Western European Refining Capability," OIL GAS—European Magazine 1 (1979), 17-20; "Prices Dilemma for Europe's Refiners," Petroleum Economist (February 1979), 42-45; Radzio, VEBA, 293-299, 313-324; "Auf Zupsitungen vorbereiten," Der Spiegel 11 (1979), 89.
76. PAAA, Zwischenarchiv, B71, 113897, "Wettbewerbsprobleme in der Mineralölwirtschaft," May 13, 1975, Heldt to State Secretary, June 3, 1975, and report from Dr. Kinkel, May 15, 1975; "Europe's Unprofitable Markets," Petroleum Economist (August 1977), 294-295; "Prices Dilemma for Europe's Refiners," Petroleum Economist (February 1979), 42-45; Michel Bacchetta, "The Crisis in Oil-Refining in the European Community," Journal of Common Market Studies 17/2 (1978), 97-119.
77. Quotation from Monopolkommission, VEBA AG, 35-36; "Auf Zupsitungen vorbereiten," Der Spiegel 11 (1979), 89; "Ziffern im Griff," Der Spiegel, 12 (1977), 78-82; Radzio, VEBA, 313-327; "The 800 Million DM Deal between Veba and BP," OIL GAS—European Magazine 2 (1978), 9-10.
78. DTM, KWU, 60, confidential Memo on the Situation in Nuclear Power Plant Regions, November 15, 1974, and "Sonderabschreibung für Kraftwerke?" Handelsblatt May 18, 1976; "K + U fragt: Prof. H. Mandel," Atomwirtschaft 2 (1973); Christian Marx, "Failed Solutions to the Energy Crisis: Nuclear Power, Coal Conversion, and the Chemical Industry in West

Germany since the 1960s," *Historical Social Research* 4/39 (2014), 251–272; Augustine, *Technocracy*, 51–74; Hatch, *Nuclear Power*, 83–86; Strunk, *AEG*, 94–98; Eichengreen, *Globalizing Capital*, 136–157; Strunk, *Die AEG*, 94–98.

79. DTM, KWU 1, "Erläuterungen zum Abschluss der KWU-Gruppe . . . ," May 20, 1976, "Notiz zur Vorbereitung des Gespräches mit dem KWU-Vorstand," January 15, 1976, and Report by Barthelt on meeting in Erlangen, July 5, 1974; DTM, FA AEG-Telefunken I.2.060 Ü, GS 5416, "Kraftwerk Union 1973: Geschäftsbericht 1973"; Strunk, *Die AEG*, 94–98; BAK, B136, 12453, Berger on KWU, January 12, 1978.
80. Quotation from BAK, B136, 12453, "Wir sind nicht unterrichtet worden," *Der Spiegel*, 132–134, September 13, 1978; Gray, "Nuclear Anxieties"; Gall, "Atoms for Brazil."
81. BBA, 122, 25, "Konstituierende Sitzung des Fachausschusses 'Energieforschung und -technik," May 12, 1975; BAK, B136, 10901, "Minister Genscher über mögliche Standorte von Kernkraftwerken," *Atom-Information*, May 1974; Augustine, *Technocracy*, 100; Stephen Milder, *Greening Democracy: The Anti-Nuclear Movement and Political Environmentalism in West Germany and Beyond, 1968–1983* (Cambridge, UK: Cambridge University Press, 2017); Michael Schüring, "West German Protestants and the Campaign against Nuclear Technology," *Central European History* 45/4 (2012), 744–762.
82. Quotations from Häfele, "Hypotheticality," 315; Häfele and Manne "Strategies"; BAK, B136, 10903, Schmitz-Wenzel, "Kernenergie als möglicher Themenbereich der Regierungserklärung," October 18, 1976.
83. Quotation from BAK, B136, 10868, Haedrich's report on Cabinet meeting, November 11, 1976.
84. Quotation from BAK, B136, 10903, Matthöfer, "Bürgerdialog Kernenergie," January 1976, and Strohl to National Delegations, September 15, 1976.
85. Kallenbach, "Kernenergie."
86. Quotation from BAK, B136, 10868, Haedrich's report on Cabinet meeting, November 11, 1976; ASD, WBA, A3, 539, "Beitrag zur Erklärung des Bundeskanzlers zu Energiefragen," November 28, 1973; Wolfgang D. Müller, "Keine Alternative zur Kernenergie," *Atomwirtschaft* 7 (1973), 385.
87. BAK, B136, 12453, Gruener to Ahrens (SPD), March 1, 1977; figures from BAK, B136, 12453, Barthelt to Schmidt, April 20, 1976; ibid., Berger on KWU's situation, January 12, 1978; DTM, KWU, 19, "Kernkraftwerke: Riesenaufträge ohne Gewinn," *Süddeutsche Zeitung*, June 30, 1976; Augustine, *Technocracy*, 116.
88. Stobaugh, "Oil Companies"; Edith Penrose, "The Development of Crisis," in Vernon (ed.), *Oil Crisis*, 39–58; Vernon, "Introduction"; Hohensee, *Ölpreisschock*, 113–118; ASD, WBA, A3, 539, Brandt to Parliament, 67. Sitzung, November 29, 1973.
89. Quotation from ASD, WBA, A3, 539, Friderichs to Parliament, 67. Sitzung, November 29, 1973; BAK, B102, 313604, "Entwurf eines Gesetzes zur Sicherung der Energieversorgungg . . . ," November 11, 1973.
90. BAK, B102, 313604, "Aktuelle Situation im Mineralölbereich," November 3, 1973; Stobaugh, "Oil Companies"; Sampson, *Seven Sisters*, 261–281; Vernon, "Introduction"; Kapstein, *Insecure Alliance*, 168–170; Türk, "Oil Crisis."
91. Mendershausen, *Oil Crisis*, 79–83; ACDP, 08-003, 021/1, Economics Ministry's report on current oil situation, December 6, 1973, and Andreas Dürr to Niegel (CSU), December 6, 1973; PAAA, Zwischenakten, B71, 105693, Department 4 to State Secretary and Foreign Minister, November 2, 1973; Hohensee, *Ölpreisschock*, 114; ASD, WBA, A 3—539, speech to Parliament on November 29, 1973, p. 3908.
92. Quotations from Hans-Wilhelm Schiffer, "Rotterdam und die deutsche Ölversorgung," *Oel—Zeitschift für Mineralölwirtschaft* (May 1979), 127; Horn, *Energiepolitik*, 35–42; Karlsch and Stokes, *Faktor Öl*, 376; BSB, "First Revision to the Energy Program" (1974).
93. PAAA, 105693, Friderichs to Brandt, December 11, 1973; "Dynamics of the Rotterdam Market," *Petroleum Economist* (February 1979), 49–52; Prodi and Clo, "Europe"; Karlsch and Stokes, *Faktor Öl*, 376.
94. Quotation from Mendershausen, *Oil Crisis*, 82; Prodi and Clo, "Europe"; "West Germany: Coping with Crisis," *Petroleum Economist* (1974), 54.

95. PAAA, 105693, Friderichs to Brandt, December 11, 1973; ACDP, 08-003, 021/1, Economics Ministry Report from December 6, 1973.
96. PAAA, Zwischenakten, B71, 113897, report on Bundestag's 7th election period, 78th meeting, February 13, 1974; PAAA, Zwischenakten, B71, 105693, Friderichs to Brandt, December 11, 1973; BAK, B102, 313604, circular for cabinet meeting on October 17, 1973.
97. Quotation from ASD, WBA, A3, Friderichs to Parliament, November 29, 1973; ACDP, 08-003, 021/1, Economics Ministry Report December 6, 1973; Mendershausen, *Oil Crisis*, 81; "West Germany: Coping with the Crisis," *Petroleum Economist* 2 (1974), 54.
98. OECD Data: https://data.oecd.org/gdp/quarterly-gdp.htm; Prodi and Clo "Europe"; Hohensee, *Ölpreisschock*, 162-163.
99. Mendershausen, *Oil Crisis*, 80-83; "How Europe's Imports Were Cut," *Petroleum Economist* (November 1974), 423-424; "Oil and Energy Notebook: West Germany," *Petroleum Economist* (May 1974), 187.
100. Prodi and Clo, "Europe"; ACDP, 08/3, 021/1, BWM to Bundestag, December 6, 1973; "Heizöl- und Kohleeinsatz in der . . . ," *DIW Wochenberichte* (January 9, 1975), 5-10; BSB, "Erste Fortschreibung des Energieprogramms" (1974); BBA, 122, 24, Peters to Koordinierungsausschusses Kohleveredlung, October 6, 1975; BBA, 122, 4, Ruhrgas to Ziegler (BMFT0), September 9, 1974; Marx, "Failed Solutions"; Strunk, *Die AEG*, 94-98.
101. Quotation from BSB, "Erste Fortschreibung des Energieprogramms" (1974).
102. Quotation from OECD, *Energy Conservation in the International Energy Agency: 1976 Review* (Paris: OECD, 1976), 8; Graf, *Petro-Knoweldge*, 148-154; Michael Graetz, *The End of Energy: The Unmaking of America's Environment, Security, and Independence* (Cambridge, MA: MIT Press, 2011), 132-137.
103. Türk, "International Energy Agency"; "BSB, "Die Energiepolitik des Bundesregierung" (1973); Meyer-Renschhausen, *Energieprogramm*, 21-22.
104. https://sites.fas.harvard.edu/~histecon/energyhistory/energydata.html; GDP and inflation data from https://data.oecd.org.
105. Quotation from Hans Matthöfer, preface to D. Ehrenstein, J. Wichert, R. A. Dickler (eds.), *Energiebedarf und Energiebedarfsforschung* (Villingen-Schwennigen: Neckar-Verlag, 1977), XXI.
106. ASD, Matthöfer Nachlass, 684, Matthöfer to Boener and Rau, September 8, 1977; Hans Diefenbacher and Jeffrey Johnson, "Energy Forecasting in West Germany: Confrontation and Convergence," in Thomas Baumgartner and Atle Midttun (eds.), *The Politics of Energy Forecasting* (Oxford: Clarendon Press, 1987), 61-84, here 74.
107. Prodi and Clo, "Europe"; Mendershausen, *Oil Crisis*, 83; Horst Meixner, "Bewusstseinsänderung als Folge der Ölkrise" and Rolf Bauerschmidt, "Energieeinsparung," in Fritz Lücke (ed.), *Ölkrise: 10 Jahre danach* (Bonn: Verlag TÜV Rheinland, 1983), 150-153 and 208-213.
108. Quotation from BSB, "Erste Fortschreibung des Energieprogramms" (1974); ASD, WBA, A3, 538, Report from Economic Advisory Council, November 27, 1973; Helmut Schmidt speech, "The Energy Crisis; A Challenge for the Western World," March 13, 1974, in Soell (ed.), *Helmut Schmidt*, 42; PAAA, Zwischenakten 113905, Kruse on the Club of Rome query: "What should be the oil optimum price?" December 11, 1975.
109. ASD, Matthöfer Nachlass, 784, Matthöfer speech at Schloss Schwanberg bei Kitzingen, March 4, 1976.
110. "Fusionen: Alles beim alten," *Der Spiegel* 40 (1978), 65.
111. BMFT, *Technologies Program, 1977-1980*, 20.
112. Meixner, "Bewusstseinsänderung"; Martin Held, "Gibt es einen Energierelevanten Wertwandel bei den Privaten Konsumenten?" in Lücke (ed.), *10 Jahre danach*, 153-170.
113. Quotation from Bauerschmidt, "Energieeinsparung"; Meixner, "Bewusstseinsänderung."
114. BSB, 7. Wahlperiode, 73th Session, January 17, 1974; Jacobs, *Panic*.
115. Quotation BSB, 7. Wahlperiode, 73th Session, January 17, 1974; Mendershausen, *Oil Crisis*, 83, FN 26.

Chapter 6

1. Research Institute for Soft Technologies, March 1979. ASD, Bechert, 110, 258.
2. Milder, *Greening Democracy*, 168, 171.
3. Civil war quotation from Augustine, *Technocracy*, 137; Andrew S. Tompkins, *Better Active than Radioactive!: Anti-Nuclear Protests in France and West Germany* (Oxford: Oxford University Press, 2015), 57–58.
4. Quotations from Hans Günter Schumacher, "Verhältnis des Bundesverbandes Buürgerinitiativen Umweltschutz zu den Umweltparteien," in Rudolf Brun (ed.), *Der grüne Protest: Herausforderung durch die Umweltparteien* (Frankfurt: Fischer, 1978), 59–73.
5. "Hesse Green List (1977)," in Margit Mayer and John Ely (eds.), *The German Greens: Paradox between Movement and Party* (Philadelphia, PA: Temple University Press, 1998), 214.
6. Nolte, "Different Sort of Neoliberalism?"
7. Peter Mair, *Ruling the Void: The Hollowing of Western Democracy* (New York: Verso, 2013); Roger Eatwell and Matthew Goodwin, *National Populism: The Revolt against Liberal Democracy* (London: Penguin Books, 2018).
8. Oliver Nachtwey, *Germany's Hidden Crisis: Social Decline in the Heart of Europe*, translated by David Fernbach and Loren Balhorn (New York: Verso, 2016/2018); Franz Walter, *Die SPD: Biographie einer Partie von Ferdinand Lassalle bis Andrea Nahles* (Reinbek bei Hamburg: Rowohlt, 2015).
9. First quotations from Gunter Schwab, *Der Tanz mit dem Teufel: Ein abenteurliches Interview* (Hannover: Sponholtz, 1958), ch. 4; last quotation cited in Raymond H. Dominick, *The Environmental Movement in Germany* (Bloomington: Indiana University Press, 1992) 153, also 148–149, 152–157; Franz-Josef Brüggemeier, *Tschernobyl, 26. April 1986. Die ökologische Herausforderung* (Munich: Deutscher Taschenbuch, 1998), 202–204.
10. Metternich cited in Dominick, *Environmental Movement*, 148–149; Adoph Metternich, *Die Wüste droht. Die gefährdete Nahrungsgrundlage menschlichen Gesellschaft* (Bremen: Friedrich Trüjen, 1947); Erich Hornsmann, *Sonst Untergang. Die Antwort der Erde auf die Missachtung ihrer Gesetze* (Rheinhausen: Verlagsanstalt Rheinhausen, 1951); Reinhard Demoll, *Bändigt den Menschen gegen die Natur oder mit ihr?* (Munich: Bruckmann, 1954); Richard H. Beyler, "Hostile Environmental Intellectuals? Critiques and Counter-Critiques of Science and Technology in West Germany after 1945," *Wissenschaftsgeschichte* 31 (2008), 393–406.
11. Schwab, *Tanz*; Joachim Radkau, *Nature and Power: A Global History of the Environment*, translated by Thomas Dunlap (New York: Cambridge University Press, 2002/2008), 272–295; Macekura, *Mismeasure*, 104–137.
12. Frank Uekoetter, *The Age of Smoke: Environmental Policy in Germany and the United States, 1880–1970* (Pittsburgh: University of Pittsburgh Press, 2009), 187–194.
13. First quotation from Kai F. Hünemörder, *Die Frühgeschichte der globalen Umweltkrise und die Formierung der deutschen Umweltpolitik (1950–1973)* (Stuttgart: Franz Steiner Verlag, 2004), 54–56; Second quotation from Uekoetter, *Smoke*, 134, also 188; Uekötter, *Greenest Nation*, 71–73, 86–88.
14. Quotation from Thomas Lekan, "Saving the Rhine: Water, Ecology, and *Heimat* in Post-World War II Germany," in Christoph Mauch and Thomas Zeller (eds.), *Rivers in History: Perspectives on Waterways in Europe and North America* (Pittsburgh: University of Pittsburgh Press, 2008), 136; Blackbourn, *Water*, 322–331; Hünemorder, *Umweltpolitik*, 80–86.
15. Quotations from Schwab, *Tanz*, 234–235; Dominic, *Environmental Movement*, 152–158; Blackbourn, *Water*, 328–329; Uekötter, *Greenest Nation*, 57, 70, 77–90; Lekan, "Rhine," 113; Brüggemeier, *Tschernobyl*, 213; Hünemorder, *Umweltpolitik*, 121; Ramachandra Guha, *Environmentalism: A Global History* (New York: Longman, 2000), 69–77.
16. Demoll, *Bändigt*, 25; also Reinhard Demoll (ed.), *Im Schatten der Technik* (Munich: Bechte, 1960).
17. Jacques Ellul, *The Technological Society*, translated by John Wilkinson (New York: Vintage, 1958/1964) 4–5, 193, 198, 208, 284; Beyler, "Environmental Intellectuals"; Stephen J. Macekura, *Of Limits and Growth: The Rise of Global Sustainable Development in the Twentieth Century* (Cambridge: Cambridge University Press, 2015), 137–146.

18. E. F. Schumacher, *Small Is Beautiful: Economics as if People Mattered* (New York: Harper Collins, 2010/1973) 22 and 68; Guha, *Environmentalism*, 67.
19. Jeremy Varon, *Bringing the War Home: The Weather Underground, the Red Army Faction, and Revolutionary Violence in the Sixties and Seventies* (Berkeley: University of California Press, 2004), 39; Andrei S. Markovitz and Philip S. Gorski, *The German Left: Red, Green and Beyond* (New York: Oxford University Press, 1993); Robert Pfalzgraff et al., *The Greens of West Germany: Origins, Strategies, and Transatlantic Implications* (Cambridge, MA: Institute for Foreign Policy Analysis, 1983), 22–23.
20. Fischer quoted in Paul Hockenos, *Joschka Fischer and the Making of the Berlin Republic: An Alternative History of Postwar* (New York: Oxford University Press, 2008), 54, also 58; Varon, *Red Army Faction*, 67; Markovitz and Gorski, *German Left*, 56.
21. 30 percent of the nation's high school and university students sympathized with Marxism or Communism; Hockenos, *Fischer*, 85, 95–96; and Gerd Koenen, *Das rote Jahrzehnt: Unsere kleine deutsche Kulturrevolution, 1967–1977* (Frankfurt: Kiepenhauer and Witsch, 2001), 184; Markovitz, *German Left*, 79–113.
22. Wolfgang Müller-Haeseler, "Blümchen statt Schlote," *Die Zeit*, September 10, 1971, https://www.zeit.de/1971/37/bluemchen-statt-schlote; "Lied vom Tod," *Der Spiegel* 51 (1971); Hünemorder, *Umweltpolitik*, 299–326.
23. Jens Ivo Engels, "Modern Environmentalism," in Frank Uekoetter (ed.), *The Turning Points of Environmental History* (Pittsburgh: University of Pittsburgh Press, 2010), 119–131; Uekotter, *Greenest Nation*, 87; Blackbourn, *Water*, 331; Brüggemeier, *Tschernobyl*, 208–211.
24. Donella H. Meadows et al., *The Limits to Growth* (Washington, DC: Potomac Associates, 1972); Robert M. Collins, *More: The Politics of Economic Growth in Postwar America* (New York: Oxford University Press, 2000), 132–145; Kai F. Hünemörder, "Kassandra im modernen Gewand. Die umweltapokalyptischen Mahnrufe der frühen 1970er Jahren," in Frank Uekötter and Jens Hohensee (eds.), *Wird Kassandra heiser? Die Geschichte falscher Ökoalarme* (Munich: Franz Steiner, 2004), 78–97.
25. Elke Seefried, "Towards the '*Limits to Growth*'? The Book and Its Reception in West Germany and Great Britain 1972/73," *Bulletin of the German Historical Institute London* 33/1 (2011): 3–37; Patrick Kupper, "'Weltuntergangs-Vision aus dem Computer'. Zur Geschichte der Studie 'Die Grenzen des Wachstums' von 1972," in Uekötter and Hohensee, *Geschichte falscher Ökoalarme*, 98–111.
26. Quotations from Erhard Eppler, "Die Qualität des Lebens" (1972), http://library.fes.de/pdf-files/akademie/online/09120.pdf; Erhard Eppler, "Dokumente der ZEIT: Eppler—Thesen zur Lebensqualität," *Die Zeit*, July 20, 1973, https://www.zeit.de/1973/29/dokumente-der-zeit-eppler-thesen-zur-lebensqualitaet; Erhard Eppler, *Links Leben: Erinnerungen eines Wertkonservativen* (Berlin: Ullstein, 2015), 162–168.
27. Brandt quoted in Steurer, *Wachstumsdiskurs*, 373–377; Collins, *More*, 132–145; Hünemorder, *Umweltpolitik*, 229–232; Erhard Eppler, *Ende oder Wende: Von der Machbarkeit des Notwendigen* (Stuttgart: W. Kohlhammer, 1975).
28. Quotations from Erhard Eppler, "Alarmruf an die menschlichen Vernunft," *SPD Pressedienst*, January 3, 1974; Introduction to Ralf Dahrendorf et al. (eds.), *Die Energiekrise: Episode oder Ende einer Ära?* (Hamburg: Hoffmann and Campe, 1974); Weizsäcker in ibid., 91–93.
29. Erhard Eppler, "Nicht Gleichheit—mehr Gerechtigkeit," *Die Zeit*, November 3, 1972, https://www.zeit.de/1972/44/nicht-gleichheit-mehr-gerechtigkeit; Eppler, *Ende oder Wende*, 22–35; Horst Ehmke, *Mittendrin: Von der Grossen Koalition zur Deutschen Einheit* (Berlin: Rowohlt, 1994), 291; Eppler, *Links Leben*, 168–177; Winkler, *Long Road*, 320.
30. Horst Ehmke, "Keine Angst vor dem Fortschritt," *Die Zeit*, April 22, 1977, https://www.zeit.de/1977/18/keine-angst-vor-dem-fortschritt; Illing, *Energiepolitik*, 125–126; Horn, *Energiepolitik*, 198; Erhard Eppler, "Umweltschutz ist kein Spleen gelangweilter Mittelständler," *Sozialdemokratischer Pressedienst* 37/248 (December 30, 1982), 8–9; Hartmut Soell, *Helmut Schmidt, 1969 bis Heute* (Munich: Deutsche Verlags-Anstalt, 2008), 339 and 356; Ulrich Herbert, *Geschichte Deutschlands im 20. Jahrhundert* (Munich: C. H. Beck, 2014), 902 and 930.
31. BAK, 136, 10868, Interview with Hans Friedrichs, November 14, 1976; and Friderichs to Westphal, November 3, 1976; BAK, B136, 10901, Matthöfer to Hans Friderichs, October

17, 1974; BAK, 136, 10903, Schmitz-Wenze, "Kernenergie als möglicher Themenbereich der Regierungserklaerung, October 18, 1976.
32. First quotation from Milder, *Greening Democracy*, 23, see also 55–57; second quotation in Hager, "Grassroots Origins," 4–5.
33. Protester cited in Augustine, *Nuclear Power*, 93–94, 96; Thompkins, *Radioactive*, 37–45; Milder, *Greening Democracy*, 39–41; Dorothy Nelkin and Michael Pollak, *The Atom Besieged: Extraparliamentary Dissent in France and Germany* (Cambridge, MA: MIT Press, 1982).
34. Winegrower cited in Augustine, *Nuclear Power*, 101, 99–105.
35. First quotation in Hockenos, *Fischer*, 141–146; Kelly cited in Milder, *Greening Democracy*, 129, also 110–114; Tompkins, *Radioactive*, 1; Sara Parkin, *The Life and Death of Petra Kelly* (New York: Pandora, 1994), 84–86
36. Holger Strohm, *Friedlich in die Katastrophe: Eine Dokumentation über Atomkraftwerke* (Frankfurt a.M.: Zweitausendeins, 1981/1973), 178–180, 241; Michael Schüring, "West German Protestants and the Campaign against Nuclear Technology," *Central European History* 45/4 (2012), 744–762; Nelkin and Pollak, *Atom Besieged*, 96–106; ASD, Bechert, 112, 262, Rezension: "Die Macht der Technik macht uns noch alle kaputt," *Abendzeitung*.
37. Petra Kelly, *Fighting for Hope*, translated by Marianne Howarth (Boston: South End, 1983/1984), 92–93; Parkin, *Kelly*, 106–107; Stephen Milder, "Thinking Globally Acting (Trans-)locally: Petra Kelly and the Transnational Roots of Green Politics," *Central European History* 43 (2010), 301–326.
38. Dieter Rucht, *Von Wyhl nach Gorleben: Bürger gegen Atomprogramm und nukleare Entsorgung* (Munich: C. H. Beck, 1980), 57.; Karl Bechert, "Gefahren der Kernkraftwerke und der Endlagerung des Atommülls," in Kurt Naumann (ed.), *Kraftwerk 2000: Ein energiepolitisches Forum* (Stuttgart: Seewald Verlag, 1975), 76–84.
39. BMFT, *Energy Research*, 92–102; Hatch, *Nuclear Power*, 80, 88–90; Rucht, *Von Wyhl*, 60–65; Illing, *Energiepolitik*, 138–140; Rucht, *Von Wyhl*, 45–50; BMFT, *Technologies Program*, 24–27.
40. BAK, 136, 10868, Dahrendorf, "Zu Fragen der Entsorgung der Kernkraftwerken," November 15, 1976; Rucht, *Von Wyhl*, 50–66.
41. Judge cited in Hatch, *Nuclear Power*, 81; Rucht, *Von Wyhl*, 65–66; Dorothy Nelkin and Michael Pollack, "French and German Courts on Nuclear Power," *Bulletin of the Atomic Scientists* 36/5 (1980), 36–42.
42. Hatch, *Nuclear Power*, 80–82; Christian Joppke, "Nuclear Power Struggles after Chernobyl: The Case of West Germany," *West European Politics* 13/2 (1990), 178–191; Klaus Humann, *Atommüll oder Der Abschied von einem teueren Traum* (Hamburg: Rowohlt, 1977/1981); Nelkin and Pollak, *Atom Besieged*, 92–93; Horst Mewes, "A Brief History of the German Green Party," in Mayer and Ely (eds.), *German Greens*, 29–48.
43. Hager, "Grassroots"; Nelkin and Pollak, *Atom Besieged*, 93–94.
44. Bess, *Light Green*, 92–104; Tompkins, *Radioactive*, 13–20; Nelkin and Pollak, *Atom Besieged*, 157–166; Albert Presas, I. Puig, and Jan-Henrik Meyer, "One Movement or Many? The Diversity of Antinuclear Movements in Europe," in Arne Kaijser et al., *Engaging the Atom: The History of Nuclear Energy and Society in Europe from the 1950s to the Present* (Morgantown: West Virginia University Press, 2021), 83–111; Hecht, *Radiance of France*.
45. Eppler, *Ende oder Wende*, 48–49; Eppler, "Qualität des Lebens"; Ronald Inglehart, *Silent Revolution: Changing Values and Political Styles among Western Publics* (Princeton, NJ: Princeton University Press, 1977); Nachtwey, *Hidden Crisis*, 19–25.
46. Gerd Langguth, *The Green Factor in German Politics: From Protest Movement to Political Party* (Boulder, CO: Westview Press, 1984), 27–30; Pfaltzgraff et al., *The Greens*, 40–43; Markovitz and Gorski, *German Left*, 79–104; Frank Bösch, "Krisenkinder. Neoliberale, die Grünen und der Wandel des Politischen in den 1970er und 1980er Jahrem" in Bösch et al., *Neoliberalismus*, 39–60; Margit Mayer and John Ely, "Success and Dilemmas of Green Party Politics," in Mayer and Ely (eds.), *German Greens*, 3–26; Werner Hülsberg, *The German Greens: A Social and Political Profile* (London: Verso, 1988), 9–16.
47. Quotation in Langguth, *Green Factor*, 33; Silke Mene, "Eine Partei nach dem Boom: Die Grünen als Spiegel und Motor Ideengeschichtlicher Wandlungsprozesse seit den 1970er Jahren," in Morten Reitmayer and Thomas Schlemmer (eds.), *Anfänge der Gegenwart: Umbrüche in*

Westeuropa nach dem Boom (Munich: Oldenbourg, 2014); Markovitz and Gorski, *German Left*, 128–129.
48. First quotation from Herbert Gruhl, "Die Grüne Notwendigkeit," in Brun (ed.), *Grüne Protest*, 117–121; second from Herbert Gruhl, *Ein Planet wird geplündert: Die Schreckensbildung unserer* Politik (Frankfurt, a.M.: Fischer, 1975/1984), 11–12.
49. Quotation from Carl Beddermann, "Die 'Grüne Liste Umweltschutz' in Niedersachsen," in Brun (ed.), *Grüne Protest*, 105–116; Manon Maren-Grisebach, *Philosophie der Grünen* (Munich: Guenter Olzog, 1982), 11–20.
50. Quotation from Rudolf Bahro, *From Red to Green: Interviews with New Left Review* (London: Verso, 1984), 119, 135; Die Grünen, *Das Bundesprogramm* (1980); see also the "Green Basic Program (1980)," "Eco-Socialists: Ecological Crisis and Social Transformation," and "The Umbau Program," in Mayer and Ely (eds.), *German Greens*, 217–218, 223, 267–293; Silke Mende, *"Nicht rechts, nicht links, sondern vorn": Eine Geschichte der Gründungsgrünen* (Munich: oldenbourg Verlag, 2011), 447–451.
51. Quotations from Beddermann, "Grüne Liste"; Marianne Gronemeyer, "Gegen die Reduzierung sozialer Wahrnehmung," in Brun (ed.), *Grüne Protest*, 34–46.
52. Quotation from Gronemeyer, "Wahrnehmung," 37; Mende, "Grünen als Spiegel."
53. Quotation from Klaus Traube, *Müssen wir umschalten? Von den politischen Grenzen der Technik* (Reinbek: Rowohlt, 1978), 14; Hülsberg, *German Greens*, 44.
54. Quotation from Theodore Ebert, "Von den Bürgerinitiativen zur Ökologiebewegung," in Theodor Ebert, Wolfgang Sternstein, and Roland Vogt (eds.), *Ökologiebewegung und ziviler Widerstand: Wyhler Erfahrungen* (Stuttgart: Umweltwissenschaft Institute, 1978), 2–17; "Kein Atomkraftwerk in Wyhl: Erklärung der badisch-elsässischen Buergerinitiativen," in Heinrich Billstein and Klaus Naumann (eds.), *Für eine bessere Republik: Alternativen der demokratischen Bewegung* (Cologne: Pahl-Rugenstein, 1981), 172–176.
55. Joschka Fischer, "Warum eigentlich nicht?" in Joschka Fischer, *Von Grüner Kraft und Herrlichkeit* (Hamburg: Rowohlt, 1984), 88–98.
56. Quotations from Hans Günter Schumacher, "Verhältnis des Bundesverbandes Bürgerinitiativen Umweltschutz zu den Umweltparteien," in Brun (ed.), *Grüne Protest*, 59–73, 62.
57. Robert Jungk, *The New Tyranny: How Nuclear Power Enslaves Us All*, translated by Christopher Trump (New York: Fred Jordan Books, 1977/1979), 161, 163, 167; Robert Jungk, *Tomorrow Is Already Here: Scenes from a Manmade World*, translated by Marguerite Waldman (London: Hart-Davis, 1954), 7–8.
58. Erhard Eppler, "Grundsatzreferate," in Forum SPD, *Fachtagung "Energie, Beschäftigung, Lebensqualität" am 28. und 29. April 1977 in Köln* (Bonn: SPD, 1977), 16–30; Klaus-Christian Wanninger, *1999: Das Atomreich* (St. Michaels: Bläschke, 1979).
59. Quotation from ASD, Matthöfer Nachlass, 237, "Ein Streitgespräch zwischen Robert Jungk und BMFT Hans Matthöfer," *Atomstaat*, January 1978; Judt, *Postwar*, 803–835.
60. Schumacher, "Bundesverbandes Buergerinitiativen Umweltschutz, 62."
61. Amory B. Lovins, "Energy Strategy: The Road Not Taken?," *Foreign Affairs* 55/1 (1976), 65–96, here 77; Amory B. Lovins, *Non-Nuclear Futures: The Case for an Ethical Energy Strategy* (New York: Harper and Row, 1975); ASD, Bechert, 101, 242; Amory B. Lovins, "Nuclear Power: Technical Bases for Ethical Concern"; Amory B. Lovins, *Soft Energy Paths: Toward a Durable Peace* (New York: Harper & Row, 1977/1979).
62. Lovins's book was quickly translated into German: Amory Lovins, *Sanfte Energie: Das Program für die energie- und industriepolitische Umrüstung unserer Gesellschaft* (Bonn: BMFT, 1978); see also his two essays in Siegfried de Witt and Hermann Hatzfeldt (eds.), *Zeit zum Umdenken! Kritik an v. Weizsäckers Atom-Thesen* (Hamburg: Rowohlt, 1979).
63. Die Grünen, *Das Bundesprogramm*, 1980; BBU, "Für eine ökologische Kreislaufwirtschaft," in Billstein and Naumann (eds.), *bessere Republik*, 164–167.
64. Quotation from Günter Altner, "Fortschritt, Umweltschutz und die Grünen," in Rudolf Brun (ed.), *Der grüne Protest: Herausforderung durch die Umweltparteien* (Frankfurt, a.M.: Fischer Taschenbuch, 1978), 18–33; Die Grünen, *Das Bundesprogram* (1980).
65. Quotation from Siegfried de Witt, "Introduction," to Siegfried de Witt and Hermann Hatfeldt (eds.), *Zeit zum Umdenken! Kritik an von Weizsäckers Atom-Thesen* (Hamburg: Rowohlt, 1979), 10–12; Altner, "Fortschritt."

66. Florentin Krause, Hartmut Bossel, and Karl Friederich Müller-Reissmann, *Energie-Wende: Wachstum und Wohlstand ohne Erdöl und Uran* (Frankfurt a.M.: Firscher, 1980).
67. Walter, *Die SPD*, 197–228; Varon, *Red Army Faction*, 30–45; Markovitz and Gorski, *German Left*, 115–121.
68. Parkin, *Kelly*, 88–90; Milder, *Greening Democracy*, 174; Joachim Hirsch, "A Party Is Not a Movement and Vice Versa," in Mayer and Ely (eds.), *The German Greens*, 180–190.
69. BAK, 136, 10903, Scheller to Schmidt, September 22, 1976; BAK, 136, 10868, Press Conference from Gerd Walter, November 2, 1976; Markovitz and Gorski, *German Left*, 189–198; Hülsberg, *German Greens*, 45–59; Hockenos, *Fischer*, 131–151; Walter, *Die SPD*, 197–228.
70. Schmidt quoted by Eppler, *Links Leben*, 174; Winkler *Long Road*, 319–322; Ehmke, *Mittendrin*, 287–302.
71. Brandt quoted in Markovitz and Gorski, *German Left*, 81; Erhard Eppler, "Warum denn nicht mit den Gruenen?" in Brun (ed.), *Grüne Protest*, 170–175; Ehkme, *Mittendrin*, 291; Soell, *Schmidt*, 351–357; Bösch, *1979*, 320.
72. SPD, *Forum SPD. Energie: Ein Letifaden zur Diskussion* (Bonn: FES, 1977), quotations from 26–27; Hans Koschnick, "Eröffnung," in SPD, *Forum SPD: Fachtagung "Energie, Beschäftigung, Lebensqualität"* (Bonn: FES, 1977), 6–13; Meyer-Renschhausen, *Energieprogramm*, 99–100, 123–126.
73. Quotation from Alois Pfeiffer in, "Wirtschaftswachstum—Arbeitsplaetze—Energiebedarf—Lebensqualität," in SPD, *Fachtagung Energie*, 45–51; Meyer-Renschhausen, *Energieprogramm*, 122–129; Knut Krusewitz, "Der Energie-Kompromiss oder die trinitarische Atom-Formel: Zu den jüngsten energiepolitsichen Gewerkschafts- und Parteibeschlüssen," *Internationale Politik* 12 (1977), 1467–1476.
74. Italics added. Erhard Eppler, "Grundsatzreferate," in SPD, *Fachtagung Energie*, 16–30; Bernd Faulenbach, *Das sozialdemokratische Jahrzehnt: Von der Reformeuphorie zur Neuen Unübersichtlichkeit. Die SPD 1969–1982* (Berlin: Vorwärts Buch, 2013), 224–229.
75. DGB, "Für wirksamen Umweltschutz," March 6, 1974, in Billstein and Naumann (eds.), *bessere Republik*, 160–163; Heinrich Siegmann, *The Conflicts between Labor and Environmentalism in the Federal Republic of Germany and the United States* (New York: St. Martin's Press, 1985), 15–16.
76. Eppler, "Grundsatzreferate"; Erhard Eppler, "Nur eine Floskel?" *Die Zeit*, August 12, 1977, https://www.zeit.de/1977/33/nur-eine-floskel.
77. Klaus Michael Meyer-Abich, "Energieplanung—Energieeinsparung—Energiequelle," SPD, *Fachtagung Energie*, 66–81.
78. ASD, Matthöfer Nachlass 0237, "Ich trete nicht zurück," August 10, 1977; ASD, Matthöfer Nachlass 0357, speech to Hessische Kreise, February 13, 1978; Meyer-Renschhausen, *Energieprogramm*, 125.
79. BAK, 136, 12453, Klaus Barthelt (KWU) to Chancellor Schmidt, April 20, 1976; Harald Legler and Eberhard Jochen, "Der Zusammenhang zwischen Energieverbrauch, Wirtschaftswachstum und Beschäftigung," *Internationale Politik* 3 (1977), 270–285.
80. BAK, 136, 12453, KWU to Schmidt, March 22, 1977; Meyer-Renschhausen, *Energieprogramm*, 126–28; Krusewitz, "Energie-Kompromiss."
81. Quotation from Carl Friedrich von Weizsäcker, "Mit der Kernenerige leben," *Die Zeit*, March 17, 1978, https://www.zeit.de/1978/12/mit-der-kernenergie-leben; Klaus Mehrens, "Energiebedarf, Wirtschaftswachstum, Beschäftigung—aus der Sicht der Gewerkschaften," in Volker Hauff (ed.), *Energie—Wachstum—Arbeitsplätze* (Villingen: Neckar Verlag, 1978), 84–98; Krusewitz, "Der Energie-Kompromiss"; BBA, 160, 231, Heinz Kühn, "Vorrang für die Kohle," *Ruhrkohle AG Werks-Zeitung*, December 5, 1977; Eppler, *Links Leben*, 189.
82. ASD, Matthöfer Nachlass, 0357, speech to Hessische Kreise, February 13, 1978; BSB, "Zweite Fortschreibung des Energieprogramms der Bundesregierung," December 12, 1977, DS 8/1357.
83. Klaus Michael Meyer-Abich, "Wirtschaftliche und politische Möglichkeiten bei der Einsparung von Energie," in D. Ehrenstein, et al., *Energiebedarf und Energiebedarfsforschung* (Villingen-Schwennigen: Neckar-Verlag, 1977), 229–262; BSB, "Zweite Fortschreibung (1977)."

84. BAK, B136, 11015, "Gesetz zur Förderung der Modernisierung von Wohnungen," August 31, 1976, and "Programm zur Förderung heizenergiesparender Investitionen in bestehenden Gebäuden," September 8, 1977; ASD, Matthöfer Nachlass, 357, Speech to "Hessischen Kreis," February 13, 1978; Meyer-Renschhaussen, *Energieprogramm*, 150–151; Bösch, *1979*, 320–326.

85. BSB, "Zweite Fortschreibung (1977)," 5; Steurer, *Wachstumsdiskurs*, 397; Meyer-Renschhausen, *Energieprogramm*, 131–134, 172.

86. Volker Hauff, "Vorwort," in Herbert Kruempelmann, *Energieversorgung und Lebensqualitaet: Argumente in der Energiediskussion—Band 6* (Villigen: Neckar Verlag, 1978), IX–X; ASD, Hauff Nachlass, 39, Speech to the Fachkongress Nukleare Entsorgung, December 2, 1978; Gunter Hofmann, "Der 'Minenhund' des Bundeskanzlers," *Die Zeit*, November 1979, https://www.zeit.de/1979/46/der-minenhund-des-bundeskanzlers.

87. Quotation from ASD, Hauff Nachlass, 39, "Strukturpolitik und Energiepolitik," October 28, 1978; also in this file, speech to the Fachkongress Nukleare Entsorgung, December 2, 1978.

88. Quotations from ASD, Hauff Nachlass, 39, Hauff to Energiekommission of IG-Chemie, October 19, 1978, Hauff, "Strukturpolitik und Energiepolitik," October 28, 1978, and Hauff to Bundestag, 8, on December 14, 1978.

89. Hans-Wilhelm Schiffer, "West Germany: Government Policies on Energy Saving," *Petroleum Economist* 7 (1980), 294–295; Heinz Jürgen Schurmann, "Die Stellung der internationalen Ölindustrie nach der zweiten Ölkrise (I)," *Oel: Zeitschrift für die Mineralölwirtschaft* 7 (1980), 174–180; Hans-Wilhelm Schiffer, "Die Entwicklung der Mineralölindustrie in der Bundesrepublik im Jahre 1979," *Oel: Zeitschrift für die Mineralölwirtschaft* 2 (1980), 44–52; "Kapstein, *Insecure Alliance*, 177–200.

90. Graetz, *End of Energy*, 61–63; Frank Bösch, "Taming Nuclear Power: The Accident Near Harrisburg and the Change in West German and International Nuclear Policy in the 1970s and early 1980s," *German History* 35/1 (2017), 71–95.

91. Quotation from "Der Unfall. Gorleben. Harrisburg," *Die Zeit* 15 (1979), https://www.zeit.de/1979/15/harrisburg-der-unfall-gorleben-die-angst; Hatch, *Nuclear Power*, 114–116; Milder, *Greening Democracy*, 210–215.

92. Quotations from Helmut Schmidt, *Kernfrage-Kernenergie: Ansprache des Bundeskanzler auf der Europäischen Nuklearkonferenz in Hamburg am 7. Mai 1979* (Bonn: Bulletin des Press- und Informationsamtes der Bundesreigung, 1979); Helmut Schmidt, "Kernenergie: unverzichtbar, aber politische nicht akzeptiert," *Atomwirtschaft* 6 (1979), 319–324; BSB, "Dritte Fortschreibung des Energieprogramms" (1981), 5; Hatch, *Nuclear Power*, 135.

93. Quotation from ASD, Hauff Nachlass, 45, "Energiepolitik für die 80er Jahre," December 17, 1979; ASD, Hauff Nachlass, 43, "Energiepolitik als Friedenspolitik," September 25, 1979; BMFT, *Energy Research*, 19–20.

94. ASD, Hauff Nachlass, 58, Hauff foreword to Ökonomie, "Ökologie, Umweltschutz: Sicherheit für die 80er Jahre," *Politik: Aktuelle Informationen der Sozialdemorkatischen partei Deutschlands*, February 1980.

95. Erhard Eppler, "Darf der Mensch, was er kann?" *Die Zeit* 40 (1979), https://www.zeit.de/1979/40/darf-der-mensch-was-er-kann; ASD, Hauff Nachlass, 45, Hauff to first session of the Enquete Kommission, January 10, 1980.

96. SPD, *Parteitag Berlin 1979: Beschlüsse zur Energiepolitik und Umweltpolitik* (Bonn: SPD, 1980); Bösch, *1979*, 321.

97. Richard Löwenthal, "Identität und Zukunft der Sozialdemokratie," *Die Zeit*, December 11, 1981, https://www.zeit.de/1981/51/identitaet-und-zukunft-der-sozialdemokratie; "Dann ist die Regierung schon 1982 am Ende," *Der Spiegel*, December 14, 1981, https://www.spiegel.de/politik/dann-ist-die-regierung-schon-1982-am-ende-a-ab850712-0002-0001-0000-000014352377; Spohr, *Global Chancellor*, 125; Eichengreen, *European Economy*, 252–276.

98. Alois Pfeiffer, "Referat zum Thema 'Energiepolitik,'" in Deutscher Gewerkschaftsbund, *Energie-Politik: Zusammenstellung wichtiger energiepolitischer Informationen des DGB* (Bonn: DGB, 1983), 1–33; ASD, Hauff Nachlass, 58, Hauff's foreword to "Ökonomie, Ökologie, Umweltschutz: Sicherheit für die 80er Jahre," 6; Wolfgang Bartels, "Energiepolitik und Umweltschutz," in Günter Arndt et al. (eds.), *DGB Programm '81* (Frankfurt: Nachrichten, 1981), 158–166; Robert A. Dickler, "Atomenergie und Arbeitsplätze: Zum Mythos des

Zielkonflikts Wirtschaftswachstum, Vollbeschäftigung und Umweltschutz," *Internationale Politik* 9 (1977), 1075–1094; Volker Hauff und Hans Christoph Binswanger, *Die ökologische Herausforderung an Wirtschaftstheorie und Wirtschaftspolitik* (Bonn: Friedrich Ebert Stiftung, 1984); Siegmann, *Labor and Environmentalism*, 40–50.

99. Eppler, "Nicht mit den Grünen?"; ASD, Hauff Nachlass, 39, Hauff at the Gesprächskreises Wirtschaft und Politik, November 15, 1978; ASD, Hauff Nachlass, 67, "Kommission für Umweltfragen und Ökologie," November 25, 1981.
100. Eppler, "Spleen," 8–9.
101. ASD, Matthöfer Nachlass, 245, "Zusammenhänge zwischen der Beeinflussung volkswirtschaftlicher Rahmenbedingungen . . ." (1982); Ehmke, *Mittendrin*, 299–300; Siegmann, *Labor and Environmentalism*, 40–44; DGB, *Energie-Politik* (1983).
102. Eckart Conze et al. (eds.), *Nuclear Threats, Nuclear Fear and the Cold War of the 1980s* (New York: Cambridge University Press, 2017); William Hitchcock, *The Struggle for Europe: The Turbulent History of a Divided Continent* (New York: Random House, 2002), 267–268.
103. SPD, *Arbeitsprogramm zur ökologischen Modernisierung der Volkswirtschaft* (Bonn: Couric-Druck, 1986); ASD, Hauff Nachlass, 72, "Kommission für Umweltpolitik beim SPD-Parteivorstand . . . ," October 4, 1984; SPD, "Die Energiepolitik der SPD: Beschluss des SPD-Parteitages, München, 19.–23. April 1982," in *Politik: Aktuelle Informationen der Sozialdemokratischen Partei Deutschlands* 4 (April 1982).
104. SPD, "Sichere Energieversorgung ohne Atomkraft: Leitlinien der SPD zur Energiepolitik," *Politik: Informationsdienst der SPD* 4 (June 1986), 1–2; Stephen Padgett, "West German Social Democrats in Opposition, 1982–1986," *West European Politics* 10/3 (1987), 333–356.
105. Bess, *Light Green*, 93–94.
106. ASD, Matthöfer Nachlass, 245, Economics Minister to Chancellor, July 28, 1982; Meixner, "Bewusstseinsänderung"; Bösch, *1979*, 320–323; Alexander Glaser, "From Brokdorf to Fukushima: The Long Journey to Nuclear Phase-out," *Bulletin of the Atomic Scientists* 68/6 (2012), 12–21.

Chapter 7

1. Hans Christoph Binswanger et al., *Arbeit ohne Umweltzerstörung: Strategien für eine neue Wirtschaftspolitik* (Frankfurt a.M.: Fischer, 1983/1988), 30.
2. Quotation in Röpke, *Free Society*, 34; Razeen, *Classical Liberalism*, 105–131.
3. KAS/ACDP 07-001-22077, CDU, "Das Berliner Program. 2. Fassung 1971."
4. SPD, "Grundsatzprogramm der Sozialdemokratischem Partei Deutschland in Bad Godesberg" (Bonn: Vorstand der SPD, 1959).
5. Alfred Müller-Armack, "Das gesellschaftspolitische Leitbild der sozialen Marktwirtschaft," *Wirtschaftspolitische Chronik* 3 (1962), 7–28; Andreas Exenberger, "Die Sozial Marktwirtschaft von Alfred Müller-Armack" (Institut für Wirtschaftstheorie, Wirtschaftspolitik und Wirtschaftsgeschichte, Universitaet Innsbruck, WP 97/01).
6. For a pioneering study on neoliberalism and the environment, see Troy Vettese, "Limits and Cornucopianism: A History of Neo-Liberal Environmental Thought, 1920–2007" (dissertation, New York University, 2019); Paul Sabin, *The Bet: Paul Ehrlich, Julian Simon, and Our Gamble over Earth's Future* (New Haven, CT: Yale University Press, 2013); Erhun Kula, *History of Environmental Economic Thought* (New York: Routledge, 1998); Pearce, "Environmental Economics"; Richard Andrews, *Managing the Environment, Managing Ourselves: A History of American Environmental Policy* (New Haven, CT: Yale University Press, 2020).
7. Martin Bemmann et al., "Einleitung," and Thomas Zeller, "Ursprung, Möglichkeiten und Grenzen des Konzepts der ökologischen Modernisierung—Kommentar," in Martin Bemmann et al. (eds.), *Ökologische Modernisierung: Zur Geschichte und Gegenwart eines Konzepts in Umweltpolitik und Sozialwissenschaften* (Frankfurt: Campus Verlag, 2014), 7–34, 127–134; Arthur P. J. Mol and Gert Spaargaren, "Ecological Modernisation Theory in Debate: A Review," *Environmental Politics* 9/1 (2000), 17–49.
8. Robert Skidelsky, *John Maynard Keynes, 1883–1946: Economist, Philosopher, Statesman* (New York: Penguin, 2003); Joanna Bockman, *Markets in the Name of Socialism: The Left-Wing Origins of Neoliberalism* (Stanford, CA: Stanford University Press, 2011); Slobodian,

Globalists; Daniel Rodgers, *Age of Fracture* (Cambridge, MA: Harvard University Press, 2011), 56–58.
9. Quotation from Binyamin Applebaum, *The Economists' Hour: False Prophets, Free Markets, and the Fracture of Society* (New York: Little Brown, 2019), 17; Gerstle, *Neoliberal Order*; Stefanie L. Mudge, *Leftism Reinvented: Western Parties from Socialism to Neoliberalism* (Cambridge, MA: Harvard University Press, 2018); Harvey, *Neoliberalism*; Frank Bösch et al., *Neoliberalismus*.
10. Quotation from Robert M. Solow, "The Economics of Resources or the Resources of Economics," *The American Economic Review* 64/2 (1974), 1–14, here 1–2.
11. Dasgupta and Heal, *Exhaustible Resources*, 1.
12. Donald A. Walker, "Early General Equilibrium Economics: Walras, Pareto, and Cassel," S. Abu Turab Rizvi, "Postwar Neoclassical Microeconomics," and Mark Blaug, "The Formalist Revolution of the 1950s," in Warren J. Samuels et al., *A Companion to the History of Economic Thought* (Malden, MA: Blackwell, 2003), 278–293, 377–394, 395–410.
13. Tom Tietenberg and Lynne Lewis, *Environmental Economics & Policy* 6th edition (Boston: Pearson, 2010), 132.
14. Harold Hotelling, "The Economics of Exhaustible Resources," *Journal of Political Economy* 39 (1931), 137–175; Sandmo, "Environmental Economics."
15. William Nordhaus, "The Allocation of Energy Resources," *Brookings Papers on Economic Activity* 3 (1973), 529–570, quotation from 530; William D. Nordhaus, "Economic Growth and Climate: The Carbon Dioxide Problem," *American Economic Association* 67/1 (1977), 341–348; Gregory F. Nemet, "Modeling Long Term Energy Futures after Nordhaus (1973)," *Journal of Natural Resources Policy Research* 7 (2015), 141–146.
16. Nordhaus, "Energy Resources," quotations from 532, 549, 553, and 570.
17. Solow, "Economics of Resources," 8–10; Robert Solow, "Comments on William Nordhaus," *Brookings Papers on Economic Activity* 3 (1973), 572–574
18. Quotation from Solow, "Economics of Resources," 12; Dasgupta and Heal, *Exhaustible Resources*, 7 and 474–478; Willam J. Baumol and Wallace E. Oates, *The Theory of Environmental Policy* (Englewood Cliffs, NJ: Prentice-Hall, 1975), 56–70; Nordhaus, "Energy Resources," 567; William Nordhaus, "Resources as a Constraint on Growth," *American Economic Review* 64/2 (1974), 22–26.
19. Arthur Pigou, *The Economics of Welfare*, 4th edition (London: Macmillan, 1920/1952), 180–184; Sandmo, "Environmental Economics," 53–56; Pearce, "Environmental Economics."
20. Quotation from Larry E. Ruff, "The Economic Common Sense of Pollution," in Robert Dorfman and Nancy S. Dorfman (eds.), *Economics of the Environment. Selected Readings*, 3rd edition (New York: W. W. Norton, 1993), 20–36, here 20. Originally in *Public Interest* 19 (1970), 69–85; Allen V. Kneese, "Water Quality Management by Regional Authority in the Ruhr Area with Special Emphasis on the Role of Cost Assessment," *Papers in Regional Science* 11/1 (1963), 229–250.
21. Allen V. Kneese, *Economics and the Environment* (New York: Penguin, 1977), quotations from 29, 77, 121–124.
22. Kneese, *Environment*, 19–30; Tietenberg, *Economics and Policy*, 27. While Kneese advocated this approach, he was not alone. Jan-Henrik Meyer, "Who Should Pay for Pollution? The OECD, the European Communities, and the Emergency of Environmental Policy in the Early 1970s," *European Review of History: Revue europeene d'histoire* 24/3 (2017), 377–398.
23. David Freeman, *A Time to Choose: America's Energy Future* (Cambridge, MA: Ballinger, 1974), Appendix F.
24. Edward A. Hudson and Dale W. Jorgenson, "U.S. Energy Policy and Economic Growth, 1975–2000," *The Bell Journal of Economics and Management Science* 5/2 (1974), 461–514, 462; Dale Jorgenson, E. R. Berndt, L. R. Christensen, and E. A. Hudson, "U.S. Energy Resources and Economic Growth," Washington, DC (September 1973); Kenneth C. Hoffman and Dale W. Jorgenson, "Economic and Technological Models for Evaluation of Energy Policy," *The Bell Journal of Economics* 8/2 (1977), 444–466.
25. James Tobin cited in Leonard Silk, "Paradox for Economists," *New York Times*, October 30, 1974, 63; Philip Mirowski, "Twelve Theses Concerning the History of Postwar Neoclassical Price Theory," *History of Political Economy* 38 (2006), 343–379.

26. Carl Christian von Weizsäcker, *Zur ökonomischen Theorie des technischen Fortschritts* (Göttingen: Vandenhoek & Ruprecht, 1966); Paul A. Samuelson (1965), "A Theory of Induced Innovation along Kennedy-Weizsäcker Lines," *Review of Economics and Statistics*, 47/4, 343–356.
27. Nathan Rosenberg, *Inside the Black Box: Technology and Economics* (New York: Cambridge University Press, 1982), 15, 14–20; Florian Brugger and Christian Gehrke, "The Neoclassical Approach to Induced Technical Change: From Hicks to Acemoglu," *Microeconomica* 68/4 (2017), 730–776; Adam Jaffe, Richard Newell, and Robert Stavins, "Technological Change and the Environment," in Karl-Göran Mäler and Jeffrey R. Vincent (eds.), *Handbook of Environmental Economics* (Amsterdam: Elsevier, 2003), 461–507.
28. Quotations from William D. Nordhaus, "How Fast Should We Graze the Global Commons?" Cowles Foundation Discussion Paper No. 615 (January 1982), 1–13; William D. Nordhaus, "Some Skeptical Thoughts on the Theory of Induced Innovation," *The Quarterly Journal of Economics* 87/2 (1973), 208–219.
29. Charles S. Maier, "'Malaise': The Crisis of Capitalism in the 1970s," in Ferguson et al. (eds.), *Shock of the Global*, 25–48.
30. Philip Mirowski, "Neoliberalism: The Movement That Dare Not Speak Its Name," *American Affairs* 2/1 (2018), 118–141; Slobodian, *Globalists*, 1–26, 224–235; Rob Van Horn and Philip Mirowski, "The Rise of the Chicago School of Economics and the Birth of Neoliberalism," in Mirowski and Plehwe, *Mont Pèlerin*, 139–179.
31. First quotation from Milton Friedman, 28, also 121–123; other quotations in Franz Kromka, "Das grüne Denken der Väter der Sozialen Marktwirtschaft," *Zeitschrit für Politik* 39/3 (1992), 264–285; Ronald Coase, "The Nature of the Firm," *Economica* 4/16 (1937), 386–405; Rodgers, *Fracture*, 41–76; Harvey, *Neoliberalism*, 67; Rob van Horn, "Reinventing Monopoly and the Role of Corporations: The Roots of Chicago Law and Economics," in Mirowski and Plehwe, *Mont Pelerin*, 204–237.
32. Reagan quoted in Sabin, *The Bet*, 141–143; Gerstle, *Neoliberal Order*; Collins, *More*, 157–186.
33. Vettese, "Limits and Cornucopianism," chapter 4; Sabin, *The Bet*, 70–71, 132–133; Julian L. Simon, "Resources, Population, Environment: An Oversupply of False Bad News," *Science* 208/4451 (1980), 1431–1437.
34. First quotations from Julian L. Simon, *The Ultimate Resource* (Princeton, NJ: Princeton University Press, 1981), 22; last quotation from Simon, "False Bad News," 1436.
35. Simon, *Ultimate Resource*, 26; Vettese, "Limits and Cornucopianism," 142–145.
36. Ronald Coase, "The Problem of Social Cost," *Journal of Law and Economics* 3 (1960), 1–44; Gary D. Libecap, *Contracting for Property Rights* (New York: Cambridge University Press, 1994); Rodgers, *Fracture*, 57–58.
37. Quotations from Alan Randall, "The Problem of Market Failure," in Dorfman and Dorfman (eds.), *Economics of the Environment*, 144–162, here 145; Ralph Turvey, "On Divergences between Social Cost and Private Cost," in ibid., 139–143.
38. Tom Tietenberg, "Cap-and-Trade: The Evolution of an Idea," *Agricultural and Resource Economics Review* 39/3 (2010), 359–367; Robert W. Hahn and Roger G. Noll, "Designing a Market for Tradeable Emission Permits," in Wesley A. Magat (ed.), *Reform of Environmental Regulation* (Cambridge, MA: Ballinger, 1982), 119–214; Robert W. Hahn and Robert N. Stavins, "Incentive-Based Environmental Regulation: A New Era from an Old Idea?" *Ecology Law Quarterly* 18/1 (1991), 1–42.
39. Quotation (italics added) from Robert W. Hahn, "Getting More Environmental Protection for Less Money: A Practitioners' Guide," *Oxford Review of Economic Policy* 9/4 (1993), 112–123; Robert W. Hahn, "Economic Prescriptions for Environmental Problems: How the Patient Followed the Doctor's Orders," *Journal of Economic Perspectives* 3/2 (1989), 95–114; Robert N. Stavins, "What Can We Learn from the Grand Policy Experiment? Lessons from SO2 Allowance Trading," *Journal of Economic Perspectives* 12/3 (1998), 69–88; Robert N. Stavins, "Harnessing Market Forces to Protect the Environment," *Environment* 31/1 (1989), 5–7, 28–35.
40. Daniel Dudek, "Emissions Trading: Environmental Perestroika or Flimflam?" *The Electricity Journal* 2/9 (November 1989), 32–40; Fred Krupp, "The Making of a Market-Minded Environmentalist," *Strategy + Business* 51 (June 10, 2008); Yergin, *The Quest*, 480–483.

41. Quotation from Tietenberg, "Cap-and-Trade"; Stavins, "What Can We Learn"; Yergin, *The Quest*, 479–484; Jonas Meckling, *Carbon Coalitions: Business, Climate Politics, and the Rise of Emissions Trading* (Cambridge, MA: MIT Press, 2011), 55; Finis Dunaway, *Seeing Green: The Uses and Abuses of American Environmental Images* (Chicago: University of Chicago Press, 2015).
42. DIW, *Sicherung der Energieversorgung für die Bundesregierung Deutschland* (Berlin: DIW, 1972); BAK, B102, 313597, Report from January 28, 1972; Graf, *Sovereignty*, 76–81.
43. Hans K. Schneider, lecture on April 9, 1975, *Die Zukunft unserer Energiebasis als ökonomisches Problem* (Opladen: Westdeutscher Verlag, 1977).
44. Figures from https://www.adac.de/verkehr/tanken-kraftstoff-antrieb/deutschland/kraftst offpreisentwicklung/.
45. Berndt Lehbert, "Untersuchung der kurz- und langfristigen Elastizitaeten der Energienachfrage in Bezug auf die Energiepreise in der Bundesrepublik Deutschland," *Kiel Working Paper, Institut für Weltwirtschaft* 59 (1977), 1–90.
46. Quotation from ADS, Hans Matthöfer NL, 684, Matthöfer to Boener, September 8, 1977; Erhard Eppler, "Grundsatzreferate" in Forum SPD, *Fachtagung "Energie, Beschäftigung, Lebensqualität*, 16–30.
47. Stefan Rath-Nagel, *Alternative Entwicklungsmöglichkeiten der Energiewirtschaft in der BRD: Untersuchung mit Hilfe eines Simulationsmodells* (Stuttgart: Birkhäuser Verlag, 1977); the studies commissioned in 1977 were published in the series Hans Matthöfer, *Argumente in der Energiediskussion* (Villingen: Neckar Verlag, 1977).
48. Quotation Dickler, "Mythos des Zielkonflikts," 1083; Robert A. Dickler, "Zum Stand der Energieanalyse in den USA," in D. Ehrenstein, J. Wichert, and R. A. Dickler (eds.), *Energiebedarf und Energiebedarfsforschung: Referate und Ergebnisse einer Tagung des BMFT* (Villingen-Schwennigen: Neckar-Verlag, 1977), 290–331.
49. Gerhard Friede, "Substitutionsansatz zur Schätzung alternativer Möglichkeiten der wirtschaftlichen Entwicklung," in Ehrenstein et al. (eds.), *Energiebedarfsforschung*, 63–76; Harald Legler and Eberhard Jochen, "Der Zusammenhang zwischen Energieverbrauch, Wirtschaftswachstum und Beschäftigung," *Internationale Politik* 3 (1977), 270–285.
50. Quotation from Hans C. Binswanger et al., *Arbeit ohne Umweltzerstörung: Strategien für eine neue Wirtschaftspolitik* (Frankfurt a.M.: Fischer Verlag, 1983), 231–232; Hermann Precht, "Arbeitsplätze durch Umweltschutz?" in Wolfgang Brinkel and Harald B. Schäfer (eds.), *Wachstum Wohin? Energie, Umweltschutz, Arbeitsplätze* (Freiburg: Dreisam Verlag, 1979), 77–88, 80; Klaus Michael Meyer-Abich, *Energieeinsparung als Neue Energiequelle: Wirtschaftspolitische Möglichkeiten und alternative Technologien* (Munich: Hanser, 1979), 57–58.
51. First quotation from Walter Schulz, "Wirtschaftstheoretische und empirische Ueberlegungen zur These der Entkopplung von Wirtschaftswachstum und Energieverbrauch," in Horst Siebert (ed.), *Erschöpfbare Ressourcen* (Berlin: Duncker & Humblot, 1980) 377–399; second from Friede, "Substitutionsansatz," 63.
52. Herbert Giersch et al., *The Fading Miracle: Four Decades of Market Economy in Germany* (Cambridge: Cambridge University Press, 1992), 139, 160, 185–189, 192–195, 207–208; Mudge, *Leftism Reinvented*, 350–355.
53. Hans-Christoph Binswanger, Werner Geissberger, and Theo Ginsburg, *Wege aus der Wohlstandsfalle: Der NAWU-Report* (Frankfurt a.M.: Fischer, 1979); Jörg Wolff (ed.), *Wirtschaftspolitik in der Umweltkrise: Strategien der Wachstumsbegrenzung und Wachstumsumlenkung* (Stuttgart: Deutsche Verlags-Anstalt, 1974).
54. Quotations from Binswanger et al., *Arbeit ohne Umweltzerstörung*, 28–29; Udo E. Simonis, "Arbeit und Umwelt: Ansatzpunkte einer integrierten Beschäftigungs- und Umweltpolitik," in Eduard Kroker (ed.), *Arbeit, Umwelt, Arbeitslosigkeit* (Koenigstein: Koenigstein, 1988), 49–67; Udo Ernst Simonis (ed.), *Ökonomie und Ökologie: Auswege aus einem Konflikt* (Karlsruhe: CF Müller Verlag, 1983); Emma Rothschild, "Maintaining (Environmental) Capital Intact," *Modern Intellectual History* 8/1 (2011), 193–212.
55. Binswanger, *NAWU*, 68, 118.
56. Binswanger, *NAWU*, 9, 20, 68, 118.
57. Carl Christian von Weizsäcker, "Leistet der Markt die optimale intertemporale Allokation der Ressourcen?" in Siebert (ed.), *Erschöpfbare Ressourcen*, 769–814, here 811.

58. Quotation from Bertram Schefold, "Ecological Problems as a Challenge to Classical and Keynesian Economics," *Metroeconomica* 37 (1985), 21–61; Bertram Schefold, *Wirtschaftsstile: Band 1: Studien zum Verhaeltnis von Ökonomie und Kultur* (Frankfurt a.M.: Fischer, 1994), 9–12; Bertram Schefold, *Wirtschaftsstile Band 2: Studien zur ökonomische Theorie und zur Zukunft der Technik* (Frankfurt: Fischer, 1995); Meyer Abich and Bertram Schefold, *Grenzen Der Atomwirtschaft* (Munich: C. H. Beck, 1986).
59. Christian Leipert and Udo Ernst Simonis, "Alternativen wirtschaftlicher Entwicklung," in Simonis (ed.), *Ökonomie und Ökologie*, 103–158.
60. Quotation from Hans K. Schneider, "Implikationen der Theorie erschöpfbarer natürlicher Ressourcen für wirtschaftspolitischen Handeln," in Siebert (ed.), *erschöpfbare Ressourcen*, 815–844, here 821.
61. Hans K. Schneider, "Implikationen."
62. Quotation from Rüdiger Pethig, "Intertemporale Allokation mit erschöpfbaren Ressourcen und endogenen Innovationen," in Siebert (ed.), *erschöpfbare Ressourcen*, 277–295; Binswanger, *Arbeit ohne Umweltzerstörung*, 92; Florian Sauter-Servaes, "Die Übergang von einer erschöpfbaren Ressource zu einem synthetischen Substitut," in Siebert (ed.), *erschöpfbare Ressourcen*, 245–258.
63. Bundestag, *Der "Schnelle Brüter" in Kalkar: Beschluss des Bundestages zur Inbetriebnahme* (Bonn: Bundestag, 1983); Klaus Michael Meyer-Abich and Robert Dickler, "Energy Issues and Policies in the Federal Republic of Germany," *Annual Revue of Energy* 7 (1982), 221–59, here 237, 250–251; W. Marth and M. Koehler, "The German Fast Breeder Program (A Historical Overview)," *Energy* 23/7–8 (1998), 593–608.
64. Meyer-Abich worked with Dickler; Schefold and Schneider submitted sub-studies or conducted interviews with the inquiry into nuclear power. Bundestag, *Der "Schnelle Brüter"*; Bundesregierung, "Materialienband zum Bericht der Enquete-Kommission 'Zukünftige Kernenergie-Politik' über den Stand der Arbeit," Deutscher Bundestag, 9. wahlperiode, Drucksache 9/2439, March 24, 1980; Stefan Cihan Aykut, "Energy Futures from the Social Market Economy to the Energiewende: The Politicization of West German Energy Debates, 1950–1990," in Andersson and Rindzeviciute, *Forging the Future*, 63–91.
65. Hans Diefenbacher and Jeffrey Johnson, "Energy Forecasting in West Germany: Confrontation and Convergence," and Thomas Baumgartner and Atle Midttun, "Energy Forecasting and Political Structure: Some Comparative Notes," in Thomas Baumgartner and Atle Midttun (eds.), *The Politics of Energy Forecasting: A Comparative Study of Energy Forecasting in Western Europe and North America* (Oxford: Clarendon, 1987), 61–84 and 267–289; Jürgen Rehm and Wolfgang Servay, "Der Intuitive Kern von Energieprognosen," in Manfred Härter (ed.), *Energieprognostik auf dem Prüfstand* (Cologne: Verlag TÜV, 1988), 31–48; Meyer-Abich and Dickler, "Energy issues."
66. Quotation from Ralf-Dieter Brunowsky and Lutz Wicke, *Der Öko-Plan: Durch Umweltschutz zum neuen Wirtschaftswunder* (Munich: Piper: 1984), 14.
67. Frank Uekötter and Kenneth Anders, "The Sum of All German Fears: Forest Death, Environmental Activism, and the Media in 1980s Germany," in Frank Uekötter (ed.), *Exploring Apocalyptica: Coming to Terms with Environmental Alarmism* (Pittsburgh: University of Pittsburgh Press, 2018), 75–106, quotation from 91; Axel Goodbody, "Anxieties, Visions, and Realities: Environmentalism in Germany," in Axel Goodbody (ed.), *The Culture of German Environmentalism: Anxieties, Visions, Realities* (New York: Berghahn Books, 2002), 32–43.
68. Quotations from the introduction to Hermann Graf Hatzfeldt (ed.), *Stirbt der Wald? Energiepolitische Voraussetzungen und Konsequenzen* Karlsruhe: C. F. Müller, 1982), 9 and 12; Karl Friedrich Wentzel, "Waldsterben: eine Bestandsaufnahme," and Dieter Deumling, "Emissionen und Immissionen," in ibid., 15–22 and 95–123; Uekötter, *Greenest Nation*, 114–116.
69. Binswanger, *Arbeit ohne Umweltzerstörung*, 21–28; Brunowsky and Wicke, *Eco-Plan*, 14–18; Uekötter, *Greenest Nation*, 114.
70. Ulrich Herbert, *A History of 20th-Century Germany*, translated by Ben Fowkes (New York: Oxford University Press, 2019), 809–812.
71. Hans K. Schneider, "Energiepolitische Alternativen fuer die 90er Jahren," *Wirtschaftsdienst* 67/3 (1987), 132–139; Lutz Wicke and Jochen Hucke, *Der Ökologische Marshallplan* (Frankfurt a.M.: Ullstein, 1989), 211–212; Meyer-Abich and Dickler, "Energy Issues."

72. Armin Frank and Franz-Josef Hinse, "Energieeinsparung durch Anwendung neuer Technologie," and Illo-Frank Primus, "Energieeinsparungspotentiale–Beispiele energiesparender Massnahmen und Technologien seit der Ölkrise," in Hauff (ed.), *Energieversorgung und Lebensqualität* (Villingen: Neckar-Verlag, 1978), 501–519 and 520–543; Thomas Bohn, "Die Entwicklung der Energietechnik als Folge der Ölkrise," and Rolf Bauerschmidt, "Energieeinsparung," in Fritz Lücke (ed.), *Ölkrise: 10 Jahre danach* (Bonn: Verlag TÜV Rheinland, 1983), 65–89 and 208–213; Frank Bösch, *1979*, 322.
73. Quotation from Gebhard Kirchgässner, "Wirtschaftswachstum, Ressourcenverbrauch und Energieknappheit," in Siebert (ed.), *Erschöpfbare Ressourcen*, 355–376, 371; K. Kriegsmann and A. Neu, "Substitutionsbeziehungen zwischen den Produktionsbeziehungen zwischen den Produktionsfaktoren Energie, Kapital und Arbeit in der Bundesrepublik Deutschland," *Zeitschrift für Energiewirtschaft* 1 (1981), 56–67; Horst Meixner, "Energieeinsparungspolitik und Marktwirtschaft: Überholte Leitbilder blockieren effiziente Strategien" in *Wirtschaftsdienst* IV (1981), 178–83; Friede, "Substitutionsansatz."
74. Rolf Peter Sieferle, *The Subterranean Forest: Energy Systems and the Industrial Revolution*, translated by Michael P. Osman (Cambridge: White Horse Press, 1982/2010); K. A. Körber, Hans Karl Schneider, and Guido Brunner (eds.), *Energiekrise—Europa im Belagerungszustand? Politische Konsequenzen aus einer eskalierenden Entwicklung* (Hamburg: Körber & Blanck, 1977); Rosenberg, *Black Box*, 3–33.
75. Hans K. Schneider, "Die Interdependenz zwischen Energieversorgung und Gesamtwirtschaft als wirtschaftspolitisches Problem," *volkswirtschaftliche Diskussionsreihe* at University of Augsburg, Nr. 15 (February 27, 1980), 1–19.
76. C. C. Weizsäcker, "Leistet der Markt"; Meixner, "Energieeinsparung"; Meixner, "Bewusstseinsänderung"; Schneider "erschöpfbarer natuerlicher Ressourcen"; Klaus Michael Meyer-Abich and Ulrich Steger, "Einleitung: Handlungsspielraume für eine bedarfsgerechte Energieversorgung . . . ," in Ulrich Steger and Klaus Michael Meyer-Abich, *Handlungsspielraeume der Energiepolitik: Mittel- und laengerfristige Perspektiven bedarfsorientierter Energiesysteme fuer die Bundesrepublik* (Villingen: Neckar-Verlag, 1980), 1–32; Udo Ernst Simonis, "Mehr Technik—Weniger Arbeit? Eine Einführung," in Udo Ernst Simonis (ed.), *Mehr Technik—Weniger Arbeit?* (Karlsruhe: C. F. Müller, 1984), 9–14.
77. Binswanger, *NAWU*, 135–137; Binswanger, *Arbeit ohne Umweltzerstörung*; Hauff and Binswanger, *Ökologische Herausforderung*.
78. Deutsche Shell AG, *Trendwende im Energiemarkt: Scenarien fuer den Bundesrepublik bis zum Jahr 2000*, Deutsche Shell Aktiengesllschaft, August 1979, Nr. 10; Krause et al., *Energie-Wende*; Meyer-Abich and Schefold, *Grenzen der Atomwirtschaft*; Baumgartner and Midttun, "Energy Forecasting."
79. Horst Siebert, "Erfolgsbedingungen einer Abgabenlösung in der Umweltpolitik," in Otmar Issing (ed.), *Ökonomische Probleme der Umweltschutzpolitik* (Berlin: Duncker & Humblot, 1976), 35–64; Horst Siebert, *Ökonomische Theorie der Umwelt* (Tübingen: JCB Mohr, 1978).
80. Siebert, *Umwelt*, 68.
81. Schefold, "Ecological Problems," 43.
82. Siebert, *Umwelt*, 68, 76–78; Siebert, "Abgabenlösung," 43.
83. Schefold, "Ecological Problems," 39.
84. Quotations from Lutz Wicke, *Umweltökonomie: Eine praxisorientierte Einführung* (Munich: Franz Vahlen, 1982), 14–16, 40–41; Mikael Skou Andersen, *Governance by Green Taxes: Making Pollution Prevention Pay* (Manchester: Manchester University Press, 1994), 24.
85. Siebert, "Umweltpolitik."
86. Wicke, *Umweltökonomie*, 371; Brunowsky and Wicke, *Eco-Plan*, 105–106.
87. The first attempt came from the United States. Thomas Neff, *The Social Costs of Solar Energy: A Study of Photovoltaic Energy Systems* (New York: Pergamon Press, 1981); Timur Ergen, *Grosse Hoffnungen und brüchige Koalitionen Industrie, Politik und die schwierige Durchsetzung der Photovoltaik* (Frankfurt a.M.: Campus-Verlag, 2015), 115, 134–145; BMFT, *Technologies Program, 1977–1980*, 74–82; Bernd Dietrich, "Neubewertung der Energieträger Kernenergie und Sonne," in Lücke (ed.), *Ölkrise*, 120–126; Duccio Basosi, "A Small Window: The Opportunities for Renewable Energies from Shock to Counter-Shock," in Basosi et al, *Oil Counter-Shock*, 336–356.

88. Lutz Wicke, *Die ökologische Milliarden: Das kostet die zerstörte Umwelt–so können wir sie retten* (Munich: Kösel, 1986); Udo E. Simonis, *Wir müssen anders wirtschaften: Ansatzpunkte einer ökologischen Umorientierung der Industriegesellschaft* (Frankfurt a.M.: Akademische Schriften, 1989).
89. Olav Hohmeyer, *Social Costs of Energy Consumption: External Effects of Electricity Generation in the Federal Republic of Germany* (Berlin: Springer, 1988); AGG, B.II.1, 6218 1/2, Olaf Hohmeyer, "Die sozialen Kosten der Elektrizitätsversorgungm I," *Eurosolar* 1 (1990), 6–10.
90. AGG, B.II.1, 6218 1/2, Voss, "Die sozialen Kosten der Elektrizitätsversorgung II," *Eurosolar* 1 (1990), 11–19.
91. Wicke, *Marshallplan*; Hohmeyer, *Social Costs*; H. C. Binswanger, Hans G. Nutzinger, Angelika Zahmt, "Umwelt-Steuern," *Bundargumente* (Bund für Umwelt und Naturschutz Deutschland) 1 (September 1990), 1–12; AGG, B.II.1 1421 2/2 Deutsche Institut für Wirtschaftsforschung, "Abschaetzung des Potentials erneuerbarer Energiequellen in der Bundesrepublik Deutschland" (1984); Dietrich "Kernenergie und Sonne" and Joachem Nitsch, "Die Mögliche Rolle der Solarenergie in der Bundesrepublic Deutschland," in Lücke (ed.), *Ölkrise*, 120–126, and 127–137; Klaus Müschen and Erika Rombert, *Strom ohne Atom: Ausstieg und Energiewende: Ein Report des Öko-Instituts Freiburg/Breisgau* (Frankfurt a.M.: Fischer, 1986), 134–135.
92. Binswanger, *NAWU*, 139.
93. Quotation from Binswanger, *Arbeit ohne Umweltzerstörung*, 194; Binswanger, *NAWU*, 142–143; Siebert, "Umweltpolitik"; Hansmeyer "Abwasserabgabe; Hans K. Schneider, "Marktwirtschaftliche Energiepolitik oder staatlicher Dirigismus?" in Hans K. Schneider (ed.), *Aufsätze aus drei Jahrzehnten zur Wirtschafts- und Energiepolitik* (Munich: Oldenbourg, 1990), 162–167; Wicke, *Umweltökonomie*, 79; Wicke, *Marshallplan*, 198; C. C. Weizsäcker, "Leistet der Markt," 813–814.
94. First quote from Brunowsky and Wicke, *Öko-Plan*, 224; second from Hans Christoph Binswanger and Alfred Jäger, "Ökonomie und Ökologie," in Forum Kirche und Gesellschaft (ed.), *Zwischen Wachstum und Lebensqualität* (Munich: Forum Kirche und Gesellschaft, 1980), 70–115, here 80; Simonis *Ökonomie und Ökologie* Udo E. Simonis, "Arbeit und Umwelt: Ansatzpunkte einer integrierten Beschäftigungs- und Umweltpolitik," in Eduard Kroker (ed.), *Arbeit, Umwelt, Arbeitslosigkeit* (Koenigstein: Koenigstein, 1988), 49–67.
95. Binswanger, *Arbeit ohne Umweltzerstörung*; Simonis, "Wir müssen anders"; Binswanger and Hauff, *Ökologische Herausforderung*; Schneider, "Energiepolitische Alternativen."
96. First quotation from Binswanger et al., "Umwelt-Steuern"; second and third from Hans C. Binswanger and Claus Wepler, "The Energy Tax as Instrument of a Sustainable Development" (1992) in (AEU), L.E.C.E., 0000106, Commission Environment; Kneese, *Economics and the Environment*, 121–134.
97. Wicke, *Eco-Plan*, 86; C.C. Weizsäcker, "Leistet der Markt."
98. Zeller, "ökologischen Modernisierung"; Arthur P. J. Mol and David A. Sonnenfeld, "Ecological Modernisation around the World: An Introduction," *Environmental Politics* 9/1 (2000), 3–14; Joseph Huber, *Die verlorene Unschuld der Ökologie. Neue Technologien und superindustriellen Entwicklung* (Frankfurt a.M.: Fischer, 1982); Joseph Huber, *Die Regenbogengesellschaft: Ökologie und Sozialpolitik* (Frankfurt a.M.: Fischer, 1985).
99. Quotation from Mikael Skou Andersen and Ilmo Massa, "Ecological Modernization—Origins, Dilemmas and Future Directions," *Journal of Environmental Policy & Planning* 2 (2000), 337–345, here 338; Mol and Sonnenfeld, "Ecological Modernisation," 5.
100. Exenberger, "Die Sozial Marktwirtschaft"; Watrin, "Müller-Armack; Kromka, "grüne Denken."
101. Quotation from Gerold Blümle, "Einige Gründe für die 'Lücke' zwischen Theorie und Praxis in der derzeitigen Umweltpolitik," in Lothar Wegehenkel (ed.), *Marktwirtschaft und Umwelt* (Tübingen: JCB Mohr, 1981), 21–26, here 24; Hans Georg Pohl, "Zukunftsfragen der Energie im Spannungsfeld von Markt und Politik," *Zeitschrift für Energiewirtschaft* 4 (1988), 285–289.
102. Fritz Holzwarth, "Marktwirtschaft und Umwelt: Anmerkungen zu dem mit diesem Titel von Lothar Wegehenkel herausgegebenen Sammelband," *Ordo* 34 (1983), 211–219.

NOTES 365

103. Quotation from Waldemar Pelz, "Ökosoziale Marktwirtschaft: Eine neu wirtschaftspolitische Konzeption?" *Ordo* 39 (1988), 295–300; Klaus W. Zimmermann, "Zur politiscshen Oekonomie von Ökosteuern," *Ordo* 47 (1996), 169–194.
104. Quotation from Holzwarth, "Marktwirtschaft"; Lothar Wegehenkel, "Marktwirtschaft und Umwelt: Eine Einleitung," Holger Bonus, "Emissionsrechte als Mittel der Privatisierung öffentlicher Ressourcen aus der Umwelt," and Alfred Schüller, "Gründe für eine Fortentwicklung der Emissionsrechtelösung," in Wegehenkel (ed.), *Marktwirtschaft und Umwelt*, 1–5, 54–77, and 78–86.
105. First quotations from Wicke, *Umweltökonomie*, 211–212; second from Simonis, "Arbeit und Umwelt"; Siebert, *Theorie der Umwelt*; Horst Siebert, *Analyse der Instrument der Umweltpolitik* (Göttingen: Schwartz Verlag, 1976), 119; Wicke, *Eco-Plan*, 92.
106. Wicke, *Eco-Plan*, 44, 157, 198–204, 227; Wicke, *Marshallplan*, 197.
107. Quotation from Peter Hennicke, "Least-Cost Planning als Methode zur Ermittlung und Umsetzung kosten-minimaler Energiedienstliestungen," in Peter Hennicke (ed.), *Den Wettbewerb im Energiesektor planen: Least-Cost Planning* (Berlin: Springer, 1991), 3–43, here 11–12; Peter Hennicke and Helmut Spitzley, "Least Cost Planning und Energiedienstleistungs unternehmen am Beispiel Bremen," in ibid., 193–212; AGG, B.II.1, 6218 1/2, Peter Hennicke, "Vom überzentralisierten Grossverbundsystem zur Kommunalisierung der Energiewirtschaft."
108. Hennicke, *Energiewende*, 15; Hennicke, "Least-Cost Planning," 35.
109. Quotation from AGG, B, II.1, 6218 1/2, Hennicke, "Vom überzentralisierten Grossverbundsystem zur Kommuanlisierung der Energiewirtschaft (II)," July 7, 1988; Hennicke, *Energiewende*, 15–29.

Chapter 8

1. Walter Wallmann (CDU), West Germany's Environmental Minister, after Chernobyl (1986), "Moralisch nicht zu verantworten," *Die Zeit*, September 19, 1986.
2. BSB, "Bedingungen und Folgen von Aufbaustrategien für eine solare Wasserstoffwirtschaft," DS 11/7993, September 24, 1990.
3. Adam Higginbotham, *Midnight in Chernobyl: The Untold Story of the World's Greatest Nuclear Disaster* (New York: Simon & Schuster, 2019), 169–172.
4. First quotation in Hockenos, *Fischer*, 210–211; second in Hülsberg, *German Greens*, 199; Joppke, "Nuclear Power Struggles; Herbert, *20th-Century*, 811–813.
5. Cited in Markovitz and Gorski, *German Left*, 212; Rita Meyhöfer and Hans-Hermann Hertle, *DGB: Ausstieg bis zum Jahr 3000? Eine gewerkschaftspolitische Bilanz* (Berlin: DGB, 1987), 3–4.
6. Beck, *Risk Society*, 21–23.
7. First quotation from Beck, *Risk Society*, 20; second from Uekötter, *Greenest Nation*, 113.
8. Udo Kords, "Tätigkeit und Handlungsempfehlungen der beiden Klima-Enquete-Kommissionen des Deutschen Bundestages (1987–1994), in H. G. Brauch (ed.), *Klimapolitik* (Berlin: Springer, 1996), 203–214; Jacobsson and Lauber, "Energy System Transformation."
9. Gerstle, *Neoliberal Order*; Slobodian, *Globalists*; Alisdair Robert, *The Logic of Discipline: Global Capitalism and the Architecture of Government* (New York: Oxford University Press, 2010); Rawi Abdelal, *Capital Rules: The Construction of Global Finance* (Cambridge, MA: Harvard University Press, 2007).
10. Winkler, *Long Road*, 369; Markovitz and Gorski, *German Left*, 169–171, 215–216; Herbert, *20th-Century*, 831; Hülsberg, *German Greens*, 180–187.
11. Peter Becker, *Aufstieg und Krise der deutschen Stromkonzerne: Zugleich ein Beitrag zur Entwicklung des Energierechts* (Bochum: Ponte Press, 2011), 42–43; Thomas P. Hughes, *Networks of Power: Electrification in Western Society, 1880–1930* (Baltimore, MD: Johns Hopkins Press, 1983), 423; Morris and Jungjohann, *Energy Democracy*, 59–60.
12. Quotation from AGG, B.II.1, 6178, 1/2, BAG Energie, "Energiewende in den Gemeinden," July 1987; ibid., "Energiewende in den Gemeinden—Ausstieg aus der Atomenergie...," January 31, 1988; BSB, "Rekommunalisierung und Demokratisierung der Energieversorgung," DS 10/5010, February 5, 1986; Becker, *Aufstieg und Krise*; Christoph H. Stefes, "Critical Junctures and the German *Energiewende*," in Hager and Stefes (eds.), *Energy Transition*, 63–89.

13. AGG, B.II.1, 6178, 1/2, BAG Energie, "Energiewende in den Gemeinden," July 1987; Winkler, *Long Road*, 372–375.
14. Quotations from AGG, B.II.1, 3987, "Die Grünen im Landtag (Stuttgart): "Thesen zum Entwurf des Energie-Dezentral-Gesetzes," October 14, 1983; BSB, "Rekommunalisierung und Demokratisierung der Energieversorgung," DS 10/5010, February 5, 1986.
15. AGG, B.II.1 6178 2/2, Hennicke, "Stadt und Energie: Eine kritische Bestandsaufnahme der Probleme kommunaler Energieversorgung," September 1985; AGG, B.II.1, 6178, 1/2, BAG Energie, "Energiewende in den Gemeinden," July 1987.
16. AGG, B.II.1 6178 2/2, Peter Hennicke "Eckpunkte, Kriterien und Leitsätze für ein 'Gesetz zur rationellen Energienutzung'"; Hennicke, "Least-Cost Planning."
17. AGG, B.II.1 6178 2/2, Hennicke, "Stadt und Energie"; BSB, "Rekommunalisierung und Demokratisierung der Energieversorgung," DS 10/5010, February 5, 1986; AGG, B.II.1 6178 2/2, Hennicke, "Stadt und Energie."
18. AGG, B.II.1, 6178, 1/2, BAG Energie, "Energiewende in den Gemeinden," July 1987.
19. AGG, B.II.1, 6178 2/2, "Energiewende in den Gemeinden—Ausstieg aus der Atomenergie . . . ," January 31, 1988; BSB, "Rekommunalisierung und Demokratisierung der Energieversorgung," DS 10/5010, February 5, 1986; AGG, B.II.1 3987, Die Grünen im Landtag (Stuttgart), "Thesen zum Entwurf des 'Energie-Dezentral-Gesetzes,'" October 14, 1983; ADS, Volker Hauff Nachlass, 73, "Sichere Energie ohne Atomkraft" (1988).
20. Quotations ACDP, 08, 003, 057/3, Hessischer Landtag, 11 Wahlperiod, 31 Sitzung, October 31, 1984; BSB, "Rekommunalisierung und Demokratisierung der Energieversorgung," DS 10/5010, February 5, 1986; AGG, B.II.1, 6178, Hennicke "Rationellen Energienutzung"; AGG, B.II.1, 6178, 1/2, BAG Energie, "Energiewende in den Gemeinden," July 1987.
21. AGG, B.II.1, 3987, "Gesetz zur Förderung dezentraliserter Energiewirtschaft," in *Staatsanzeiger . . .*, November 19, 1983.
22. Markovitz and Gorski, *German Left*, 203–206; Hockenos, *Fischer*, 199–210.
23. ACDP, 08–003, 076/1, Gesetzentwurf der Fraktionen der SPD und der Grünen in Hessischer Landtag, October 2, 1984, and Stellungnahme der VDEW, October 25, 1984; Hockenos, *Fischer*, 202–203.
24. Quotation from von Büllesheim, BSB, Plenarprotokoll 10/222, June 19, 1986, 17227; BSB, "Beschlussempfehlung und Bericht des Ausschusses für Wirtschaft," DS 10/6677; BSB, Antrag der Grünen, "Rekommunalisierung und Demokratisierung der Energieversorgung," DS 10/5010, February 5, 1986.
25. ACDP, 08–003, 057/3, Hessischer Landtag, 11. Wahlperiode, 31. Sitzung, October 31, 1984; Walter, *Die SPD*, 231–237.
26. ADS Hauff NL, 000072, Hauff to Brandt, June 20, 1986; ADS, Hauff NL, 000072, Kommission für Umweltpolitik, Protokoll der konstitutierenden Sitzung, October 4, 1984; Hauff und Binswanger, *Ökologische Herausforderung*; Volker Hauff, "Wir müssen uns radikal für die Umwelt entscheiden," *SPD Pressedienst* 40/162 (August 27, 1985), 5–6.
27. First quotation from Alfred Dregger, second from Heiner Geissler, in *CDU Bundesparteitag Mainz*, October 7–8, 1986; BSB, "Unterrichtung durch die Bundesregierung: Energiebericht der Bundesregierung," DS 10/6073, September 26, 1986; Illing, *Energiepolitik*, 169–171.
28. Trampert cited in Markovitz and Gorski, *German Left*, 212; Hockenos, *Fischer*, 210–215; Hülsburg, *German Greens*, 172–173, 191–201; Jürgen-Friedrich Hake et al., "The German Energiewende—History and Status Quo," *Energy* 92 (2015), 532–546.
29. Quotation from ADS, Hauff NL, 000073, "Sichere Energieversorgung ohne Atomkraft," (1988); Meyhöfer and Hertle, *Ausstieg*, 2–4; ADS, Hauff, NL, 000092, Vorlage des Zwischenberichts der Kommission "Sichere Energieversorgung ohne Atomkraft," August 11, 1986.
30. Volker Hauff, "Atomenergie überfordert die Menschen," *Die Zeit* 38 (1986), https://www.zeit.de/1986/38/atomenergie-ueberfordert-die-menschen; Volker Hauff, *Energiewende: Von der Empörung zur Reform* (Munich: Knaur, 1986); SPD, *Die Lehren aus Tschernobyl: von der Empörung zur Reform* (Bonn: Vorstand der SPD, 1986); ADS, Hauff NL, 000092, Jürgen Schreiber, "Umsteiger des Jahres," *Natur* 12 (1986).
31. Quotation from Hauff, *Energiewende*, 87.
32. Hauff, "Atomenergie."

33. Hauff, *Energiewende*, 102–103; SPD, *Ohne Atomkraft: Argumente der SPD zur Energiepolitik* (Bonn: Vorstand der SPD, 1986); ADS, Hauff NL, 000073, Hauff to PV-Kommission Energie und Umweltpolitik, June 1, 1988; Meyer-Abich and Schefold, *Atomwirtschaft*; Jarausch, *Out of Ashes*, 620–634.
34. BSB, "Rekommunalisierung und Demokratisierung . . . ," (1986); ADS, Hauff NL, "Hauff legt Ausstiegsplan der SPD vor," *Die Welt*, August 12, 1986; "SPD-Ausstiegsszenario unrealistisch," *Handelsblatt*, August 18, 1986; "IG Chemie: Hauff-Papier unrealistich und falsch," *Süddeutsche Zeitung*, August 16, 1986.
35. Meyer-Abich and Schefold, *Atomwirtschaft*; Bertram Schefold, "Die Bilanz bei einem Verzicht ist günstig" *Die Zeit* 40 (1986), https://www.zeit.de/1986/38/atomenergie-ueberfordert-die-menschen; Müschen and Rombert, *Energiewende*; ADS, Hauff NL, 000092, "Bonn hält an Kernenergie fest," *Stuttgarter Zeitung*, September 4, 1986.
36. Hauff, *Energiewende*; SPD, *Die Lehren aus Tschernobyl*; BSB, Plenarprotokoll 10/222, June 19, 1986.
37. BSB, "Energiebericht der Bundesregierung," 1986; BSB, "Antrag der Fraktion der SPD: Programm: Energieeinsparung und rationelle Energienutzung," DS 11/2242, May 4, 1988; Hanns W. Maull, "Energy Security: A European Perspective," *Energy Papers: International Energy Program* 6 (November 1984), 1–21.
38. Spencer Weart, *The Discovery of Global Warming* (Cambridge, MA: Harvard University Press, 2008), 114–118, 140–147; Bonneuil and Fressoz, *Shock of the Anthropocene*, 72–79.
39. Quotation from Deutsche Physikalische Gesellschaft E.V., "Warnung vor einer drohenden Klimakatastrophe" (1986), 1–13; Jacobsson and Lauber, "Energy System Transformation."
40. "Das Weltklima gerät aus den Fugen," *Der Spiegel* 33 (1986), https://www.spiegel.de/politik/das-weltklima-geraet-aus-den-fugen-a-fa7f2e33-0002-0001-0000-000013519133; Tod im Treibhaus," *Der Spiegel* 4 (1986), https://www.spiegel.de/wissenschaft/tod-im-treibhaus-a-bf06be9c-0002-0001-0000-000013517345; Kords, "Klima-Enquete."
41. World Commission on Environment and Development, *Our Common Future* (New York: Oxford University Press, 1987); Ergen, *Grosse Hoffnungen*, 175–180.
42. Quotation from BSB, "Dritter Bericht der Enqute-Kommission: Vorsorge zum Schutz der Erdatmosphäre zum Theme Schutz der Erde," DS 11/8030, May 24, 1990, 66; BSB, "Erster Zwischenbericht der Enquete Kommission," DS 11/3246, November 2, 1988; Monika Ganseforth, "Politische Umsetzung der Empfehlungen der beiden Klima-Enquete-Kommissionen (1987–1994)—eine Bewertung," in Brauch (ed.), *Klimapolitik*, 215–224; Elke Bruns, Dörte Ohlhorst, Bernd Wenzel, and Johann Köppel, *Erneuerbare Energien in Deutschland: Eine Biographie des Innovationsgeschehens* (Berlin: BM UNR, 2009), 83–92.
43. AGG, B.II.1, 6178 1/2, BAG, "Energiewende in den Gemeinden," January 31, 1988; Michael Huber, "Leadership and Unification: Climate Change Policies in Germany," in Ute Collier and Ragnar E. Löfstedt (eds.), *Cases in Climate Change Policy: Political Reality in the European Union* (London: Earthscan, 1997), 65–86; Willy Leonhardt, "Kommunale Starthilfe für Photovoltaik," in Hennicke (ed.), *Least-Cost Planning*, 283–295; Morris and Jungjohann, *Energy Democracy*, 74–78.
44. "The Umbau Program: The Transformation of Industrial Society" (1986), in Mayer and Ely (eds.), *German Greens*, 267–270; ADS, Hauff NL, 000073, "Ökologische Erneuerung der Industriegesellschaft," June 16, 1988; Volker Hauff, "Arbeitsprogramm zur ökologischen Modernisierung der Volkswirtschaft," *SPD Nachrichten* (July 1985).
45. AGG, B.II.1, 793 2/2, "Energiespar- und strukturgesetz," February 12, 1990; BSB, Antrag der SPD, "Forschungs- und Entwicklungsprogramm Solarenergie und Wasserstoff," DS 11/1175, November 12, 1987; BSB, Gesetzentwurf der Fraktion der SPD, "Entwurf eines Energiegesetzes," Drucksache 11/7322, June 1, 1990.
46. Quotation from ADS, Hauff NL, 000073, "Ökologische Erneuerung der Industriegesellschaft," May 31, 1988; ibid., "Leitantrag zur ökologischen Erneuerung der Volkswirtschaft," June 16, 1988, and Hauff to members of the "PV Kommission Energie und Umweltpolitik," June 1, 1988.
47. ACDP, 07-001, 3552, Bundesfachausschuss Energie und Umwelt. Arbeitskreis I: "Wirtschaftswachstum, Energiebedarf und ökologische Grenzen," August 25, 1977; CDU, "Grundsatzprogramm 'Freiheit, Solidarität, Gerechtigkeit,'" October 23–25, 1978; ACDP,

Riesenhuber NL, 27/2, "Die Energie-Jahre 1973–1980" (1982); Dörte Ohlhorst, *Windenergie in Deutschland: Konstellationen, Dynamik und Regulierungspotenziale im Innovationsprozess* (Wiesbaden: VS Verlag, 2009), 212–214

48. Rothschild, "(Environmental) Capital"; World Commission, *Common Future*, 16; Elke Seefried, "Rethinking Progress: On the Origin of the Modern Sustainability Discourse, 1970–2000," *Journal of Modern European History* 13/3 (2015), 377–400.
49. Quotation from CDU, *Politik auf der Grundlage des christlichen Menschenbildes*, June 13–15, 1988, 46.
50. Quotation from CDU, *Protokoll 37. Bundesparteitag, Bremen*, September 11–13, 1989, 156; Dietmar Czok, *Nutzen und Haushalten: Christlichen Demokraten für Landeskultur und Umweltschutz* (Bonn: CDU, 1988).
51. Christian Hübner, *History of Energy and Climate Energy Policy in Germany: Christian Democratic Union Perspectives 1958–2014* (Lima: KAS, 2014); Uekötter, *Greenest Nation*, 122.
52. Winkler, *Long Road*, 366–367, 372–374; Herbert, *20th-Century*, 797, 802–804; Jacobsson and Lauber, "Energy System Transformation"; B.II.1. 796 1/2, Report from Winfried Damm, May 10, 1988.
53. Quotations from CDU, CDU, *Politik auf der Grundlage des christlichen Menschenbildes*, June 13–15, 1988, 53–57, 63–66; Wicke and Hucke, *Marshallplan*.
54. ADS, Hauff NL, 000073, UPI, "Ökosteuern als marktwirtschaftliches Instrument," April 1988.
55. ADS, Hauff NL, 000073, Hauff to PV-Kommission Energie und Umweltpolitik, June 1, 1988; Jarausch, *Out of Ashes*, 619, 628–631; Kershaw, *Global Age*, 280–293, 352.
56. Quotation from Harald B. Schäfer, "Steuerung durch Steuern: Zum Erfordernis der ökologischen Modernisierung der Volkswirtschaft," *SPD Pressedienst* 44/71 (1989), 5–6; ADS, Hauff NL, 000073, Hauff to PV-Kommission Energie und Umweltpolitik, June 1, 1988; AGG, B.II.1, 6219 1/2, "Fortschritt 90 Regierungsprogramm," *Frankfurter Rundschau*, March 5, 1990; SPD, "Der Neue Weg: Ökologisch, Sozial und Wirtschaftlich star," September 28, 1990.
57. Quotations from CDU, "Protokoll 37. Bundesparteitag in Bremen," September 11–13, 1989, 162; BSB, "Ankündigung von Mitgliedern der Bundesregierung zur Einführung einer CO2-Abgabe/Klimaschutzsteuer," DS 11/8479, November 23, 1990; AGG, B.II.1 6219 1/2, "Der Frist Läuft," *Wirtschaftswoche* Nr. 27, July 5, 1990; Ganseforth, "Klima-Enquete"; Huber, "Climate Change."
58. Quotations from AGG, B.II.1, 6219 1/2, CDU Wirtschaftsrat, "Kursbestimmung—Energiepolitik für die 90er Jahre," December 1990; CDU Plenum in Wiesbaden: *Politik auf der Grundlage des christlichen Menschenbildes*, June 6, 13–15; Huber, "Climate Change."
59. Quotation from AGG, B.II.1, 6219 1/2, "Einigung im Streit um Kohlendioxid," *Süddeutsche Zeitung*, 130, June 1990; Paul H. Suding, "Zehn Jahre Energiesparpolitik in der BRD," *Zeitschrift für Energiepolitik* 3 (1988), 191–203; BSB, "Energiebericht der Bundesregierung"; ACDP, 08-003, 057/3, May 17, 1984, "Stellungnahme der Vereinigung Imdustrielle Kraftwirtschaft . . ."; BSB, SPD Antrag, "Program: Energieeinsparung und rationelle Energienutzung," May 4, 1988; Huber, "Climate Change."
60. Mary Sarotte, *The Collapse: The Accidental Opening of the Berlin Wall* (New York: Basic Books, 2014); Herbert, *20th-Century*, 887–906.
61. Wolfgang Schäuble cited in Becker, *Aufstieg und Krise*, 63–69.
62. Quotation from AGG, B.II.1, 6219, 1/2, "Die Frist Läuft"; BSB, "Dritter Bericht der Enquete-Kommission: Vorsorge zum Schutz der Erdatmosphäre zum Theme Schutz der Erde," DS 11/8030, May 24, 1990, 30–32; AGG, B.II.1, 6219 1/2, "Einigung im Streit um Kohlendioxid," June 1990; Ganseforth, "Klima-Enquete"; Huber, "Climate Change"; Franz Josef Schafhausen, "Klimavorsorgepolitik der Bundesregierung," in Brauch, *Klimapolitik*, 237–249.
63. Walter, *SPD*, 230–240; Herbert, *20th-Century*, 904–905.
64. Antje Vollmer cited in Herbert, *20th-Century*, 905; Markovitz and Gorski, *German Left*, 231–234; Hockenos, *Fischer*, 220–222.
65. Quotation from Roberts, *Discipline*, 5; Gerstle, *Neoliberal Order*; Slobodian, *Globalists*.
66. Commission of the European Communities, "Completing the Internal Market: White Paper," June 14, 1985, 8; Laurent Warlouzet, *Governing Europe in a Globalizing World: Neoliberalism and its Alternatives Following the 1973 Oil Crisis* (New York: Routledge, 2018).

67. Kohl cited in Herbert, *20th-Century*, 786; Bösch, *1979*, 288–292; Otto Graf Lambsdorff, "Konzept für eine Politik zur Überwindung der Wachstumsschwäche und zur Bekämpfung der Arbeitslosigkeit," September 9, 1982; Winkler, *Long Road*, 359–360.
68. CDU, *Stuttgarter Leitsätze für die 80er Jahre*, May 9–11, 1984, 2–4; Frank Bösch, Thomas Hertfeldter, and Gabriele Metzler, "Einführung," and Thomas Handschumacher, "Eine 'neoliberale' Verheissung: Das politische project der 'Entstaatlichung' in der Bundesrepublik der 1970er und der 1980er Jahre," in Bösch et al., *Neoliberalismus*, 13–29, 149–178.
69. Quotation from CDU, *Stuttgarter Leitsätze für die 80er Jahre*, May 9–11, 1984; Heinrich Geissler in *Tagesprotokoll, 32. Bundesparteitag in Stuttgart*, May 10, 1984, 124; BSB, "Energiebericht der Bundesregierung" (1986).
70. Quotation from CDU, *Protokoll 37. Bundesparteitag*, September 11–13, 1989, Bremen, 161; CDU, *Politik auf der Grundlage des christlichen Menschenbildes*, June 13–15, 1988, 53–55.
71. Quotation from BSB, Antwort der Bundesregierung: "Förderung und Nutzung 'Erneuerbarer Energiequellen'" DS 11/2684, July 20, 1988; BSB, "Energiebericht der Bundesregierung," 1986; Stefes, *"Energiewende"*; Ohlhorst, *Windenergie*, 112–114; Jacobsson and Lauber, "Energy System Transformation."
72. First and third quotations from CDU, *Stuttgärter Leitsätze für die 80er Jahre*, May 9–11, 1984, 5; second from ACDP, 01–560 Riesenhuber, 027/2, "Zur Energiepolitik der CDU/CSU," March 22, 1982; Franz-Josef Strauss's comments Bundesparteitag in Weisbaden, June 13–15, 1988, 68.
73. Quotation from Riesenhuber cited in "Sonne Statt Kernkraft"; also Kohl's preamble to CDU, *Politik auf der Grundlage des christlichen Menschenbildes*, June 13–15, 1988; CDU, *Stuttgarter Leitsätze für die 80er Jahre*, May 9–11, 1984, 5–7; Franz-Josef Strauss in CDU Bundesparteitag Mainz, October 7–8, 1986, 83.
74. Ergen, *Grosse Hoffnungen*, 134–135; Jacobsson and Lauber, "Energy System Transformation"; BSB, SPD: "Forschungs- und Entwicklungsprogramm Solarenergie und Wasserstoff," DS 11/1175, November 12, 1987; BSB, "Unterrichtung durch die Bundesregierung: Faktenbericht 1990 zum Bundesbericht Forschung 1988," DS 11/6886, April 5, 1990; BSB, Plenarprotokill 11/216, June 20, 1990, 16909–17004; Illing, *Energiepolitik*, 177.
75. BSB, Plenarprotokoll 11/128, Discussion of Renewable Energy, February 23, 1989; Ohlhorst, *Windenergie*, 112–114; Bruns et al., *Erneuerbare Energien*, 331–344; Ergen, *Grosse Hoffnungen*, 180–182; BSB, "Rekommunalisierung und Demokratisierung der Energieversorgung," DS 11/6484, February 14, 1990.
76. Odell, *World Power*, 249–252; Parra, *Oil Politics*, 247–249; Garavini, 301–360.
77. Guiliano Garavini, "Thatcher's North Sea: The Return of Cheap Oil and the 'Neo-liberalisation' of European Energy," *Contemporary European History* (December 2022), doi.org/10.1017/S0960777322000686, 1–16; Tyler Priest, "Shifting Sands: The 1973 Oil Shock and the Expansion of Non-OPEC Supply," in Bini et al. (eds.), *Oil Shock*, 236–251; Peter R. Odell, "Prospects for West European Energy Markets," *International Energy Program: Energy Papers* No. 14 (March 1987); Parra, *Oil Politics*, 249, 258–261; Bradford, *Energy System*, 723–742; Odell, *World Power*, 260–263.
78. Duccio Basosi, Giuliano Garavini, and Massimiliano Trentin, *Counter-Shock: The Oil Counter-Revolution of the 1980s* (London: I. B. Taurus, 2018); Parra, *Oil Politics*, 257–259, 276–291; Odell, *World Power*, 248–261; Garavini, "North Sea"; Yergin, *The Prize*, 236–245; Garavini, *OPEC*, 352–359; Jacobs, *Panic*, 288–289.
79. B. A. Rahmer, "West Germany: Difficult Readjustments," *Petroleum Economist*, April 1982, 151–152; Bernhard Hnat, *Strukturwandel in der mineralölverarbeitenden Industrie der Bundesrepublik Deutschland* (Göttingen: Vandenhoeck & Ruprecht, 1991).
80. BSB, "Energiebericht der Bundesregierung," 1986; ACDP, 08–003, 057/3, BDI to Energieausschuss, "Aktuelle Entwicklung und Ausblick," May 8, 1985; Hnat, *Strukturwandel*; Baum, "Continuing Troubles"; Parra, *Oil Politics*, 276–280; Illing, *Energiepolitiik*, 171–174.
81. Garavini, "North Sea"; BSB, "Energiebericht der Bundesregierung," 1986; Ingo Hensing and Wolfgang Ströble, "Der Ölmarkt zu Beginn der neunziger Jahre," *Zeitschrift für Energiewirtschaft* 4 (1991), 225–231; Paul Horsnell and Robert Mabro, *Oil Markets and Prices: The Brent Market and the Formation of World Oil Prices* (Oxford: Oxford University Press, 2000); Jacobs, *Panic*, 283; Yergin, *The Prize*, 702–709; Giovanni Favero and Agnelo Faloppa, "Price Regimes,

Price Series and Price Trends: Oil Shocks and Counter-Shocks in Historical Perspective," David E. Spiro, "The Role of the Dollar and the Justificatory Discourse of Neoliberalism," and Catherine R. Schenk, "The Oil Market and Global Finance in the 1980s," in Basosi, Garavini, and Trentin, *Counter-Shock*, 15–34, 35–54, 55–75.

82. BSB, "Energiebericht der Bundesregierung," 1986; Odell, "Energy Markets"; Ulf Lantzke, "The Role of Emergency Oil Stocks: A European Perspective," *International Energy Program: Energy Papers* 7 (July 1985), 1–22; Illing, *Energiepolitik*, 171–174; BSB, "Rekommunalisierung und Demokratisierung" (1986).

83. Quotation from AGG, B.II.1, 6178 1/2, "Energiewende in den Gemeinden," January 31, 1988; BSB, "Ersatz des Kohlepfennigs durch eine Primärenergie- und Atomstromsteuer" (1988).

84. Bruns et al., *Erneurbare Energie*, 215–226; Hans Halter, "Das Undenkbare denken! Sonne und Wasserstoff, das neue Zeitalter der Energieerzeugung," *Der Spiegel*, August 17, 1987, https://www.spiegel.de/politik/das-undenkbare-denken-a-7f2df12a-0002-0001-0000-000013525 271; Morris and Jungjohann, *Energy Democracy*, 148; Hermann Scheer, *A Solar Manifesto* (London: James & James, 2001/1993), 44.

85. Morris and Jungjohann, *Energy Democracy*, 37; Bruns et al., *Erneuerbare Energien*, 331–339; Hake et al., "Energiewende"; Müschen and Rombert, *Energiewende*, 131–134.

86. First quotation in Morris and Jungjohann, *Energy Democracy*, 37; BSB, "Rekommunalisierung und Demokratisierung" (1986); BSB, Bericht der Enquete Kommission: "Gestaltung der technischen Entwicklung . . . ," DS 11/7993, September 24, 1990; AGG, B.II.1 1421 2/2, DIW: "Abschätzung des Potentials erneuerbarer Energiequellen in der BRD" (1984); BSB, Antwort der Bundesregierung: "Förderung und Nutzung 'Erneuerbarer Energiequellen," DS 11/2684, July 20, 1988.

87. Quotation from AGG, B.II.1, 1421/1, "Die Sonne braucht einen Anwalt," *Energiewende Magazine* III (1989); Hager, "Grassroots Origins"; Ergen, *Grosse Hoffnungen*, 134–140; Morris and Junjoghann, *Energy Democracy*, 148.

88. Morris and Jungjohann, *Energy Democracy*, 55–62, quotation from 61; Ohlhorst, *Windenergie*, 112–114; Robert W. Righter, *Windfall: Wind Energy in America Today* (Norman: University of Oklahoma Press, 2011); Bruns et al., *Erneuerbare Energien*, 334–343; AGG, B.II.1, 796 1/2, Fördergesellcahft Windenergie e.V. "Perspektiven der Windenergienutzung in der Bundesrepublik Deutschland," July 1988.

89. Figures from Scheer, *Solar Manifesto*, 98–100; *Der Spiegel*, "Sonne statt Kernkraft"; Bradford, *Solar Revolution*, 108–110.

90. Quotation from *Der Spiegel*, "Sonne Statt Kernkraft," https://www.spiegel.de/politik/sonne-statt-kernkraft-a-7b6d6358-0002-0001-0000-000013517822; AGG, B.II.1, 1421 2/2, DIW: "Abschätzung des Potentials erneuerbarer Energiequellen in der BRD" (1984); BSB, "Bericht der Enquete Kommission 'Gestaltung der technischen Entwicklung; . . . ," DS 11/7993, September 24, 1990; Bruns et al., *Erneuerbarer Energien*, 226–231; BSB, "Förderung und Nutzung 'Erneuerbarer Energiequellen" ' (1988); John Perlin, *From Space to Earth: Solar Electricity* (Ann Arbor, MI: Aatec, 1999).

91. Quotation from H. Hauser, "Solarstromvergütung zum Selbstkostenpreis: Memorandum zum Aufbau einer nennenswerten Solarstromerzeugung," February 15, 1989, https://www.sfv.de/lokal/mails/wvf/kostendeckende_Verguetung_bis_hin_zum_EEG_2004; *Der Spiegel*, "Sonne statt Kernkraft."

92. Quotation from AGG, B.II.1, 796 1/2, Fördergesellschaft Windenergie e.V. "Perspektiven der Windenergienutzung in der Bundesrepublik Deutschland," July 1988; BSB, "Gestaltung der technischen Entwicklung; Technikfolgen-Abschätzung und -Bewertung" (1990); Jacobsson and Lauber, "Energy System Transformation."

93. Hans Halter, "Das Undenkbare denken," https://www.spiegel.de/politik/das-undenkbare-denken-a-7f2df12a-0002-0001-0000-000013525271.

94. BSB, "Dritter Bericht der Enquete-Kommission" (1990); BSB, "Gestaltung der techischen Entwicklung; Technikfolgen-Abschätzung und -Bewertung" (1990).

95. Cited in Morris and Jungjohann, *Energy Democracy*, 61–62; Peter Salje, *EEG 2014. Gesetz für den Aufbau erneuerbarer Energien* (Cologne: Carl Heymanns Verlag, 2015), 144.

96. Volkmar Lauber and Lutz Mez, "Three Decades of Renewable Electricity Policies in Germany," *Energy and Environment* 15/4 (2004), 599–623; Jacobsson and Lauber, "Energy System Transformation"; BSB, "Förderung und Nutzung 'Erneuerbarer Energiequellen'" (1988); Bruns et al., *Erneuerbare Energien*, 98, 231–233; Jan Oelker and Christian Hinsch, *Windgesichter: Aufbruch der Windenergie in Deutschland* (Dresden: Sonnenbuch, 2005), 220–225.
97. BSB, "Förderungs- und Markteinführungsmassnahmen für erneuerbare Energiequellen in industrialiserten Ländern," DS 11/2755, August 4, 1988; AGG, B.II.1, 6178 2/2, Hennicke, "Stadt und Energie"; BSB, "Rekommunalisierung und Demokratisierung der Energieversorgung" (1986); SPD, *Die Lehren aus Tschernobyl*; AGG, B.II.1, 5099, Öko-Institut, "Das Grüne Energiewende—Szenario 2010," 1988; H. Hauser, "Solarstromvergütung zum Selbstkostenpreis."
98. Quotation from BSB, "Förderung und Nutzung "Erneuerbare Energiequellen," DS 11/2029, March 18, 1988; Andreas Berchem, "Das unterschätzte Gesetz," *Die Zeit*, September 22, 2006, https://www.zeit.de/online/2006/39/EEG; Ergen, *Grosse Hoffnungen*, 180–185; Bruns et al., *Erneuerbare Energien*, 98–99; Stephen Milder, "A Struggle to Remake the Market: Feed-in Rates and Alternative Energy in 1980s West Germany," *Contemporary European History* 31/4 (2022), 593–609.
99. Quotation from F. J. Strauss, Bundesparteitag in Wiesbaden, June 13–15, 1988; BSB, Plenarprotokoll 11/128, Discussion of Renewables, February 23, 1989; Lauber and Mez, "Renewable Electricity"; BSB, Antwort der Bundesregierung: "Förderung kleiner Wasserkraftwerke," DS 11/5025, July 31, 1989.
100. Quotation from BSB, Plenarprotokoll, 11/128, Discussion of Renewables, February 23, 1989; BSB, Plenarprotokoll, 11/216, Discussion of Feed-In, June 20, 1990, 17092; BSB, "Entwurf eines Gesetzes über die Einspeisung von Strom aus erneuerbaren Energien in das öffentliche Netz (Stromeinspeisungsgesetz)," DS 11/7816, September 7, 1990.
101. AGG, B.II.1, 796 1/2, Report from Winfried Damm May 10, 1988, and "Regeneratives 1989" (1989); AGG, B.II.1, 6218 1/2, Wolfgang Daniels, "Interfraktioneller Erfolg bei den Stromeinspeisevergütungen," *Das Solarzeitalter* 3 (1990), 31; BSB, Plenarprotokoll 11/128, Discussion of Renewables, February 23, 1989; Berchem. "Das unterschätze Gesetz"; BSB, Plenarprotokoll, 11/216, Discussion of Feed-In, June 20, 1990, 17092.
102. BSB, Plenarprotokoll 11/224, Discussion of Feed-In Legislation, September 13, 1990.
103. BSB, "Forschungs- und Entwicklungsprogramm Solarenergie und Wasserstoff" (1987); BSB, "Beschlussempfehlung und Bericht: 'Forschungs- und Entwicklungsprogramm Solarenergie und Wasserstoff'" DS 11/6857, March 30, 1990.
104. BSB, Plenarprotokoll, 11/216, Discussion of Feed-In, June 20, 1990; AGG, B.II.1, 6218 1/2, Wolfgang Daniels, "Interfraktioneller Erfolg bei den Stromeinspeisevergütungen," *Das Solarzeitalter* 3 (1990), 31; Morris and Jungjohann, *Energy Democracy*, 123–128.
105. BSB, "Entwurf eines Gesetzes über die Einspeisung von Strom aus erneuerbaren" (1990).
106. AGG, B.II.1, 6218 1/2, Interessenverband Windkraft Binnenland EV to Bundestag, June 19, 1990; Stefes, "*Energiewende*"; Bruns et al., *Erneuerbare Energien*, 98; Ohlhorst, *Windenergie*, 112–114; BSB, Plenarprotokoll, 11/216, Discussion of Feed-In, June 20, 1990.
107. "Primärenergieverbrauch nach der Substitutionsmethode" AG Energiebilanzen e.V.
108. BSB, "Aufbaustrategien für eine solare Wasserstoffwirtschaft" (1990).
109. Quotation from Berchem, "Das unterschätze Gesetz"; BSB, "Entwurf eines Gesetzes über die Einspeisung von Strom" (1990); BSB, Plenarprotokoll 11/224, Feed-In Debate, September 13, 1990.

Chapter 9

1. Leonid Brezhnev, Premier of the Soviet Union. AAPBD, 1973, "Gespräch Brandt mit Breschnew," May 18, 1973, doc. 145.
2. Vladimir Putin, "Protocol of Speech to the German Bundestag," September 25, 2001, https://www.bundestag.de/parlament/geschichte/gastredner/putin/putin_wort-244966; Jeannette Prochnow, "Fossilized Memory: The German-Russian Energy Partnership and the Production of Energy-Political Knowledge," *Global Environment* 11 (2013),

94–129; Dietmar Bleidick, *Die Ruhrgas 1926 bis 2013: Aufstieg und Ende seines Marktführers* (Munich: De Gruyter, 2017), 393–396; Simon Pirani, *Changes in Putin's Russia: Power, Money, and People* (New York: Pluto Press, 2010).

3. Daniel Yergin, *The New Map: Energy, Climate, and the Clash of Nations* (New York: Penguin, 2021); Jeronim Perovic (ed.), *Cold War Energy: A Transnational History of Soviet Oil and Gas* (Cham: Palgrave Macmillan, 2017); Margarita Balcameda, *Russian Energy Chains: The Remaking of Technopolitics from Siberia to Ukraine to the European Union* (New York: Columbia University Press, 2021).

4. Thane Gustafson, *The Bridge: Natural Gas in a Redivided Europe* (Cambridge, MA: Harvard University Press, 2020); Per Högselius, *Red Gas: Russia and the Origins of European Energy Dependence* (New York: Palgrave Macmillan, 2013).

5. Jeronim Perovic, "The Soviet Union's Rise as an International Energy Power: A Short History," and Falk Flade, "Creating a Common Energy Space: The Building of the Druzhba Oil Pipeline," in Perovic (ed.), *Cold War Energy*, 1–43, 321–344; Oscar Sanchez-Sibony, *Red Globalization: The Political Economy of the Soviet Cold War from Stalin to Khruschchev* (New York: Cambridge University Press, 2014).

6. Cantoni, "NATO and the EEC," 134.

7. Antony Blinken, *Ally versus Ally: Europe, America, the Siberian Pipeline Crisis* (New York: Praeger, 1987), 72–82; Cantoni, "NATO"; Bruce Jentleson, *Pipeline Politics: The Complex Political-Economy of East-West Energy Trade* (Ithaca, NY: Cornell University Press, 1986), 87–91; Andreas Metz, "50 Years of Pipes for Gas," *Mittel und Osteuropa Jahrbuch* (2020), 1–18.

8. BAK, B136, 7702, Report from Oil Sector meeting, July 7, 1961, Washington Embassy to Chancellor, August 29, 1961, Prass, "Oil Imports from Eastern Bloc States," August, 31, 1961, "Coordination of Trade Policy among EEC," July 20, 1962, Kling, "Oil Imports from the Eastern Bloc," March 1, 1962; Cantoni, "NATO"; Jentelson, *Pipeline Politics*, 97, 114–116; Blinken, *Ally*, 78–81.

9. Quotation from Angela Stent, *From Embargo to Ostpolitik: The Political Economy of West-German Soviet Relations, 1955–1980* (Cambridge: Cambridge University Press, 1981), 104–109; Perovic, "Energy Power"; Jentleson, *Pipeline Politics*, 116.

10. Yergin, *The Quest*, 318; Gustafson, *The Bridge*, 19.

11. "Gas Club" quotation in Gustafson, *The Bridge*, 2–5, 11–19; BSB, "Energiebilanz des Bundesgebietes," DS 3665; Bleidick, *Ruhrgas*, introduction, 207–214, 239–264; Horn, *Energiepolitik*, 56–70; BAK, NS Schiller, 176, "Report on Energy Policy from BWM," November 12, 1964.

12. Gerber, *Energiekonsums*, 166–172; Bleidick, *Ruhrgas*, 261–269; Gustafson, *Bridge*, 72–79; Martin Quinlan, "Gas Markets Poised for Further Growth," *Petroleum Economist* 52/2 (February 1985), 45–48; Jonathan Stern, *European Gas Markets: Challenges and Opportunities in the 1990s* (Dartmouth: Royal Institute, 1990), 6–14.

13. Official quoted in Hogselius, *Red Gas*, 76; ACDP, 08-003 CDU/CSU Fraktion, Diebäcker to CDU/CSU, January 8, 1969; BAK, B136, 2506, Fischer-Menshausen to Chancellor's Office, October 26, 1960; Odell, *World Power*, 131.

14. Brandt cited in Gottfried Niedhardt, "Transformation through Communication and the Quest for Peaceful Change," *Journal of Cold War Studies* 18/3 (2016), 14–59; second quotation from AAPBD, 1967, Doc. 238, "Aufzeichnung des Referenten Bahr," June 28, 1967, 966–967; Timothy Garton Ash, *In Europe's Name: Germany and the Divided Continent* (New York: Random House, 1993), 56–70, 365–369; Carole Fink, *Cold War: An International History* (Boulder, CO: Westview, 2017), 149–156; William Glenn Gray, "Paradoxes of 'Ostpolitik': Revisiting the Moscow and Warsaw Treaties, 1970," *Central European History* 49/3–4 (2016), 409–440.

15. Thane Gustafson, *Crisis Amid Plenty: The Politics of Soviet Energy under Brezhnev and Gorbachev* (Princeton, NJ: Princeton University Press, 1991), 35; Högselius, *Red Gas*, 31–43; Michael De Groot, "The Soviet Union, CMEA, and the Energy Crisis of the 1970s," *Journal of Cold War Studies* 22/4 (2020), 4–30; Paul Josephson, *Conquest of the Russian Arctic* (Cambridge, MA: Harvard University Press, 2014).

16. Högselius, *Red Gas*, 45–63, 80–83; Metz, "50 Years"; Frank Bösch, "Energiewende nach Osten," *Die Zeit* 42, October 10, 2013, https://www.zeit.de/2013/42/1973-gas-pipeline-sowjetunion-gazprom; Gustafson, *Bridge*, 41–64.
17. Ash, *Europe's Name*, 55–57; Fink, *Cold War*, 149–151.
18. Högselius, *Red Gas*, 106–119; Dunja Krempin, "Rise of Siberia and the Soviet-West German Energy Relationship during the 1970s," in Perovic (ed.), *Cold War Energy*, 253–281; Metz, "50 Years"; Carter, "Osthandel," 30–38.
19. AAPBD, 1970, "Aufzeichung des Min. Dir. Herbst," Doc. 23, January 26, 1970; "Salto am Trapez," *Der Spiegel*, February 8, 1970, https://www.spiegel.de/politik/salto-am-trapez-a-23649de1-0002-0001-0000-000045202633; "Auf kleiner Flamme," *Der Spiegel*, October 10, 1969, https://www.spiegel.de/politik/auf-kleiner-flamme-a-409ba6ba-0002-0001-0000-000045562725; "'Mr. Deutsche Bank' ist Tot," *Manager Magazin*, May 25, 2004, https://www.manager-magazin.de/unternehmen/karriere/a-301360.html; Niedhardt, "Peaceful Change"; Blinken, *Ally*, 2, 27–28; Högselius, *Red Gas*, 106–122.
20. AAPBD, 1970, "Gespräch Bahr und Gromyko," doc. 33, February 3, 1970, "Aufzeichnung Boschafters Emmel," doc. 49, February 11, 1970; "Metz, "50 Years"; Bösch, "Energy Diplomacy"; Gustafson, *Bridge*, 68–85; Högeselius, *Red Gas*, 129.
21. Quotation from "Jetz geht es mit den dicken Hunden los," *Der Spiegel*, May 20, 1973, https://www.spiegel.de/politik/jetzt-geht-es-mit-den-dicken-hunden-los-a-ca910515-0002-0001-0000-000042001288; Gray, "Ostpolitik"; Herbert, *20th-Century*, 710–711; Stent, *Embargo*, 180–195; Niedhardt, "Peaceful Change."
22. Quotation in James Mark et al., *1989: A Global History of Eastern Europe* (New York: Cambridge University Press, 2019), 33; AAPBD, 1972, "Gespräch zwischen Brandt und Patolitschew," July 5, 1972, doc. 198; Metz, "50 Years"; Martin Malia, *The Soviet Tragedy: A History of Socialism in Russia, 1917–1991* (New York: Free Press, 1994), 351–366.
23. De Groot, "Energy Crisis"; Perovic, "Energy Power"; Blinken, *Ally*, 26; Gustafson, *Bridge*, 49–50; Bösch, "Energy Diplomacy"; Gustafson, *Crisis Amid Plenty*, 35, 145–146; Högselius, *Red Gas*, 31–43.
24. Quotations from AAPBD, 1973, "Gespräch Brandt mit Breschnew," May 18, 1973, doc. 145; Ash, *Europe's Name*, 251; Krempin, "Siberia."
25. Quotation from "Ich hoffe, wir bekommen mehr Erdgas und Öl," *Der Spiegel*, May 27, 1973, https://www.spiegel.de/politik/ich-hoffe-wir-bekommen-mehr-erdgas-und-oel-a-11956035-0002-0001-0000-000041986672; BAK, B136, 7668, "finanziellen Wirkungen des Energiekonzepts," July 4, 1973; BSB, "Die Energiepolitik des Bundesregierung (1973)"; Werner Lippert, *The Economic Diplomacy of Ostpolitik: The Origins of NATO's Energy Dilemma* (New York: Berghahn, 2011), 101–120.
26. Quotation from "Die Russen sind Da," *Die Zeit*, October 12, 1973, https://www.zeit.de/1973/41/die-russen-sind-da.
27. First quotation from ADS, WB Archive, A-3, 538, Press Conference on November 15, 1973; second from Graf, *Petro-Knowledge*, 260; PAAA, 113897, Parliamentary proceedings 7. Wahlperiode, 78 Sitzung, February 13, 1974, and Heinz Reintges, "Energiewirtschaft und Energiepolitik," June 11, 1975; Bösch, "Energy Diplomacy."
28. AAPBD, 1973, "Gespräch des Brandt mit Falin," March 23, 1973, doc. 87, and "Gespräch Scheel mit Gromyko," November 2, 1973, doc. 354; Hans Otto Eglau, "Energie Versorgung: Der Schah lockt mit Öl und Gas," *Die Zeit*, October 19, 1973, https://www.zeit.de/1973/43/der-schah-lockt-mit-oel-und-gas/seite-2; "Erdgas aus dem Iran," *Mineralölzeitschrift* (March 1974), 77–79; Hogselius, *Red Gas*, 173–177; Carter, "Osthandel," 79–82.
29. BSB, "Erste Fortschreibung des Energieprogramms der Bundesregierung," (1974); BAK, B102, 281077, Friderichs to Schmidt, March 10, 1974; Krempin, "Siberia."
30. First quotations from Klaus Liesen, "Erdgas: Seine gegenwärtige und zukunftige Bedeutung in Westeuropa," *Oel: Zeitschrift für die Mineralölwirtschaft* 15 (September 1977), 242–245; last quotations from Ruhrgas, *Natural Gas on the Road to the Next Century* (Essen: Ruhrgas AG, 1979), 3; Bleidick, *Ruhrgas*, 305; "Trauer zum Ex-Ruhrgas-Chef Liesen," *RP-Online*, March 31, 2017.

31. Jeffrey Segal and Frank Niering, "Special Report on World Natural Gas Pricing," *Petroleum Economist* (September 1980), 373-378; Thane Gustafson, *Soviet Negotiating Strategy: The East-West Gas Pipeline Deal, 1980-1984* (Santa Monica, CA: Rand, 1984); E. Salinger, "Zur Entwicklung der Erdgasversorgung," *Oel: Zeitschrift für die Mineralölwirtschaft* 18 (December 1980), 316-322; Erdöl und Erdgas im Westhandel der Sowjetunion," *DIW Wochenberichte* 41 (1974), 193-197; Gustafson, *Bridge*, 26-27.
32. David S. Painter, "From Linkage to Economic Warfare: Energy, Soviet-American Relations, and the End of the Cold War," in Perovic (ed), *Cold War Energy*, 283-318; De Groot "Energy Crisis"; Krempin, "Siberia."
33. Brezhnev quotation from AAPBD, 1978, "Gespräch zwischen Schmidt und Breschnew," May 4, 1978, doc. 135; "Breschnew-Besuch: Auf der Nadelspitze," *Der Spiegel*, May 7, 1987, https://www.spiegel.de/politik/breschnew-besuch-auf-der-nadelspitze-a-47937017-0002-0001-0000-000040617745; Stephan Kieninger, "Diplomacy Beyond Deterrence: Helmut Schmidt and the Economic Dimension of Ostpolitik," *Cold War History* 20/2 (2020), 179-196; Jentleson, *Pipeline Politics*, 140-141; Blinken, *Ally*, 28-29; Hogselius, *Red Gas*, 179.
34. First quotation from Margaret Thatcher Foundation (MTF), TNA, PREM 16/1655, Callaghan-Schmidt Meeting, April 24, 1978; second quotation, Schmidt in Kieninger, "Schmidt"; TNA, Cold War in Eastern Europe, FCO 28/3982, "Long-Term Programme for the Main Trends of Economic and Industrial Cooperation between the FRG and the USSR. May 1978," June 27, 1980; Ash, *In Europe's Name*, 90-95.
35. Hogselius, *Red Gas*, 181-187; Jentleson, *Pipeline Politics*, 164; Bösch, *1979*, 9-46.
36. First quotations from MTA, TNA, PREM 19/59 f39, Chancellor conversation with Schmidt, October 31, 1979; Last quotation from MTA, TNA, COI Transcript, "Joint Press Conference with West Germany Chancellor," May 11, 1979; Helmut Schmidt, *Perspectives on Politics*, edited by Wolfram F. Hanrieder (Boulder, CO: Westview Press, 1982), 165.
37. Segal and Niering, "Special Report"; "Ruhrgas Attacks Algeria Deals," *Financial Times*, October 5, 1982; Gustafson, *Gas Pipeline Deal*, 7; BSB, "Dritte Fortschreibung des Energieprogramms der Bundesregierung," November 5, 1981, DS 9/983; Jentleson, *Pipeline Politics*, 169-172; Hogselius, *Red Gas*, 170-173, 191.
38. AAPBD, 1980, "Gespräch des Schmidt mit Vertretern der Wirtschaft . . . ," January 3, 1980, doc. 32; TNA, Cold War in Eastern Europe, FCO 28/3982, Palmer, "FRG/USSR Economic Commission," June 3, 1980; Gustafson, *Pipeline Deal*, 28; Perovic, "Energy Power"; Blinken, *Ally*, 31-42.
39. Quotation from AAPBD, 1980, "Deutsche-Sowjetische Regierungsgespräch in Moskau," July 1, 1980, doc. 193; Axel Lebahn, "Yamal Gas Pipeline from the USSR to Western Europe in the East-West Conflict," *Aussenpolitik* 3/34 (1983), 257-276; Kieninger "Schmidt."
40. Josef Joffe, "Europe and America: The Politics of Resentment," *Foreign Affairs* 61/3 (1983), 569-590; AAPBD, 1981, Gespräch Genscher mit Frydenlund," May 21, 1981, doc. 145; AAPBD, 1982, "Aufzeichnung des Staatssekretärs Lautenschlager," July 5, 1982, doc. 200; Claudia Wörmann, *Osthandel als Problem der Atlantischen Allianz: Erfahrungen aus dem Erdgas-Röhren-Geschäft mit der UdSSR* (Bonn: DGfAP, 1986), 137; Gustafson, *Gas Pipeline Deal*, 17-22; William Glenn Gray, "Learning to Recycle: Petrodollars and the West, 1973-1975," in Bini et al. (eds.), *Oil Shock*, 172-198.
41. Carter quoted in quoted in Fink, *Cold War*, 195.
42. Richard Perle in Jentleson, *Pipeline Politics*, 172-173; Joffe, "Resentment."
43. *Der Spiegel* quoted in Blinken, *Ally*, 105; Lebahn, "Yamal"; Fink, *Cold War*, 198-207.
44. Schmidt cited in Jentleson, *Pipeline Politics*, 172; Högselius, *Red Gas*, 188-190.
45. AAPBD, 1982, "Pfeffer to AA," September 28, 1982, doc. 250; BSB, von Würzen of BWM on natural gas, DS 9/35, December 1, 1980; Wörmann, *Osthandel*, 121.
46. AAPBD, 1981, "Gespräch Schmidt mit Mitterand," July 13, 1981, doc. 201, "Gespräch Schmidt und Mitterand," July 20, 1981, doc. 212, "Hermes an das AA," May 22, 1981, doc. 149, "Gespräch Genscher mit Haig," March 9, 1981, doc. 62; Helmut Schmidt, *Menschen und Mächte* (Berlin: Siedler Verlag, 1987), 79; BSB, Dritte Fortschribung.
47. OECD, *Economic Survey 1981-1982: Germany* (Paris: OECD, 1982); Brenner, *Global Turbulence*, 304; Eichengreen, *European Economy*, 264.
48. Schmidt, *Mächte*, 379; BSB, "Dritte Fortschribung."

49. Quotation from AAPBD, 1981, "Deutsch-amerikanische Regierungsgespräch in Montebello," July 19, 1981, doc. 209; Josef Esser and Werner Väth, "Overcoming the Steel Crisis in the Federal Republic of Germany," in Vincent Meny and Yves Wright (eds.), *The Politics of Steel: Western Europe and the Steel Industry in the Crisis Years (1974-1984)* (Berlin: De Gruyter), 1986), 623–691; Josef Esser and Wolfgang Fach, "Crisis Management: Made in Germany: The Steel Industry," in Peter Katzenstein (ed.), *Industry and Politics in West Germany: Toward the Third Republic* (Ithaca, NY: Cornell University Press, 1989), 221–248; Jarausch, *Out of Ashes*, 620–624.
50. "Mannesmann Sharply Lower," *Financial Times*, March 23, 1980; Mannesmann Demag Expects Downturn," *Financial Times*, August 14, 1981; Horst Wessel, *Kontinuität im Wandel: 100 Jahre Mannesmann, 1890–1990* (Düsseldorf: Mannesmann-Archiv, 1990), 459–460; Gustafson, *Gas Pipeline Deal*, 25–26; Jentleson, *Pipeline Politics*, 183.
51. "AEG's Difficulties Put Coalition in Dilemma," *Financial Times*, July 8, 1982; *Financial Times*; "AEG Crisis Heightens German Bank Fears," *Financial Times*, July 9, 1982; Gustafson, *Gas Pipeline Deal*, 8, 25–26.
52. First quotation in Jentleson, *Pipeline Politics*, 187; "AEG Faces Tough Decision on US Embargo," *Financial Times*, August 6, 1982; Grüner responding to SPD on AEG inquiry, June 28, 1982, Drucksache 9/1821; Lebahn, "Yamal"; Lippert, *NATO's Energy Dilemma*, 169.
53. AAPBD, 1982, "Gespräch Genscher mit Bush," March 9, 1982, doc. 78; AAPBD, 1981, "Gespräch Genscher mit Frydenlund," May 21, 1981, doc. 145; Heinrich Bechtoldt, "Bonn after the Decision on the Polish Crisis," *Aussenpolitik: German Foreign Affairs Review* 33/2 (1982), 118–122; Gustafson, *Gas Pipeline Deal*, 17–22.
54. Jentleson, *Pipeline Politics*, 197; Blinken, *Ally*, 31–56.
55. Högselius, *Red Gas*, 197; Metz, "50 Years"; Painter, "Economic Warfare."
56. TNA, Cold War in Eastern Europe, FCO 28/4401, Lawrence Brady remarks on November, 12, 1981.
57. Ruhrgas AG, *Natural Gas Today and Tomorrow* (Essen: Ruhrgas AG, 1987); Stern, *European Gas Markets*, 5–8; Högselius, *Red Gas*, 198.
58. BSB, "Energiebericht der Bundesregierung," September 26, 1986, DS 10/6073; Bradford, *Energy Systems*, 957–958; Farrenkopf and Przigoda (eds.), *Glück auf*, table 9, p. 548; Quinlan, "Gas Markets"; Vladimir Baum, "Oil Industry's Continuing Troubles," *Petroleum Economist* 52 (April 1985), 125–126; Parra, *Oil Politics*, 243; Odell, *World Power*, 250–251.
59. Helmut Kohl, *Errinerrungen* (Munich: Droemer, 2004), 441; Illing, *Energiepolitik*, 161–162; CDU, *Politik auf der Grundlage des christlichen Meschenbildes*, June 13–15, 1988.
60. Quotation from Quinlan, "Gas Markets"; Stern, *European Gas Markets*, 14; Bleidick, *Ruhrgas*, 276–279.
61. First quotation from ACDP, 08–003 CDU/CSU Fraktion, 057/03, "Wirtschaftsrat Kommission Energiepolitik," February 7, 1985; second from AGG, B.II.1, 6219 1/2, "Koalitionsvereinbarung für die 12. Legislaturperiode," January 16, 1991; Bösch, Thomas, and Metzler, *Grenzen des Neoliberalismus*.
62. Quotation from G. Rexrodt, BSB, Energy Discussion, April 17, 1997, Plenarprotokoll 13/169; see comments from Wolfgang Schäuble on liberalization, March 30, 1995, Plenarprotokoll 13/31; AGG, B.II.1 621 "Energiepolitik für die 90er Jahre," December 1990.
63. First quotation from BSB, "Dritte Fortschreibung"; second quotation in BSB, DS 11/2760, August 5, 1988; BSB, "Dritter Bericht der Enquete-Kommission: Vorsorge zum Schutz der Erdatmosphäre...," May 24, 1990, DS 11/8030.
64. First two quotations from Ruhrgas, *Natural Gas Today*, 14; last quotation from Ruhrgas, *Next Century*, 10; BSB, "Erste Fortschreibung des Energieprogramms der Bundesregierung" (1974).
65. Gas lobbyist quoted in "Ein Vergehen an der Umwelt," *Der Spiegel*, July 10, 1988, https://www.spiegel.de/politik/ein-vergehen-an-der-umwelt-a-d97d007b-0002-0001-0000-00001 3528748.
66. First quotations from A. Haeberlin, "Natural Gas and the Climate Debate—An Interim Balance," *GWF Gas—Erdgas* 138/10 (1997), 616–620; bridge quotation from J. Christiansen, "Going in the Right Direction—Natural Gas as a Bridge to an Ideal Energy Supply," *Gas Journal* 48/3 (1997), 28–32; Alfred Voss, "Energie und Klima–Wirtschaftsverträglicher

Klimaschutz, die Quadratur des Kreises?" *Wechselwirkungen Jahrbuch* (1996), n.p.; Alfred Voss, Ulrich Faul, and Peter Schaumann, "Strategien zur Minderung energiebedingter Treibhausgasemissionen–Die Rolle der Kernenergie und des Erdgases," in H. G. Brauch (ed.), *Energiepolitik* (Berlin: Springer, 1997), 389–400; Ortwin Renn, D. Oesterwind, and A. Voss, "Sanfte Energieversorgung–eine neue Utopie?" *Energiewirtschaftliche Tagesfragen* 30/2 (1980), 111–117.

67. Quotation from BSB, "Umweltbericht 1988," May 20, 1998, DS 13/10735; BSB, "Beschluss der Bundesregierung zur Reduzierung der energiebedingten CO2-Emissionen...," February 12, 1992, DS 2081.
68. AGG, B.II.1, 6219 1/2, "Koalitionsvereinbarung für die 12. Legislaturperiode," January 16, 1991; Astrid Eckert, *West Germany and the Iron Curtain: Environment, Economy, and Culture in the Borderlands* (New York: Oxford University Press, 2019), 144–153.
69. Quotation from BSB, Plenarprotokoll 12/8, February 20, 1991; BSB, "Dritte Fortschreibung"; IEA, *Energy Policies of IEA Countries: Germany 2002 Review* (Paris: IEA, 2002); Gustafson, *Bridge*, 84–91.
70. Stern, *European Gas*; Helm, *British Energy Policy*, ch 8; Bradford, *Energy Systems*, 269–270; Fabio Domanico, "Concentration in the European Electricity Industry: The Internal Market as Solution?" *Energy Policy* 35 (2007), 5064–5076.
71. IEA, *Germany 2002*; Huber, "Climate Change Policies"; Ute Ballay, "Natural Gas Set to Boom in East Germany," *Petroleum Economist* 63/3 (1996); Ute Ballay, "Gas Makes Inroads into German Energy Market," *Petroleum Economist* 63/5 (1996).
72. David Knott, "Germany to Open Gas Distribution," *Oil and Gas Journal* 94 (December 30, 1996), 23; Högselius, *Red Gas*, 205–208.
73. Bush cited in Fritz Bartel and Nuno P. Monteiro, "Introduction," in Nuno P. Montiero and Fritz Bartel (eds.), *Before and after the Fall: World Politics and the End of the Cold War* (New York: Cambridge University Press, 2021), 1. Also in this volume, Mary Elise Sarotte, "The Historical Legacy of 1989: The Arc to Another Cold War?"; V. M. Zubock, *Collapse: The Fall of the Soviet Union* (New Haven, CT: Yale University Press, 2021).
74. Nadejda M. Victor and David G. Victor, "Bypassing Ukraine: Exporting Russian Gas to Poland and Germany," in David G. Victor, Amy M. Jaffe, and Mark H. Hayes (eds.), *Natural Gas and Geopolitics from 1970 to 2040* (New York: Cambridge University Press, 2009), 122–169; Jonathan P. Stern, *The Russian Natural Gas "Bubble": Consequences for European Gas Markets* (London: Royal Institute for European Affairs, 1995); Bonneuil and Fressoz, *Shock of the Anthropocene*, 103.
75. German law allowed TPA, but a company that owned a network could refuse to grant TPA if this would affect supply to its own customers. Quotations from Knott, "Gas Distribution"; Högselius, *Red Gas*, 205–208; Victor and Victor, "Bypassing Ukraine"; Stern, *Gas "Bubble."*
76. Primitive quotation from Gustafson, *Bridge*, 222; blood quotation from David Knott, "Vying for Germany's Gas Market," *Oil and Gas Journal* (September 14, 1992), 29; David Knott, "Truce Between German Gas Suppliers," *Oil and Gas Journal* 92 (February 14, 1994), 36.
77. Quotation from Knott, "Gas Distribution"; Gustafson, *Bridge*, 218–226; Victor and Victor, "Bypassing Ukraine"; IEA, *Germany 2002*.
78. Quotations from AEI, Commission of the European Communities, "Green Paper: For a European Energy Policy" (February 1995); AEI, Competitiveness Advisory Group, "Enhancing European Competitiveness" (December 1995); AEI, "Completing the Internal Market. White Paper" (June 1985); John Gillingham, *European Integration, 1950–2003: Superstate or New Market Economy* (New York: Cambridge University Press, 2003), 251; Yergin, *Quest*, 386.
79. Gustafson, *Bridge*, 224–229; IEA, *Germany 2002*; Knott, "Gas Distribution"; Desmond Dinan, *Ever Closer Union: An Introduction to European Integration*, 4th edition (London: Lynne Rienner, 2010), 468.
80. Gustafson, *Bridge*, 229–231; IEA, *Energy Policies of IEA Countries: Germany 2007 Review* (Paris: IEA, 2007).
81. Stern, *European Gas Markets*; IEA *Germany 2002*; Knott, "Gas Distribution"; Bleidick, *Ruhrgas*, 280; Victor and Victor, "Bypassing Ukraine"; AGG, B.II.1, 6178 1/2, "Energiewende in den Gemeinden," July 1987; Domanico, "Concentration."

82. Högselius, *Red Gas,* 171–176, 197–198; Sanchez-Sibony, *Red Globalization,* 191–192; Gregory Vilchek and Olga Bykova, "The Origin of Regional Ecological Problems within the Northern Tyumen Oblast, Russia," *Arctic and Alpine Research* 24/2 (1992), 99–107.
83. BSB, "Erster Bericht der Enquete-Kommission "Schutz der Erdatmosphäre," March 31, 1992, DS 12/2400; BSB, "Dritter Bericht der Enquete-Kommission: Vorsorge zum Schutz der Erdatmosphäre," May 24, 1990, DS 11/8030.
84. Quotation from Knott, "Vying"; Stern, *Gas "Bubble"*; Yergin, *Quest,* 332–337; IEA, *Germany 2002*; Victor and Victor, "Bypassing Ukraine."
85. Prochnow, "Fossilized Memory"; Gustafson, *Bridge,* 362–368; Ben Knight, "The History of Nord Stream," *Deutsche Welle,* July 23, 2021, https://www.dw.com/en/the-history-of-nord-stream/a-58618313; Dietmar Student, "Jenseits von Aden," *Manager Magazin* (June 24, 2004); "Nordstream: Pipeline Manufacturing of Large Steel Components," https://www.wermac.org/nordstream/nordstream_part3.html.

Chapter 10

1. Wolf von Fabeck, solar engineer, "Leitartikel," *Solarbrief* 4 (1998), https://www.sfv.de/solarbr/1998-4.
2. First quotation from United Nations, *United Nations Framework Convention on Climate Change* (1992), 9; Bush cited in Yergin, *The Quest,* 462–463; Topfer label in "Der legendäre 'Erdgipfel' von Rio 1992," *Hamburger Abendblatt,* June 13, 2012; John Vidal, "Rio+20," *The Guardian,* June 19, 2012.
3. Kohl cited in "Klimakonferenz: Ernst nehmen," *Der Spiegel,* April 9, 1995, https://www.spiegel.de/politik/ernst-nehmen-a-15fa6c8a-0002-0001-0000-000009181644.
4. Hustedt cited in Gerd Stadermann, "Die Mutter des EEG," *Sonnenenergie* 3 (2017) 48–51; Kate Connolly, "Hermann Scheer Obituary," *The Guardian,* October 18, 2010, https://www.theguardian.com/world/2010/oct/18/hermann-scheer-obituary; Hermann-Scheer-Stiftung and Eurosolar, *Im Gedenken an Hermann Scheer. Sonderdruck* (Berlin: Hermann-Scheer-Stiftung, 2012).
5. Quotations from Scheer, *Solar Manifesto,* 3, 5, 149, 157.
6. Scheer, *Manifesto,* 1; second quotation is Bert Bolin in Yergin, *Quest,* 491; Jonas Meckling, *Carbon Coalitions: Business, Climate Politics, and the Rise of Emissions Trading* (Cambridge, MA: MIT Press, 2011); Hannah Ritchie and Max Roser, "CO2 Emissions," *Our World in Data,* https://ourworldindata.org/co2-emissions, accessed January 11, 2023.
7. Kate Aronoff, *Overheated: How Capitalism Broke the Planet—And How We Fight Back* (New York: Verso, 2019); Michael Mann, *The New Climate War: The Fight To Take Back Our Planet* (New York: Public Affairs, 2021).
8. Quotation from William F. Grover and Joseph G. Peschek, *The Unsustainable Presidency: Clinton, Bush, Obama, and Beyond* (New York: Palgrave Macmillan, 2014), 38; Klein, *Capitalism vs. the Climate,* 19 and 69; Colin Crouch, *Post-Democracy* (Cambridge, UK: Polity, 2004); Joseph Stiglitz, *The Roaring Nineteens: A New History of the World's Most Prosperous Decade* (New York: W. W. Norton, 2003); Gerstle, *Neoliberal Order.*
9. Quotation from Herbert, *20th-Century Germany,* 930; Philipp Ther, *Europe since 1989: A History* translated by Charlotte Hughes-Kreutzmüller (Princeton, NJ: Princeton University Press, 2016), 264.
10. Quotation from Morris and Jungjohann, *Energy Democracy,* 351; Hager, "Grassroots Origin."
11. Quotations Ernst von Weizsäcker, Amory B. Lovins, and L. Hunter Lovins, *Factor Four: Doubling Wealth and Halving Resource Use* (London: Earthscan, 1998/2001), 143, 229, 215–222. German edition in 1995; Hans Günter Brauch, *Ernst Ulrich von Weizsäcker: A Pioneer on Environmental, Climate and Energy Policies* (Heidelberg: Springer, 2014).
12. Merkel cited in Bert Bolin, *A History of the Science and Politics of Climate Change: The Role of the Intergovernmental Panel on Climate Change* (New York: Cambridge University Press, 2007), 108; Christiane Beuermann and Jill Jäger, "Climate Change Politics in Germany: How Long Will the Double Dividend Last?" in Tim O'Riordan and Jill Jäger (eds.), *The Politics of Climate Change: A European Perspective* (New York: Routledge, 1996), 186–227.

13. Liesel Hartenstein, "Warum der Erdgipfel vln Rio folgenlos blieb—Wege für eine Überlebensstrategie," in H. G. Brauch (ed.), *Klimapolitik* (Berlin: Springer, 1996), 225–236; Yergin, *Quest*, 750–760; Herbert, *20th-Century Germany*, 944; Kundnani, *Paradoxes*, 46–47.
14. "Feldzug der Moralisten: Vom Umweltschutz zum Öko-Wahn," *Der Spiegel*, September 24, 1995; Olaf Stampf, "Der Weltuntergang fällt aus," *Der Spiegel*, December 14, 1997, https://www.spiegel.de/spiegel/print/index-1995-39.html; Directorate General for Energy, *European Opinion and Energy Matters 1997* (Brussels: European Commission, 1997); Mann, *Climate Wars*, 136; Uekoetter, *Greenest Nation*, 139–145.
15. Herbert, *20th-Century*, 943–947; Ther, *Europe since 1989*, 259–265.
16. Roman Herzog, "Renewal of Confidence," April 26, 1997. https://ghdi.ghi-dc.org/sub_document.cfm?document_id=3946; Ther, *Europe since 1989*, 259–265; Brenner, *Global Turbulence*, 229–236; Herbert, *20th-Century Germany*, 930–940; Nachtwey, *Hidden Crisis*.
17. AGG, B.II.3 406, "Einstieg ins Solarzeitalter," March 12, 1996; BSB, Plenarprotokoll 13/44, June 22, 1995; CDU Grundsatzprogramm-Kommission, "Grundsatzprogramm der Christlich Demokratischen Union Deutschlands—Entwurf," January 1993; Beuermann and Jäger, "Climate Change Politics"; Michael Huber, "Leadership and Unification: Climate Change Policies in Germany," in Ute Collier and Ragnar E. Löfstedt, *Cases in Climate Change Policy: Political Reality in the European Union* (London: Earthscan, 1997), 65–86.
18. CDU Grundsatzprogramm-Kommission, "Grundsatzprogramm" (1993), 30; AGG, B.II.I 6219, 1/2, CDU/CSU Koaitionsvereinbarung für die 12. Legislaturperiode, January 16, 1991; CDU, *Für Wachstum und Beschäftigung: Beschluss des Bundesvorstandes ...*, January 14, 1994; Huber; "Climate Change Policies."
19. BSB, "Das energiepolitische Gesamtkonzept der Bundesrepublik," December 11, 1991. DS 12/1799; CDU Grundsatzprogramm-Kommission, "Grundsatzprogramm" (1993); CDU, CSU, FDP, "Koalitionsvereinbarung für die 13. Legislaturperiode," (1994); AGG, B.II.1 6219 1/2, Wirtschaftsrat der CDU, "Kursbestimmung—Energiepolitik für die 90er Jahre," December, 1990.
20. Merkel cited in AGG, B.II.3 439 2/2, "Einfaches Instrument," *Der Spiegel* 22 (1997), 95; "Schäuble, passen Sie Auf," *Der Spiegel* 16 (1998), https://www.spiegel.de/politik/schaeuble-passen-sie-auf-a-b222db88-0002-0001-0000-000007860966?context=issue; Barbara Schmidt-Mattern und Stephan Detjen, "Angela Merkel und der Kampf gegen die Erderwärmung" *Deutschlandfunk*, September 13, 2019, https://www.deutschlandfunk.de/klimapolitik-angela-merkel-und-der-kampf-gegen-die-100.html; Ganseforth, "Klima-Enquete-Kommissionen," 215–224.
21. Quotation from Ursula Weidenfeld, *Die Kanzlerin: Porträt einer Epoche* (Berlin: Rowohlt, 2021), 243; Ralph Bollmann, *Angela Merkel: Die Kanzlerin und ihre Zeit* (Munich: C. H. Beck, 2021), 172–174; AGG, B.II.1, 6219 1/2, Wirtschaftsrat der CDU, "Energiepolitik für die 90er Jahre"; CDU, "Protokoll 37. Bundesparteitag, 11.–13. September 1989 in Bremen"; Eckert, *Iron Curtain*, 201–244; "Verwegener Plan," *Der Spiegel* 15 (1995); AGG, B.II.3 439 2/2, "Einfaches Instrument," *Der Spiegel* 22 (1997), https://www.spiegel.de/wirtschaft/ver wegener-plan-a-358b7874-0002-0001-0000-000009181974.
22. CDU, CSU, FDP, "Koalitionsvereinbarung fuer die 13. Legislaturperiode des Deutsches Bundestages," (1994); Vladimir Baum, "Chancellor Kohl and the New Power Generation," *Petroleum Economist* 62/9 (1995), 10–11; AG Energiebilanzen, "Auswertungstabellen zur Energiebilanz Deutschland, 1990 bis 2017" (2018).
23. Quotations from BSB, Plenarprotokoll 12/195, Braunkohle Discussion, December 1, 1993.
24. AGG, B.II.1, 6219 1/2, Wirtschaftsrat der CDU, "Energiepolitik für die 90er Jahre"; BSB, "Das energiepolitische Gesamtkonzept der Bundesrepublik," DS 12/1799, December 11, 1991; Bundesverband Braunkohle, *Braunkhole im Zeitraum 1985–2010* (Berlin: Debriv, 2010), 20–31; Vladimir Baum, "Praise for the Country's Energy Sector Is Hard-Earned," *Petroleum Economist* 66/1 (1999), 11.
25. Becker, *Aufstieg und Krise*; Andrew Mollet, "Gas and Electricity Reforms," *Petroleum Economist* 3/61 (1994), 39; BSB, "Entwurf eines Gesetztes zur Neuordnung der Energiewirtschaft," DS 13/5352, July 25, 1996; AGG, B.II.3, 695 1/2 "Zukunft des Stromeinspeisungsgesetzes und der Erneuerbaren Energien," October 15, 1996.

26. Rainer Eising, *Liberalisierung und Europäisierung: Die regulative Reform der Elektrizitätsversorgung in Grossbritannien, der Europäischen Gemeinschaft und der Bundesrepublik Deutschland* (Wiesbaden: Springer, 2000), 106, 268–290; BSB, Plenarprotokoll 13/169, Energy Liberalization Debates, April 17, 1997; Dieter Helm, "The Assessment: European Networks—Competition, Interconnection, and Regulation," *Oxford Review of Economic Policy* 17/3 (2001), 297–312; David Buchan, *Energy and Climate Change: Europe at the Cross Roads* (Oxford: Oxford University Press, 2009), 1–28.
27. Quotation from AGG, B.II.3 698, Merkel, "Ökologische Herausforderungen für Strominvestoren," *Handelsblatt* (1997); BSB, Plenarprotokoll 13/169, Energy Liberalization Debates, April 17, 1997; AGG, B.II.3 439 1/2, Cronauge, "Energierechtsreform und kommunale Selbstverwaltung," March 1995.
28. Cited in Vladimir Baum, "Market Participants Will Define the Rules," *Petroleum Economist* 67/4 (2000), 3; Mollet, "Gas and Electricity Reforms"; BSB, Plenarprotokoll 13/169, Energy Liberalization Debates, April 17, 1997; Morris and Jungjohann, *Energy Democracy*, 136–137; AGG, B.II.3 698, "Anpassung des Stromeinspeisungsgesetztes . . . ," n.d.; "Plugging into European Electricity," *Petroleum Economist* 67/3 (2000), 8; Helen Avati, "Market Opening Gathers Pace," *Petroleum Economist* 66/4 (1999), 31; Fereidoon P. Sioshani, "Opportunities and Perils of the Newly Liberalized European Electricity Markets," *Energy Policy* 29 (2001), 419–427.
29. AGG, Fell 135, Hustedt and Fell, "Für eine Novellierung des Stromeinspeisegesetztes...," August 23, 1999; AGG, B.II.4 713, Erneuerbare Energien—eine Erfolgsgeschichte," December 21, 1999.12.21; AGG B.II.3 699, Werner Kleinkauf, "Wirtschaftlichkeit von Windenergieanlagen" 1998; Hager, "Grassroots Origins."
30. AGG, B.II.3 439 2/2, Schmidt to Meinecke, January 16, 1996, "Conference Program for Sunny Times," June 29, 1996; Morris and Jungjohann, *Energy Democracy*, 128–130, 150–151; Staffan Jacobsson and Volkmar Lauber, "The Politics and Policy of Energy System Transformation—Explaining the German Diffusion of Renewable Energy Technology," *Energy Policy* 34 (2006), 256–276.
31. AGG, B.II.3 702 1/2, "Stromeinspeisungsgesetz erneut auf dem Prufstand," March 4, 1996; AGG, B.II.3 695 2/2, BDI to Friedhelm Ost, March 11, 1996 and "Die Selbstjustiz von Energieversorgungsunternehmen . . . ," June 26, 1995; AGG, B.II.3, 702 1/2, "Politische Lösung der Stromeinspeisungsproblematik gefordert," March 6, 1996; Huber, "Climate Change Policies."
32. Quotation from AGG, B.II.3 698, Bundesverband Windenergie to Schäuble and CDU, November 21, 1997; ibid., Arbeitsgemeinschaft Wasserkraftwerke BW to Hauser, November 24, 1997; AGG, Heyne 104 1/2, "Für eine Novellierung des Stromeinspeisegesetztes...," August 23, 1999; AGG, B.II.4 713, BMWT, "Bericht zur Härteklausels . . . ," December 9, 1999; AGG, B.II.3 695 1/2, "Zukunft des Stromeinspeisungsgesetzes und der Erneuerbaren Energien," October 15, 1996; AGG, B.II.3, 702 2/2, Harig of Preussen-Elektra AG, March 6, 1996; Huber, "Climate Change Policies"; Morris and Jungjohann, *Energy Democracy*, 153–158.
33. BSB, Plenarprotokoll 13/169, Liberalization Debates, April 17, 1997; "Gegenwind aus Bonn," *Der Spiegel* 37 (September 7, 1997), https://www.spiegel.de/politik/gegenwind-aus-bonn-a-c201a5f7-0002-0001-0000-000008777815; AGG, B.II.3 698, Max Straubinger to Bundesverband Windenergie, July 14, 1997; "Historisches zur kostendeckenden Vergütung..."; Jacobsson and Lauber, "Diffusion."
34. First quotation from AEI, EU Commission, "White Paper," 1995; second from AEI, EU Commission, "On the State of Liberalisation of the Energy Markets," May 4, 1999, AEI; "Van Miert verlangt Reformen . . . ," *Solarbrief* 4 (1998), https://www.sfv.de/sob98406; AGG, Heyne 104 1/2, "Für eine Novellierung des Stromeinspeisegesetzes . . . ," August 23, 1999; AGG, Heyne 104 1/2, Fell to Künast and Roth, March 16, 2001; Jacobsson and Lauber, "Diffusion."
35. AGG, B.II.4 713, Ursula Fuentes, "Erneurbaren Energien—Eine Erfolgsgeschichte," December 21, 1999; AGG, B.II.3 695, 1/2, Pressemitteilung Nr. 303/95, May 10, 1995; AGG, B.II.3, 454 1/2, "10-Punkte Program für Einsteig ins Solarzeitalter," BSB Drucksache 13/4481, April 26, 1996.

36. Morris and Jungjohann, *Energy Democracy*, 156–158; Jacobsson and Lauber, "Diffusion"; AGG, B.II.3 406, "8-Punkte Programm für den Einstieg ins Solarzeitalter," March 12, 1996.
37. AGG, Heyne 104 1/2, Eco Institute, "Strom ohne Grenzen" (1997); AGG, B.II.3 698, Bundesverband Windenergie to Hustedt, November 6, 1997; ibid., Bundesverband Windenergie, circular, July 31, 1997; AGG, B.II.4 713, Fuentes, "Erneuerbare Energien"; AGG, B.II.4 4144, "The New German Renewable Energy Law," n.d.; AGG, B.II.4 714, Herdan to Wissman, February 9, 2000.
38. Wicke and Hucke, *Ökologische Marshallplan*, 229–231; Leonhardt, "Kommunale Starthilfe."
39. Bürgerantrag in Aachen, December 2, 1991, https://www.sfv.de/lokal/mails/wvf/kostendeckende_Verguetung_bis_hin_zum_EEG_2004; "Historisches zur kostendeckenden Vergütung...," *Solarenergie Förderverein*; "Gewinnmaximierung durch Zentralisierung," *Solarbrief* 1 (1997), https://www.sfv.de/briefe/brief97_1/sob97131; "Feuer und Flamme," *Der Spiegel* (January 9, 1994), https://www.spiegel.de/wirtschaft/feuer-und-flamme-a-0b653d19-0002-0001-0000-000013683091?context=issue; Morris and Jungjohann, *Energy Democracy*, 73–94, 128–130.
40. Quotation from Edgar Wolfrum, *Die Geglückte Demokratie* (Stuttgart: Klett-Cotta, 2006), 474; "Nicht so Richtig," *Der Spiegel* (July 23, 1995), https://www.sfv.de/lokal/mails/wvf/kostendeckende_Verguetung_bis_hin_zum_EEG_2004; Herbert, *20th-Century Germany*, 981–992.
41. Quotation from Bahro, *Red to Green*, 120; Hubert Kleinert, "Politics and Progress: The German Green Party 1983–2003," in Frank Zelko and Caroline Brinkmann (eds.), *The Origins of Green Parties: Reflections on the First Three Decades* (Washington, DC: Heinrich Boll Foundation, 2006), 77–88; Jachnow, "German Greens."
42. Quotation from Hockenos, *Joschka Fischer*, 236–237, also 202–210 and 250–251; Roland Roth and Detlef Murphy, "From Competing Factions to the Rise of the Realos," in Mayer and Ely, *German Greens*, 49–72, here 57; Hülsburg, *German Greens*, 186–186; Markovitz and Gorski, *German Left*, 215–216.
43. Röpke cited in Foucault, *Biopolitics*, 157; Eucken cited in Razeen, *Classical Liberalism*, 112.
44. "Einigkeit und Grün und Freiheit—Ökolibertäre Grüne Gründungserklärung," http://oekolibertaere.xobor.de/t1f2-Gruendungserklaerung.html; "Eco-Libertarian's Founding Statement," in Mayer and Ely, *German Greens*, 240-243.
45. Joschka Fischer, *Das Ende der ökologischen Bescheidenheit: Von der Krise der Umweltpolitik zum Der Umbau der Industriegesellschaft* (Frankfurt a.M.: Eichborn, 1989), 59–61; Roth and Murphy, "Competing Factions," 59; Walter Boehlich, "Gedächtnistrübungen: Thomas Schmid," *TAZ-Archiv*, September 7, 1995, 10.
46. First quotation from AGG, B.II.3 406, "8-Punkte Programm für den Einstieg ins Solarzeitalter," March 12, 1996; second from ibid., "Solarzeitalter: eine konkrete machbare Vision...," 1996; AGG, Heyne 104 1/2, "Entwurf eines Gesetzes zur Förderung der Stromerzeugung aus erneuerbaren Energien," BSB DS 14/2341, December 13, 1999; Stadermann, "Mutter des EEG."
47. AGG, B.II.3, 695 1/2, Pressemitteilung 303/95, May 10, 1999; Stadermann, "Mutter des EEG."
48. Hockenos, *Fischer*, 252–253; Jachnow, "German Greens."
49. "Leitartikel," *Solarbrief* 1 (1998), 2; Huber, "Climate Change Policies"; Ganseforth, "Klima-Enquete-Kommissionen"; BSB, "Programm für Klimaschutz, Wirtschaftsmodernisierung...," DS 13/187, January 11, 1995; Andersen and Massa, "Ecological Modernization"; Ehmke, *Mittendrin*, 297; Edgar Wolfrum, *Rot-Grün an der Macht: Deutschland 1998–2005* (Munich: C. H. Beck, 2013), 214.
50. Walter, *Die SPD*, 271–275; Hockenos, *Fischer*, 278–280.
51. Tony Blair and Gerhard Schröder, "Europe: The Third Way/ Die Neue Mitte," Friedrich Ebert Stiftung Working Documents No. 2 (June 1998); Ther, *Europe since 1989*, 263–265.
52. AGG, Heyne, 104 2/2, Fell, "Rede zur Debatte über die Umwelt- und Energiepolitik," Oct. 5, 1999; AGG, B.II.3 454 1/2, "10-Punkte Programm für den Einstieg ins Solarzeitalter," BSB, DS, 13/4481.
53. First quotation from AGG, B.II.3, 454 1/2, Hustedt, "Mehr Jobs tanken:...," April 1998; second from AGG, B.II.3 454 ½, "Einstieg in eine ökoklogisch-sozial Steuerreform," and "10-Punkte Programm für den Einstieg ins Solarzeitalter," BSB, DS, 13/4481.
54. Quotation from AGG, B.II.3, 454 ½, Hustedt, "Mehr Jobs tanken:...," April 1998.

55. Hermann Scheer, *The Solar Economy: Renewable Energy for a Sustainable Global Future*, translated by Andrew Ketley (London: Earthscan, 2005/1999), xvi; Stadermann, "Mutter des EEG."
56. BSB, Plenarprotokoll 13/110, June 13, 1996, Stromeinspeisungsgesetz discussions.
57. AGG, B.II.3, 454 1/2, Hustedt, "Mehr Jobs tanken: . . . ," April 1998; AGG, B.II.3, 4554 1/2, "Mit der Energiesteuer Lohnnebenkosten Senken," BSB Drucksache 13/7750, May 16, 1997; SPD und Bündnis 90/Die Grünen, "Aufbruch und Erneuerung . . . ," October 20, 1998; Hermann Scheer, "Das deutsche 100.000-Dächer-Photovolkait-Programm," *Solarzeitalter* 4 (1998), 1–5.
58. Hockenos, *Fischer*, 252–262; Herbert, *20th-Century Germany*, 992–993.
59. Quotations from Scheer, *Solar Economy*, 3 and 49; Scheer, *Energy Imperative*, 22; AGG, B.II.3 695 2/2, "Die Selbstjustiz von Energieversorungsunternehmen . . . ," June 26, 1995.
60. Quotations from Scheer, *Energy Imperative*, 8, 10, 35, 114–116; and Scheer, *Solar Economy*, xiv; AGG, B.II.3 6952/2, "Die Selbstjustiz von Energieversorungsunternehmen...," June 26, 1995; AGG, B.II.3 695 1/2, BSB Plenarprotokoll 13/110, "Discussion of Stromeinspeisungsgesetzt," June 13, 1996.
61. AGG, B.II.3, 406, "Der Staat der 'Stromer,'" *Der Spiegel* 46 (1995), https://www.spiegel.de/wirtschaft/der-staat-der-stromer-a-d16944a4-0002-0001-0000-000009230206; Scheer, *Solar Economy*, 48; Becker, *Aufstieg und Krise*, 56–75; AGG, B.II.3 439 1/2, Least Cost Planning Conference, October 21, 1995; AGG, B.II.3 454 1/2, BSB, "Entwurf eines Gesetztes zur Neuordnung der Energiewirtschaft," DS 13/5352, July 25, 1996.
62. Quotation from AGG, B.II.3, 406, "Der Staat der 'Stromer,'" *Der Spiegel* (1995).
63. Quotation from B.II. 3, 406, "Eckpunktepapier: Ersatz des Energiewirtschaftsgesetzes . . . ," n.d.; BSB, Plenarprotokoll 13/208, "Discussion of Stromeinspeisungsgesetz," November 28, 1997; AGG, B.II.3, 454 1/2, BSB Drucksache, 13/5352, "Entwurf eines Gesetzes zur Neuordnung der Energiewirtschaft," July 25, 1996; BSB, B.II.3 454 1/2, Drucksache 13/4481, "10-Punkte Program für den Einstieg ins Solarzeitalter," April 26, 1996; "Der Staat der 'Stromer,'" *Der Spiegel* 46 (1995), 77; Radkau, *Aufstieg und Fall*, 349; Michael Lindemann, "Mannesmann und Veba in Telecoms Alliance," *Financial Times*, January 18, 1996.
64. Scheer, *Solar Economy*, 25, 44, 77–85, and 167–169; Scheer, *Energy Imperative*, 115.
65. AGG, B.III.3 406, "Eckpunktepapier: Ersatz des Energiewirtschaftsgesetzes durch ein Energiegesetz," n.d.
66. Quotation from AGG, B.II.3 695, 1/2, BSB Plenarprotokoll 13/110, Discussion of Stromeinspeisungsgesetz, June 13, 1996; Morris and Jungjohann, *Energy Democracy*, 165–168, 194–197; Toke and Lauber, "Neoliberalism and Environmental Policy."
67. Buchan, *Energy and Climate Change*, 20–35; Domanico, "Concentration"; Helm "European Networks."
68. Quotation from AGG, B.II.3 439 1/2, "Grüne sorgen für Stromstoss," November 14, 1996; AGG, B.II.3 699, Pressemitteilung Nr. 0257/98, April 29, 1998; AGG, B.II.3, 698, Pressemitteilung Nr. 544/96, Hustedt "EU-Energie-Binnenmarkt verhindert Einstieg ins Solarzeitalter," June 21, 1996; AGG, B.II.3 454 1/2, BSB, "Entwurf eines Gesetztes zur Neuordnung der Energiewirtscahft," DS 13/5352, July 25, 1996.
69. Baum, "Market Participants", 3; Helm, "European Networks"; Becker, *Aufstieg und Krise*, 110–146.
70. Colin J. Campbell and Jean H. Laherrère, "The End of Cheap Oil," *Scientific American* 278/3 (March 1998), 78–84; Hans Kronberger, *Blüt für Öl: Der Kampf um die Ressourcen* (Vienna: Uranus, 1998); SFV 98/3.
71. Quotation from AGG, Heyne, 104 2/2, "Beschluss des Parteirats von Bündnis 90/Die Grünen," October 9, 2000 and Pressemitteilung Nr. 62/99, "Regenerative Energieträger . . . ," February 11, 1999; AGG, Fell 16, C. J. Campbell, "A New Energy Crisis," September 12, 2000; AGG Heyne 106 1/2, Pressemitteilung Nr. 616/2000, "Weg vom Öl!" October 17, 2000, and MWT Aktuell, "Hohe Rohölpreise sichern Ölversorgung . . . ," October 2, 2000; H. J. Fell, "100% Erneuerbare Energien!" https://hans-josef-fell.de/ueber-mich/; Klaus Rüfer, "Energieautarkes Wohnen in Hammelburg," *Hier ist Bayern*, November 19, 2021, https://www.br.de/nachrichten/bayern/energieautarkes-wohnen-in-hammelburg-raps oel-sonne-und-co,SoHVXyy.

72. AGG, Heyne 106 1/2, Pressemitteilung Nr. 524/2000, "Das Ende des Erdölzeitalters ist in Sicht," September 6, 2000; AGG, Heyne 104 2/2, Pressemitteilung 534/2000, Reinhard Loske, "Energieeinsparung fördern...," September 10, 2000; Scheer, *Energy Economy*, 9, 109–112.
73. Hockenos, *Fischer*, 258, 281–282; Jachnow, "German Greens"; Herbert, *20th-Century Germany*, 1001–1007; Lauber and Mez, "Renewable Electricity."
74. Quotation from AGG, Heyne, 109 1/2, Trittin, "Die Energiewende vollenden...," March 18, 2000; ibid., "Kernenergie-Ausstieg im Konsens mit der Wirtschaft," *HIB*, February 15, 2000.
75. Quotation from AGG, Heyne 109 2/2, Trittin, "Neue Energie ist Grün...," August 2000; "Der verpatzte Ausstieg," *Der Spiegel* 4 (1999), https://www.spiegel.de/politik/der-verpatzte-ausstieg-a-bf5f7fd8-0002-0001-0000-000008541420; Wolfrum, *Rot-Grüne*, 230–237; Hockenos, *Fischer*, 281–284; SPD and Bündnis 90/Die Grünen, "Aufbruch und Erneuerung"; "Stiller Abschied," *Der Spiegel* 18 (1999), https://www.spiegel.de/wirtschaft/stiller-abschied-a-83032c63-0002-0001-0000-000017069022; AGG, Heyne 109, 1/2, "Stromerzeugung aus Kernenergie 'geordnet beenden,'" September 19, 2001.
76. AGG, Heyne 109, 1/2, "Stromerzeugung aus Kernenergie 'geordnet beenden,'" September 19, 2001, "Vom Stand des Ausstiegs: Ein Zwischenbericht," March 8, 2000, and Trittin, "Die Energiewende vollenden...," March 18, 2000; AGG, Heyne 109 2/2, "Atomausstieg—wo stehen wir...," March 16, 2000, letter from Trittin and Baake, June 15, 2000, and "Durchbruch in den Konsensverhandlungen:...," June 14, 2000; Wolfrum, *Rot-Grüne*, 238–239.
77. Quotation from AGG, Heyne 104 2/2, "Rede zur Debatte über die Umwelt- und Energiepolitik," May 10, 1999; Hockenos, *Fischer*, 282; Wolfrum, *Rot-Grüne*, 239–243.
78. Quotation AGG, Heyne 109, 1/2, BSB, DS, 14/6890, September 11, 2001.
79. AGG, Heyne 104 1/2, Hustedt and Fell, "Für eine Novellierung des Stromeinspeisegesetzes in diesem Jahr...," August 23, 1999; AGG, B.II.3 439 1/2, Conference on Revising the Energy Law, November 30, 1995, Peter Hennicke, "Zehn Thesen...," (September 1995), and "Die Grünen wollen Marktkräfte entfesseln, *Neue Zürchner Zeitung*, January 25, 1996; AGG B.II.3 454 1/2, Hustedt, "Mehr Jobs tanken." (April 1998); Jacobsson and Lauber, "Diffusion."
80. AGG, B.II.3 406, "Solarzeitalter: eine konkrete machbare Vision...."
81. Quotation from AGG, B.II.3 439 1/2, Eckpunktepapeir: Ersatz des Energiewirtschaftsgesetzes durch ein Energiegesetzt; BSB, "Entwurf eines Gesetzes zur Förderung der Stromerzeugung aus erneuerbaren Energien," DS 14, 2341, December 13, 1999.
82. Quotation from AGG, B.II.3, 695 1/2, BSB Plenarprotkoll 13/110, June 13, 1996; Morris and Jungjohann, *Energy Democracy*, 151.
83. AGG, Heyne 104 2/2, *HIB*, "100,000 Dächer," May 11, 2000; AGG, Heyne 104, 1/2, "Das 100,000 Dächer Programm Photovoltaik," November 19, 1999; Scheer, *Solar Economy*, 245–256; Scheer, "100,000 Dächer Photolotaik-Programm"; Jacobsson and Lauber, "Diffusion."
84. AGG, B.II.4 4144, Finance Ministry to Scheer, February 17, 2000; AGG, Heyne 104, 1/2, "Das 100,000 Dächer Programm Photovoltaik," November 19, 1999; AGG, Heyne 104 2/2, Pressemitteilung 62/99, "Regenerative Energieträger...," February 11, 1999.
85. IEA, *Energy Policies of IEA Countries: Germany 2002 Review* (Paris: OECD/IEA, 2002); Stefany Griffith-Jones, "National Development Banks and Sustainable Infrastructure: The Case of the KfW," Global Economic Governance Initiative Working Paper 006 (July 2016), 1–33.
86. BSB, "Nationales Klimaschutzprogram...," DS 14/4729, November 14, 2000; BSB, "Dritter Bericht der Enquete-Kommission," DS 11/8030, May 24, 1990; IEA, *Energy Efficiency Market Report 2013: Market Trends and Medium-Term Prospects* (Paris: IEA/OECD, 2013), 154–160; Damian Carrington, "How a Green Investment Bank Really Works," *The Guardian Environment Blog*, May 24, 2012; Bremer Energie Institut, "Ermittlung von Effekten des KfW-CO2-Gebäudesanierungsprogramms," July 2007.
87. AGG, Heyne 104 2/2, "Ein Jahr 100.000-Dächer-Programm," March 8, 2000; AGG, B.II.4, 714, VDMA to Wissman, February 9, 2000; AGG, Heyne 104 1/2, "Das 100.000 Dächer Programm Photovoltaik," November 19, 1999; Morris and Jungjohann, *Energy Democracy*, 214–215.
88. Scheer, *Energy Imperative*, 114.

89. AGG, Heyne 104 1/2, Hustedt and Fell, "Für eine Novellierung des Stromeinspeisegesetzes," August 23, 1999; AGG, B.II.4 713, Fuentes, "Erneuerbaren Energien," December 21, 1999; AGG, Heyne 104 2/2, Rede zur Debatte über die Umwelt- und Energiepolitik, May 10, 1999.
90. Fabeck, "Leitartikel," Solarbrief 1 (1998), 2; AGG, B.II.3 699, Hustedt, "Heisser Herbste," Tageszeitung, September 27, 1997; AGG, B.II.3 698, BWE, "Leitantrag des BWE-Vorstandes zur Energiepolitik," April 25, 1997.
91. AGG, Heyne 104 2/2, Rede zur Debatte über die Umwelt- und Energiepolitik Parteirat," May 10, 1999; AGG, B.II.4 713, "Entschliessung des Präsidiums des Deutschen Bauernverbandes," December 7, 1999.
92. Quotation from AGG, Fell 135, Vahrenholt of Shell AG to Werner Müller, June 15, 1999; AGG, B.II.4 713, "Clement eröffnet Europas Grösste Solarfabrik...," November 16, 1999; AGG, Heyne 104 1/2, Hustedt and Fell, "Für eine Novellierung des Stromeinspeisegesetzes in diesem Jahr," August 23, 1999.
93. Quotation from AGG, Heyne 104 2/2, "Solarwirtschaft und Wissenschaft unterstützen Rot-Grüne Energiepolitik," Pressemitteilung 629/2000, October, 24, 2000; AGG, B.II.4 714, position papers by Eurosolar, VDMA, IG Metall, February 14, 2000; AGG, Heyne 104 2/2, "Energiewende: Jetzt!" Pressemitteilung 458/99, October 31, 1999, "Rede zur Debatte über die Umwelt- und Energiepolitik Parteirat," May 10, 1999, and Fell, "Regenerative Energieträger...," Pressemitteilung 62/99, February 11, 1999.
94. AGG, Heyne 104 1/2, Hustedt and Fell, "Für ein Novellierung des Stromeinspeisegesetzes in diesem Jahr," August 23, 1999; AGG, B.II.4 713, Siefken to Hustedt, December 29, 1999; AGG, Heyne 109 2/2, "Atomausstieg–Wo Stehen Wir, Wie Geht es Weiter," March 16, 2000; AGG, Fell 135, Fell to Müller, September 15, 1999; Morris and Jungjohann, Energy Democracy, 212–214.
95. BSB, Plenarprotokoll 14/79, "Erneuerbare Energien-Gesetz," December 16, 1999; BSB, Plenarprotokoll 14/91, "Erneuerbare Energien-Gesetz," February 25, 2000; AGG, Heyne 104 1/2, BSB "Entwurf eines Gesetzes zur Förderung der Stromerzeugung...," DS 14/2341, December 13, 1999; AGG, B.II.4 713, Hustedt, "Der Einstieg im das Solarzeitalter hat begonnen," December 16, 1999; AGG, B.II.4 4144, "Bündnis 90/Die Grünen, "Startschuss für Strom aus Erneuerbaren Energien," March 2000.
96. Beatrice Cointe and Alain Nadai, Feed-in Tariffs in the European Union (Cham, Switzerland: Springer, 2018).
97. AGG, Heyne 104 1/2, Hustedt and Fell, "Für eine Novellierung des Stromeinspeisegesetzes in diesem Jahr," August 23, 1999.
98. AGG, B.II.3 454 1/2, Hustedt, "Mehr Jobs tanken," April 1998; 2.
99. AGG, B.II.3 454 1/2, BSB, "Einstieg in eine ökologisch-soziale Steuerreform," DS 13/3555, January 22, 1996; AGG, B.II.4 4144, BSB, "SPD und Bündnis 90/Die Grünen Entwurf eines Gesetzes zur Förderung der Stromerzeugung aus erneuerbaren Energien," DS 14/2341, February 14, 2000; AGG, B.II.3, 454 1/2, Hustedt, "Ökologisch-sozial Steuerreform," June 1998.
100. Fabeck cited in "Keine Baugenehmigungen," Solarbrief 3 (1998), 23.
101. Quotation from AGG, Heyne 104 1/2, BSB, "Entwurf eines Gesetzes zur Förderung der Stromerzeugung aus erneuerbaren Energien," DS 14/2341; AGG, B.II.4 7134.
102. Quotation from Walter Hamm, "Entartung des politischen Wettbewerbs," ORDO: Jahrbuch für die Ordnung von Wirtschaft und Gesellscahft 56 (2005), 19–37; Walter Hamm, "Die Ökosteuer—eine ordnungspolitische Fehlleistung," ORDO: Jahrbuch für die Ordnung von Wirtschaft und Gesellschaft 52 (2001), 1–13; Alfred Voss and Stefan Rath-Nagel, "Konzeption eines effizienten und marktkonformen Fördermodells für erneuerbare Energien," Stuttgart, February 28, 2000.
103. BSB, "Entwurf eines Gesetzes zum Einstieg in die ökologische Steuerreform," DS 14/40, Nov 17, 1998.
104. Quotations and figures from AGG, B.II.3 454 1/2, Hustedt, "Ökologisch-Sozial Steuerreform"; AGG, B.II.3 454 1/2, BSB, "Mit der Energiesteuer Lohnebenkosten senken," DS 13/7750, May 16, 1997, and BSB, "Beschlussempfehlung und Bericht des Finanzausschusses," DS 13/10924, June 8, 1998.

105. Cited in Wolfrum, *Rot-Grüne*, 218, 214–221; AGG, B.II.3, 454 2/2 "5 Mark statt 5 Millionen Arbeitslose," March 23, 1998, "Brief an die Partei Bündnis 90/Die Grünen," March 19, 1998, and and Hustedt and Schmidt, "Grünes Licht für neue Arbeit," May 4, 1998; AGG, B.II.3 454 1/2, "Jeder Gewinnt! Warum Benzin . . . ," March 25, 1998; Hockenos, *Fischer*, 244–246; Scheer, *Solar Economy*, 257.
106. Quotation from AGG, B.II.3 439 2/2, "Einfaches Instrument," *Der Spiegel* 22 (1997); SPD and Bündnis 90/Die Grünen, "Aufbruch und Erneuerung," October 20, 1998; AGG, Heyne 129, "Deutschlands international Verwantwortung," July 2, 2001; AGG, B.II.3 454 1/2, BSB, "Einstieg in eine ökologisch-soziale Steuerreform," Drucksache 13/3555, January 22, 1996; Beuermann and Jäger, "Climate Change Politics."
107. Hamm, "Ökosteuer"; Zimmermann, "Ökosteuern,"; AGG, B.II.3 454 1/2, BSB, "Beschlussempfehlung und Bericht des Finanzausschuss . . . ," DS 13/10924, June 8, 1998; Wolfrum, *Rot-Grüne*, 217–230; Huber; "Climate Change Policies."
108. AGG, Heyne 104, Hustedt, "Rede zur Debatte über die Umwelt- und Energiepolitik Parteirat," May 10, 1999; AGG, Heyne 129, "Grüne: Ökosteuer muss steigen," *Die Welt*, December 28, 2000, *Pressemitteilung* 67/2001, February 4, 2001, and Loske to Trittin, Müller, and Scheel, December 5, 2001; Wolfrum, *Rot-Grüne*, 217–230; Hockenos, *Fischer*, 283.
109. Quotation from AGG, Heyne 129, Polling Report from March 24, 2001.
110. AGG, Heyne 129, *Pressemitteilung*, March 31, 2001; DIW Wochenberichte about Eco Tax, April 5, 2001, and "Deutschlands international Verantwortung," July 2, 2001; Christiane Beuermann and Tilman Santarius, "Ecological Tax Reform in Germany: Handling Two Hot Potatoes at the Same Time," *Energy Policy* 34 (2006), 917–929.
111. AGG, Heyne 106 1/2, "Das Ende des Erdölzeitalters ist in Sicht," *Pressemitteilung* 524/2000, September 6, 2000.
112. Griffith-Jones, "National Development Banks."

Coda

1. First quotation in Tom Morton and Katja Müller, "Lusatia and the Coal Conundrum: The Lived Experience of the German *Energiewende*," *Energy Policy* 99 (2016), 277–287; second and third in Morris and Jungjohann, *Energy Democracy*, 175, 348; final quotation from "Germany's Energy Transformation," *The Economist*, July 28, 2012.
2. IEA, *France 2009 Review* (Paris: IEA, 2009); IEA, *The United Kingdom 2012 Review* (Paris: IEA, 2012); Yergin, *The Quest*, 329–330, 382–395, 409–413; Graetz, *End of Energy*; Smil, *Energy Transitions*, 99–107.
3. Bollmann, *Angela Merkel*, 310–311; Jonas Sonnenschein and Peter Hennicke, *The German Energiewende: A Transition towards an Efficient, Sufficient Green Energy Economy* (Lund: Lund University, 2015), 15–16; Joyce Mushaben, *Becoming Madam Chancellor: Angela Merkel and the Berlin Republic* (Cambridge: Cambridge University Press, 2017), 216, 225–228.
4. Jürgen-Friedrich Hake et al., "The German Energiewende—History and Status Quo," *Energy* 92 (2015), 532–546; Hager, "Green Energy Revolution."
5. Quotation in Joern Hoppmann et al., "The Two Faces of Market Support—How Deployment Policies Affect Technological Exploitation in the Solar Photovoltaic Industry," *Research Policy* 42 (2013), 989–1003, 996; Ann-Kathrin Blankenberg and Ulrich Dewald, "Public Policy & Industry Evolution: The Evolution of the Photovoltaic Industry in Germany," Druid Academy Working Paper 2013.
6. Quotation in Morris and Jungjohann, *Energy Democracy*, 174–175, also 151 and 217; Wilfried Lütkenhorst and Anna Pegels, "Germany's Green Industrial Policy" GSI and IISD Research Report, January 2014; David Buchan, "The *Energiewende*—Germany's Gamble," Oxford Institute for Energy Studies, SP 26, June 2012; David Nelson et al., "Policy and Investment in German Renewable Energy," *Climate Policy Initiative*, Working Paper (April 2016); Lauber and Jacobsson, "Lessons"; AGG, Heyne 104 1/2 Fell to Künast and Roth, March 16, 2001; Scheer, *Energy Imperative*, 116–118.
7. Quotation from Blue Green Alliance, "Overview of the Solar Energy Industry and Supply Chain," Working Paper, January 2011; Bradford, *Solar Revolution*, 102–107.

8. Uwe Nestle, "Does the Use of Nuclear Power Lead to Lower Electricity Prices?" *Energy Policy* 41 (2012), 152–160; Federal Ministry of Economics and Technology, "Energy Concept for an Environmentally Sound, Reliable, and Affordable Energy Supply," September 28, 2010; Ortwin Renn and Jonathan Paul Marshall, "Coal, Nuclear and Renewable Energy Policy in Germany," *Energy Policy* 99 (2016), 224–232.
9. First quotation from Mushaben, *Madam Chancellor*, 231; last quotation from Florian Coulmas and Judith Stalpers, *Fukushima: Von Erdbeben zur atomaren Katastrophe* (Munich: Beck, 2011), 127–130.
10. Merkel cited in Christian von Hirschhausen et al. (eds.), *Energiewende: "Made in Germany"* (Cham: Springer, 2018), 37.
11. Quotation from Miranda Schreurs, "The Ethics of Nuclear Energy: Germany's Energy Politics after Fukushima," *Journal of Social Science* 77 (2014), 9–29, here 24; Friedrich Kunz and Hannes Weigt, "Germany's Nuclear Phase Out," *Economics of Energy & Environmental Policy* 3/2 (2014), 13–27; Peter Hennicke and Paul Welfens, *Energiewende nach Fukushima. Deutsche Sonderweg oder weltweites Vorbild* (Munich: Oekom, 2012), 7–11.
12. FAZ cited in Sebastian Strunz, "The German Energy Transition as a Regime Shift," *Ecological Economics* 100 (2014), 150–158, here 154; second quotation from Hennicke and Welfens, *Energiewende*, 7–8.
13. IEA, *Energy Efficiency Market Report 2013*, 149; Barbara Schlomann and Wolfgang Eichhammer, "Energy Efficiency Policies and Measures in Germany," Fraunhofer ISI Working Paper, November 2012; AG Energiebilanzen, "Auswertungstabellen zur Energiebilanz Deutschland, 1990 bis 2017."
14. Jeremy Rifkin, *The Third Industrial Revolution: How Lateral Power Is Transforming Energy, The Economy, and the World* (New York: St. Martin's Griffin, 2013), 68–70; Federal Ministry of Economics and Technology, "Energy Concept."
15. Lars Borchert, "The People's Energiewende: Germany between Citizens' Energy and Nimbyism," *Clean Energy Wire*, March 10, 2015, https://www.cleanenergywire.org/news/citizen-participation-key-energy-transition-success; Paul Hockenos, "Jobs Won, Jobs Lost," *Clean Energy Wire*, March 30, 2015, https://www.cleanenergywire.org/dossiers/energy-transitions-effect-jobs-and-business; Umweltbundesamt, "Fast 2 Millionen Beschäftigte im Umweltschutz," November 5, 2014; Lütkenhorst and Pegels, "Green Industrial Policy."
16. "Germany's Energy Transformation," *The Economist*; Buchan, "Germany's Gamble"; Borchert, "The People's Energiewende"; Agora, "Understanding the Energiewende."
17. Peter Sopher, "Political Lessons from Germany's Energiewende," EDF Report, November 17, 2014; Kerstine Appunn and Ruby Russel, "German Utilities and the Energiewende," *Clean Energy Wire*, January 12, 2015, https://www.cleanenergywire.org/factsheets/german-utilities-and-energiewende.
18. Scheer, *Energy Imperative*, 7 and 22.
19. Buchan, "Germany's Gamble"; Aaron Wiener, "Made in the Shade," *Foreign Policy*, July 9, 2012, https://foreignpolicy.com/2012/07/09/made-in-the-shade/; Hubert Schmitz and Rasmus Lema, "The Global Green Economy: Competition or Cooperation Between Europe and China?" in Fagerberg et al. (eds.), *Triple Challenge*, 119–142; Keith Bradsher, "When Solar Panels Became Jobs Killers," *NYT*, April 8, 2017, https://www.nytimes.com/2017/04/08/business/china-trade-solar-panels.html.
20. Arne Jungjohann and Craig Morris, "The German Coal Conundrum," *Heinrich Böll Foundation North America*, June 2014; Jonathan Elkind and Damian Bednarz, "Warsaw, Brussels, and Europe's Green Deal," *Columbia SIPA Center on Global Energy Policy*, July 30, 2020; Frank Dohmen et al., "German Failure on the Road to a Renewable Future," *Der Spiegel International*, May 13, 2019, https://www.spiegel.de/international/germany/german-failure-on-the-road-to-a-renewable-future-a-1266586.html.
21. Sonnenschein and Hennicke, *Germany's Energiewende*; "The Energiewende and Energy Prices: Public Support and Germany's Long-Term Vision," *Carbon Brief*, July 26, 2013; Dimitri Pescia et al., "Understanding the Energiewende," Agora Energiewende (2015).
22. Federal Ministry of Economics and Technology, "Energy Concept"; Christoph Stefes, "Bypassing Germany's Reformstau," *German Politics* 19/2 (2010), 148–163; Morris and Jungjohann, *Energy Democracy*, 348.

23. Clement cited in James Kanter, "EU Carbon Trading System brings Windfalls for Some," *NYT*, December 9, 2008, https://www.nytimes.com/2008/12/09/business/worldbusiness/09iht-windfall.4.18536167.html.
24. Pescia, "Energiewende"; Hockenos, "Jobs Won, Jobs Lost."
25. Spiegel cited in Morris and Jungjohann, *Energy Democracy*, 363–366; Merkel cited in "Germany's Energy Transformation," *The Economist*.
26. Etienne Benson, *Surroundings: A History of Environments and Environmentalism* (Chicago: University of Chicago Press, 2020), 179, 185–8; Yergin, *The Quest*, 516–517; Julian Wettengel, "Polls Reveal Citizens' Support for Climate Action and Energy Transition," *Clean Energy Wire*, December 14, 2021, https://www.cleanenergywire.org/factsheets/polls-reveal-citizens-support-energiewende.
27. Figures from *Our World in Data*, "Germany: CO2 Country Profile"; Kerstine Appunn et al., "Germany's Greenhouse Gas Emissions and Energy Transition Targets," *Clean Energy Wire*, December 21, 2021, https://www.cleanenergywire.org/factsheets/germanys-greenhouse-gas-emissions-and-climate-targets.
28. Quotation in Sören Amelang, "Germany's Energy Transition Revamp Stirs Controversy over Speed, Participation," *Clean Energy Wire*, June 29, 2016, https://www.cleanenergywire.org/dossiers/reform-renewable-energy-act.
29. Jungjohann and Morris, "Coal Conundrum"; Kerstine Appunn et al., "Germany's Energy Consumption and Power Mix in Charts," *Clean Energy Wire*, December 21, 2021.
30. Gareth Bryant, "Creating a Level Playing Field? The Concentration and Centralisation of Emissions in the European Union Emissions Trading System," *Energy Policy* 99 (2016), 308–318; Andrea Brock and Alexander Dunlap, "Normalising Corporate Counterinsurgency," *Political Geography* 62 (2018), 33–47; Patrick Graichen et al., "The Role of Emissions Trading in the Energy Transition," Agora Energiewende (2015); Kanter, "EU Carbon Trading."
31. Gabriel cited in Morton, "Lusatia"; Uwe Jun, "Die SPD in der Ära Merkel: Eine Partei auf der Such nach sich selbst," and Annette Elisabeth Töller, "Kein Grund zum Feiern! Die Umwelt- und Energiepolitik der dritten Regierung Merkel," in Reimut Zohlnhöfer and Thomas Saalfeld (eds.), *Zwischen Stillstand, Politikwandel und Krisenmanagement. Eine Bilanz der Regierung Merkel, 2013–2017* (Wiesbaden: Springer, 2019), 39–62, 569–590.
32. Quotations from Ministry of Economics and Energy, "Making a Success"; Adam Tooze, "Germany's Unsustainable Growth: Austerity Now, Stagnation Later," *Foreign Affairs* (Sept/Oct 2012), 23–30; Markus Brunnermeier, Harold James, Jean-Pierre Landau, *The Euro and the Battle of Ideas* (Princeton, NJ: Princeton University Press, 2016), 137–155.
33. Morris and Jungjohann, *Energy Democracy*, 370–374; Lauber and Jacobson, "Lessons"; Buchan, "Germany's Gamble."
34. Quotation from Marlies Uken, "Energiewende ohne Bürger," *Die Zeit*, February 13, 2018, https://www.zeit.de/wirtschaft/2018-02/windenergie-energiewende-union-spd-buergerprojekte; Gunther Latsch et al., "Gone with the Wind," *Der Spiegel International* 1/30 (2014), https://www.spiegel.de/international/business/wind-power-investments-in-germany-proving-riskier-than-thought-a-946367.html.
35. AGG, Hans Josef Fell, 2, BSB, "Für eine sichere Energieversorgung im 21. Jahrhundert," DS 16/579, February 7, 2006.
36. Johannes Teyssen of E.ON paraphrased in "Business Worries as Russian Sanctions Mount," *Deutsche Welle*, March 18, 2014; Yergin, *The Quest*, 339–343.
37. Constanze Stelzenmüller, "Germany's Russia Question: A New Ostpolitik for Europe," *Foreign Affairs* (March/April 2009), 89–100; Gustafson, *The Bridge*, 362–369; David Buchan and Malcom Keay, *Europe's Long Energy Journey: Towards an Energy Union?* (Oxford: Oxford University Press, 2015), 160–181.
38. Frank Dohmen and Alexander Jung, "How Germany Could End Russian Gas Dependency," *Der Spiegel International*, May 6, 2014, https://www.spiegel.de/international/business/german-alternatives-to-russian-gas-numerous-but-pricey-a-967682.html; Tatiana Mitrova and Tim Boersma "The Impact of US LNG on Russian Natural Gas Export Policy," *Columbia SIPA Center on Global Energy Policy* (December 2018).
39. Pierre Noël and Chi Kong Chyong, "How to Break the Logjam over Nord Stream 2," *Foreign Policy* (April 29, 2021), https://foreignpolicy.com/2021/04/29/nord-stream-2-russia-germ

any-gas-pipeline-lng-baltic-biden-europe-ukraine-energy/; Gustafson, *The Bridge*, 279–383; Jürgen Dahlkamp et al., "The Anatomy of Germany's Reliance on Russian Natural Gas," *Der Spiegel International* June 29, 2022, https://www.spiegel.de/international/business/anatomy-of-germany-s-reliance-on-russian-natural-gas-decades-of-addiction-a-ad156813-3b24-424f-a51e-3ffbd7b6385c.

40. Quotation from BSB, "Europäische Energieunion," DS 18/1461, May 21, 2014; BSB, "Kleine Anfrage," DS 18/2828, October 8, 2014.

41. Dirk Jansen and Dororthea Schubert, *Zukunft statt Braunkohle: 30 Jahre Widerstand gegen den Braunkohlentagebau Garzweiler II* (Düsseldorf: BUND, 2014); Brock and Dunlap, "Counterinsurgency"; Andreas Malm and the Zetkin Collective, *White Skin, Black Fuel: On the Danger of Fossil Fascism* (London: Verso, 2021), 7, 99; Christian Stöcker, "The Kids Aren't Nearly Angry Enough," *Der Spiegel International*, April 16, 2019, https://www.spiegel.de/international/europe/our-climate-catastrophe-the-kids-are-not-nearly-angry-enough-a-1262697.html. Wettengel, "Polls Reveal Citizens' Support"; Ende Gelände, "Keep It in the Ground!" https://2015.ende-gelaende.org/en.html.

42. Kommission Wachstum, Strukturwandel, und Beschäftigung, *Abschlussbericht* (Berlin: BMWi, 2019), 2–4.

43. Katharina Baudisch and Dörte Fouquet, "Germany's Coal Exit Plan," *Renewable Energy Law and Policy Review* 45 (2019), 46–59; Alexander Rietzenstein and Rebekka Popp, "The German Coal Commission—A Role Model for Transformative Change?" E3G Briefing Paper, April 2019.

44. IPCC, *Sixth Assessment Report: The Physical Science Basis*, August 2021; Mann, *Climate War*, 148–150.

45. Quotation from Constanze Stelzenmüller, "Germany Is Now the Fulcrum for Vladimir Putin's Pressure," *Brookings Institute*, August 11, 2022, https://www.brookings.edu/blog/order-from-chaos/2022/08/11/germany-is-now-the-fulcrum-for-vladimir-putins-pressure/; Constanze Stelzenmüller, "Putin's War and European Energy Security" *Brookings Institute*, June 7, 2022, https://www.brookings.edu/testimonies/putins-war-and-european-energy-security-a-german-perspective-on-decoupling-from-russian-fossil-fuels/; Dahlkamp, "Anatomy"; Judy Dempsey, "Europe's Continued Commitment to Ukraine Hinges on Germany," *Carnegie Europe*, July 19, 2022, https://carnegieeurope.eu/strategiceurope/87531; Allison Meakem, "Germany Confronts Its Nuclear Demons," *Foreign Policy*, June 20, 2022, https://foreignpolicy.com/2022/06/20/germany-nuclear-power-energy-weapons-nato-russia-ukraine-war-energy-crisis-greens/; Zia Weise, "Why Germany Won't Give up on Giving Up Nuclear," *PoliticoPro*, April 28, 2022, https://www.politico.eu/article/politics-behind-germany-refusal-reconsider-nuclear-phaseout/.

46. Federal Government, "Policy Statement by Olaf Scholz," February 27, 2022.

INDEX

For the benefit of digital users, indexed terms that span two pages (e.g., 52–53) may, on occasion, appear on only one of those pages.

Abu Dhabi
  oil embargo (1967) and, 90–91
  oil industry in, 96
acid rain, 199–201, 200f
Adenauer, Konrad
  generally, 54
  coal industry and, 25–26, 27, 39, 40, 41, 43
  energy security and, 45
  nuclear energy industry and, 102–4, 105–6, 129–30
  oil industry and, 34–35, 41, 42–43
AEG
  generally, 170
  natural gas industry and, 263
  nuclear energy industry and, 112, 114, 118–19, 121–22
  photovoltaic cells (PV) and, 231–32
AEG Kanis, 251–52, 253, 255
Aerospace Center, 231–32
Afghanistan, Soviet invasion of, 252
Agip, 133
air pollution, 158, 184
Algeria
  natural gas production in, 251
  oil embargo (1967) and, 90
  oil industry in, 65–66, 92
Allais, Maurice, 65–66
Allied occupation of Germany
  coal industry during, 22
  International Authority of the Ruhr (IAR), 25
  nuclear energy industry and, 102
  Post-War rebuilding of Germany, 22
  rebuilding during, 22
Altmaier, Peter, 304
Altner, Günter, 198–99
American Economic Association, 58–59

Amery, Carl, 169
Amoco, 230
Andersson, Jenny, 52
ARAMCO, 73–74, 90
Argentina, electricity demand in, 119–20
Armand, Louis, 52–54, 101–2, 105, 106
Association of German Automobile Clubs, 76–77
Association of German Industry, 112
Atomic Affairs Ministry, 102–3
Atomic Commission, 102–3, 107, 109, 110–11, 112, 113–14, 123
Atomic Law (1960), 109, 166–67, 301
Atomic Ministry, 114–15
"Atomic State," 171–72
Austria, natural gas and, 245
Autobahn, 75–76, 126f
automobile industry, 75–77

Baade, Fritz, 32, 52, 62
backstop technology, 187, 197–98
Bad Godesberg program, 108
Baerbock, Annalena, 307–8
Bahr, Egon, 244, 245, 246, 247
Bahrain, oil embargo (1967) and, 91
Bahro, Rudolf, 169–70
Balke, Siegfried
  generally, 107
  energy security and, 66–67
  energy transitions and, 332n.1
  nuclear energy industry and, 104, 106, 108–9, 110–11, 113
  state intervention, on, 116
Bartelt, Klaus, 119–20
BASF, 260, 261, 263, 307, 310
Basic Energy Research Program, 141
Bauer, Walter, 84

Bayernwerk, 113–14
Beck, Ulrich, 212
Beddermann, Carl, 170, 173–74
Belarus, independence of, 260
Bell Labs, 115
Bennigsen-Foerder, Rudolf von, 133–35
Benz, Karl, 75–76
Bergbau-AG Lothringen, 33
Bergmann, Burckhard, 238–39, 239f
Berlin Climate Conference (1995), 270–71
Berlin Wall, 224–25, 260
Binswanger, Hans Christoph, 205–6
   Ecological Modernization and, 185, 291
   emissions trading, on, 208
   energy efficiency and, 216
   malleability of energy economy and, 202
   neoclassical economics and, 196–98
   oil industry and, 358n.1
   renewable energy and, 205
   taxation, on, 209
biogas, 289, 290, 298–99, 303
Blair, Tony, 281
Bloch, Ernst, 101–2
BMFT, 141
Bohm, Franz, 48–49
Boulding, Kenneth, 58–59
BP. *see* British Petroleum (BP)
Brandt, Leo, 109, 338n.2
Brandt, Willy
   generally, 93, 163–64, 171–72, 174
   air pollution and, 158
   coal industry and, 140
   environmental issues and, 161, 162–63
   jobs and, 179
   natural gas industry and, 244, 246, 247–49, 265
   nuclear energy industry and, 117, 119, 138, 174
   Oil Shock (1973) and, 125–27, 129–31, 147
   Ostpolitik and, 244, 246, 247, 250–51, 263
   research and development and, 141
Brazil
   electricity demand in, 119–20
   nuclear energy industry in, 138–39, 144, 254, 307
Brezhnev, Leonid
   natural gas and, 247–48, 249, 250–52, 255–56, 263, 371n.1
   nuclear weapons and, 245–46
Brigitta, 242–43
British Petroleum (BP)
   concentration of power and, 30–31, 97–98, 143–44
   mercantilism and, 141–43
   Middle East, in, 74, 75
   natural gas and, 251–52, 264
   oil and, 67, 230
   oil prices and, 36, 37
   pipelines, 78–79
   power of, 83
   Social Market Economy and, 34–35, 79
   taxation and, 43
Brown, Boveri, and Cie (BBC), 112
Brundtland, Gro Harlem, 222
Brundtland Report, 220, 222
Burchard, Hans Joachim, 63
Burckhardt, Helmut, 33–34, 41, 43, 66, 93
Burgbacher, Fritz, 54–56, 84
Bush, George H.W., 192, 260, 267–68
Bush, George W., 268–69, 294, 297
Business Association for Ruhr Mining (UvRb), 28–29, 37, 39, 40, 41–42

Campbell, C.J., 284–85
carbon emissions
   carbon tax, 204, 223–24, 273
   CDU and, 223–24
   difficulty in establishing accountability, 4
   emissions trading, 191–92, 208, 291–92, 294, 297
   GDR, in, 225
   historical growth in, 44f
   Reunification and, 225
   targets, 304
carbon tax, 204, 223–24, 273
car-free Sundays, 125, 126f, 130f
Carson, Rachel, 158
Carter, Jimmy, 139, 189, 252
CDU. *see* Christian Democratic Union (CDU)
cheap energy paradigm, 10–11
chemical industry, 111–12, 161
Chernobyl nuclear accident (1986), 211–12, 216–17, 220, 222–23
Chevron, 74, 75, 230
China
   economic growth in, 293
   solar energy in, 303
   WTO and, 284–85
Christian Democratic Union (CDU)
   carbon emissions and, 223–24
   climate change and, 220, 236, 237, 270–71
   coal industry and, 39–40, 85–86, 273
   competition and, 273–75
   coupling paradigm and, 228, 273, 305–6
   divisions within, 223
   Ecological Modernization and, 209, 257–58
   economic growth and, 60
   electoral success of, 87
   electricity sector and, 216, 234–36
   energy forecasts and, 54–56
   energy liberalization and, 284
   energy policy, 236
   energy price policy, 34, 228
   energy tax and, 293
   environmental issues and, 222–23, 272
   FDP, coalition with, 213–14, 216–17, 257, 273, 300

First Grand Coalition, 89, 117, 118, 160
natural gas industry and, 240–41, 247, 257–59, 261, 307
neoliberalism and, 227, 305–6
nuclear energy industry and, 109, 113, 115, 116, 117, 118, 121–22, 216–17, 218–19, 273, 287, 300
oil industry and, 20, 92
Oil Shock (1973) and, 151
Ordoliberalism and, 107–8, 227, 228, 236
Peace Marches and, 181
reform gridlock and, 278–79
renewable energy and, 228, 233, 276, 277
Reunification and, 224–25
Second Grand Coalition, 298, 305–6
Social Market Economy and, 183–84, 227
Soviet oil and, 242
SPD, coalitions with, 89, 117, 118, 160, 298, 305–6
state intervention and, 227–28
sustainability and, 222
technology and, 228
VEBA and, 133–34
Christians, Wilhelm, 246
Christian Socialist Union (CSU), 102–3
Clement, Wolfgang, 239f, 303
climate change
 CDU and, 220, 236, 237, 270–71
 coal industry and, 308–9
 difficulty in establishing accountability, 4
 Ecological Modernization and, 220, 221–22, 259
 energy security and, 308, 311
 failures of call to action, 269
 Framework Convention on Climate Change (UNFCCC), 267–68, 270–71
 Green Party and, 221–22
 natural gas industry and, 264–65
 neoliberalism and, 269
 politicization of energy and, 219–22, 237
 projections of, 219–20
 Reunification and, 269, 271
 SPD and, 221–22
 unintended consequences of, 5
Climate Inquiry, 223–24, 234, 236
Climate March, 309f
Clinton, Bill, 269, 287–88
Club of Rome, 139, 162, 163, 184, 186, 187, 219, 232
Coal Commission, 308–10
coal industry
 generally, 19, 20
 capital flight and, 85
 cartels in, 27, 28–29, 34, 37–38, 40, 41–43
 CDU and, 39–40, 85–86, 273
 climate change and, 308–9
 closure of mines, 85–86, 86f, 88f

collapse of market in, 19–20
comparison with oil industry, 34–35
competition in, 28
decline of, 43, 85–86, 87–89
Economic Miracle and, 29
Economics Ministry and, 32–33, 39, 40, 41
ECSC and, 27–28, 39–40, 42–43, 51–52, 85
energy crises and, 23–24, 38–43, 38f
energy forecasts for, 31–32
FDP and, 273
Garzweiler II mine, 273, 274f, 308
gasification and liquifaction, 140–41
Graf Bismarck coal mine, 86, 86f, 88f
high point of, 29
imports, 32
increased use of coal, 3–4
Korean War as energy crisis in, 25
OEEC and, 33–34
oil, transition to, 19–20, 29–30, 43, 46, 60–61, 73
Oil Shock (1973), mercantilist responses to, 140–41
Ordoliberalism and, 51–52
oversupply in, 38–39
persistence of, 304–5
phasing out of, 309–10
Post-War period, in, 22, 23
pricing regimes in, 32–33, 34–35, 38, 85
productivity in, 35
protests in, 41–43, 42f
rationing, 26
recession, effect of, 87
reorganization of, 93–94
retraining program in, 43
Social Market Economy and, 26, 34–35, 36–37
SPD and, 86, 140, 273
strikes in, 89
structural change in, 40
subsidies in, 85
substitution for oil as result of Oil Shock (1973), 148
surface mining, 273, 274f
Third Reich and, 22, 23
US, tensions with regarding, 25–26
wages in, 29
Coase, Ronald, 190, 191–92
Cockfield, Arthur, 226–27
Cold War, 181, 241–42, 247, 252–54
commercial combined cycle gas turbines (CCGT), 259, 263
Commission for Environmental Questions and Ecology, 180
Communist League, 173–74
Communist Party, 173
Compagnie Française des Pétroles (CFP), 92, 97, 141–42, 143–44

competition
  CDU and, 273–75
  coal industry, in, 28
  electricity sector, in, 273–76, 275f
  oil industry, in, 63–64
  Ordoliberalism and, 49, 64
conservation
  energy saving stamp, 180f
  jobs and, 180–81
  Oil Shock (1973) and, 148
consumer society, 76–77
Copenhagen Climate Summit (2009), 304
Council of Economic Experts, 60, 129, 195–96, 202
coupling paradigm
  CDU and, 228, 273, 305–6
  cracks in, 193–94
  decoupling process, 182
  Economics Ministry and, 257–58
  emergence of, 47–48
  energy as complementing other production, 62–63
  energy security and, 67
  environmental movement and, 170, 172–73
  Interior Ministry and, 195
  natural gas industry and, 240–41, 255–56, 263, 264
  nuclear energy industry and, 116–17, 136–37, 175, 176
  oil price decline and, 236
  Oil Shock (1973) and, 127, 149, 150–51
  Ordoliberalism and, 47–48, 183–84
  persistence of, 303
  problems with, 68–70
  resource shortages, impact of, 163–64
  SPD and, 175, 176, 194, 305–6
Croatia, independence of, 271
Cuban Missile Crisis (1962), 241–42
Czechoslovakia, Soviet intervention in, 245–46

Dahrendorf, Ralf, 166–67
Dance with the Devil (Schwab), 158
Daniels, Wolfgang, 235
Data Resources Incorporated, 188
De Gaulle, Charles, 168
Democratic National Party, 89
Democratic Republic of Congo, energy consumption in, 4–5
Demoll, Reinhard, 159–60
Denmark, 131–32, 185, 232–33
détente, 252
Deutsche Bank
  natural gas industry and, 246, 247–48, 250–52, 253, 263, 265
  oil industry and, 30
Deutsche BP, 36, 63, 80
Deutsche Erdoelversorgungsgesellschaft mbH (DEMINEX), 95–96, 98, 124, 135, 142–43

Deutsche Erdöl AG (DEA), 86, 89–90, 93
Deutsche ESSO, 36, 57, 78, 146
Deutsche Shell, 36, 84, 143, 146
Deutsche Texaco AG, 90
De Witt, Siegfried, 167
Dickler, Robert, 194–95
Diesel, Rudolf, 75–76
digital revolution, 272
Dittfurth, Jutta, 211–12
divergence of German energy policy, 10–12
  energy transitions, role of, 15–16
  foreign energy sources, reliance on, 11
  Green *Sonderweg* (special path), 11–12
  market mechanisms, 12
  multiple actors involved in, 12
  neoliberalism, divergence from, 11–12
  nuclear energy industry and, 167–68, 182
  triumphalism and, 12
Dolinski, Urs, 67
Dumont, René, 167

Earth Summit (1992), 267–69, 270–71
Eco-Alliance (SPD–Green Party)
  generally, 1–2, 15, 270
  achievements of, 293–94
  divisions in, 285–87
  Ecological Modernization and, 297
  Electricity Feed-in Law (1990) and, 289
  end of, 298
  energy security and, 307–8
  externalities and, 291
  formation of, 270, 282
  Kosovo War and, 285
  legacy of, 298
  nuclear energy industry and, 287
  Ordoliberalism and, 283–84, 290–91
  renewable energy and, 280–82, 285, 287, 289–90, 294
  solar energy and, 287–90
  taxation and, 292–93
  wind energy and, 289–90
eco-catastrophism, 158, 169
ecocide, 3–4
Eco-Institute in Freiburg
  generally, 177, 198–99, 215, 232
  eco-catastrophism and, 169
  Electricity Feed-in Law (1990) and, 234
  energy efficiency and, 172–73
  energy policy, 209
  formation of, 167
  jobs and, 218
  nuclear energy industry and, 198
eco-libertarianism, 279–80
ecological economics, 59
Ecological Modernization
  generally, 15, 216
  carbon tax and, 273

CDU and, 209, 257–58
climate change and, 220, 221–22, 259
Eco-Alliance and, 297
electricity sector and, 209–10, 235
energy tax and, 205–7
environmental economics and, 203–4
externalities and, 291
Green Party and, 209, 259, 282
Keynesianism versus, 185
limitations of, 209–10
natural gas industry and, 237, 240–41, 256, 264
neoclassical economics versus, 185
neoliberalism versus, 185
nuclear energy industry and, 217, 218
Ordoliberalism versus, 185, 207–8
origins of, 185
price-induced technological development and, 202, 204
renewable energy and, 205, 263, 270, 289
rise of, 185
rolling back of, 305–6
SPD and, 209, 259, 280, 282
taxation and, 292
ecological tax, 205–7, 293
ecology, 6
Economic Commission for Europe (UNECE), 35–36
Economic Cooperation Administration (ECA), 30
economic growth
  CDU and, 60
  coupling with energy policy (see coupling paradigm)
  energy and, 58–59
  energy security, importance of compared, 48
  environmental movement, concerns of, 170
  GNP, fixation on, 57–59
  growth paradigm in BDR, 60–61
  neoclassical economics and, 58–60
  Ordoliberalism and, 60
  quantitative economic growth, 57–58
Economic Miracle
  coal industry and, 29
  energy crises and, 13
  energy pricing regimes, effect of, 43–44
  free market and, 44–45
  inexpensive energy, role of, 7
  Ordoliberalism and, 123
Economics and the Environment (Kneese), 188
Economics Ministry
  carbon emissions and, 272
  Cartel Office, 146
  coal industry and, 32–33, 39, 40, 41
  coupling paradigm and, 257–58
  electricity sector and, 214, 234, 235–36
  energy crises and, 193
  energy forecasts, 32, 54, 61
  energy liberalization and, 262

energy planning and, 128
energy policy, 34–35
energy security and, 83
natural gas industry and, 246, 249
neoliberalism and, 227
"new order" in energy and, 92
nuclear energy industry and, 102, 107, 108–11, 113–14, 118
oil industry and, 34–35, 36, 43, 82, 83–84, 93, 95, 134–35
Oil Shock (1973) and, 131
renewable energy and, 228, 277
SPD and, 89, 93
Suez Crisis (1956) and, 80–81
US, relationship with, 29–30, 31
The Economics of Welfare (Pigou), 187–88
Egypt
  1973 War and, 125–26
  oil embargo (1967) and, 90
  Suez Crisis (1956), 13, 33–34, 64, 71, 80–83, 128
Ehmke, Horst, 163–64
Eisenhower, Dwight D., 64, 81, 101–2, 111
Electricité de France (EDF), 164–65, 167–68, 182
Electricity Feed-in Law (1990), 234–37, 276, 277–78, 280, 289, 290
Electricity Law (1965), 86
electricity sector
  auctions, 306
  CDU and, 216, 234–36
  competition in, 273–76, 275f
  consolidation in, 117
  Ecological Modernization and, 209–10, 235
  Electricity Feed-in Law (1990) (see Electricity Feed-in Law (1990))
  energy efficiency in, 176–77, 215–16
  generation of electricity, 299f
  Green Party and, 214–16, 234, 235–36
  increase in electricity consumption, 116–17
  increase in electricity demand, 119–20
  nuclear energy industry and, 111
  power surges, 303
  recommunalization of, 214–16
  Renewable Energy Act (2000), 290, 291, 294, 298–99, 303, 305–6
  solar energy (see solar energy)
  SPD and, 216, 234, 235
  surcharge, 303
  third party access (TPA), 274–75
  wind energy (see wind energy)
ELF-ERAP, 92, 97, 143–44
Ellul, Jacques, 159–60, 170
Eltville Program, 110–11
emissions trading, 191–92, 208, 291–92, 294, 297
Energy and Mining Union (IGBE), 82–83, 85, 89, 92–94

energy consumption
    energy forecasts of, 142f, 194–95, 194f, 202–3
    GDR, in, 225
    global differences in, 4–5
    graphic representation, 2f, 13f
    historical growth in, 4, 5–6, 43–44, 61
    naturalization of, 6–7
energy crises
    generally, 221f
    Chernobyl nuclear accident (1986) as, 211–12, 216–17, 220
    coal industry and, 23–24, 38–43, 38f
    Economic Miracle and, 13
    Economics Ministry and, 193
    effect on German energy policy, 44–45
    energy forecasts and, 193–94
    energy transitions, role in, 10, 20–21
    Iranian Revolution (1979) as, 178–79, 251
    Korean War as, 23–24, 25–26
    Oil Shock (1973) (see Oil Shock (1973))
    Ordoliberalism and, 194
    Suez Canal closure (1967), 67, 90–91, 128
    Suez Crisis (1956) as, 33–34, 64, 71, 80–83, 128
    Three Mile Island nuclear accident (1979) as, 178, 179–80
Energy Economic Institute (EWI), 47, 54–56, 61, 62, 67, 193, 197
Energy Economic Law (1935), 214–15
energy efficiency
    decreases in, 224
    electricity sector, in, 176–77, 215–16
    energy saving stamp, 180f
    energy transitions and, 12–13
    environmental movement and, 172–73
    fossil fuels and, 8
    jobs and, 180–81
    modernization and, 301–2
    Oil Shock (1973) and, 12–13, 148, 296
    price elasticity and, 201–2
    productivity increase, 201
    promotion of, 177
    Second Revision to the Energy Program and, 176–77
    SPD and, 177, 180, 182, 296
energy forecasts
    CDU and, 54–56
    coal industry, for, 31–32
    diversity in, 198–99
    Economics Ministry, 32, 54, 61
    energy consumption, of, 142f, 194–95, 194f, 202–3
    energy crises and, 193–94
    Energy Inquiry, 54–58, 60–61, 63, 68–69
    energy security concerns and, 64
    graphic representation, 55f
    growth of, 63

nuclear energy industry and, 116–17
OECD and, 53
OEEC and, 31, 47, 52–54, 63
oil industry and, 57, 61
Oil Shock (1973) and, 149
Ordoliberalism and, 54, 56, 68–69
SPD and, 54–56
Energy Industry Liberalization Act (1998), 262
Energy Inquiry, 54–58, 60–61, 63, 68–69
energy intensity, 3f, 301–2
energy liberalization, 257–58, 260, 262
energy planning
    Economics Ministry and, 128
    FDP and, 128–29
    SPD and, 128–29
Energy Program (1973), 96–97, 128–29, 134–35
energy security
    climate change and, 308, 311
    close international integration and, 66–67
    coupling paradigm and, 67
    Eco-Alliance and, 307–8
    economic growth, importance of compared, 48
    Economics Ministry and, 83
    energy forecasts and, 64
    energy transitions and, 306, 307–8
    Green Party and, 310–11
    Liberal Party and, 310–11
    low cost supply as incompatible with, 69
    market approach to, 67
    mixed strategy, 67–68
    natural gas industry and, 240, 249–50
    politicization of energy and, 14
    "renaissance" in BDR policy, 66
    SPD and, 82–83, 307–8, 310–11
    state intervention and, 67–68
    Suez Crisis (1956) and, 82–83
energy tax, 205–7, 293
energy transitions (Energiewende)
    generally, 3, 298, 311–12
    apparent triumph of, 298–302
    coal to oil, 19–20, 29–30, 43, 46, 60–61, 73
    competing energy paradigms and, 295–96
    costs of, 303, 304
    divergence of German energy policy, role in, 15–16
    Ecological Modernization (see Ecological Modernization)
    energy crises, role of, 10, 20–21
    energy efficiency and, 12–13 (see also energy efficiency)
    energy markets, role of, 8
    energy security and, 306, 307–8
    environmental movement and, 172–73
    future transitions, 310–12
    Green Party and, 15, 157
    ideas, role of, 9–10
    intermittancy, problem of, 310

linkage to other goals, 296–97
Marxian perspective on, 8
mobilization, role of, 9
multiple causes of, 9, 10
natural gas industry and, 12–13, 240–41 (*see also* natural gas industry)
non-state actors, role of, 9
nuclear energy industry and, 12–13, 14–15, 100–1 (*see also* nuclear energy industry)
obstacles to, 302–5, 311
oil industry and, 12–15 (*see also* oil industry)
oil to natural gas, 229
Ordoliberalism and, 46
policy linkage, role of, 9
political coalitions, role of, 9–10
price and, 8
renewable energy and, 12–13, 270, 298–99, 301–2 (*see also* renewable energy)
SPD and, 15, 157
stalling of, 7–8
state intervention and, 294
states, role of, 9
technology and, 8
Engelsberger, Matthias, 234–36
Enrico Fermi nuclear power plant (US), 122
ENRON Corporation, 261
Ente Nazionale Idrocarburi (ENI), 78–79, 133, 135, 143–44
environmental economics, 187–88, 203–4
environmental issues
 CDU and, 222–23
 Green Party and, 237
 Keynesianism, problematic nature of, 195–96
 neoclassical economics, problematic nature of, 195–96
 Ordoliberalism, problematic nature of, 195–96
 Social Market Economy, environmental concerns with, 184
 SPD and, 161, 162–64, 177–78, 237
 sustainability, 222
Environmental Ministry, 211–12
environmental movement. *see also* Green Party
 air pollution and, 158
 coupling paradigm and, 170, 172–73
 demographics of, 169
 eco-catastrophism and, 158, 169
 economic growth, concerns regarding, 170
 energy efficiency and, 172–73
 energy transitions and, 172–73
 jobs, juxtaposition with, 179–80
 neoliberalism, effect of, 213
 nuclear energy industry, protests against, 155–56, 156f, 164–67, 171–73, 198, 296
 Oil Shock (1973) and, 163
 origins of, 158–61
 Reunification, effect of, 213, 225–26
 river pollution and, 159

student mobilization and, 160–61
technology, concerns regarding, 159–60, 170–71
Environmental Program, 203
Environmental Protection Party, 173–74
E.ON, 264, 265, 284, 302, 304–5, 306, 307
Eppler, Erhard
 generally, 163f, 198
 conservation, on, 164
 energy efficiency and, 180, 308–9
 energy security and, 178–79
 environmental movement and, 174
 jobs and, 175, 179
 Limits to Growth and, 162–63
 nuclear energy industry and, 175–76, 177–78, 181–82
 reform wing of SPD, in, 305
 restructuring of energy and, 177
 "Silent revolution" and, 169
 state intervention, on, 163–64
Equipment and Machinery Producers Association (VDMA), 289–90
Erhard, Ludwig
 generally, 20–21, 93
 coal industry and, 25–26, 39, 40, 41–43, 85–86, 87, 93
 coalition government and, 89
 coupling paradigm and, 303
 economic growth and, 60
 energy pricing regimes and, 32–33, 34–35, 36, 38, 87, 228
 Korean War and, 24, 25–26, 28–29
 nuclear energy industry and, 102
 oil industry and, 43, 82–83, 296
 Ordoliberalism and, 46, 50–51, 97
 Social Market Economy and, 43
 Suez Crisis (1956) and, 81–82
ESSO-Exxon
 concentration of power and, 30–31, 81, 98, 124, 128
 infrastructure, 78–79
 mercantilism and, 140, 141–42
 Middle East, in, 74, 75
 national interests of, 95
 natural gas industry and, 242–43, 247, 264
 oil and, 67
 oil prices and, 36–37
 pipelines, 78–79
 power of, 83
 Social Market Economy and, 34–35, 79
 taxation and, 43
Estonia, independence of, 260
Ethics Commission, 301
Etzel, Franz, 105
Eucken, Walter
 neoliberalism versus, 190, 227
 Ordoliberalism and, 48–49, 50–51, 52
 state intervention, on, 207–8

European Association for Renewable Energy (EUROSOLAR), 276–77, 289
European Atomic Energy Community (Euratom)
　breeder reactors and, 121–22
　creation of, 82, 106
　debates in, 106–7
　energy forecasts and, 116–17
　financial assistance from, 113–14
　non-military vision for nuclear energy, 105–7
　resource shortages and, 120
　US, cooperation with, 111–12
　"Wise Men," 105, 106–7
European Coal and Steel Community (ECSC)
　generally, 82, 91–92, 128, 131–32
　creation of, 20
　energy policy, 85
　High Authority, 27–28, 39–40, 43, 85, 91–92
　integration of Europe and, 105
　pricing regimes of, 27–28, 51–52
　Suez Crisis (1956) and, 39–40
　wages in coal industry and, 42–43
European Commission
　electricity sector in, 274–75
　energy liberalization and, 262
　neoliberalism and, 10–11, 208, 213, 226–27, 269
　oil industry and, 91–92
　renewable energy and, 277
　Single Europe Act, 226–27
European Community
　coal industry in, 43
　Council of Ministers, 85, 131–32
　energy security in, 66–67
　natural gas industry in, 259
　neoliberalism in, 226–27
　nuclear energy industry in, 121–22
　Oil Shock (1973) and, 131–32
　Soviet oil and, 241–42
　tension with, 7
European Council
　coal industry and, 27–28, 39–40, 43
　energy liberalization and, 262
European Court of Justice, 294
European Economic Community (EEC)
　creation of, 82
　oil industry and, 92
　stockpiling of reserves and, 91–92
European Investment Bank, 259
European Nuclear Congress, 178
European Parliament, energy liberalization and, 262
European Reconstruction Program. see Marshall Plan
European Recovery Program, 109–10, 113–14
European Union
　generally, 3–4
　creation of, 262
　Emissions Trading Scheme (ETS), 294, 297, 304–5

　energy liberalization in, 262
　renewable energy in, 278
Expert Committee for Energy Research and Technology, 144
externalities, 191–92, 204, 205, 291
Exxon. see ESSO-Exxon

Fabeck, Wolf von, 276–77, 278, 289, 291, 294, 377n.1
Falkenheim, Ernst, 34–35
fast breeder reactors, 121–22, 139–40, 174–75
FDP. see Free Democratic Party (FDP)
Federal Association of Citizens' Initiatives for Environmental Protection (BBU), 161, 165, 171, 173–74
Federal Association of Solar Energy, 234
Federal Association of Young Entrepreneurs, 292
Federal Cartel Office, 134–35
Federal Environmental Agency, 204
Federal Statistical Agency, 60
Federal Wind Association, 289
Federation of German Industry, 148
Fell, Hans Josef, 285, 289, 290–91, 293–94, 300
Fermi, Enrico, 121
Finance Ministry, nuclear energy industry and, 118
First Grand Coalition (SPD–CDU), 89, 117, 118, 160
First Revision to the Energy Program (1974), 133
Fischer, Joschka
　generally, 286
　decentralization, on, 279–80
　Eco-Alliance and, 279, 280
　energy liberalization, on, 284
　environmental movement and, 160–61, 170–71
　Kosovo War and, 285
　nuclear energy industry and, 217
Fischer-Menshausen, Herbert, 43
Flick Affair, 140, 223
Ford, 76
Ford Foundation, 188
Forest Death, 199–201, 200f
fossil fuels
　benefits of, 4–5
　coal industry (see coal industry)
　energy efficiency and, 8
　historical growth in use of, 4, 5–6
　inexpensive nature of, 7
　Marxian perspective on, 8
　oil industry (see oil industry)
　renewable energy versus, 295
　waste and, 5
Fourth Atomic Research Program, 141
Framework Convention on Climate Change (UNFCCC), 267–68, 270–71
France
　Allied occupation of Germany (see Allied occupation of Germany)
　coal industry in, 39–40, 85

Communism in, 173
dirigisme in, 11, 12, 128
ECSC and, 27–28
energy consumption in, 2f, 2, 29–30, 157
energy efficiency in, 148
energy intensity in, 3f
energy policy in, 51
energy security concerns in, 65–66, 92
Marginalism, 65–66
mercantilism in, 141–42
Ministry of Justice, 167–68
natural gas industry in, 251
1970s, new energy policy in, 157
nuclear energy industry in, 102–3, 106, 108, 112, 113, 122–23, 164, 182
oil industry in, 33, 63–64, 77–78, 82, 84, 97, 133
Oil Shock (1973) and, 131–32
protests against nuclear energy in, 164–65, 167–68
recent developments in, 297
renewable energy in, 297
state intervention in energy in, 51
Suez Crisis (1956) and, 71
Fraunhofer Institute for Solar Energy Systems, 198, 232
Fraunhofer Institute for Systems Technology and Innovation Research, 205
Free Democratic Party (FDP)
  CDU, coalition with, 213–14, 216–17, 257, 273, 300
  coal industry and, 273
  electoral performance of, 173–74
  energy planning and, 128–29
  energy tax and, 293
  neoliberalism and, 227
  nuclear energy industry and, 216–17, 273
  oil, taxation of, 181
  Oil Shock (1973) and, 129–30
  Peace Marches and, 181
  SPD, coalition with, 95, 128–29, 161, 171, 181
Fridays for Future, 308
Friderichs, Hans, 126–27, 129–31, 134–35, 146, 147
Friedman, Milton, 190–91, 227
Fukushima Daiichi nuclear accident (2011), 300–1, 304
future studies, 47

Gabriel, Sigmar, 298, 305, 307
Gaddafi, Muammar, 253–54
Galbraith, John Kenneth, 59
Garzweiler II coal mine, 273, 274f, 308
Gazprom, 238–39, 260–61, 263, 265
Gelsenberg Benzin AG (GBAG), 95, 134–35, 143–44
General Agreement on Tariffs and Trade (GATT), 226–27

General Electric, 113
General Theory of Employment, Interest, and Money (Keynes), 189
Geneva Convention, 111
geothermal energy, 290
German Democratic Republic (GDR)
  Basic Treaty, 247
  carbon emissions in, 225
  energy consumption in, 225
  fall of, 224–25
German Economic Research Institute (DIW), 47, 67, 193
German Farmers Association, 289
German Fishing Association, 159
German Physics Society (DPG), 219–20
German Research Council, 102
Giordani, Francesco, 105
globalization, 6
global warming. *see* climate change
Goetzberger, Adolf, 232, 233
"Golden Girdle," 72–73
Gore, Al, 304
Göttingen Manifesto, 104f, 105, 106
Graf Bismarck coal mine, 86f, 86, 88f
Grass, Günter, 268f
Greece, Communism and, 73–74
Green Budget Germany, 283–84
green energy. *see* renewable energy
Green List for Environmental Protection, 173–74
Green Party
  climate change and, 221–22
  creation of, 173–74
  decoupling process and, 182
  divisions in, 279
  Eco-Alliance (*see* Eco-Alliance (SPD–Green Party))
  Ecological Modernization and, 209, 259, 282
  electoral success of, 213–14
  Electricity Feed-in Law (1990) and, 280
  electricity sector and, 214–16, 234, 235–36
  energy security and, 310–11
  energy transitions and, 15, 157
  environmental issues and, 237
  environmental movement (*see* environmental movement)
  natural gas industry and, 264
  nuclear energy industry and, 217, 219
  oil price decline and, 231
  origins of, 157
  renewable energy and, 270, 282–83
  Reunification and, 225–26
  Sonderweg (special path), 12, 157
  SPD, coalition with (*see* Eco-Alliance (SPD–Green Party))
  taxation and, 292
  threat to SPD, as, 157, 174, 182
  wind energy and, 280

Greenpeace, 308
Green *Sonderweg* (special path), 12, 157
Group of Seven (G-7), 137–38
Gruhl, Herbert, 169–70, 173–74, 198
Gulf, 74, 230
Gulf War, 271
Gundremmingen nuclear power plant, 113–14, 114f, 118, 119
Gutermuth, Heinrich, 29, 40–43, 44–45

Häfele, Wolf, 121–22, 139–40, 144–45
Hamm, Walter, 291–92
Hanover Trade Fair (1969), 246
Hansen, Jim, 219
Harrod, Roy, 59
Hartley, Harold, 31–32, 35, 53–54, 328n.40
Harvey, Hal, 299
Hatzfeldt, Hermann, 199–201
Hauff, Volker
    Ecological Modernization and, 181–82, 221–22
    Electricity Feed-in Law (1990) and, 234, 235
    energy efficiency and, 180
    environmental policy, 216, 217–19
    nuclear energy industry and, 178–79
    "Path of Oil," 177–78
    reform wing of SPD, in, 305
    restructuring of energy and, 177
Hayek, Friedrich, 190–91, 203, 226, 227
Heath, Edward, 141–42
Heisenberg, Werner, 102–3, 104, 108
Hennicke, Peter, 209–10, 215, 220, 234, 301
Herzog, Roman, 272
Hicks, John, 189
Hiroshima bombing (1945), 101–2
Hirsch, Etienne, 344n.96
historical background, 4–7
    energy consumption, growth in, 4, 5–6
    fossil fuels, growth in use of, 4, 5–6
    naturalization of energy consumption, 6–7
Hitler, Adolf, 20, 21, 23, 75–76
Hobsbawn, Eric, 5
Höchst, 113, 123–24
Hohmeyer, Olav, 205
Holocaust, 5, 6
Holzwarth, Fritz, 207–8
Hotelling, Harold, 186–87
Hudson, Edward A., 188, 194–95
Hustedt, Michaela
    competition, on, 290–91
    Electricity Feed-in Law (1990) and, 280, 289, 290
    energy efficiency and, 308–9
    nuclear energy industry and, 300
    power of utilities, on, 283, 284
    renewable energy and, 270, 281–82, 285, 286–87, 298–99
    solar energy and, 288
    state intervention, on, 294

IG Farben, 30, 260
IG Metall, 289–90
India, electricity demand in, 119–20
Industrial Mining Union (IGB), 27, 29, 40, 41–43
inflation, 24–25
Institute for Applied Ecology. *see* Eco-Institute
Institute for Economic Research, 292
Institute for Energy Economics and Rational Energy Use, 258
Institute for Global Economics, 32
Interior Ministry
    coupling paradigm and, 195
    Division U, 161
    nuclear energy industry and, 137
internal combustion engine (ICE), 72, 75–77
International Authority of the Ruhr (IAR), 25
International Conference on Atomic Energy, 102–3
International Energy Agency (IEA), 128, 131–32, 231–32, 255
International Institute for Applied Systems Analysis (IIASA), 139
International Institute for Society and the Environment, 197, 198
International Panel on Climate Change (IPCC), 4, 269, 271, 304, 310, 311
International Petroleum Exchange, 230–31
Investment Aid Law (1952), 26, 29
Iran
    Islamic Republic, 178
    natural gas production in, 256
    oil exports from, 80
    oil industry in, 96
    revolution in, 178–79, 251
Iraq
    Kuwait, invasion of, 271
    oil embargo (1967) and, 90–91
    oil exports from, 80
    oil price increase, 96
Iraq Petroleum Company (IPC), 75
Ireland in European Community, 131–32
Israel
    1973 War and, 125–26
    Six Day War and, 90
    Suez Crisis (1956) and, 71
Italy
    coal industry in, 39–40, 85
    Communism in, 173
    energy consumption in, 29–30
    natural gas industry in, 251
    oil industry in, 63–64, 92, 133–34

Japan
    decoupling in, 229

INDEX

digital revolution in, 272
Fukushima Daiichi nuclear accident (2011), 300–1, 304
nuclear weapons, use of in, 101–2
renewable energy in, 287–88
steel industry in, 254
Jaruzelski, Wojciech, 253
Jevons, Stanley, 58
jobs
  conservation and, 180–81
  energy efficiency and, 180–81
  environmental movement, juxtaposition with, 179–80
  nuclear energy industry and, 218
  solar energy and, 218
  SPD and, 179, 218
  wind energy and, 218
Johnson, Lyndon B., 64–65
Jorgenson, Dale W., 188, 194–95
Jungjohann, Arne, 11
Jungk, Robert, 171–72, 268

Karlsruhe Institute, 108–9
Kelly, Petra, 165, 166, 173–74, 198
Kennedy, John F., 64–65, 244
Keynes, John Maynard, 185, 189
Keynesianism
  generally, 59
  coal industry and, 129–30
  energy policy, 176–77
  energy tax and, 206
  environmental issues, problematic nature of, 195–96
  neoclassical economics and, 189
  neoliberalism versus, 185, 190
  Ordoliberalism versus, 185
Keyser, Theobald
  coal industry and, 27, 28–29, 34, 40–41, 87, 89
  oil industry and, 37, 43
Khomeini, Ayatollah, 178
Khrushchev, Nikita, 247–48
Kiel Institute for the World Economy, 52, 193–94
Kissinger, Henry, 131
Kneese, Allen, 188, 203
Kohl, Helmut
  carbon emissions and, 273
  Chernobyl nuclear accident (1986) and, 211–12
  climate change and, 220, 267–69
  coupling paradigm and, 257
  electoral success of, 213–14
  energy efficiency and, 272
  energy policy, 236
  environmental issues and, 223
  neoliberalism and, 227
  nuclear energy industry and, 216–17
  Reunification and, 224–25, 269, 271
Köhler, Horst, 298

Kontinental Öl AG, 30
Korean War, 20, 23–24, 25–26, 62, 64
Kosovo War, 285
Kosygin, Alexei, 247–48, 251–52
Kraftwerk Union (KWU)
  coal gasification and liquifaction and, 140
  nuclear energy and, 99, 119–20, 124, 138, 144, 145, 150, 175–76
Kratzmüller, Emil, 78
Kreditanstalt für Wiederaufbau (KfW), 288, 290–91, 294, 298–99
Krekeler, Heinz, 338n.1
Kriele, Rudolf, 107
Krupp, 19, 251–52
Kuwait
  Iraqi invasion of, 271
  oil embargo (1967) and, 90, 91
Kyoto Protocols (1997), 267–69, 294, 304–5

Lafontaine, Oskar, 225–26, 280–81, 285, 292
Lambsdorff, Otto Graf, 227
Lantzke, Ulf, 128–30, 133–34, 337n.111
Large Wind-Energy Installation (GROWIAN), 231–33
Latvia, independence of, 260
Law Promoting Stabilization and Growth in the Economy (1967), 60
Lawrence Livermore Labs, 115, 165–66
Law to Secure the Energy Supply (1973), 129
Least Cost Planning (LCP), 215
Lebanon, oil embargo (1967) and, 90
Lehbert, Berndt, 193–95, 201–2
Liberal Party, 310–11
Libya
  oil embargo (1967) and, 90–91
  oil industry in, 95, 253–54
Liebrucks, Manfrfed, 67
Liesen, Klaus, 249–51, 258
Limits to Growth (Club of Rome), 162, 164, 172, 184, 186, 190–91, 197–98, 204
Lithuania, independence of, 260
Los Alamos Laboratory, 115
Lovins, Amory, 172–73, 175, 215

malleability of energy economy, 202–3
Malthus, Thomas, 59
Mandel, Heinrich, 113–14, 345n.1
Manhattan Project, 115
Mannesmann AG
  coal gasification and liquifaction and, 140
  natural gas industry and, 245, 246–48, 251–52, 253, 255, 263, 265
market failures, 187–88
market pricing
  oil industry, in, 63–64
  Ordoliberalism and, 50–51
  resource economics and, 186–87

Marshall Plan, 20, 30–31, 102, 288
Marxist Student Association Spartacus, 280
Massachusetts Institute of Technology, 162, 186
Mattei, Enrico, 133–34, 135
Matthöfer, Hans
  coupling paradigm and, 149–50, 177, 194
  Ecological Modernization and, 182
  nuclear energy industry and, 136–37, 138, 140–41, 174–76
  taxation, on, 180–81
Max-Planck-Institute, 175
McCloy, John J, 25–26
McKelvey, Victor, 65
McNeill, J.R., 57–58
Menne, Wilhelm Alexander, 112–13, 123–24
mercantilism
  coal industry, in, 140–41
  failure of, 141–45
  natural gas industry and, 135–36, 145
  nuclear energy industry and, 137–40, 144–45
  oil industry and, 133–35, 141–44
  Oil Shock (1973), mercantilist responses to, 133–45
  SPD and, 141
Merkel, Angela
  generally, 272
  carbon emissions and, 273
  carbon tax and, 273
  climate change and, 270–71
  competition, on, 274–75
  energy policy, 304
  energy security and, 310–11
  natural gas industry and, 307
  nuclear energy industry and, 273, 301
  Renewable Energy Act and, 305–6
  renewable energy and, 277, 298, 304
Metternich, Anton, 158
Mexico
  oil production in, 229
  steel industry in, 254
Meyer-Abich, Klaus Michael, 175, 198–99, 220
Middle East Emergency Committee (MEEC), 74
Mining and Energy Union, 175–76
Mittelstand companies, 2
Mobil, 74, 81
Moldova, independence of, 260
Monnet, Jean, 27–28, 105–6, 128
Mont Pèlerin Society, 49
Morris, Craig, 11
Moscow Treaty (1970), 247
Müller, Werner, 286, 293
Müller, Wolfgang D., 347n.41
Müller-Armack, Alfred
  air pollution and, 158
  coal industry and, 51–52

  concentration, on, 185
  oil industry and, 81
  Social Market Economy and, 49, 50–51, 183–84
Mumford, Lewis, 58–59
Mussolini, Benito, 133

Nagasaki bombing (1945), 101–2
Nasser, Gamal Abdel, 71, 80–81, 96
National Iranian Oil Company (NIOC), 96
natural gas industry
  CDU and, 240–41, 247, 257–59, 261, 307
  climate change and, 264–65
  commercial combined cycle gas turbines (CCGT), 259, 263
  coupling paradigm and, 240–41, 255–56, 263, 264
  Ecological Modernization and, 237, 240–41, 256, 264
  Economics Ministry and, 246, 249
  emergence of, 242–43
  energy liberalization and, 260, 262
  energy security and, 240, 249–50
  energy transitions and, 12–13, 240–41
  Green Party and, 264
  hard energy path, 264
  high price of natural gas, 243
  increased reliance on, 3–4
  Liquefied Natural Gas (LNG), 250–51
  mercantilism and, 135–36, 145
  neoliberalism and, 263
  Nord Stream pipelines, 265–66, 293–94, 307
  oil, transition from, 229
  Oil Shock (1973), mercantilist responses to, 135–36, 137–40, 145
  OPEC, as alternative to, 240
  Ostpolitik and, 244, 246, 247, 250–51, 263
  problems in, 264–65
  Russia, imports from, 239–41, 264–65, 306
  Soviet Union, imports from, 12–13, 14, 135–36, 239–40, 244–56, 263, 264, 296, 306
  SPD and, 240–41, 249, 256, 264, 296, 307
  taxation, 258
  third party access (TPA), 261, 263
  Yamal pipeline, 251–54, 255–56, 307
neoclassical economics
  generally, 47
  Ecological Modernization versus, 185
  economic growth and, 58–60
  environmental issues, problematic nature of, 195–96
  Keynesianism and, 189
  neoliberalism and, 190
  resource economics and, 186
  skepticism regarding, 197–98
  technology and, 197–98

neoliberalism
  generally, 6
  CDU and, 227, 305–6
  cheap energy paradigm versus, 10–11
  climate change and, 269
  divergence of German energy policy from, 11–12
  Ecological Modernization versus, 185
  Economics Ministry and, 227
  emissions trading and, 191–92
  environmental movement, effect on, 213
  externalities and, 191–92
  FDP and, 227
  GATT and, 226–27
  Keynesianism versus, 185, 190
  natural gas industry and, 263
  neoclassical economics and, 190
  Ordoliberalism compared, 184, 190, 291–92
  producer versus consumer behavior, 192–93
  SPD and, 305–6
neo-mercantilism, 127
Netherlands
  natural gas production in, 251
  oil industry in, 30–31, 63–64, 97, 133
  Oil Shock (1973) and, 131–32
New Analyses for Growth and the Environment (NAWU), 196, 202, 204
New York Mercantile Exchange, 230–31
Nixon, Richard, 245–46
Nolte, Paul, 11
Nordhaus, William, 186–87, 189, 197–98
Nord Stream pipelines, 265–66, 293–94, 307
North Atlantic Treaty Organization (NATO), 102–3, 241–42, 260
Norway
  natural gas production in, 251, 256
  oil production in, 229
Nossener Brücke CCGT plant, 259
nuclear energy industry
  attempts to move away from, 216–19
  Bad Godesberg program, 108
  CDU and, 109, 113, 115, 116, 117, 118, 121–22, 216–17, 218–19, 273, 287, 300
  chemical industry and, 111–12
  Chernobyl nuclear accident (1986), 211–12, 216–17, 220, 222–23
  concentration in, 99–100, 124
  coupling paradigm and, 116–17, 136–37, 175, 176
  dangers of, 122, 137
  debates regarding, 106–7, 110
  detachment from military, 104–5
  divergence of BDR energy policy, 167–68, 182
  Eco-Alliance and, 287
  Ecological Modernization and, 217, 218
  economic development and, 106–7
  Economics Ministry and, 102, 107, 108–11, 113–14, 118
  economies of scale and, 118
  electricity sector and, 111
  Eltville Program, 110–11
  energy forecasts and, 116–17
  energy transitions and, 12–13, 14–15, 100–1
  entrepreneurial state and, 112–15, 123
  fast breeder reactors, 121–22, 139–40, 174–75
  FDP and, 216–17, 273
  financial risk in, 109–10
  Fourth Atomic Research Program, 141
  Fukushima Daiichi nuclear accident (2011), 300–1, 304
  Green Party and, 217, 219
  Interior Ministry and, 137
  jobs and, 218
  lack of investment in, 111
  Marshall Plan and, 102
  mercantilism and, 137–40, 144–45
  military considerations in, 103
  nuclear waste, problem of, 166–67
  Oil Shock (1973), mercantilist responses to, 137–40
  OPEC, as counterstrategy against, 164
  Ordoliberalism and, 102, 118, 123
  Parliamentary Inquiry into Nuclear Power, 198–99, 202–3
  phasing out of, 300
  private enterprise and, 109–10, 112
  problematic policies in, 124
  promise of, 101–2
  protests against, 155–56, 156f, 164–67, 171–73, 198, 296
  rapid growth of, 118–20, 122–23
  research and development in, 115–16
  resource shortages in, 120
  role of state in creation in, 100–1
  Social Market Economy and, 108–9, 116
  SPD and, 100–1, 103, 105–6, 108, 109, 116, 117, 118, 123, 124, 164, 174–76, 178–79, 217, 219, 287
  Spitzingsee Agenda, 115
  state intervention in, 107–8, 115–16
  technocratic approach, 124
  Technology Ministry and, 137
  Third Atomic Program, 121
  Three Mile Island nuclear accident (1979), 178, 179–80
  uranium enrichment, 120–21
Nuclear Research Institute, 258
nuclear weapons, use in Japan, 101–2

Oak Ridge National Laboratory, 121–22
Obama, Barack, 297
Ohnesberg, Benno, 160–61

oil industry
  automobile industry and, 75–77
  buyouts in, 90, 91
  CDU and, 20, 92
  coal, transition from, 19–20, 29–30, 43, 46, 60–61, 73
  comparison with coal industry, 34–35
  competition in, 63–64
  concentration in, 94–95, 98, 134–35
  divergence of German policy and, 13–15
  Economics Ministry and, 34–35, 36, 43, 82, 83–84, 93, 95, 134–35
  energy forecasts and, 57, 61
  energy transitions and, 13–15
  foreign producers, reliance on, 30–31
  "Golden Girdle," 72–73
  increased consumption of oil, 77–78, 84
  increased production of oil, 74, 74f, 84
  infrastructure, development of, 31, 75, 78–80
  internal combustion engine (ICE) and, 72, 75–77
  liberalization of, 82
  market pricing in, 63–64
  Marshall Plan and, 30
  mercantilism and, 133–35, 141–44
  natural gas, transition to, 229
  neo-mercantilism in, 127
  Oil Shock (1973), mercantilist responses to, 133–35
  OPEC, oil production outside of, 229
  pipelines, 78–79, 80
  price decline as countershock, 228–31, 230f, 236
  price increases and, 285
  pricing as policy, 151
  pricing regimes in, 35–36, 37
  productivity in, 35
  refineries, 78
  scarcity in, 62
  Social Market Economy and, 34–35
  Soviet Union, imports from, 33, 241–42
  SPD and, 92, 94–95, 273–74
  stockpiling of reserves, 91–92
  substitution of coal as result of Oil Shock (1973), 148
  supply chain in, 72–73
  tariffs, 33
  taxation of oil, 180–81
  tax policy and, 84
Oil Shock (1973)
  car-free Sundays and, 125, 126f, 130f
  CDU and, 151
  coal industry, mercantilist responses in, 140–41
  conservation and, 148
  coupling paradigm and, 127, 149, 150–51
  Economics Ministry and, 131
  effects of in BDR, 125–27, 129, 130f
  embargo of oil and, 125–26, 129, 132

  energy efficiency and, 12–13, 148, 296
  energy forecasts and, 149
  energy transitions and, 12–13
  environmental movement and, 163
  FDP and, 129–30
  mercantilist responses to, 133–45
  natural gas industry, mercantilist responses in, 135–36, 137–40, 145
  nuclear energy industry, mercantilist responses in, 144–45
  oil industry, mercantilist responses in, 133–35, 141–44
  price increase and, 125–26, 129, 131, 146–47
  pricing as policy, 151
  recession resulting from, 137–38, 147
  resource shortages and, 186
  Social Market Economy and, 151
  solar energy, impact on, 205
  SPD and, 129–30, 151
  spot market, reliance on, 146–47
  substitution of coal as result of, 148
  technocratic agenda, as revealing weaknesses of, 150
  wind energy, impact on, 205
100,000 Roofs, 1–2, 288–89, 294, 298–99
100 MW Wind Demonstration Program, 228
OPEC. see Organization of Petroleum Export Countries (OPEC)
Ordoliberalism
  generally, 15
  overview, 46
  CDU and, 107–8, 227, 228, 236
  coal industry and, 51–52
  competition and, 49, 64
  conservative nature of, 49–50
  coupling paradigm and, 47–48, 183–84
  decentralization and, 49–50, 92, 279–80
  Eco-Alliance and, 283–84, 290–91
  Ecological Modernization versus, 185, 207–8
  economic growth and, 60
  Economic Miracle and, 123
  energy crises and, 194
  energy forecasts and, 54, 56, 68–69
  energy policy, 46–47, 51, 61
  environmental issues, problematic nature of, 195–96
  free market and, 128
  Keynesianism versus, 185
  market pricing and, 50–51
  neoliberalism compared, 184, 190, 291–92
  neo-mercantilism versus, 127
  nuclear energy industry and, 102, 118, 123
  origins of, 48–49
  Social Market Economy and, 50–51, 184
  technocratic corporatism versus, 73
  Third Path and, 49–50
  Third Reich versus, 49

Organisation for Economic Co-operation and
    Development (OECD)
  conservation and, 148
  energy forecasts and, 53
  Oil Shock (1973) and, 147
Organisation for European Economic
    Co-operation (OEEC)
  coal industry and, 33–34
  energy forecasts and, 31, 47, 52–54, 63
  Oil Committee, 91–92
  stockpiling of reserves and, 91–92
  Suez Crisis (1956) and, 81, 82
Organization of Arab Petroleum Exporting Countries
    (OAPEC), 125–26, 129, 132, 140, 146, 249
Organization of Petroleum Export
    Countries (OPEC)
  generally, 140, 151
  contracts with, 307
  counterstrategies against, 131–32
  creation of, 75
  decrease in oil production, 229
  natural gas as alternative to, 240, 256
  nuclear energy as counterstrategy against, 164
  oil production outside of, 229
  price decline as countershock, 228–29, 231
  price increases, 95, 96, 125–26, 129, 131, 146–
    47, 178
  volatility of, 253–54
Ostpolitik, 244, 246, 247, 250–51, 263

Pahlavi, Mohammad Reza, 178
Paley Commission, 59
Paley Report, 59, 64
Paris Climate Accords (2015), 4
Paris Climate Conference (2015), 308
Paris Treaty (1951), 27–28
Parliamentary Inquiry into Nuclear Power, 198–
    99, 202–3
"Path off Oil," 177–78
Peace Marches (1981-1982), 181
photovoltaic cells (PV), 231–32, 233, 276–77,
    299–300, 300f
Pigou, Arthur Cecil, 184, 187–88, 204–5, 209
Poland, martial law in, 253
politicization of energy
  climate change and, 219–22, 237
  energy economics and, 14
  energy security and, 14
  origins of, 15–16, 21
  price and, 13–14
pollution
  air pollution, 158, 184
  emissions trading and, 191–92
  environmental economics and, 204
  externality, as, 191–92
  market failure, as, 187–88
  river pollution, 159, 184

Post-War rebuilding of Germany, 22
power surges, 303
Preussen-Elektra, 118–19, 273–74, 277
price elasticity, 195, 201–2
price-induced technological development, 188,
    189, 202, 204
primary energy, 328n.36
Program for the Ecological Modernization of the
    Economy, 181–82
Prometheus, 159
Promoting the Rationalization of Hard Coal
    Mining (1963), 85
Putin, Vladimir
  generally, 3–4, 239f, 306
  natural gas exports and, 238–39, 265–66, 307
  Nord Stream pipelines and, 265–66, 307
  Ukraine War and, 310, 311

Qatar
  natural gas production in, 256
  oil embargo (1967) and, 91

Radford, Arthur, 103
Rationalization Type I, 202
Rationalization Type II, 202–3
Reagan, Ronald
  free market and, 226, 227
  neoliberalism and, 11, 184–85, 190–91
  optimism of, 201, 213
  Soviet Union and, 252–53
Reconstruction Credit Organization, 109–10
Red Army Faction, 279–80
reform gridlock, 278–79
refugees, 305
renewable energy
  CDU and, 228, 233
  Eco-Alliance and, 280–82, 285, 287,
    289–90, 294
  Ecological Modernization and, 205, 263, 270, 289
  energy transitions and, 12–13, 270, 298–99,
    301–2
  externalities and, 205
  financing of, 287–88
  fossil fuels versus, 295
  Green Party and, 270, 282–83
  modernization and, 301–2
  oil price decline, effect of, 231–32
  100,000 Roofs, 1–2, 288–89, 294, 298–99
  photovoltaic cells (PV), 231–32, 233, 276–77
  price gap in, 205
  Renewable Energy Act (2000), 290, 291, 294,
    298–99, 303, 305–6
  rise of, 231–33
  risks to, 277–78
  solar energy (see solar energy)
  SPD and, 270
  wind energy (see wind energy)

Renewable Energy Act (2000), 290, 291, 294, 298–99, 303, 305–6
Research and Technology Ministry, 194–95
resource economics
  backstop technology and, 187
  flawed markets and, 187
  market pricing and, 186–87
  neoclassical economics and, 186–87
Resources for the Future (RFF), 59, 64, 203
resource shortages
  coupling paradigm, impact on, 163–64
  nuclear energy industry, in, 120
  Oil Shock (1973) and, 186
Reunification
  carbon emissions and, 225
  climate change and, 269, 271
  costs of, 271–72
  environmental movement, effect on, 213, 225–26
  Green Party and, 225–26
  recession resulting from, 272
  SPD and, 225–26
Rheinische Energie AG, 54–56
Rheinisch-Westfälisches Elektrizitätswerke (RWE), 134–35
  coal gasification and liquifaction and, 140
  coal industry and, 273–74, 304–5, 308
  concentration of power and, 214
  electricity sector and, 111, 117
  losses of, 302
  nuclear energy and, 112, 113–14, 118–19, 124, 144
  power of, 283–84
  renewable energy and, 278
Ricardo, David, 59
Riesenbauer, Heinz, 228
Rio Earth Summit (1992), 267–69, 270–71
river pollution, 159, 184
Robinson, Austin, 53–54, 56, 58
Röpke, Wilhelm
  concentration, on, 185, 227
  decentralization and, 279
  ECSC and, 51–52
  market pricing, on, 183–84
  neoliberalism versus, 190
  Ordoliberalism and, 49–51, 52, 60
  state intervention, on, 207–8
Royal Dutch Shell
  concentration of power and, 30–31, 97
  mercantilism and, 140, 142–43
  Middle East, in, 75
  natural gas industry and, 242–43, 247, 264
  oil and, 67, 230
  oil prices and, 36–37
  pipelines, 78–79
  power of, 34–35, 83, 128
  Social Market Economy and, 79
  taxation and, 43

Ruhr. see coal industry
Ruhrgas AG
  generally, 265
  concentration of power and, 214
  natural gas and, 238, 242–43, 246–48, 249–52, 253, 257, 258, 260–61, 262–64, 265
  oil gasification and liquifaction and, 140
  power of, 307
Ruhrkohle A.G. (RAG), 94, 98, 124, 133–34, 140
Russia. see also Soviet Union
  generally, 3–4
  Crimea, annexation of, 305, 306, 307–8, 310
  natural gas imports from, 239–41, 264–65, 306
  post-Soviet emergence of, 260
  sanctions against, 307
  Ukraine War and, 306, 310–11
RWE. see Rheinisch-Westfälisches Elektrizitätswerke (RWE)

Sadat, Anwar, 96
Sandia Laboratory, 115
Sasbach Sundays, 232
Saudi Arabia
  decrease in oil production, 229
  1973 War and, 125–26
  oil embargo (1967) and, 90–91
  oil exports from, 80
  oil industry in, 73–75, 92
  Suez Crisis (1956) and, 71
Scheer, Hermann
  generally, 304, 317n.1
  competing energy paradigms and, 295–96
  competition, on, 290–91, 294
  death of, 302
  energy efficiency and, 308–9
  energy transitions and, 16
  nuclear energy industry and, 300
  Ordoliberalism and, 291–92
  reform wing of SPD, in, 305
  renewable energy and, 1–3, 235, 268–69, 268f, 270, 276–77, 282, 285, 287–88, 289, 290, 294, 298–99
  solar energy and, 268–69
Schefold, Bertram, 185, 197, 203–4, 205–6, 207, 209
Schiller, Karl, 89, 93–94, 95, 129–30
Schmid, Thomas, 279–80
Schmidt, Helmut
  generally, 163f, 174, 180–81
  arms race and, 181
  coal industry and, 89
  energy policy, 94
  environmental issues and, 163–64
  G-7 and, 137–38
  jobs and, 179
  loss of influence, 181–82
  natural gas industry and, 250–52, 253–54, 265

nuclear energy industry and, 136–37, 174–76, 178–79, 182
oil industry and, 95, 251
Oil Shock (1973) and, 129–31, 134, 147
Ostpolitik and, 250–51
Schneider, Hans Karl, 62–63, 67–68, 185, 197–98, 202, 330n.79
Schöller, Heinrich, 111
Scholz, Olaf, 311
Schönau Power Rebels, 220, 278
Schröder, Gerhard
　generally, 280, 298
　automobile industry and, 293
　energy policy, 293
　free markets and, 280
　labor market reform and, 293–94
　natural gas industry and, 265–66
　Nord Stream pipelines and, 307
　nuclear energy industry and, 286
　taxation and, 292
Schumacher, Ernst Friedrich, 159–60, 170, 171, 268
Schumann, Robert, 27
Schwab, Günther, 158, 159–60
Scientific Research Ministry, 115
Second Grand Coalition (SPD–CDU), 298, 305–6
Second Parliamentary Inquiry into Climate Change, 258
Second Revision to the Energy Program, 176–77
shale gas, 297
Shell. *see* Royal Dutch Shell
Shelley, Mary, 79
Siebert, Horst, 203, 204, 207
Sieferle, Rolf Peter, 202
Siemens, 99, 112, 118–19, 121–22
"Silent revolution," 169
Simon, Julian, 190–91, 192–93
Simonis, Udo Ernst
　generally, 208
　Ecological Modernization and, 185, 291
　energy tax and, 205–6, 207, 209
　free markets, on, 197
Single Europe Act, 226–27
Sinn, Hans Werner, 303
Six Day War (1967), 90–91
Slovenia, independence of, 271
Small is Beautiful (Schumacher), 159–60
Social Democratic Party (SPD)
　Bad Godesberg program, 108
　carbon tax and, 223
　CDU, coalitions with, 89, 117, 118, 160, 298, 305–6
　climate change and, 221–22
　coal industry and, 86, 140, 273
　coupling paradigm and, 175, 176, 194, 305–6
　decoupling process and, 182

　divisions within, 174–75, 177–78, 281–82
　Eco-Alliance (*see* Eco-Alliance (SPD–Green Party))
　Ecological Modernization and, 209, 259, 280, 282
　Economics Ministry and, 89, 93
　ECSC and, 27
　electricity sector and, 216, 234, 235
　end of rule, 181–82
　energy efficiency and, 177, 180, 182, 296
　energy forecasts and, 54–56
　energy planning and, 128–29
　energy security and, 82–83, 307–8, 310–11
　energy tax and, 292
　energy transitions and, 15, 157
　environmental issues and, 161, 162–64, 177–78, 237
　FDP, coalition with, 95, 128–29, 161, 171, 181
　First Grand Coalition, 89, 117, 118, 160
　Green Party, coalition with (*see* Eco-Alliance (SPD–Green Party))
　internal reform in, 157
　jobs and, 179, 218
　lack of leftist alternative to, 173
　mercantilism and, 141
　natural gas industry and, 240–41, 249, 264, 296, 307
　natural gas production and, 256
　neoliberalism and, 305–6
　new generation of leadership in, 181–82
　nuclear energy industry and, 100–1, 103, 105–6, 108, 109, 116, 117, 118, 123, 124, 164, 174–76, 178–79, 217, 219, 287
　oil, taxation of, 181
　oil industry and, 92, 94–95, 273–74
　oil pricing regimes and, 36, 37
　Oil Shock (1973) and, 129–30, 151
　Peace Marches and, 181
　reform gridlock and, 278–79
　renewable energy and, 270
　research and development and, 141
　Reunification and, 225–26
　Second Grand Coalition, 298, 305–6
　Social Market Economy and, 183–84
　Sonderwag and, 12
　Soviet oil and, 242
　supply-side programs in energy sector and, 127–28
　technocratic corporatism and, 73, 97
　threat from Green Party, 157, 174, 182
Social Market Economy
　basic principles of, 183–84
　CDU and, 183–84, 227
　coal industry and, 26, 34–35, 36–37
　energy policy and, 7
　environmental concerns with, 184
　limits of, 20–21

Social Market Economy (*cont.*)
  nuclear energy industry and, 108–9, 116
  oil industry and, 34–35
  Oil Shock (1973) and, 151
  Ordoliberalism and, 50–51, 184
  SPD and, 183–84
  triumphalism and, 20–21
solar energy
  generally, 1–2
  challenges of, 295
  Eco-Alliance and, 287–90
  Electricity Feed-in Law (1990) and, 234–36
  growth of, 231–33, 294, 299
  jobs and, 218
  Oil Shock (1973), impact of, 205
  100,000 Roofs, 1–2, 288–89, 294, 298–99
  photovoltaic cells (PV), 231–32, 233, 276–77, 299–300, 300*f*
  price of, 213, 276–78
  pricing regimes, 290–91, 298–99
  Renewable Energy Act (2000) and, 290, 303, 305–6
  return on investment, 298–99
  Technology Ministry and, 228
Solar Energy Development Association, 276–77
Solar Manifesto (Scheer), 268–69
Solow, Robert M., 60, 62, 186–87
South Korea, steel industry in, 254
Soviet Union. *see also* Russia
  Afghanistan, invasion of, 252
  arms race and, 181
  carbon emissions in, 43–44
  Chernobyl nuclear accident (1986), 211–12, 216–17, 220, 222–23
  collapse of Communism in, 213, 260–61
  Ministry for Gas, 244–45
  Ministry of Gas, 260–61, 263
  natural gas imports from, 12–13, 14, 135–36, 239–40, 244–56, 263, 264, 296, 306
  oil imports from, 33, 241–42
  pipelines and, 241–42
  Post-War occupation of Germany, 22
  Sputnik, 115
  US, relations with, 252–54, 255
Spaak, Paul-Henri, 105
SPD. *see* Social Democratic Party (SPD)
Spitzingsee Agenda, 115
Sputnik, 115
state action paradigm, 10–11
steel industry, 254
Stoltenberg, Gerhard, 116, 118–19, 121, 155–56
Strauss, Franz Josef, 102–3, 107, 108, 109, 199
Strohm, Holger, 165–66, 174
student mobilization, 160–61
Subterranean Forest (Sieferle), 202
Suez Canal closure (1967), 67, 90–91, 128
Suez Crisis (1956), 13, 33–34, 64, 71, 80–83, 128

supply-side programs in energy sector, 127–28
sustainability, 222
Sweden, National Defense Institute, 211
Switzerland, nuclear energy industry in, 164
Syria
  1973 War and, 125–26
  oil embargo (1967) and, 90

Taiwan, steel industry in, 254
taxation
  carbon tax, 204, 223–24, 273
  CDU and, 293
  Eco-Alliance and, 292–93
  Ecological Modernization and, 292
  ecological tax, 205–7
  energy tax, 205–7, 293
  FDP and, 293
  Green Party and, 292
  natural gas tax, 258
  oil, taxation of, 180–81
  oil industry, tax policy and, 84
  SPD and, 292
Technical and Scientific Association, 116
technocratic corporatism, 73, 97–98
The Technological Society (Ellul), 159–60
technology
  backstop technology, 187, 197–98
  CDU and, 228
  energy transitions and, 8
  environmental movement, concerns regarding technology, 159–60, 170–71
  neoclassical economics and, 197–98
  price-induced technological development, 188, 189, 202, 204
Technology Ministry
  nuclear energy industry and, 137, 145
  oil industry and, 150
  renewable energy and, 231–32
  solar energy and, 228
  wind energy and, 228
Texaco, 74, 75, 81, 90, 91, 93, 95
Thatcher, Margaret, 11, 213, 226–27, 230–31, 262
Third Atomic Program, 121
Third Path, 49–50
Third Reich
  generally, 6
  attempts to come to terms with, 7
  automobile industry in, 75–76
  coal industry and, 22, 23
  destruction of, 21
  food supply in, 66
  1970s, reminders of during, 171–72
  oil industry during, 30
  Ordoliberalism versus, 49
Third Revision to the Energy Program, 178, 253–54
Three Mile Island nuclear accident (1979), 178, 179–80

Thyssen AG, 19, 245, 246–47, 263
Töpfer, Klaus, 222–24, 225, 227–28, 267–68
Transalpine Pipeline (TAP), 78–79
Trans-Arabian Pipeline (TAP), 75
The Transformation of Industrial Society (Fischer), 279–80
Traube, Klaus, 170, 174, 181–82, 216
Treaty of Evian, 92
Treaty of Maastricht (1996), 262
Treaty of Rome (1957), 82, 105–6
Trittin, Jürgen, 1–4, 285–86, 293, 306, 307–8
Truman, Harry S., 64, 73–74
Turkey, Communism and, 73–74
Turkish Petroleum Company, 92

Ukraine
 independence of, 260
 natural gas industry and, 265
Ukraine War, 306, 310–11
The Ultimate Resource (Simon), 190–91
unemployment. *see* jobs
United Kingdom
 Allied occupation of Germany (*see* Allied occupation of Germany)
 Central Electricity Authority, 108
 coal industry in, 8
 energy consumption in, 2, 2f, 157
 energy efficiency in, 148
 energy intensity in, 3f
 energy liberalization in, 257
 European Community, in, 131–32
 mercantilism in, 141–42
 National Coal Board, 159–60
 natural gas industry in, 297
 neoliberalism in, 11, 208, 213, 226–27
 1970s, new energy policy in, 157
 nuclear energy industry in, 102–3, 106, 108, 111, 112, 113, 122–23, 164, 182
 oil embargo (1967) and, 90–91
 oil industry in, 30–31, 33, 63–64, 84, 97, 133
 oil production in, 229, 230–31
 Oil Shock (1973) and, 131–32, 147
 recent developments in, 297
 Soviet oil and, 241–42
 state intervention in energy in, 51
 Suez Crisis (1956) and, 71, 80–81
United Nations
 Brundtland Report, 220, 222
 Climate Conference (2017), 308
 Conference on Environment and Development (Earth Summit) (1992), 267–68, 270–71
 Economic Commission for Europe (UNECE), 35–36
 Framework Convention on Climate Change (UNFCCC), 267–68, 270–71
United States
 air pollution in, 158

Allied occupation of Germany (*see* Allied occupation of Germany)
Atomic Energy Commission (AEC), 111–12
automobile industry in, 75–76
carbon emissions in, 43–44
Clean Air Act of 1990, 192
coal industry in, 20
Committee on Energy Supplies and Resources, 64–65
Council of Economic Advisors, 189
de-cartelization and, 25
Defense Advanced Research Projects Agency (DARPA), 115
dematerialization of economy in, 14, 47, 59, 62, 69
digital revolution in, 272
economic models in, 202
energy consumption in, 2, 2f, 4–5, 157
energy intensity in, 3f
energy liberalization in, 257, 297
energy security concerns in, 64–65
environmental economics in, 203
Euratom, cooperation with, 111–12
Export-Import Bank, 111–12, 113–14
fossil fuels in, 22–23
Geological Survey (USGS), 65
German oil industry, role in building, 29–30
global power of, 5
Justice Department, 90–91
National Petroleum Council, 64–65
natural gas industry in, 297
neoclassical economics in, 47, 58–60
neoliberalism in, 10–11, 184, 185, 190–93, 198, 208, 213, 226, 227
New Deal, 159–60
1970s, new energy policy in, 157
1973 War and, 125–26
nuclear energy industry in, 102–3, 105, 106, 111–12, 113, 115, 116, 118, 137, 164, 182
nuclear weapons, use of, 101–2
oil embargo (1967) and, 90–91
oil industry in, 30–31, 63–64, 73–74, 92, 95, 133
oil production in, 229
Oil Shock (1973) and, 20, 125, 131, 147, 151
Public Utility Regulatory Policies Act of 1978 (PURPA), 234
recent developments in, 297
solar energy in, 226
Soviet Union, relations with, 252–54, 255
Three Mile Island nuclear accident (1979), 178, 179–80
uranium supply in, 120–21
Vietnam War and, 160–61
University of Bremen, 167
University of Chicago, 190
uranium enrichment, 120–21

Usher, Bruce, 8

VEBA (Vereinigte Elektrizitäts und Bergwerks Aktiengesellschaft)
  generally, 95
  chemicals and, 161
  coal and, 87, 273–74
  diversification of, 94
  electricity sector and, 117
  natural gas and, 264, 265
  nuclear energy and, 118–19, 144
  oil and, 98, 127, 133–35, 142–44, 146
  power of, 124, 283–84
Velvet Revolutions, 224–25
Vietnam War, 160–61
Vögler, Alfred, 242
Volkswagen, 76–77, 77f, 272, 293
Voss, Alfred, 258

Wallmann, Walter, 365n.1
Walter Eucken Institute, 207–8
Warsaw Treaty (1970), 247
Washington Energy Conference, 131
Wawersik, Rudolf, 335n.64
Wegenhekel, Lothar, 207–8
Weimar Republic
  automobile industry in, 75–76
  collapse of, 48–49
Weiszäcker, Carl Christian von, 197
Weiszäcker, Carl Friedrich von, 104, 104f, 139, 163, 175
Weiszäcker, Ernst Ulrich von, 270–71
Wessels, Theodore
  coupling paradigm and, 68–69
  energy forecasts and, 56, 61–62
  energy security and, 66, 67, 326n.1
  Ordoliberalism and, 64
Westinghouse, 112
Westrick, Ludwig, 39
Wicke, Lutz
  Ecological Modernization and, 185, 208–9, 291
  energy tax and, 206–7, 209, 223
  environmental economics and, 204–6

wind energy
  generally, 1, 2
  challenges of, 295
  Eco-Alliance and, 289–90
  Electricity Feed-in Law (1990) and, 234, 235–36, 276, 277, 278
  emergence of, 231–33
  expansion of, 299
  financing of, 277–78
  Green Party and, 280
  growth of, 231–33, 276, 299
  jobs and, 218
  Large Wind-Energy Installation (GROWIAN), 231–33
  Oil Shock (1973), impact of, 205
  100 MW Wind Demonstration Program, 228
  price of, 213, 277–78
  pricing regime, 298–99
  Renewable Energy Act (2000) and, 290, 303, 305–6
  return on investment, 298–99
  Technology Ministry and, 228
Wingas, 261, 262
Winnacker, Karl, 113
Wintershall, 261
World Economic Summit (1990), 223–24
World Trade Organization, 10–11, 269, 284–85
World War II
  generally, 5
  Allied occupation of Germany (see Allied occupation of Germany)
  defeat in, 21
  destruction caused by, 21
  energy constraints during, 23
  Post-War rebuilding of Germany, 22
Wuppertal Institute for Climate, Environment, and Energy, 198, 270–71

Yamal pipeline, 251–54, 255–56, 307
Yugoslavia, civil war in, 271

Ziesing, Hans-Joachim, 67